Koichi Tanaka

Solvent-free Organic Synthesis

Related Wiley-VCH titles:

Microwaves in Organic Synthesis
A. Loupy (ed.)

2002. 523 pages.
Hardcover. ISBN 3-527-30514-9

Ionic Liquids in Synthesis
P. Wasserscheid and T. Welton (eds.)

2003. 380 pages.
Hardcover. ISBN 3-527-30515-7

Chemical Synthesis Using Supercritical Fluids
P. G. Jessop and W. Leitner (eds.)

1999. 500 pages.
Hardcover. ISBN 3-527-29605-0

Organic Synthesis on Solid Phase
F. Zaragoza Dörwald (ed.)

Supports, Linkers, Reactions
2nd Edition, 2002. 553 pages.
Hardcover. ISBN 3-527-30603-X

Koichi Tanaka

Solvent-free Organic Synthesis

WILEY-VCH GmbH & Co. KGaA

Prof. Dr. Koichi Tanaka
Department of Applied Chemistry
Faculty of Engineering
Ehime University
Matsuyama, Ehime, 790-8577
Japan

1st Reprint 2004

Library of Congress Card No.: applied for

A catalogue record for this book is available from the British Library

Bibliographic information published by Die Deutsche Bibliothek
Die Deutsche Bibliothek lists this publication in the Deutsche Nationalbibliografie; detailed bibliographic data is available in the Internet at <http://dnb.ddb.de>.

ISBN 3-527-30612-9

Typesetting: K+V Fotosatz GmbH, Beerfelden
Printing: Strauss Offsetdruck GmbH, Mörlenbach
Bookbinding: J. Schäffer GmbH & Co. KG, Grünstadt

Foreword

Waste prevention and environmental protection are major requirements in an overcrowded world of increasing demands. Synthetic chemistry continues to develop various techniques for obtaining better products with less environmental impact. One of the more promising approaches is solvent-free organic synthesis; this book of Koichi Tanaka collects recent examples in this field in a concise way so that their performance and merits can be easily judged. This endeavor is very welcome, as most recent syntheses and educational textbooks largely neglect solvent-free techniques.

The field of solvent-free organic synthesis covers all branches of organic chemistry. It includes stoichiometric solid–solid reactions and gas–solid reactions without auxiliaries yielding single products in pure form that do not require solvent-consuming purification steps after the actual reaction. It also includes some stoichiometric melt reactions that occur without auxiliaries and with quantitative yield due to direct crystallization of the product. Although such reactions are by far the best choices for application of solvent-free chemistry, the advantages of avoiding solvents should not be restricted to them. Solvent-free conversions can be profitably applied even when unfavorable crystal packing and low melting points impede solid-state reaction and when melt reactions without direct crystallization do not provide 100% yield of one product. The higher concentration of reactants in the absence of solvents usually leads to more favorable kinetics than in solution. In some cases auxiliaries such as catalysts or solid supports may be required. Solid supports and microwave heating, instead of cooling or convection heating, are frequently used in solvent-free reaction steps. However, costly procedures should always be compared with inexpensive, waste-free techniques that do not require steps such as recrystallization, extraction, chromatography, or disposal of distillation residues.

The attitude implied in most current publications restricts (or extends) the term 'solvent-free' to the stoichiometric application of solid or liquid reagents, with less than a 10% excess of a liquid or soluble reagent and/or less than 10% of a liquid or soluble catalyst. It seems widely accepted in the field that solvents used for pre-adsorption of reagents to a support or for desorption, purification, and isolation of the products are not counted in 'solvent-free' syntheses. On the other hand, photolysis of insoluble solids in (usually aqueous) suspensions undoubtedly qualifies for inclusion as a solvent-free technique, but not the taking up of reagents from a liquid for reaction with a suspended solid.

Reacting gases may be in excess if they react with solids and do not condense in liquid phases, but supercritical media are clearly not the subject of solvent-free chemistry and deserve their own treatment. For practical reasons, this book does not deal with homogeneous or contact-catalyzed gas-phase reactions. Furthermore, very common polymerizations (except for solid-state polymerizations), protonations, solvations, complexations, racemizations, and other stereoisomerizations are not covered, to concentrate on more complex chemical conversions. This strategy allowed for presenting diverse reaction types and techniques, including those that proceed only in the absence of liquid phases, in one convenient volume.

The performance and scalability of the various techniques is most easily compared in a side-by-side format. With respect to experimental procedures, it is now recognized that many chemical conversions (e.g., formation of C–N or C–C bonds) that were reported to require solid supports with catalytic activity and microwave irradiation (and thus introduced environmental concerns) do not require such auxiliaries or irradiation. They occur exothermally at low temperatures with quantitative yields and without solvent-consuming workups even on a large scale.

This valuable compilation will become a useful resource for the development of improved, environmentally benign syntheses in industry and academia with the aims of avoiding catalysts and saving resources wherever possible and of preventing all the waste that is produced by using auxiliaries and by unnecessarily creating nonuniform reactions with less than 100% yield. This clearly designed and structured book on solvent-free organic synthesis will be of great value for the broader application of better synthetic techniques and thus for a better environment.

Gerd Kaupp
University of Oldenburg

Contents

Preface

The elimination of volatile organic solvents in organic syntheses is a most important goal in 'green' chemistry. Solvent-free organic reactions make syntheses simpler, save energy, and prevent solvent wastes, hazards, and toxicity.

The development of solvent-free organic synthetic methods has thus become an important and popular research area. Reports on solvent-free reactions between solids, between gases and solids, between solids and liquid, between liquids, and on solid inorganic supports have become increasingly frequent in recent years.

This volume is a compilation of solvent-free organic reactions, covering important papers published during the past two decades. It contains graphical summaries of 537 examples of solvent-free organic reactions and is divided into 14 chapters:

1. Reduction,
2. Oxidation,
3. Carbon–Carbon Bond Formation,
4. Carbon–Nitrogen Bond Formation,
5. Carbon–Oxygen Bond Formation,
6. Carbon–Sulfur Bond Formation,
7. Carbon–Phosphorus Bond Formation,
8. Carbon–Halogen Bond Formation,
9. Nitrogen–Nitrogen Bond Formation,
10. Rearrangement,
11. Elimination,
12. Hydrolysis,
13. Protection,
14. Deprotection.

Each summary includes a structure scheme, an outline of the experimental procedure, and references to help the reader.

I hope that this volume will contribute to the studies of organic chemists in industry and academia and will encourage the pursuit of further research into solvent-free organic synthesis.

Koichi Tanaka
Ehime University

1 Reduction

1.1 Solvent-Free Reduction

Type of reaction: reduction
Reaction condition: solid-state
Keywords: ketone, NaBH$_4$, alcohol

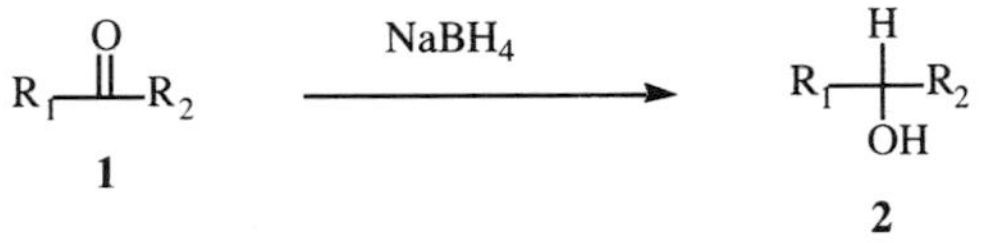

$$R_1 \overset{O}{\underset{}{\text{—C—}}} R_2 \quad \xrightarrow{\text{NaBH}_4} \quad R_1 \overset{H}{\underset{OH}{\text{—C—}}} R_2$$

1   **2**

a: R$_1$; R$_2$ = Ph
b: R$_1$=*trans*-PhCH=CH; R$_2$=Ph
c: R$_1$=2-naphthyl ; R$_2$=Me
d: R$_1$=PhCH(OH); R$_2$=Ph
e: R$_1$=PhCH$_2$; R$_2$=Ph
f: R$_1$, R$_2$=4-*t*-Bu-cyclohexyl

Exerimental procedures:

When a mixture of the powdered ketones and a ten-fold molar amount of NaBH$_4$ was kept in a dry box at room temperature with occasional mixing and grinding using an agate mortar and pestle for 5 days, the corresponding alcohols were obtained in good yields.

References: F. Toda, K. Kiyoshige, M. Yagi, *Angew. Chem. Int. Ed. Engl.,* **28**, 320 (1989).

Type of reaction: reduction
Reaction condition: solid-state
Keywords: ketone, enantioselective reduction, BH$_3$-ethylenediamin complex, inclusion complex, alcohol

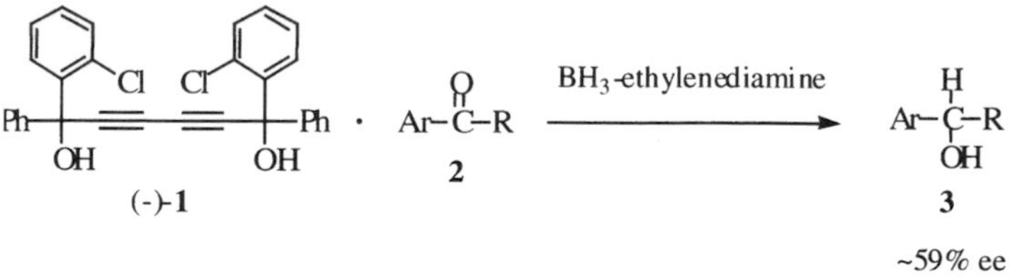

a: Ar=Ph; R=Me
b: Ar=2-MeC$_6$H$_4$; R=Me
c: R$_1$= 1-naphthyl; R$_2$=Me
d: R$_1$= 2-naphthyl; R$_2$=Me
e: R$_1$=Ph; R$_2$=Et
f: R$_1$=2-MeC$_6$H$_4$; R$_2$=Et

Exerimental procedures:

A mixture of finely powdered inclusion complex of (−)-**1** with **2** was kept under N$_2$ at room temperature for 24 h by occasional stirring. The reaction mixture was decomposed with water and extracted with ether. The ether solution was washed with dilute HCl, dried, and evaporated to give crude alcohols. Distillation of the crude alcohols in vacuo gave pure alcohols.

References: F. Toda, K. Mori, *Chem. Commun.,* **1245** (1989).

Type of reaction: reduction
Reaction condition: solid-state
Keywords: cage diketone, sodium borohydride, alcohol

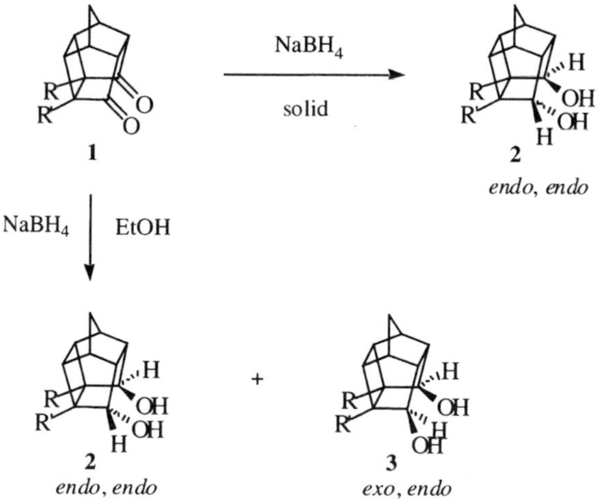

a: R=R'=H; b: R,R'=-CH$_2$CH$_2$-; c: R,R'=-CH=CH-

Exerimental procedures:
Cage diketone **1a** (87 mg, 0.50 mmol) and $NaBH_4$ (400 mg, excess) were ground together under an argon atmosphere into a fine powder, thereby producing an intimate solid mixture. The resulting powdery mixture was agitated under argon at room temperature for 7 days. Water (15 mL) then was added, and the resulting mixture was extracted with $CHCl_3$ (3×20 mL). The combined extracts were washed with water (30 mL), dried (Na_2SO_4), and filtered, and the filtrate was concentrated in vacuo to afford pure *endo,endo*-diol **2a** (89 mg, 100%) as a colorless microcrystalline solid: mp 275–276 °C.

References: A. P. Marchand, G. M. Reddy, *Tetrahedron*, **47**, 6571 (1991).

Type of reaction: reduction
Reaction condition: solid-state
Keywords: 7-norbornenone, $NaBH_4$, π-face selectivity, alcohol

a: $R_1=R_2=COOCH_3$
b: $R_1=H$; $R_2=COOCH_3$
c: $R_1=H$; $R_2=CN$
d: $R_1=R_2=CH_2OCH_3$

Exerimental procedures:
A mixture of **1a** and $NaBH_4$ (excess) was fully ground and left aside in a sample vial (1–2 days, sonication reduces the reaction time to a few hours). Usual workup led to the formation (80%) of **2a** and **3a** in 87:13 ratio.

References: G. Mehta, F. A. Khan, K. A. Lakshmi, *Tetrahedron Lett.*, **33**, 7977 (1992).

Type of reaction: reduction
Reaction condition: solvent-free
Keywords: ketone, aldehyde, carboxylic acid chloride, butyltriphenylphosphonium tetraborate, alcohol

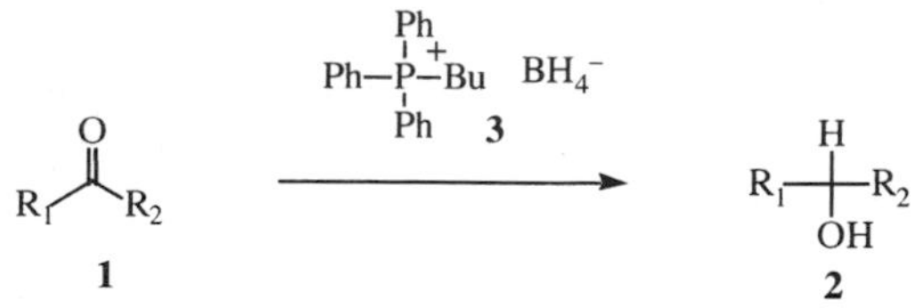

$$Ph-\overset{\overset{\displaystyle Ph}{|}}{\underset{\underset{\displaystyle Ph}{|}}{P}}\overset{+}{-}Bu \quad BH_4^- \quad \textbf{3}$$

$$R_1 \overset{O}{\underset{}{\bigwedge}} R_2 \quad \xrightarrow{\hspace{3cm}} \quad R_1 \overset{H}{\underset{OH}{-}} R_2$$

1   **2**

a: R_1=Ph; R_2=H
b: R_1=2-MeOC$_6$H$_4$; R_2=H
c: R_1=3-MeOC$_6$H$_4$; R_2=H
d: R_1=4-MeOC$_6$H$_4$; R_2=H
e: R_1=2-BrC$_6$H$_4$; R_2=H
f: R_1=4-BrC$_6$H$_4$; R_2=H
g: R_1=2-MeC$_6$H$_4$; R_2=H
h: R_1=4-MeC$_6$H$_4$; R_2=H
i: R_1=2-ClC$_6$H$_4$; R_2=H
j: R_1=3-ClC$_6$H$_4$; R_2=H
k: R_1=4-ClC$_6$H$_4$; R_2=H
l: R_1=2-NO$_2$C$_6$H$_4$; R_2=H
m: R_1=3-NO$_2$C$_6$H$_4$; R_2=H
n: R_1=4-NO$_2$C$_6$H$_4$; R_2=H

o: R_1=2,5-(MeO)$_2$C$_6$H$_3$; R_2=H
p: R_1=3,4-(MeO)$_2$C$_6$H$_3$; R_2=H
q: R_1=3,5-(MeO)$_2$C$_6$H$_3$; R_2=H
r: R_1,R_2=-(CH$_2$)$_5$-
s: R_1,R_2=-(CH$_2$)$_4$-
t: R_1=Ph; R_2=Me
u: R_1=-(CH$_2$)$_6$-; R_2=H
v: R_1=PhCH=CH; R_2=H
w: R_1=PhCH=CH; R_2=Me
x: R_1=Ph; R_2=Cl
y: R_1=PhCH$_2$; R_2=Cl
z: R_1=4-NO$_2$C$_6$H$_4$; R_2=Cl

Experimental procedures:

A mortar was charged with aldehyde, ketone, or carboxylic acid chloride and re-
ducing reagent **3**. The mixture was ground at room temperature with a pestle until TLC showed complete disappearance of the starting material. The mixture
was then extracted with CCl$_4$ (2×10 mL). Evaporation of the solvent gave the
corresponding alcohols. The product was purified by column chromatography on
silica gel using a mixture of ethyl acetate/n-hexane (10:90) as eluent.

References: A. R. Hajipour, S. E. Mallakpour, *Synth. Commun.*, **31**, 1177 (2001).

Type of reaction: reduction
Reaction condition: solid-state
Keywords: N-vinylisatin, hydrogenation, gas-solid reaction, *N*-ethyldioxindole,
N-ethylisatin

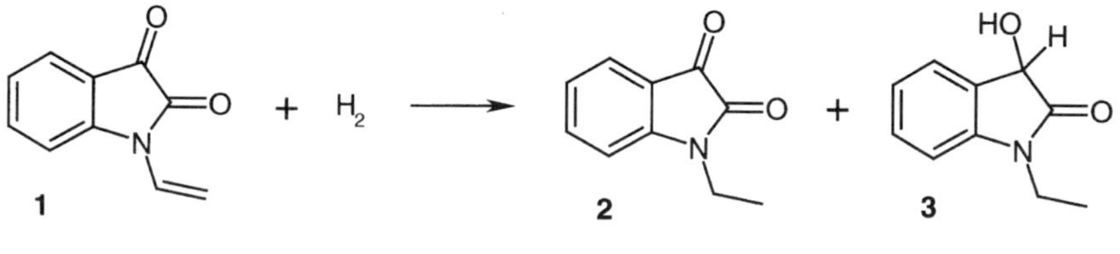

1   +   H$_2$ $\longrightarrow$ **2**   +   **3**

Experimental procedure:
Powdered crystals of **1** (670 mg, 3.9 mmol) that were recrystallized from *n*-hexane were evacuated in a 1 L flask and heated to 45 °C. Hydrogen gas was fed in from a steel cylinder (1 bar, 45 mmol) and the system kept at 45 °C for 2 days. The crystals changed their appearance and contained 67 mg (10%) unreacted **1**, 502 mg (74%) **2** and 110 mg (16%) **3**. The products were separated by preparative TLC on 200 g SiO_2 with dichloromethane.

If sublimed **1** was equally treated with H_2, no hydrogenation occurred. Thus, residual Pd impurities from the synthesis of **1** appear to activate the hydrogen in these solid-state reactions.

References: G. Kaupp, D. Matthies, *Chem. Ber.,* **120**, 1897 (1987).

Type of reaction: reduction
Reaction condition: solid-state
Keywords: cinnamic acid, hydrogenation, gas-solid reaction, 3-phenylpropionic acid

Experimental procedure:
Cinnamic acid crystals **1** were doped by inclusion of some Pd (compound) upon recrystallization from methanol with $Na_2[PdCl_4]$ (10^{-4} mol L^{-1}). Such crystals **1** were hydrogenated with excess H_2 at 1 bar and 30 °C for 6 days and yielded 48% of **2**.

References: G. Kaupp, D. Matthies, *Mol. Cryst. Liq. Cryst.,* **161**, 119 (1988).

Type of reaction: reduction
Reaction condition: solid-state
Keywords: epoxide, disodium *trans*-epoxysuccinate, palladium catalyst, disodium malate, alkene

Experimental procedures:
A supported palladium catalyst (0.50 g) was prepared with hydrogen gas at 200 °C for 30 min. The catalyst was mixed with disodium *trans*-epoxysuccinate (0.10 g), and the mixture was ground well with a mortar and pestle at room temperature. The mixture was placed in an autoclave and then shaken in the presence of hydrogen gas (9.0 MPa) at 100 °C for 14 h.

References: T. Kitamura, T. Harada, *J. Mol. Catal.*, **148**, 197 (1999).

Type of reaction: reduction
Reaction condition: solvent-free
Keywords: 2-vinylnaphthalene, hydrogenation, hydroformylation, subcritical CO_2

Experimental procedures:
The hydrogenation of vinylnaphthalene **1** was performed by mixing solid chlorotris(triphenylphosphine)rhodium catalyst (7.0 mg, 7.6 µmol) with solid 2-vinylnaphthalene (350 mg, 2.27 mmol, substrate:Rh = 300:1), both fine powders. The mixture was placed, with a stirring bar, into a 22 mm diameter flat-bottomed glass liner in a 160-mL high-pressure vessel, which was then sealed and warmed to 33 °C in a water bath. The vessel was flushed and pressurized with H_2 to 10 bar. This was considered the start of the reaction. Carbon dioxide was then added to a total pressure of 67 bar. After 30 min, the vessel was removed from the water bath and vented. The product mixture was dissolved in $CDCl_3$ and characterized by ^1H NMR spectroscopy.

References: P. Jessop, D.C. Wynne, S. DeHaai, D. Nakawatase, *Chem. Commun.*, 693 (2000).

Type of reaction: reduction
Reaction condition: solvent-free
Keywords: epoxide, *trans*-epoxysuccinic acid, hydrogenation, hydrogenolysis, alcohol

$$YO_2C \underset{H^{\prime\prime\prime}}{\overset{CO_2Y}{\diagup\!\!\diagup}} H \quad \xrightarrow[\text{Pd catalyst}]{H_2} \quad YO_2C \underset{OH}{\diagup\!\!\diagdown} CO_2Y \quad + \quad YO_2C \diagdown\!\!\diagup CO_2Y$$

Y=H; H$_2$TES; Y=Na; Na$_2$TES   +   YO$_2$CH$_2$C CH$_2$CO$_2$Y

Experimental procedures:
A supported Pd catalyst (0.1 g) was pretreated at 200 °C for 30 min with a H$_2$ stream. The resulting catalyst was mixed with H$_2$TES or Na$_2$TES (0.1 g), and the mixture was ground to a fine powder using a mortar and pestle. The mixture was placed in a Schlenk tube, and then the air in the tube was replaced by hydrogen gas. The reaction vessel was allowed to stand at 30 °C in the pressure of hydrogen (0.1 MPa) for 2 days.

References: T. Kitamura, T. Harada, *Green Chem.*, **3**, 252 (2001).

1.2   Solvent-Free Reduction under Microwave Irradiation

Type of reaction: reduction
Reaction condition: solid-state
Keywords: ketone, aldehyde, NaBH$_4$, alumina, microwave irradiation, alcohol

$$R-\!\!\left\langle\bigcirc\right\rangle\!\!-\overset{O}{\underset{}{C}}-R_1 \quad \xrightarrow[\text{MW}]{NaBH_4-Al_2O_3} \quad R-\!\!\left\langle\bigcirc\right\rangle\!\!-\overset{OH}{\underset{H}{C}}-R_1$$

$$\mathbf{1} \qquad\qquad\qquad\qquad\qquad \mathbf{2}$$

a: R=Me ; R$_1$=H
b: R=Cl; R$_1$=H
c: R=NO$_2$; R$_1$=H
d: R=H; R$_1$=Me
e: R=R$_2$=Me
f: R=H; R$_1$=Ph
g: R=H; R$_1$=CH(OH)Ph
h: R=OCH$_3$; R$_1$=CH(OH)C$_6$H$_4$-OCH$_3$-*p*

Experimental procedures:

Freshly prepared NaBH$_4$-alumina (1.13 g, 3.0 mmol of NaBH$_4$) is thoroughly mixed with neat acetophenone **1d** (0.36 g, 3.0 mmol) in a test tube and placed in an alumina bath inside the microwave oven and irradiated (30 s). Upon completion of the reaction, monitored on TLC (hexane-EtOAc, 8:2, v/v), the product is extracted into ethylene chloride (2×15 mL). Removal of solvent under reduced pressure essentially provides pure *sec*-phenethyl alcohol **2d** in 87% yield. No side product formation is observed in any of the reactions investigated and no reaction takes place in the absence of alumina.

References: R.S. Varma, R.K. Saini, *Tetrahedron Lett.*, **38**, 4337 (1997).

Type of reaction: reduction
Reaction condition: solid-state
Keywords: ketone, aldehyde, deuteriation, alumina, sodium borodeuteride, microwave irradiation, alcohol

a: R$_1$=Ph, R$_2$=H
b: R$_1$=PhCH=CH, R$_2$=H
c: R$_1$=4-NO$_2$C$_6$H$_4$, R$_2$=H
d: R$_1$=2,4,6-(MeO)$_3$C$_6$H$_2$, R$_2$=H
e: R$_1$=PhCH$_2$, R$_2$=H
f: R$_1$=1-naphthyl, R$_2$=H
g: R$_1$=2-naphthyl, R$_2$=H
h: R$_1$=Ph, R$_2$=Me
i: R$_1$=4-NO$_2$C$_6$H$_4$, R$_2$=Me
j: R$_1$=3-ClC$_6$H$_4$, R$_2$=Me

Experimental procedures:

For solid carbonyl compounds, the substrate e.g. *p*-nitroacetophenone **1i** (50 mg, 0.3 mmol) was thoroughly mixed with alumina doped NaBD$_4$ (0.126 g, 0.3 mmol of NaBD$_4$) using a pestle and mortar. The mixture was transferred to a loosely capped glass vial and irradiated in the microwave oven for 1 min at full power (750 W). The sample was allowed to cool to room temperature. The product was extracted using CHCl$_3$ (2 mL). The solvent was removed by rotary evaporation before being re-dissolved in CHCl$_3$ or CDCl$_3$ prior to NMR analysis. For liquid carbonyl compounds, thoroughly mixing was achieved by shaking the substrate with alumina doped NaBD$_4$ in the glass vial.

References: W.T. Erb, J.R. Jones, S. Lu, *J. Chem. Res. (S)*, 728 (1999).

Type of reaction: reduction
Reaction condition: solvent-free
Keywords: aldehyde, Cannizzaro reaction, barium hydroxide, microwave irradiation, alcohol, carboxylic acid

$$R-CHO \quad + \quad (CH_2O)_n \quad \xrightarrow[\text{MW}]{\text{Ba(OH)}_2 \cdot 8H_2O} \quad R-CH_2OH \quad + \quad R-COOH$$

$$\mathbf{1} \qquad\qquad\qquad\qquad\qquad\qquad\qquad \mathbf{2} \qquad\qquad \mathbf{3}$$

a: R=Ph
b: R=4-ClC$_6$H$_4$
c: R=4-BrC$_6$H$_4$
d: R=4-FC$_6$H$_4$
e: R=2-FC$_6$H$_4$
f: R=2-HOC$_6$H$_4$
g: R=4-MeC$_6$H$_4$
h: R=PhCH=CH

Experimental procedures:

In a typical experiment, benzaldehyde (106 mg, 1 mmol) was added to the finely powdered paraformaldehyde (60 mg, 2 mmol). To this mixture, powdered barium hydroxide octahydrate (631 mg, 2 mmol) was added in a glass test tube and the reaction mixture was placed in an alumina bath (neutral alumina: 125 g, mesh ~ 150, Aldrich; bath: 5.7 cm diameter) inside a household microwave oven and irradiated for the specified time at its full power of 900 W intermittently or heated in an oil bath at 100–110 °C. On completion of the reaction, as indicated by TLC (hexane-EtOAc, 4:1, v/v), the reaction mixture was neutralized with dilute HCl and the product extracted into ethyl acetate. The combined organic extracts were dried over anhydrous sodium sulfate and the solvent removed under reduced pressure. The pure benzyl alcohol (99 mg, 91%), however, is obtained by extracting the reaction mixture with ethyl acetate prior to neutralization and subsequent removal of the solvent under reduced pressure.

References: R.S. Varma, K.P. Naicker, P.J. Liesen, *Tetrahedron Lett.,* **39**, 8437 (1998).

Type of reaction: reduction
Reaction condition: solvent-free
Keywords: aldehyde, cross-Cannizzaro reaction, microwave irradiation, alcohol

$$R-CHO \xrightarrow[\text{NaOH, MW}]{\text{HCHO}} R-CH_2OH$$

1   **2**

a: R=Ph
b: R=4-ClC$_6$H$_4$
c: R=4-MeOC$_6$H$_4$
d: R=4-Me$_2$NC$_6$H$_4$
e: R=4-MeC$_6$H$_4$
f: R=4-O$_2$NC$_6$H$_4$

g: R=3-O$_2$NC$_6$H$_4$
h: R=2-O$_2$NC$_6$H$_4$
i: R=PhCH=CH
j: R=2-furyl
k: R=2-thienyl

Experimental procedures:
A mixture of benzaldehyde **1a** (0.53 g, 5 mmol), paraformaldehyde (1 g, 30 mmol) and solid sodium hydroxide (0.16 g, 4 mmol) were taken in an Erlenmeyer flask and placed in a commercial microwave oven operating at 2450 MHz frequency. After irradiation of the mixture for 25 s (monitored by TLC), it was cooled to room temperature, extracted with chloroform and dried over anhydrous sodium sulfate. Then the solvent was evaporated to give the corresponding benzylalcohol **2a** in 90% yield exclusively without the formation of any side products. Preparative column chromatography with silica gel was used for further purification of the alcohols, eluting with petroleum ether (60/80)-CHCl$_3$ (1:1).

References: J. A. Thakuria, M. Baruah, J. S. Sandhu, *Chem. Lett.,* 995 (1999).

Type of reaction: reduction
Reaction condition: solvent-free
Keywords: aromatic nitro compound, sodium hypophosphite, microwave irradiation, aromatic amine

$$R\underset{R}{\overbrace{}}NO_2 \xrightarrow[\text{MW}]{\text{NaH}_2PO_2 / \text{FeSO}_4 \cdot 7H_2O} R\underset{R}{\overbrace{}}NH_2$$

1   **2**

R, R$_1$= H, Me , OH, CONH$_2$, Ph, COOH, CN, NH$_2$

Experimental procedures:
Nitrobenzene (1 mmol) dissolved in minimum amount of dichloromethane, adsorbed over the neutral alumina (substrate:alumina=1:2, w/w), dried and mixed with ferrous sulfate (1.2 mmol) and sodium hydrogen phosphite (5 mmol). It was transfered into a test tube and subjected to microwave irradiation (BPL make, BMO 700T, 650 W, power 80%). Reaction was monitored by TLC (hexane-ethyl acetate, 70:30). After completion of the reaction (50 s), it was leached with di-

chloromethane (3×20 mL). Evaporation of the solvent under reduced pressure gave the amino product in good yield (78%). The product was further purified by passing through a column of silica gel (60–120 mesh) using hexane-ethyl acetate (8:2) as eluent.

References: H. M. Meshram, Y. S. S. Ganesh, K. C. Sekhar, J. S. Yadav, *Synlett,* 993 (2000).

Type of reaction: reduction
Reaction condition: solvent-free
Keywords: aromatic nitro compound, hydrazine hydrate, alumina, microwave irradiation, aromatic amine

a: $R_1=R_2=H$
b: $R_1=4\text{-OMe};\ R_2=H$
c: $R_1=4\text{-Me};\ R_2=H$
d: $R_1=4\text{-Cl};\ R_2=H$
e: $R_1=4\text{-I};\ R_2=H$
f: $R_1=3\text{-OH};\ R_2=H$
g: $R_1=2\text{-OH};\ R_2=H$
h: $R_1=2\text{-NH}_2;\ R_2=H$
i: $R_1=4\text{-NH}_2;\ R_2=H$
j: $R_1=2\text{-NH}_2;\ R_2=5\text{-Me}$
k: $R_1=2\text{-NH}_2;\ R_2=5\text{-CF}_3$

Experimental procedures:
Aromatic nitro compound (10 mmol) was mixed with inorganic solid support or alumina (10 g) and the mixture was added to hydrazine hydrate (30 mmol) and $FeCl_3 \cdot 6H_2O$ (0.5 mmol). The solid homogenized mixture was placed in a modified reaction tube which was connected to a removable cold finger and sample collector to trap excess hydrazine hydrate. The reaction tube was placed in a Maxidigest MX 350 (Prolabo) microwave reactor fitted with a rotational mixing system. After irradiation for a specified period, the contents were cooled to room temperature and the product extracted into ethyl acetate (2×20 mL). The solid inorganic material was filtered and the solvent was removed under reduced pressure to afford the product that was further purified by crystallization.

References: A. Vass, J. Dudas, J. Toth, R.S. Varma, *Tetrahedron Lett.,* **42**, 5347 (2001).

Type of reaction: reduction
Reaction condition: solid-state
Keywords: ester, potassium borohydride-lithium chloride, microwave irradiation, alcohol

$$R_1\text{-COOR}_2 \xrightarrow[\text{MW}]{\text{KBH}_4\text{-LiCl}} R_1\text{-CH}_2\text{OH} \ + \ R_2\text{-OH}$$

$$\mathbf{1} \qquad\qquad\qquad\qquad \mathbf{2}$$

a: R_1=Ph; R_2=Et   g: R_1=m-MeC$_6$H$_4$; R_2=Et
b: R_1=p-ClC$_6$H$_4$; R_2=Et   h: R_1=PhCH$_2$; R_2=Et
c: R_1=p-BrC$_6$H$_4$; R_2=Et   i: R_1=PhCH$_2$; R_2=Me
d: R_1=p-MeC$_6$H$_4$; R_2=Et   j: R_1=4-pyridyl; R_2=Et
e: R_1=o-ClC$_6$H$_4$; R_2=Et   k: R_1=3,4-(PhCH$_2$O)$_2$C$_6$H$_3$; R_2=Et
f: R_1=o-NO$_2$C$_6$H$_4$; R_2=Et   l: R_1=p-EtOCOC$_6$H$_4$; R_2=Et

Experimental procedures:

Potassium borohydride (1.0 g, 20 mmol), anhydrous lithium chloride (0.8 g, 20 mmol) were thoroughly mixed in a mortar and transferred to a flask (100 mL) connected with reflux equipment, then dry THF (10 mL) was added and the mixture was heated to reflux for 1 h. After cooling, the ester (10 mmol) was added and stirred for 0.5 h at room temperature, then the THF was removed under reduced pressure. After the mixture was irradiated by microwave for 2–8 min, the mixture was cooled to room temperature, water (20 mL) was added, extracted with ether (3×15 mL), dried with magnesium sulfate, and evaporated to give the crude product, which was purified by crystallization, distillation or column chromatography.

References: J.-C. Feng, B. Liu, L. Dai, X.-L. Yang, S.-J. Tu, *Synth. Commun.*, **31**, 1875 (2001).

2   Oxidation

2.1   Solvent-Free Oxidation

Type of reaction: oxidation
Reaction condition: solid-state
Keywords: ketone, Baeyer-Villiger reaction, *m*-chloroperbenzoic acid, ester

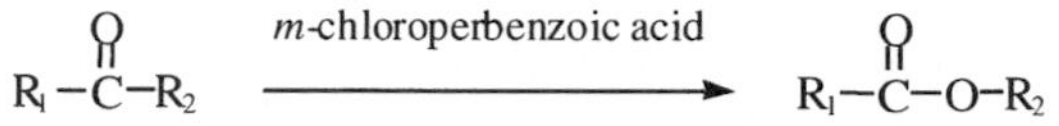

$$R_1-\overset{\overset{\textstyle O}{\|}}{C}-R_2 \xrightarrow{\textit{m-chloroperbenzoic acid}} R_1-\overset{\overset{\textstyle O}{\|}}{C}-O-R_2$$

$$\mathbf{1} \qquad\qquad\qquad\qquad\qquad \mathbf{2}$$

a: R_1=4-BrC$_6$H$_4$; R_2=Me
b: R_1=Ph; R_2=CH$_2$Ph
c: R_1=R_2=Ph
d: R_1=Ph; R_2=4-MeC$_6$H$_4$
e: R_1=Ph; R_2=2-MeC$_6$H$_4$

Experimental procedures:
The oxidations were carried out at room temperature with a mixture of powdered ketone and 2 mol equiv. of powdered *m*-chloroperbenzoic acid. When the reaction time was longer than 1 day, the reaction mixture was ground once a day with agate pestle and mortar. The excess of peroxy acid was decomposed with aqueous 20% NaHSO$_4$, and evaporated. The crude product was chromatographed on silica gel (benzene-CHCl$_3$).

References: F. Toda, M. Yagi, K. Kiyoshige, *Chem. Commun.*, 958 (1988).

Type of reaction: oxidation
Reaction condition: solid-state
Keywords: decalone, Baeyer-Villiger oxidation, norsesquiterpenoid, lactone

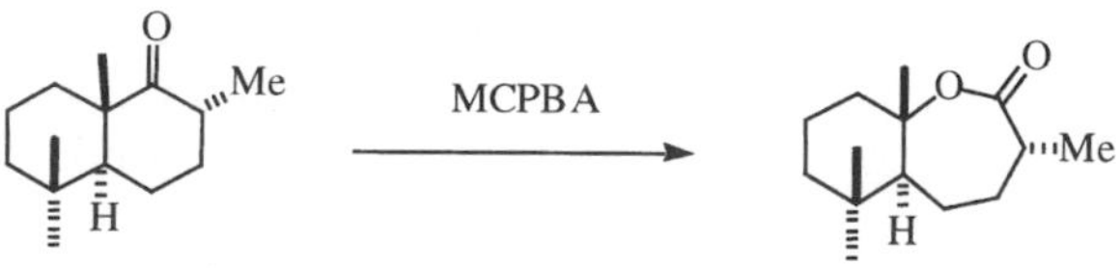

$$\mathbf{1} \qquad\qquad\qquad\qquad\qquad \mathbf{2}$$

Experimental procedures:

A mixture of the decalone **1** (40.8 mg, 196 µmol) and MCPBA (127 mg, 80%, 589 µmol) was left to stand at room temperature for 8 h and at 60 °C for 12 h. The resulting mixture was diluted with EtOAc and the organic layer was washed with sat. NaHCO$_3$ (2×), water and brine. Evaporation of the solvent followed by MPLC purification of the residue (EtOAc-*n*-hexane, 1:10) gave lactone **2** (31.5 mg, 72%) as a colorless oil.

References: H. Hagiwara, H. Nagatome, S. Kazayama, H. Sakai, T. Hoshi, T. Suzuki, M. Ando, J. Chem. Soc., *Perkin Trans. 1,* 457 (1999).

Type of reaction: oxidation
Reaction condition: solvent-free
Keywords: alcohol, ammonium chlorochromate, montmorillonite K-10, ketone

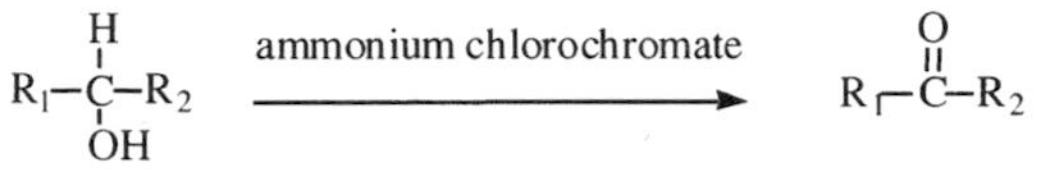

$$R_1-\overset{\overset{\displaystyle H}{|}}{\underset{\underset{\displaystyle OH}{|}}{C}}-R_2 \quad \xrightarrow{\text{ammonium chlorochromate}} \quad R_1-\overset{\overset{\displaystyle O}{||}}{C}-R_2$$

1   **2**

a: R$_1$=Ph; R$_2$=H
b: R$_1$=4-MeC$_6$H$_4$; R$_2$=H
c: R$_1$=5-Me- 2-NO$_2$C$_6$H$_3$; R$_2$=H
d: R$_1$=4-NO$_2$C$_6$H$_4$; R$_2$=H
e: R$_1$=2-HOC$_6$H$_4$; R$_2$=H
f: R$_1$=Ph; R$_2$=Me
g: R$_1$=R$_2$=Ph
h: R$_1$=Ph; R$_2$=COPh
i: R$_1$, R$_2$=-(CH$_2$)$_5$-

Experimental procedures:

Preparation of Ammonium Chlorochromate/Montmorillonite K-10. To a solution of chromium trioxide (40 g, 0.4 mol) in water (100 mL) was added ammonium chloride (21.4 g, 0.4 mol) within 15 min at 40 °C. The mixture was cooled until a yellow-orange solid formed. Reheating to 40 °C gave a solution. Montmorillonite K-10 (200 g) was then added with stirring at 40 °C. After evaporation in a rotary evaporator, the orange solid was dried in vacuo for 2 h at 70 °C. It can be kept for several months in air at room temperature without losing its activity.

Oxidation of Alcohols in the Solventless System. The above reagent (1.7 g, 2.6 mmol) was added to an appropriate neat alcohol (1.3 mmol). This mixture was thoroughly mixed using a pestle and mortar. An exothermic reaction ensued with darkening of the orange reagent and was complete almost immediately as confirmed by TLC (hexane-AcOEt, 8:2). The product was extracted into CH$_2$Cl$_2$ and passed through a small bed of silica gel (1 cm) to afford the corresponding pure carbonyl compounds.

References: M.M. Heravi, R. Kiakojoori, K.T. Hydar, *J. Chem. Res. (S),* 656 (1998).

Type of reaction: oxidation
Reaction condition: solvent-free
Keywords: allylic alcohol, manganese dioxide, barium manganate, aldehyde, ketone

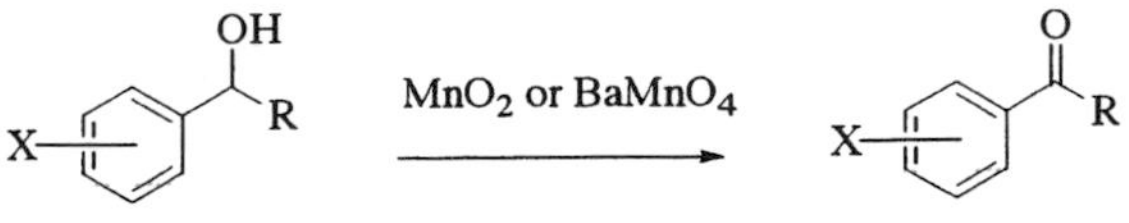

a: X=R=H	f: X=3,4-OCH$_2$O-; R=H
b: X=4-MeO; R=H	g: X=H; R=Me
c: X=4-Me; R=H	h: X=4-Br; R=Me
d: X=3-Me; R=H	i: X=H; R=Et
e: X=3-Cl; R=H	j: X=H; R=Ph

a: R=Ph
b: R=4-MeOC$_6$H$_4$
c: R=furyl

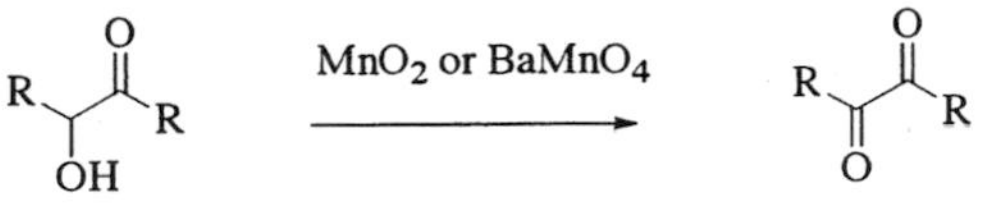

Experimental procedures:

Oxidation of Benzoin to Benzil by MnO$_2$ as a Typical Procedure for the Oxidation of Biaryl Acyloins. A mixture of benzoin **3a** (0.212 g, 1 mmol) and MnO$_2$ (0.174 g, 2 mmol) was prepared and magnetically agitated in an oil bath at 90 °C for 4 h. The progress of the reaction was monitored by TLC. The reaction mixture was applied on a silica gel pad (3 g) and washed with Et$_2$O (20 mL) to afford pure benzil **4a** quantitatively (mp 94 °C). The same reaction with BaMnO$_4$ proceeded to completion after 2 h using 1.5 mmol of the oxidant.

References: H. Firouzabani, B. Karimi, M. Abbassi, *J. Chem. Res. (S)*, 236 (1999).

Type of reaction: oxidation
Reaction condition: solvent-free
Keywords: olefin, allylic alcohol, epoxidation, tungstic acid, fluoroapatite, urea-H$_2$O$_2$, epoxide

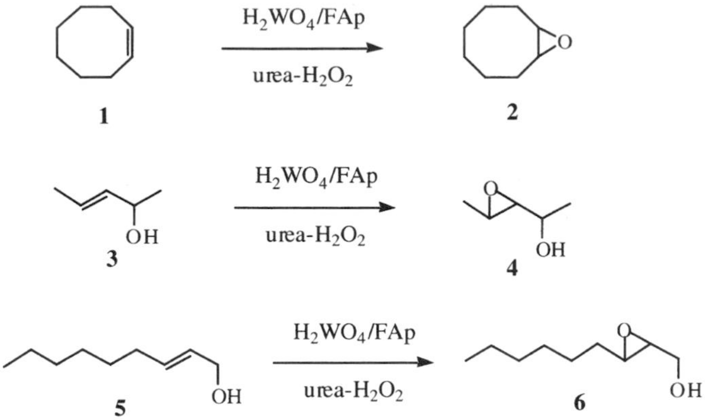

1 → **2**

H₂WO₄/FAp / urea-H₂O₂

3 → **4**

H₂WO₄/FAp / urea-H₂O₂

5 → **6**

H₂WO₄/FAp / urea-H₂O₂

Experimental procedures:
To a solid mixture of FAp powder (0.50 g) with urea-H_2O_2 powder (0.235 g, 2.5 mmol) was added tungstic acid powder (0.025 g, 0.10 mmol) in a test tube with a screw-cap, and mixed sufficiently. The solid mixture was permeated by a cyclooctene liquid **1** (0.110 g, 1.0 mmol), and the mixture was left without stirring at room temperature. After 48 h the reaction smoothly proceeded to afford epoxycyclooctane **2** in 90% yield.

References: J. Ichihara, *Tetrahedron Lett.*, **42**, 695 (2001).

Type of reaction: oxidation
Reaction condition: solid-state
Keywords: 2-hydroxybenzaldehyde, sulfide, nitrile, pyridine, urea-hydrogen peroxide complex, catechol, sulfoxide, sulfinic ester, amide, pyridine-*N*-oxide

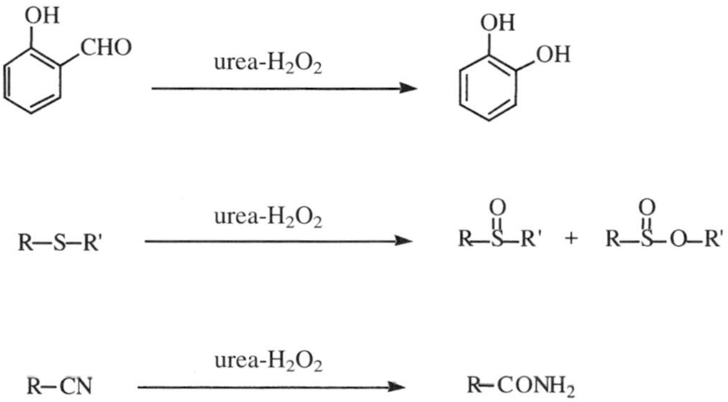

urea-H_2O_2

R—S—R' → R—S—R' + R—S—O—R'   urea-H_2O_2

R—CN → R—CONH₂   urea-H_2O_2

Experimental procedures:
The starting material (2 mmol) was added to the finely powdered urea-hydrogen peroxide adduct (376 mg, 4 mmol) in a glass test tube, and the reaction mixture was placed in an oil bath at 85 °C. After completion of the reaction, monitored by TLC, the reaction mixture was extracted into ethyl acetate and the combined extracts were washed with water and dried over anhydrous sodium sulfate. The solvent was removed under reduced pressure to afford the crude product, which was purified by chromatography to deliver pure product, as confirmed by the spectral analysis.

References: R.S. Varma, K.P. Naicker, *Org. Lett.,* **1**, 189 (1999).

Type of reaction: oxidation
Reaction condition: solid-state
Keywords: nitroxyl, verdazyl, nitrogen dioxide, bromine, xenon difluoride, gas-solid reaction

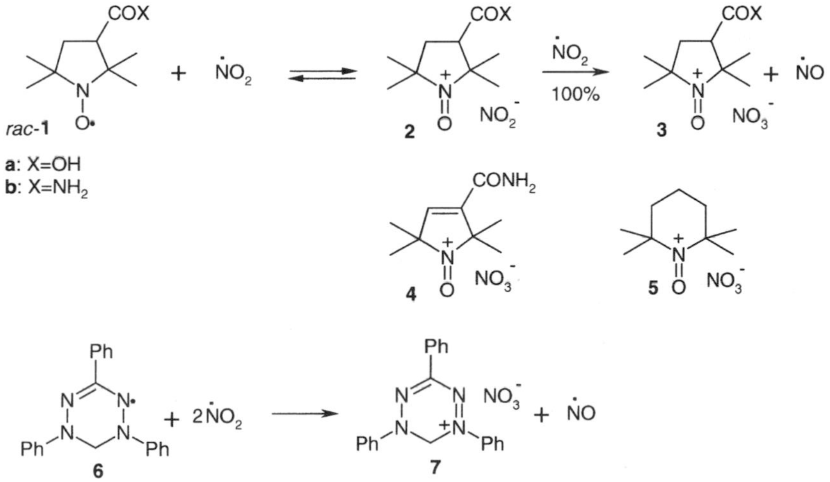

Experimental procedures:
An evacuated 100-mL flask was filled with N_2O_4/NO_2 to a pressure of 650 mbar (296 mg, 6.4 mmol NO_2). The sampling flask was connected to an evacuated 1-L flask which was then connected to an evacuated 10-mL flask that was cooled to 5 °C and contained the nitroxyl **1a**, or **1b**, or the nitroxyl precursor to **4** (500 mg, 2.70 mmol). After 1 h, the cooling bath was removed and excess NO_2 and NO were condensed to a cold trap at −196 °C for further use. The yield was 665 mg (100%) of pure **3a**, or **3b**, or **4**. (Ref. 1)

Similarly, 2 g quantities of tetramethylpiperidine-*N*-oxyl (TEMPO) were reacted at −10 °C (initial pressure of NO_2 0.03 bar) in 12 h with a quantitative yield of pure **5**. (Ref. 1)

Similarly, triphenylverdazyl **6** (200 mg, 0.64 mmol) was oxidized with NO_2 (3.2 mmol) at an initial pressure of 0.2 bar at 0 °C with a quantitative yield of pure **7**. (Ref. 1)

Similarly, the corresponding bromides of **3, 4, 5, 7** were quantitatively obtained if the free radical precursors were oxidized with bromine vapor in evacuated flasks. (Ref. 2)

Similarly, the corresponding fluorides of **3, 4, 5, 7** were quantitatively obtained by oxidation with gaseous XeF_2. (Ref. 3)

References: (1) G. Kaupp, J. Schmeyers, *J. Org. Chem.,* **60**, 5494 (1995).
(2) S. Nakatsuji, A. Takai, M. Mizumoto, H. Anzai, K. Nishikawa, Y. Morimoto, N. Yasuoka, J. Boy, G. Kaupp, *Mol. Cryst. Liq. Cryst.,* **334**, 177 (1999).
(3) G. Kaupp, *Comprehensive Supramolecular Chemistry*, Vol. 8, 381 (Ed. J.E.D. Davies), Elsevier, Oxford (1996).

Type of reaction: oxidation
Reaction condition: solvent-free
Keywords: alcohol, alumina-supported permanganate, ketone, aldehyde

$$
\underset{\textbf{1}}{\underset{R_2}{\overset{R_1}{>}}CH-OH} \quad \xrightarrow{\ KMnO_4/alumina\ } \quad \underset{\textbf{2}}{\underset{R_2}{\overset{R_1}{>}}=O}
$$

a: R_1=Ph; R_2=H
b: R_1=4-$NO_2C_6H_4$; R_2=H
c: R_1=3,4-$(MeO)_2C_6H_3$; R_2=H
d: R_1=4-PhC_6H_4; R_2=Me
e: R_1=2-pyridyl; R_2=Ph
f: R_1=Ph; R_2=Me
g: R_1=4-$MeOC_6H_4$; R_2=H
h: R_1=2-$MeOC_6H_4$; R_2=H
i: R_1=R_2=Ph
j: R_1=3-$MeOC_6H_4$; R_2=H
k: R_1=4-ClC_6H_4; R_2=H

l: R_1=2-ClC_6H_4; R_2=H
m: R_1=Ph; R_2=$PhCH_2$
n: R_1=4-BrC_6H_4; R_2=Me
o: R_1=4-ClC_6H_4; R_2=Me
p: R_1=Ph; R_2=PhCO
q: R_1=2,3-$(MeO)_2C_6H_3$; R_2=H
r: R_1=PhCH=CH; R_2=Ph
s: R_1=PhCH=CH; R_2=Me
t: R_1=4-$NO_2C_6H_3$CH=CH; R_2=H
u: R_1=PhCH=CH; R_2=H

Experimental procedures:
The alumina supported permanganate is prepared by combining $KMnO_4$ (4 g, 25.3 mmol) and alumina (neutral, 5 g) in a mortar and grinding with a pestle until a fine, homogeneous, purple powder is obtained. Benzoin (0.42 g, 2 mmol) is added to $KMnO_4/Al_2O_3$ (1 g, 2.8 mmol), the mixture was grinded with a pestle in a mortar until TLC showed complete disappearance of starting material, which

required 5 min (Table 1). Acetone (15 mL) was added to the reaction mixture and after evaporated by rotary evaporator. The residue is taken up into ether and washed with H_2O (10 mL), dried ($MgSO_4$) and the ether evaporated to give crude material.

References: A.R. Hajipour, S.E. Mallakpour, G. Imanzadeh, *Chem. Lett.*, 99 (1999).

Type of reaction: oxidation
Reaction condition: solvent-free
Keywords: silyl ether, tetrahydropyranyl ether, deprotection, wet alumina, chromium oxide, ketone, aldehyde

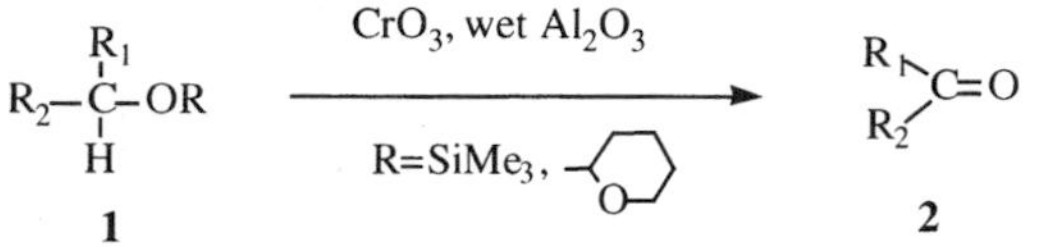

a: R_1=Ph; R_2=H
b: R_1=4-MeC$_6$H$_4$; R_2=H
c: R_1=2-NO$_2$-5-MeC$_6$H$_3$; R_2=H
d: R_1=Ph; R_2=Me

e: R_1=R_2=Ph
f: R_1,R_2=C$_5$H$_{10}$
g: R_1=PhCH=CH; R_2=H

Experimental procedures:
In a watch glass, neat trimethylsilyl ether **1** (1 mmol), or tetrahydropyranyl ether **2** (mmol), was mixed with the above catalyst (1 mmol) with a spatula. An exothermic reaction started with darkening of the orange color of the reagent and completion was confirmed by TLC (hexane-EtOAc, 8:2). The product was extracted into CH_2Cl_2 (2×20 mL) and is passed through a small bed of alumina (1 cm) to afford the corresponding carbonyl compound.

References: M.M. Heravi, D. Ajami, M. Ghassemzadeh, *Synthesis,* 393 (1999).

Type of reaction: oxidation
Reaction condition: solvent-free
Keywords: 1-hydroxyphosphonate, Arbuzov reaction, acylphosphonate

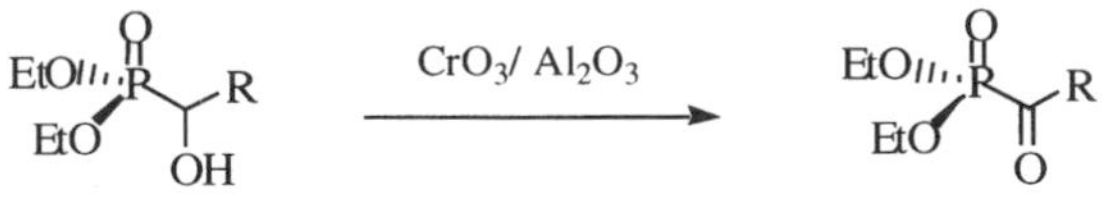

$$\text{EtO}_{\prime\prime\prime}\overset{\overset{O}{\parallel}}{P} \text{R} \quad \xrightarrow{\text{CrO}_3/\text{Al}_2\text{O}_3} \quad \text{EtO}_{\prime\prime\prime}\overset{\overset{O}{\parallel}}{P} \text{R}$$

1

2

a: R=Ph
b: R=4-MeC$_6$H$_4$
c: R=4-ClC$_6$H$_4$
d: R=4-NO$_2$C$_6$H$_4$
e: R=4-MeOC$_6$H$_4$
f: R=2-ClC$_6$H$_4$
g: R=2-NO$_2$C$_6$H$_4$
h: R=3-NO$_2$C$_6$H$_4$
i: R=2,4-Cl$_2$C$_6$H$_3$

Experimental procedures:

Thirty mmol of the reagent is prepared by the combination of CrO$_3$ (30 mmol, 3 g, finely ground) and alumina (Al$_2$O$_3$, neutral, 5.75 g) in a mortar and by grinding them together with a pestle until a fine, homogeneous, orange powder is obtained (5–10 min). The 1-hydroxyphosphonate (10 mmol) is added to this reagent. After 2.5–12 h of vigorous stirring, the reaction mixture is washed with CH$_2$Cl$_2$ (200 mL), dried (Na$_2$SO$_4$), and the solvent evaporated to give the crude products. Pure product is obtained by distillation under reduced pressure in 65–90% yields.

References: B. Kaboudin, *Tetrahedron Lett.,* **41**, 3169 (2000); H. Firouzabadi, N. Iranpoor, S. Sobhani, A. Sardarian, *ibid.,* **42**, 4369 (2001).

Type of reaction: oxidation
Reaction condition: solvent-free
Keywords: alcohol, pyridinium chlorochromate, microwave oven, ketone, aldehyde

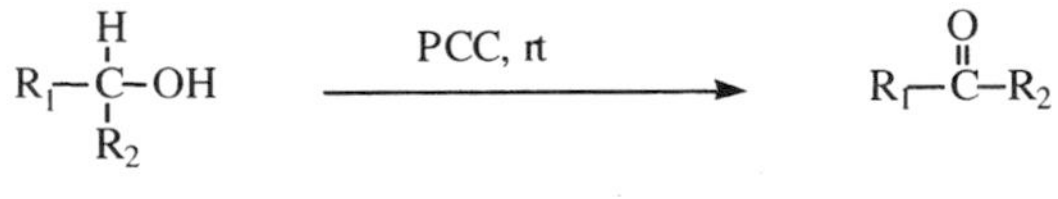

$$\text{R}_1-\overset{\overset{H}{|}}{\underset{\underset{R_2}{|}}{C}}-\text{OH} \quad \xrightarrow{\text{PCC, rt}} \quad \text{R}_1-\overset{\overset{O}{\parallel}}{C}-\text{R}_2$$

1

2

a: R$_1$=Ph; R$_2$=H
b: R$_1$=4-BrC$_6$H$_4$; R$_2$=H
c: R$_1$=4-MeOC$_6$H$_4$; R$_2$=H
d: R$_1$=4-ClC$_6$H$_4$; R$_2$=H
e: R$_1$=4-NO$_2$C$_6$H$_4$; R$_2$=H
f: R$_1$=4-MeC$_6$H$_4$; R$_2$=H
g: R$_1$=4-PhCH$_2$OC$_6$H$_4$; R$_2$=H
h: R$_1$=2-thienyl; R$_2$=H
i: R$_1$=2-furyl; R$_2$=H
j: R$_1$=1-naphthyl; R$_2$=H
k: R$_1$=R$_2$=Ph
l: R$_1$=Ph, R$_2$=Me

Experimental procedures:
PCC (3–15 mmol) was added to the substrate (3 mmol) in a mortar. Starting materials were instantly mixed and then stored for the appropriate period at room temperature or in an microwave oven without any further agitation. The progress of the reaction was monitored by dissolving a sample in acetone and using TLC on silica gel (hexane-Et_2O, 3:1). Upon completion of the reaction, HCl (20%, 30 mL) was added and extracted with Et_2O (3×25 mL). The organic layer was separated and dried ($MgSO_4$). Evaporation of the solvent gave the corresponding carbonyl compounds in 75–96% yields.

References: P. Salehi, H. Firouzabadi, A. Farrokhi, M. Gholizadeh, *Synthesis*, 2273 (2001).

Type of reaction: oxidation
Reaction condition: solid-state
Keywords: *a,β*-unsaturated acyl hydrazo compound, $Fe(NO_3)_3 \cdot 9H_2O$, *a,β*-unsaturated acyl azo compound

$$PhCH{=}CH{-}\overset{\overset{O}{\|}}{C}{-}NHNHAr \quad \xrightarrow{Fe(NO_3)_3 \cdot 9H_2O} \quad PhCH{=}CH{-}\overset{\overset{O}{\|}}{C}{-}N{=}N{-}Ar$$

1   **2**

a: $Ar{=}4\text{-}ClC_6H_4$
b: $Ar{=}2\text{-}ClC_6H_4$
c: $Ar{=}4\text{-}BrC_6H_4$
d: $Ar{=}4\text{-}IC_6H_4$
e: $Ar{=}2,4\text{-}Cl_2C_6H_3$
f: $Ar{=}4\text{-}MeC_6H_4$

Experimental procedures:
A mixture of **1** (1 mmol) and $Fe(NO_3)_3 \cdot 9H_2O$ (2 mmol, 0.808 g) was ground in an agate mortar at room temperature until the solid mixture changed from white powder into red or brown dope with balmy odor releasing (5–10 min). The resulting mixture was extracted with acetone (10 mL). Afterwards cold water (15 mL) was added and brown, red, orange or violet-red crystals were precipitated. The products were isolated by filtrating, washed with water until the washings became neutral. The crude products were chromatographed on a column of silica (60–100 mesh) and eluted with the mixed solvent of petroleum ether (bp 60–90°C) and acetone (5:1). The pure products were dried in vacuum below 50°C.

References: Y.-W. Zhao, Y.-I. Wang, H. Wang, Z.-F. Dun, X.-Z. Yao, *Synth. Commun.*, **31**, 2563 (2001).

Type of reaction: oxidation
Reaction condition: solid-state
Keywords: diaryl carbonohydrazide, $K_3Fe(CN)_6$, diaryl carbazone

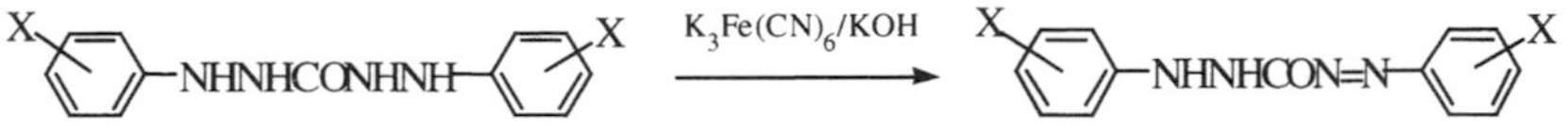

1   **2**

a: X=H
b: X=2-Me
c: X=3-Me
d: X=4-Me
e: X=2,3-Me$_2$
f: X=2,6-Me$_2$
g: X=4-NO$_2$

Experimental procedures:
A mixture of 0.01 mol of **1** and 0.041 mol of $K_3Fe(CN)_6$ and 0.3–0.4 mmol of KOH was ground in an agate mortar. After 20–30 min, the color of the mixture turned orange-yellow, orange or dark green. Then 20 mL of water was added, then yellow, orange and orange-red and dark green products were precipitated. The product was isolated by filtration, and washed with water four times. The products were recrystallized from a mixture of ethanol and water, and dried under vacuum. Structures of these products were characterized by elemental analysis, IR and ^1H NMR spectroscopy.

References: J-P. Xiao, Y.-L. Wang, Q.-X. Zhou, *Synth. Commun.*, **31**, 661 (2001).

Type of reaction: oxidation
Reaction condition: solvent-free
Keywords: sulfide, manganese dioxide, H_2SO_4/silica gel, sulfoxide

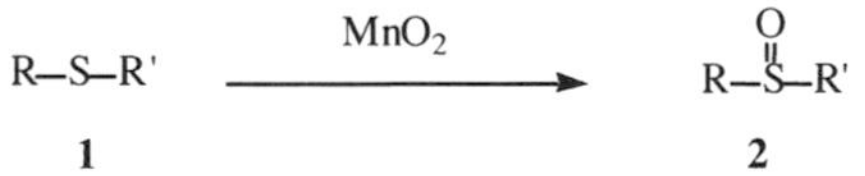

1   **2**

a: R=Ph; R'=Me
b: R=4-Me COC$_6$H$_4$; R'=Me
c: R=Ph; R'=Et
d: R=Ph; R'=Bu
e: R=Ph; R'=i-Pr
f: R=3-Me C$_6$H$_4$; R'=*i*-Pr
g: R=Ph; R'=cyclopentyl
h: R=Ph; R'=cyclohexyl
i: R=R'=Pr
j: R=R'=Bu

Experimental procedures:

A mixture of MnO_2 (0.52 g, 6 mmol) and the catalyst (1.8 g) was added to methyl phenyl sulfide **1a** (0.37 g, 3 mmol) and the resulting solid mixture was magnetically agitated for 1 h at 35–40 °C. The progress of the reaction was monitored by TLC using hexane/Et_2O (1:1) as eluent. After completion of the reaction, the solid mixture was applied directly on a silica gel column and eluted with $CHCl_3$. Evaporation of the solvent afforded almost pure methyl phenyl sulfoxide **2a** in 80% yield (0.33 g) whose physical data are consistent with those reported in the literature.

References: H. Firouzabadi, M. Abbassi, *Synth. Commun.,* **29**, 1485 (1999).

Type of reaction: oxidation
Reaction condition: solvent-free
Keywords: sulfide, *tert*-butyl hydroperoxide, silica gel, sulfoxide

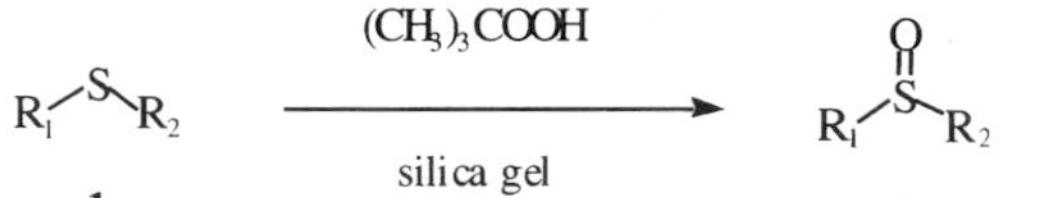

$$R_1 \overset{S}{\diagup} R_2 \quad \xrightarrow[\text{silica gel}]{(CH_3)_3COOH} \quad R_1 \overset{\overset{O}{\|}}{\underset{S}{\diagup}} R_2$$

1 **2**

a: $R_1 = R_2$ *n*-Bu
b: $R_1 = Ph$, $R_2 = CH_2Ph$
c: $R_1 = R_2 = Ph$

Experimental procedures:

Into a 25-mL round-bottomed flask was weighed 2.5 g of Merck 10181 chromatographic silica gel that had been equilibrated with the atmosphere at 120 °C for at least 48 h. The flask was stoppered and the contents allowed to cool to 25 °C. The substrate was added without solvent and the resulting mixture tumbled on a rotary evaporator at atmospheric pressure until uniformly free-floating. The oxidant was then added and the mixture again tumbled. After being heated, the mixture was allowed to cool to 25 °C and was washed with 100–200 mL of ethyl acetate. The adsorbent was collected by vacuum filtration and the filtrate concentrated under reduced pressure.

References: P.J. Kropp, G.W. Breton, J.D. Fields, J.C. Tung, B.R. Loomis, *J. Am. Chem. Soc.,* **122**, 4280 (2000).

Type of reaction: oxidation
Reaction condition: solvent-free
Keywords: amine, surface-mediated reaction, silica gel, hydroxylamine

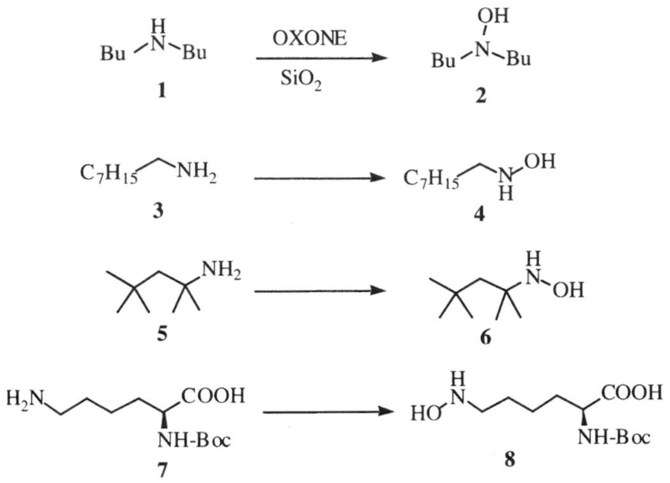

Experimental procedures:
Into a 25-mL round-bottomed flask was weighed 2.5 g of Merck 10181 silica gel or Fisher A540 alumina that had been equilibrated with the atmosphere at 120 °C for at least 48 h. The flask was stoppered, and the content was allowed to cool to 25 °C. The amine was added without solvent and the resulting mixture tumbled until uniformly free-flowing. The oxidant was then added and the mixture again tumbled. After being heated, the mixture was allowed to cool to 25 °C and was stirred overnight with 100 mL of MeOH. The adsorbent was collected by vacuum filtration and washed with an additional 50 mL of MeOH, and the combined filtrates were concentrated under reduced pressure.

References: J. D. Fields, P. J. Kropp, *J. Org. Chem.,* **65**, 5937 (2000).

Type of reaction: oxidation
Reaction condition: solvent-free
Keywords: thiol, ferriprotoporphyrin(IX) chloride, peroxidase, oxidative coupling, disulfide

$$\text{AcHN-}\underset{\underset{CH_2SH}{|}}{\overset{\overset{H}{|}}{C}}\text{-COOH} \quad \xrightarrow[\text{air}]{\text{activated HSC}} \quad \begin{array}{c} H \\ | \\ \text{AcHN-C-COOH} \\ | \\ CH_2 \\ | \\ S \\ | \\ S \\ | \\ CH_2 \\ | \\ \text{AcHN-C-COOH} \\ | \\ H \end{array}$$

1   **2**

Experimental procedures:

Reaction mixtures were prepared by mechanical dispersion of the thiol (200 mg g^{-1}) on the activated support. Water (0.5 mL g^{-1}) was added and the mixture was carefully mixed until an homogeneous (hydrated) loose solid was obtained. This solid was left at room temperature or heated in an open vessel for the indicated time. Products were eluted from the support either with diluted hydrochloric acid (cystine, oxidized glutathione, penicylamine) or methanol (*N*-acetylcystine, cystine methyl ester). The solution was decolorized by addition of activated carbon, filtered and evaporated.

References: E. Guibe-Jampel, M. Therisod, J. Chem. Soc., *Perkin Trans. 1*, 3067 (1999).

Type of reaction: oxidation
Reaction condition: solvent-free
Keywords: thiol, oxidative coupling, active manganese dioxide, barium manganate, disulfide

$$\text{R-SH} \quad \xrightarrow{\text{MnO}_2 \text{ or BaMnO}_4} \quad \text{R-S-S-R}$$

1   **2**

a: R=Pr
b: R=Bu
c: R=*cyclo*-C$_6$H$_{11}$
d: R=Ph
e: R=3-MeC$_6$H$_4$
f: R=PhCH$_2$

Experimental procedures:

The reaction of the thiols (3 mmol) were carried out with active manganese dioxide or barium manganate with the aid of magnetic stirrer at room temperature. 100% conversion was obtained in <5 min in each experiment. Application of the mixture on a silica gel pad followed by washing by an appropriate solvent afforded almost pure disulfide in 91–98% yields. In addition, the synthesis of diphenyldisulfide (97%) for the laboratory scale synthesis was performed successfully by using thiophenol (1.49 g, 12 mmol), active manganese dioxide or barium manganate (14 mmol) by strong magnetic agitation in about 5 min at room temperature.

References: H. Firouzabadi, M. Abbassi, B. Karimi, *Synth. Commun.*, **29**, 2527 (1999).

Type of reaction: oxidation
Reaction condition: solvent-free
Keywords: thiol, pyridinium chlorochromate, oxidative coupling, disulfide

$$\text{R—SH} \xrightarrow[\text{rt}]{\text{PCC}} \text{R—S—S—R}$$

$$\quad \mathbf{1} \qquad\qquad\qquad\qquad \mathbf{2}$$

a: R=Ph
b: R=4-MeC$_6$H$_4$
c: R=4-ClC$_6$H$_4$
d: R=3-MeC$_6$H$_4$
e: R=2-MeOCOC$_6$H$_4$
f: R=PhCH$_2$

g: R=cyclohexyl
h: R=n-Bu
i: R=CH$_3$(CH$_2$)$_7$
j: R=HOCH$_2$CH$_2$
k: R=HOOCCH$_2$

Experimental procedures:
Pyridinium chlorochromate (1 mmol) was added to thiol (1 mmol) in a mortar. Starting materials were mixed and stood together for the appropriate period at room temperature. Progress of the reaction was followed by dissolving a sample in acetone and using thin layer chromatography on silica gel (MeOH-HOAc, 5:1). After completion of the reaction hydrochloric acid (20%) was added and the mixture extracted with ether (2×20 mL) and chloroform (2×20 mL). The organic layer was separated and dried (MgSO$_4$). Final evaporation of solvent gave the products in 47–94% yields.

References: P. Salehi, A. Farrokhi, M. Gholizadeh, *Synth. Commun*, **31**, 2777 (2001).

2.2 Solvent-Free Oxidation under Microwave Irradiation

Type of reaction: oxidation
Reaction condition: solvent-free
Keywords: arene, $KMnO_4$-alumina, microwave irradiation, aromatic ketone

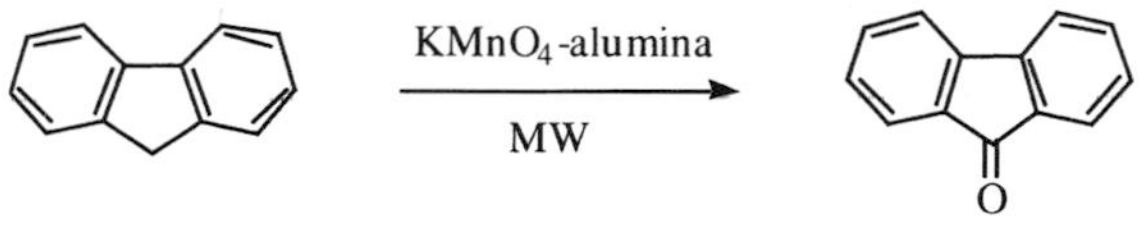

Experimental procedures:
Preparation of oxidative system: Finely ground potassium permanganate (50 g) dissolved in water (100 mL) was added to alumina (acidic or neutral, Merck activity I, 63–200 nm; 200 g). After shaking for 15 min, the majority of the water was removed by evaporation under reduced pressure and the obtained powder was dried under microwave irradiation for 5 min.
Fluoren-9-one: In a Pyrex matrix adapted to a Synthewave 402 monomode reactor, fluorene **1** (2 mmol, 0.332 g) was added to the $KMnO_4$-alumina mixture (6 mmol, 4.74 g). After 5 min of mechanical stirring, the mixture was irradiated (under stirring) at 150 W for 10 min. At the end of exposure to microwaves, the mixture was cooled to room temperature and eluted with diethyl ether (50 mL). After filtration and solvent removal, the crude product was identified by comparison (GC and NMR) with an authentic sample.

References: A. Oussaid, A. Loupy, *J. Chem. Res. (S)*, 342 (1997).

Type of reaction: oxidation
Reaction condition: solvent-free
Keywords: olefin, SeO_2/t-BuOOH, microwave irradiation, *trans-α,β*-unsaturated aldehyde

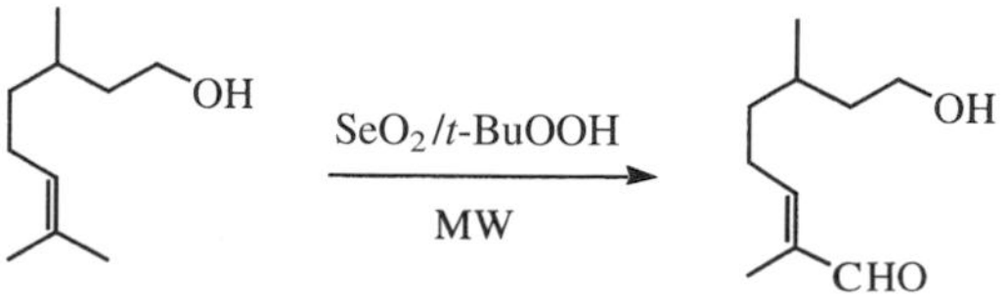

Experimental procedures:

Citronellol **1** (3.2 mmol, 0.500 g), SeO_2 (1.6 mmol, 0.176 g) and *t*-BuOOH (70%, 4.48 mmol, 0.576 g) were dissolved in a small amount of dichloromethane. Silica (1 g) was then added to form a slurry and excess solvent was evaporated off to obtain free flowing silica which was then exposed to microwave irradiation at power level 9 (640 W) for 10 min. Diethyl ether was then added and the mixture filtered. The filtrate was then washed (10% KOH, brine) and then dried (anhydrous Na_2SO_4). The solvent was evaporated to furnish the pure product in 83% yield.

References: J. Singh, M. Sharma, G.L. Kad, B.R. Chhabra, *J. Chem. Res. (S)*, 264 (1997).

Type of reaction: oxidation
Reaction condition: solvent-free
Keywords: alcohol, clayfen, microwave irradiation, ketone, aldehyde

$$\begin{array}{c} R_1 \\ | \\ CH\text{-}OH \\ | \\ R_2 \end{array} \quad \xrightarrow[\text{MW}]{\text{clayfen}} \quad \begin{array}{c} R_1 \\ \diagdown \\ C=O \\ \diagup \\ R_2 \end{array}$$

1   **2**

a: R_1Ph; R_2=H
b: R_1=Ph; R_2=Et
c: R_1=Ph; R_2=PhCO
d: R_1=4-MeC$_6$H$_4$; R_2=H
e: R_1=4-MeOC$_6$H$_4$; R_2=H
f: R_1=4-MeOC$_6$H$_4$; R_2=4-MeOC$_6$H$_4$
g: R_1= tetrahydrofurfuryl; R_2=H
h: R_1,R_2=-(CH$_2$)$_5$-

Experimental procedures:

Clayfen (0.125 g) was thoroughly mixed with neat benzoin **1c** (0.106 g, 0.5 mmol) in the solid state using a vortex mixer and the material was placed in an alumina bath inside the microwave oven and irradiated. Upon completion of the reaction, monitored on TLC (hexane-AcOEt, 10:1, v/v), the product **2c** was extracted into methylene chloride. That the effect is not purely thermal is supported by the fact that this reaction could be completed in 18 h in an oil bath at a comparable temperature of 65 °C.

References: R.S. Varma, R. Dahiya, *Tetrahedron Lett.*, **38**, 2043 (1997).

Type of reaction: oxidation
Reaction condition: solvent-free
Keywords: alcohol, manganese dioxide, silica gel, microwave irradiation, ketone, aldehyde

$$\underset{\textbf{1}}{\overset{R_1}{\underset{R_2}{\diagup}}CH\text{-}OH} \quad \xrightarrow[\text{MW}]{\text{MnO}_2\text{-silica}} \quad \underset{\textbf{2}}{\overset{R_1}{\underset{R_2}{\diagup}}C\text{=}O}$$

a: R_1=Ph; R_2=H
b: R_1=Ph; R_2=Et
c: R_1=R_2=Ph
d: R_1=Ph; R_2=PhCO
e: R_1=4-MeC$_6$H$_4$; R_2=H
f: R_1=4-MeOC$_6$H$_4$; R_2=H
g: R_1=4-MeOC$_6$H$_4$; R_2=4-MeOC$_6$H$_4$
h: R_1=PhCH=CH; R_2=H

Experimental procedures:
MnO$_2$ "doped" silica (1.25 g, 5 mmol of MnO$_2$ on silica gel, Selecto Scientific, 230–400 mesh with large surface area of 600 m^2 g^{-1}) is thoroughly mixed with benzyl alcohol 1a (108 mg, 1 mmol) and the material is placed in an alumina bath inside the microwave oven and irradiated for 20 s. Upon completion of the reaction, monitored on TLC (hexane-AcOEt, 10:1), the product is extracted into methylene chloride, solvent removed and the residue passed through a bed of silica gel (4 cm) to afford exclusively benzaldehyde **2a**. The overoxidation to carboxylic acid is not observed. The same reaction could be completed in 2 h at a comparable temperature of 55 °C in an oil bath.

References: R. S. Varma, R. K. Saini, R. Dahiya, *Tetrahedron Lett.*, **38**, 7823 (1997).

Type of reaction: oxidation
Reaction condition: solvent-free
Keywords: alcohol, iodobenzene diacetate, alumina, microwave irradiation, ketone, aldehyde

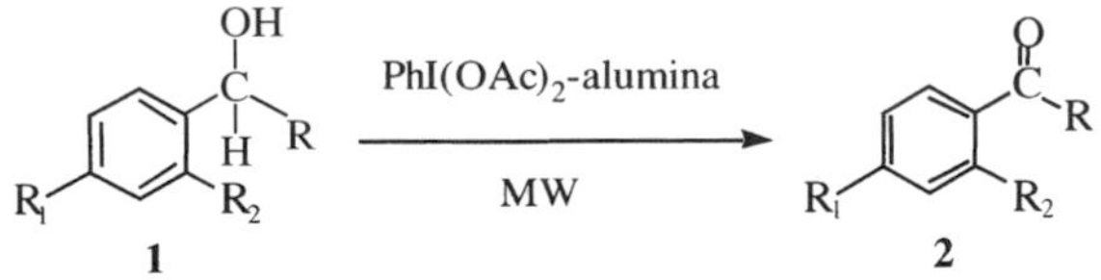

a: $R_1=R_2=R=H$
b: $R_1=4\text{-Me}$; $R_2=R=H$
c: $R_1=4\text{-MeO}$; $R_2=R=H$
d: $R_1=H$; $R_2=CH_2OH$; $R=H$
e: $R_1=R_2=H$; $R=Et$
f: $R_1=R_2=H$; $R=PhCO$
g: $R_1=4\text{-MeO}$; $R_2=H$; $R=4\text{-MeOC}_6H_4CO$

Experimental procedures:

The oxidation of benzyl alcohol **1a** to benzaldehyde **2a** is representative of the general procedure employed. Benzyl alcohol **1a** (0.108 g, 1 mmol) and IBD (0.355 g, 1.1 mmol) doped on neutral alumina (1 g) are mixed thoroughly on a vortex mixer. The reaction mixture is placed in an alumina bath inside an un-modified household microwave oven and irradiated for a period of 1 min. On completion of the reaction, followed by TLC examination (hexane-AcOEt, 9:1, v/v), the product is extracted into dichloromethane and is neutralized with aque-ous sodium bicarbonate solution. The dichloromethane layer is separated, dried over magnesium sulfate, filtered, and the crude product thus obtained is purified by column chromatography to afford pure benzaldehyde **2a** in 94% yield. Alter-natively, the crude products are charged on a silica gel column that provides io-dobenzene on elution with hexane followed by pure carbonyl compounds in sol-vent system (hexane-ethyl acetate, 9:1, v/v).

References: R. S. Varma, R. Dahiya, R. K. Saini, *Tetrahedron Lett.,* **38**, 7029 (1997).

Type of reaction: oxidation
Reaction condition: solvent-free
Keywords: alcohol, wet alumina, chromium(VI) oxide, microwave irradiation, ketone, aldehyde

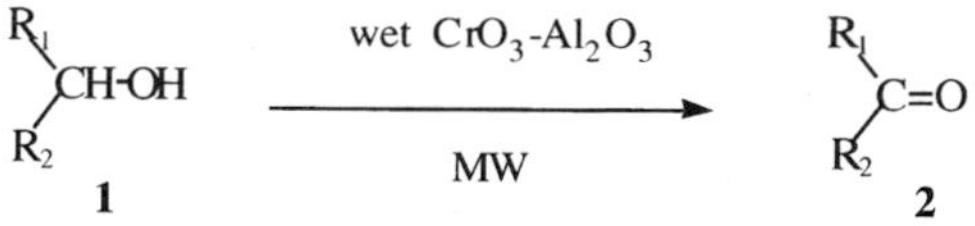

$$R_1\text{-CH-OH} \quad \xrightarrow[\text{MW}]{\text{wet } CrO_3\text{-}Al_2O_3} \quad R_1\text{-C=O}$$

R₂ → R₂

1 → **2**

a: R=Ph; R_2=H
b: R=4-MeC$_6$H$_4$; R_2=H
c: R=4-MeOC$_6$H$_4$; R_2=H
d: R=4-NO$_2$C$_6$H$_4$; R_2=H
e: R=R_2=Ph
f: R=Ph; R_2=Me
g: R=PhCO; R_2=Ph

Experimental procedures:

Wet alumina is prepared by shaking neutral alumina oxide (10 g, Aldrich, Brockmann I, ~ 150 mesh) with distilled water (2 mL). The reagent is prepared by mixing CrO_3 (0.8 g, 8 mmol) with wet alumina (2.4 g) using a pestle and mortar. This reagent is gradually added to the benzyl alcohol **1a** (0.432 g, 4 mmol) and mixed with a spatula. An exothermic reaction ensues with darkening of the orange color of the reagent and is completed almost immediately as confirmed by TLC (hexane-AcOEt, 8:2). The product is extracted into methylene chloride (2×25 mL) and is passed through a small bed of alumina (1 cm) to afford pure benzaldehyde **2a**. In some cases, brief microwave irradiation (inside an alumina bath in an unmodified household microwave oven) completes the reaction.

References: R.S. Varma, R.K. Saini, *Tetrahedron Lett.*, **39**, 1481 (1998).

Type of reaction: oxidation
Reaction condition: solvent-free
Keywords: alcohol, zeolite HZSM-5, chromium trioxide, microwave irradiation, ketone, aldehyde

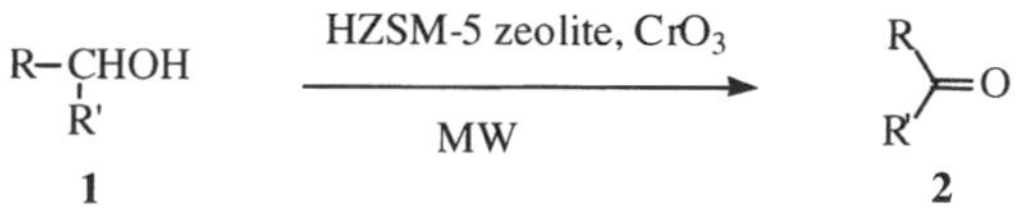

$$\text{R-CHOH} \quad \xrightarrow[\text{MW}]{\text{HZSM-5 zeolite, } CrO_3} \quad \text{R}$$
R' → R' =O

1 → **2**

a: R=Ph; R'=H
b: R=4-MeC$_6$H$_4$; R'=H
c: R=5-Me-4-NO$_2$C$_6$H$_3$; R'=H
d: R=Ph; R'=Ph
e: R=Ph; R'=COPh
f: R=PhCH=CH; R=H
g: R,R'=-(CH$_2$)$_5$-

Experimental procedures:

Chromium trioxide (2 mmol) and a indicated equivalent of HZSM-5 zeolite were crushed together in a mortar so as to form an intimate mixture. A neat alcohol was added to this intimate mixture. The resulting mixture was mixed thoroughly using a spatula. This mixture was placed on a microwave oven and irradiated (900 W) for an indicated time. The crude product was subjected to column chromatography using hexane-EtOAc (8:2) as eluent affording the corresponding carbonyl compound.

References: M.M. Heravi, D. Ajami, K. Tabar-Hydar, M. Ghassemzadeh, *J. Chem. Res. (S),* 334 (1999).

Type of reaction: oxidation
Reaction condition: solvent-free
Keywords: benzoin, alumina, copper(II) sulfate, microwave irradiation, benzil

a: $R_1 = R_2 = Ph$
b: $R_1 = Ph$; $R_2 = p\text{-MeC}_6H_4$
c: $R_1 = Ph$; $R_2 = p\text{-MeOC}_6H_4$
d: $R_1 = R_2 = p\text{-ClC}_6H_4$
e: $R_1 = R_2 = p\text{-MeC}_6H_4$
f: $R_1 = p\text{-MeC}_6H_4$; $R_2 = p\text{-MeOC}_6H_4$
g: $R_1 = R_2 = 2\text{-furyl}$

Experimental procedures:

Benzoin **1** (1 mmol) and $CuSO_4\text{-}Al_2O_3$ (1.5 g, 0.85 mmol of $CuSO_4 \cdot 5H_2O$) were mixed thoroughly on a vortex mixer. The reaction mixture contained in glass tubes was placed in an alumina bath (heat sink) inside the microwave oven and irradiated for a specified time. On completion of the reaction, followed by TLC examination (hexane-ethyl acetate, 9:1), the product was extracted into methylene chloride (3×10 mL). The solvent was removed under reduced pressure and the residue crystallized from an appropriate solvent to afford nearly quantitative yields of benzils **2**.

References: R.S. Varma, D. Kumar, R. Dahiya, *J. Chem. Res. (S)*, 324 (1998).

Type of reaction: oxidation
Reaction condition: solvent-free
Keywords: benzoin, zeolite, microwave irradiation, benzil

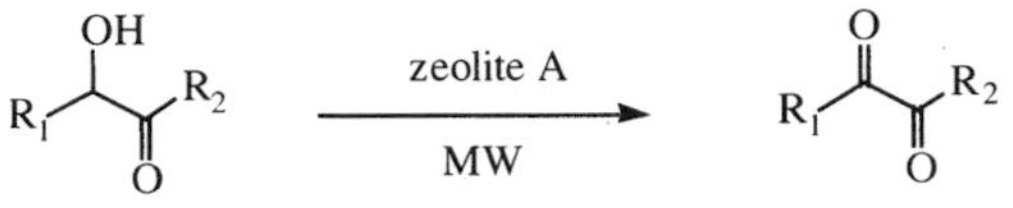

a: $R_1=R_2=Ph$
b: $R_1=R_2=p\text{-}ClC_6H_4$
c: $R_1=R_2=p\text{-}MeC_6H_4$
d: $R_1=R_2=p\text{-}MeOC_6H_4$
e: $R_1=p\text{-}MeOC_6H_4$; $R_2=Ph$
f: $R_1=R_2=furyl$

Experimental procedures:
Benzoin **1a** (2 mmol) and zeolite A (2 g) were mixed thoroughly in a mortar and the mixture was transfered to a beaker and irradiated for 6 min. The progress of the reaction was monitored by TLC using CH_2Cl_2 as solvent. The mixture was extracted into methylene chloride and then filtered. The solvent was removed under reduced pressure to afford pure benzil **2a** (340 mg, 80%). Further purification was carried out by column chromatography (CH_2Cl_2-light petroleum, 80:20, v/v) and crystallization in EtOH.

References: S. Balalaie, M. Golizeh, M. S. Hashtroudi, *Green Chem.,* **2**, 277 (2000).

Type of reaction: oxidation
Reaction condition: solvent-free
Keywords: tetrahydropyranyl ether, deprotection, montmorillonite K-10, microwave irradiation, ketone, aldehyde

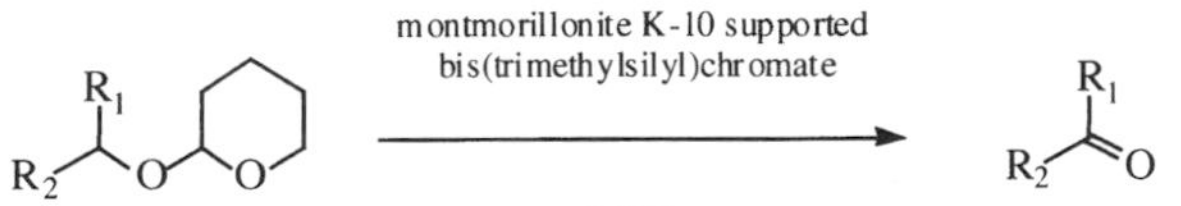

a: $R_1=H$; $R_2=Ph$
b: $R_1=H$; $R_2=4\text{-}MeC_6H_4$
c: $R_1=H$; $R_2=2\text{-}NO_2\text{-}5\text{-}MeC_6H_3$
d: $R_1=Me$; $R_2=Ph$
e: $R_1=H$; $R_2=CH=CHPh$
f: $R_1=R_2=Ph$

Experimental procedures:
Montmorillonite K-10 supported bis(trimethylsilyl)chromate (0.93 g, equivalent to 1.5 mmol) was thoroughly mixed with 0.192 g benzyltetrahydropyranyl ether (1 mmol), and the material was placed in a beaker inside the microwave oven (900 W) for 20 s. After completion of the reaction as monitored by TLC (hexane-AcOEt, 8:2) the product was extracted into CH_2Cl_2, the solvent was removed, and the residue was chromatographed on a silica gel column using hexane-AcOEt (8:2) to afford 92% benzaldehyde.

References: M.M. Heravi, D. Ajami, *Monatsh. Chem.,* **130,** 709 (1999); M.M. Eravi, R. Hekmatshoar, Y.S. Beheshtiha, M. Ghassemzadeh, *Monatsh. Chem.,* **132,** 651 (2001).

Type of reaction: oxidation
Reaction condition: solid-state
Keywords: trimethylsilyl ether, microwave irradiation, oxidative deprotection, ketone, aldehyde

a: R_1=Ph; R_2=H
b: R_1=2-MeOC$_6$H$_4$; R_2=H
c: R_1=4-MeOC$_6$H$_4$; R_2=H
d: R_1=4-MeC$_6$H$_4$; R_2=H
e: R_1=2-NO$_2$-5-MeC$_6$H$_3$; R_2=H
f: R_1,R_2=-(CH$_2$)$_5$-
g: R_1=C$_6$H$_{13}$; R_2=H
h: R_1=PhCH=CH; R_2=H

Experimental procedures:
Zeofen (equivalent to 1.2 mmol of ferric nitrate) was gradually added to benzyl trimethylsilyl ether **1a** (0.18 g, 1 mmol) in a beaker and mixed with spatula. The beaker was placed under microwave irradiation for 20 s. The product was extracted with methylene chloride and was passed through a small bed of silica gel to afford pure benzaldehyde **2a**.

References: M.M. Heravi, D. Ajami, M. Ghassemzadeh, K. Tabar-Hydar, *Synth. Commun.,* **31,** 2097 (2001).

Type of reaction: oxidation
Reaction condition: solvent-free
Keywords: benzoin, $Fe(NO_3)_3 \cdot 9H_2O$, microwave irradiation, benzil

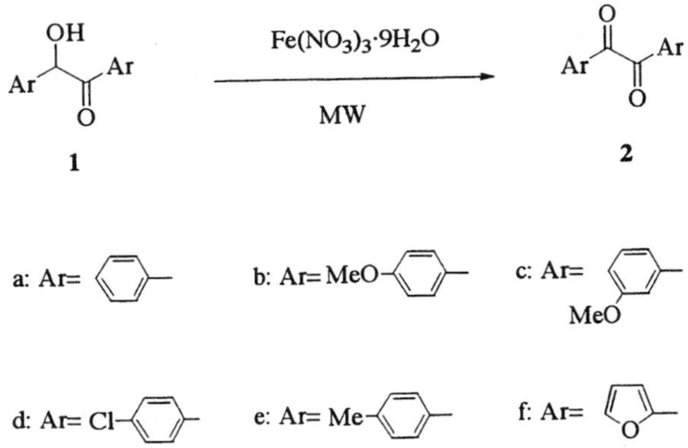

Experimental procedures:
Benzoin (1 mmol) and $Fe(NO_3)_3 \cdot 9H_2O$ (1 g) was mixed thoroughly and then irradiated in a microwave oven for 0.5–1 min. Acetone (10 mL) was added to the crude mixture, then mixed thoroughly. After that, 30 mL of cold water was added and faint yellow products were precipitated. The products were isolated by filtration, recrystallized with methanol, and dried under vacuum.

References: Y. Zhao, Y. Wang, *J. Chem. Res. (S)*, 70 (2001).

Type of reaction: oxidation
Reaction condition: solvent-free
Keywords: sulfide, sodium periodate, wet silica, microwave irradiation, sulfoxide

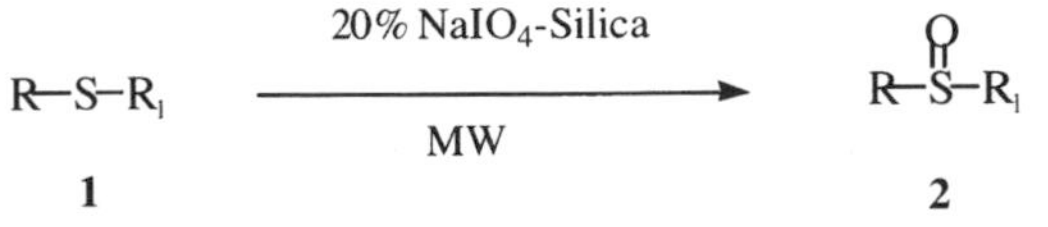

a: $R=PhCH_2$, $R_1=Ph$
b: $R=R_1=Ph$
c: $R=R_1=PhCH_2$
d: $R=Ph$; $R_1=Me$
e: $R=R_1=Bu$
f: $R=C_{12}H_{25}$, $R_1=Me$
g: R, $R_1=-(CH_2)_4-$

Experimental procedures:
The sulfide **1** (0.75 mmol) is dissolved in dichloromethane (2–3 mL) and adsorbed over silica supported sodium periodate (20%, 1.36 g, 1.28 mmol) that is
wetted with 0.3 mL of water by thoroughly mixing on a vortex mixture. The adsorbed powdered material is transferred to a glass test tube and is inserted in an
alumina bath (alumina: 100 g, mesh 65–325, Fisher scientific; bath: 5.7 cm diameter) inside the microwave oven. The compound is irradiated for the time specified in the table and the completion of the reaction is monitored by TLC examination. After completion of the reaction, the product is extracted into ethyl acetate (2×15 mL). The removal of solvent at reduced pressure affords crude sulfoxide **2** that contains less than 5% sulfone. The final purification is achieved by
column chromatography over silica gel column or a simple crystallization.

References: R. S. Varma, R. K. Saini, H. M. Meshram, *Tetrahedron Lett.*, **38**, 6525 (1997).

Type of reaction: oxidation
Reaction condition: solvent-free
Keywords: sulfide, iodobenzene diacetate, microwave irradiation, sulfoxide

$$R_1\text{—S—}R_2 \xrightarrow[\text{MW}]{\text{PhI(OAc)}_2\text{-alumina}} R_1\text{—}\overset{\overset{O}{\|}}{S}\text{—}R_2$$

1   **2**

a: $R_1=R_2=i\text{-Pr}$
b: $R_1=R_2=n\text{-Bu}$
c: $R_1=R_2=Ph$
d: $R_1=CH_2Ph$; $R_2=Ph$
e: $R_1=R_2=CH_2Ph$
f: $R_1=CH_2Ph$; $R_2=Me$
g: $R_1=n\text{-}C_{12}H_{25}$; $R_2=Me$
h: $R_1, R_2=\text{-}(CH_2)_4\text{-}$
i: $R_1, R_2=\text{-}(CH_2)_2CO(CH_2)_2\text{-}$

Experimental procedures:
Neutral alumina (1.5 g) was thoroughly mixed with iodobenzene diacetate (532
mg, 1.65 mmol) and benzyl phenyl sulfide **1d** (300 mg, 1.5 mmol) using a pestle
and mortar. The adsorbed material was placed in an alumina bath inside the microwave oven and irradiated at 50% power for two successive intervals of 45 s
each (with time interval of 3–4 min; bath temperature rose to 80–85 °C). The
progress of the reaction was monitored by TLC (hexane-ethyl acetate, 7:3, v/v).
When the reaction was complete the whole material was directly charged onto a
silica gel column which provided iodobenzene on elusion with hexane (100 mL).
The fractions eluted by chloroform-hexane (1:1 v/v) provided sulfone (<7% by

NMR of crude product) and finally elution by chloroform afforded pure benzyl phenyl sulfoxide **2d** in 86% yield, mp 123 °C.

References: R. S. Varma, R. K. Saini, R. Dahiya, *J. Chem. Res. (S),* 120 (1998).

Type of reaction: oxidation
Reaction condition: solvent-free
Keywords: sulfide, clayfen, iron(III) nitrate, microwave irradiation, sulfoxide

$$R_1\text{—S—}R_2 \quad \xrightarrow[\text{MW}]{\text{clayfen}} \quad R_1\text{—}\overset{\overset{\textstyle O}{\|}}{S}\text{—}R_2$$

$$\mathbf{1} \qquad\qquad\qquad\qquad \mathbf{2}$$

a: $R_1=R_2=Pr$
b: $R_1=R_2=Bu$
c: $R_1=C_{12}H_{25}$, $R_2=Me$
d: $R_1=Ph$; $R_2=Me$
e: $R_1=R_2=Ph$
f: $R_1=Ph$, $R_2=PhCH_2$
g: $R_1=R_1=PhCH_2$
h: $R_1,R_2=-(CH_2)_4-$
i: $R_1,R_2=-CH_2CH_2COCH_2CH_2-$

Experimental procedures:
Clayfen (1.0 g) is thoroughly mixed with the sulfide (2 mmol) in a test tube. The reaction mixture is placed in an alumina bath inside the microwave oven and is irradiated for the stipulated time. On completion of the reaction, monitored by TLC, the products were extracted with CH_2Cl_2 (3×10 mL). The solvent was removed on a rotary evaporator and the crude product, thus obtained, was charged on a silica gel column. The fractions eluted by chloroform-hexane (1:1) provided sulfone and final elution in chloroform afforded pure sulfoxide.

References: R. S. Varma, R. Dahiya, *Synth. Commun.,* **28**, 4087 (1998).

Type of reaction: oxidation
Reaction condition: solid-state
Keywords: sulfide, *tert*-butyl hydroperoxide, silica gel, microwave irradiation, sulfoxide, sulfone

$$1 \xrightarrow[\text{silica gel, MW}]{(CH_3)_3COOH} 2$$

$$2 \xrightarrow[\text{silica gel, MW}]{(CH_3)_3COOH} 3$$

Experimental procedures:

Into a 25-mL round-bottomed flask was weighed 2.5 g of Merck 10181 chromatographic silica gel that had been equilibrated with the atmosphere at 120 °C for at least 48 h. The flask was stoppered and the contents allowed to cool to 25 °C. The substrate was added without solvent and the resulting mixture tumbled on a rotary evaporator at atmospheric pressure until uniformly free-flowing. The oxidant was then added and the mixture again tumbled. After being heated, the mixture was allowed to cool to 25 °C and was washed with 100–200 mL of ethyl acetate. The adsorbent was collected by vacuum filtration and the filtrate concentrated under reduced pressure. The residue was weighed and analyzed by ^{1}H NMR spectroscopy.

References: P.J. Kropp, G.W. Breton, J.D. Fields, J.C. Tung, B.R. Loomis, *J. Am. Chem. Soc.*, **122**, 4280 (2000).

Type of reaction: oxidation
Reaction condition: solvent-free
Keywords: thiol, Hyflo Super Cel, microwave irradiation, oxidative coupling, disulfide

$$2 \underset{1}{R{-}SH} \xrightarrow[\text{MW}]{\text{air, Hyflo Super Cel}} \underset{2}{R{-}S{-}S{-}R}$$

a: R=4-NO$_2$C$_6$H$_4$
b: R=2-pyridyl
c: R=2-HOOCC$_6$H$_4$
d: R=2-ClC$_6$H$_4$

Experimental procedures:

Reaction mixtures were prepared by mechanical dispersion of the thiol on the support (Hyflo Super Cel, Fulka) with a vortex mixer, or by impregnation of the support with a thiol solution in either, methylene dichloride or methanol. Oxidation was performed in an open vessel, in a microwave reactor Synthewave 402 (Prolabo) or in a preheated oil bath. After the reaction, the coated support was eluted with an appropriate solvent. After drying, the support could be reused without loss of activity. The products were identified by NMR, MS and by comparison with authentic samples. There was no evidence of the presence of side-products.

References: L. S. Marie, E. G. Jampel, M. Therisod, *Tetrahedron Lett.,* **39**, 9661 (1998).

Type of reaction: oxidation
Reaction condition: solid-state
Keywords: 1,4-dihydropyridine, phenyliodine(III) bis(trifluoroacetate), microwave irradiation, pyridine

a: R=H
b: R=Me
c: R=Et
d: R=Ph
e: R=PhCH$_2$
f: R=4-CH$_3$C$_6$H$_4$
g: R=4-NO$_2$C$_6$H$_4$
h: R=2-furyl

Experimental procedures:
1,4-Dihydroxypyridines **1** (1 mmol) and phenyliodine(III)bis(trifluoroacetate) (PIFA) (1.2 mmol) were mixed thoroughly using a vortex mixer at room temperature. After 15 min the reaction mixture was poured into water and the product extracted into methylene chloride (2×10 mL). The combined extract was dried over anhydrous sodium sulfate and the solvent was removed under reduced pressure. The residue was passed through a small bed of silica gel using hexane-EtOAc (9:1) as eluent to afford the corresponding pyridine derivative **2**.

References: R. S. Varma, D. Kumar, *J. Chem. Soc., Perkin Trans. 1,* 1755 (1999).

Type of reaction: oxidation

Reaction condition: solvent-free

Keywords: 1-hydroxyphosphonate, CrO_3/alumina, microwave irradiation, α-keto phosphonate

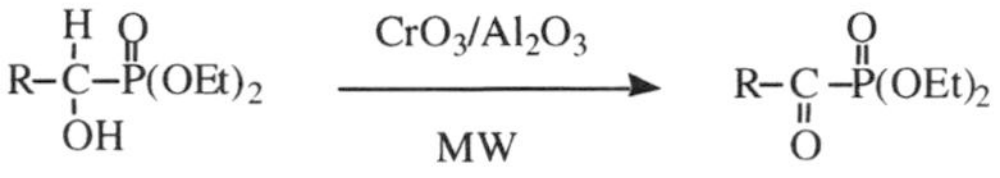

$$R-\overset{\underset{\displaystyle OH}{|}}{\underset{\underset{}{}}{\overset{H}{\underset{|}{C}}}}-\overset{O}{\overset{\|}{P}}(OEt)_2 \quad \xrightarrow[\text{MW}]{CrO_3/Al_2O_3} \quad R-\overset{O}{\underset{\|}{C}}-\overset{O}{\overset{\|}{P}}(OEt)_2$$

$$\textbf{1} \qquad\qquad\qquad\qquad \textbf{2}$$

a: R=n-Bu   j: R=p-NO$_2$C$_6$H$_4$
b: R=n-pentyl   k: R=p-MeOC$_6$H$_4$
c: R=PhCH$_2$CH$_2$ l: R=o-ClC$_6$H$_4$
d: R=PhCH=CH   m: R=o-O$_2$NC$_6$H$_4$
e: R=MeCH=CH   n: R=2,4-Cl$_2$C$_6$H$_3$
f: R=Ph   o: R=furfuryl
g: R=p-MeC$_6$H$_4$ p: R=α-naphthyl
h: R=p-ClC$_6$H$_4$ q: R=β-naphthyl

Experimental procedures:

Thirty mmol of the reagent is prepared by the combination of CrO_3 (30 mmol, finely ground) and alumina (Al_2O_3, neutral, 5.75 g) in a mortar and by grinding until a fine, homogeneous, organic powder is obtained (5–10 min). The 1-hydroxyphosphonate (10 mmol) was added to this reagent and was irradiated by microwave. The reaction mixture was washed with CH_2Cl_2 (200 mL), dried (Na_2SO_4), and the solvent evaporated under reduced pressure in 65–90% yields.

References: B. Kaboudin, R. Nazari, *Synth. Commun.*, **31**, 2245 (2001).

3 Carbon–Carbon Bond Formation

3.1 Solvent-Free C–C Bond Formation

Type of reaction: C–C bond formation
Reaction condition: solid-state
Keywords: phenol, oxidative coupling reaction, $FeCl_3 \cdot 6H_2O$

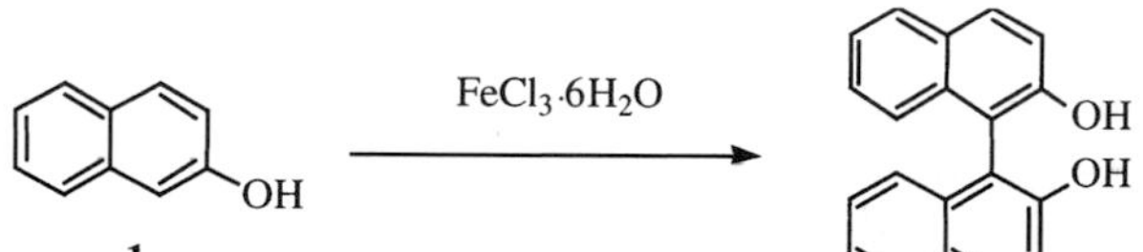

Experimental procedures:
A mixture of **1** (1 g, 7 mmol) and $FeCl_3 \cdot 6H_2O$ (3.8 g, 14 mmol) was finely powdered by agate mortar and pestle. The mixture was then put in a test tube and
kept at 50 °C for 2 h. Decomposition of the reaction mixture with dilute HCl
gave **2** in 95% yield.

References: F. Toda, K. Tanaka, S. Iwata, *J. Org. Chem.*, **54**, 3007 (1989).

Type of reaction: C–C bond formation
Reaction condition: solid-state
Keywords: ketone, Grignard reaction, alcohol

Ph$_2$C=O   +   R-MgX   ⟶   Ph$_2$RCOH   +   Ph$_2$CHOH

1   **2**   **3**   **4**

a: R=Me; X=I
b: R=Et; X=Br
c: R=*i*-Pr; X=Br
d: R=Ph; X=Br

Experimental procedures:

One mole of ketone and three moles of the dried Grignard reagent were finely powdered and well mixed with agate mortar and pestle, and the mixture was kept at room temperature for 0.5 h. The reaction mixture was decomposed with aqueous NH$_4$Cl, extracted with ether, and the extract dried over Na$_2$SO$_4$. Evaporation of the solvent gave products.

References: F. Toda, H. Takumi, H. Yamaguchi, *Chem. Exp.,* **4**, 507 (1989).

Type of reaction: C–C bond formation
Reaction condition: solid-state
Keywords: acetylenic compound, CuCl$_2$, Glaser coupling, diacetylenic compound

$$\underset{\overset{|}{OH}}{\overset{\overset{R'}{|}}{R-C}}-\!\!\equiv\!\!-H \quad\xrightarrow{\text{CuCl}_2\cdot 2\text{ pyridine}}\quad \underset{\overset{|}{OH}}{\overset{\overset{R'}{|}}{R-C}}-\equiv-\equiv-\underset{\overset{|}{OH}}{\overset{\overset{R'}{|}}{C}}-R$$

1   **2**

a: R=Ph; R'=Ph
b: R=Ph; R'=2-ClC$_6$H$_4$
c: R=4-MeC$_6$H$_4$; R'=4-MeC$_6$H$_4$
d: R=Ph; R'=2,4-Me$_2$C$_6$H$_3$
e: R=2,4-Me$_2$C$_6$H$_3$; R'=2-ClC$_6$H$_4$
f: R=Ph; R'=Me
g: R=Ph; R'=*n*-Bu
h: R, R'=fluorenyl

Experimental procedures:

When a mixture of powdered propargyl alcohol **1a** and CuCl$_2$·2 pyridine complex was reacted at 50 °C for 20 h, the coupling product **2a** was obtained in 65% yield.

References: F. Toda, Y. Tokumaru, *Chem. Lett., 987* (1990).

Type of reaction: C–C bond formation
Reaction condition: solvent-free
Keywords: ethynylbenzene, ketone, tertiary alkynol

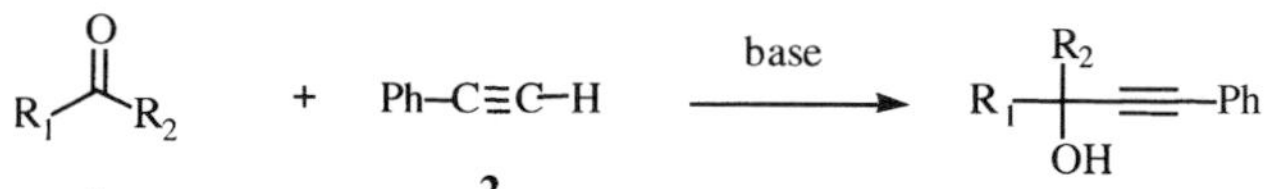

a: R_1=Me; R_2=Me
b: R_1=Me; R_2=Et
c: R_1=Me; R_2=n-Pr
d: R_1=Et; R_2=Et
e: R_1=i-Pr, R_2=i-Pr
f: R_1=Ph; R_2=Me
g: R_1=Ph; R_2=Et
h: R_1=Ph; R_2=n-Pr
i: R_1=Ph; R_2=i-Pr
j: R_1,R_2=-(CH$_2$)$_4$–
k: R_1,R_2=-(CH$_2$)$_5$–
l: R_1,R_2=-(CH$_2$)$_6$–

Experimental procedures:

After acetone **1a** (1.0 g, 17.2 mmol), ethynylbenzene **2** (1.8 g, 17.2 mmol) and potassium *t*-butoxide (1.9 g, 17.2 mmol) were well-mixed with agate mortar and pestle, the mixture was kept at room temperature for 20 min. The reaction product was mixed with 10% aqueous sodium chloride, filtered, washed with water, and dried to give **3a** as colorless crystals (2.6 g, 94% yield).

References: H. Miyamoto, S. Yasaka, K. Tanaka, *Bull. Chem. Soc. Jpn.,* **74**, 185 (2001).

Type of reaction: C–C bond formation
Reaction condition: solvent-free
Keywords: ketone, aldehyde, pinacol coupling reaction, Zn, ZnCl$_2$, α-glycol

$$\underset{\textbf{1}}{\overset{\overset{\displaystyle O}{\|}}{Ar-C-R}} \quad \xrightarrow{Zn-ZnCl_2} \quad \underset{\underset{\displaystyle OH \ \ OH}{|\ \ \ \ |}}{\overset{\overset{\displaystyle R \ \ \ \ R}{|\ \ \ \ |}}{Ar-C-\!\!-\!\!C-Ar}}$$

1   2

a: Ar=Ph; R=H
b: Ar=4-MeC$_6$H$_4$; R=H
c: Ar=4-ClC$_6$H$_4$; R=H
d: Ar=4-BrC$_6$H$_4$; R=H
e: Ar=4-CNC$_6$H$_4$; R=H
f: Ar=4-PhC$_6$H$_4$; R=H
g: Ar=Ph; R=Me
h: Ar=4-BrC$_6$H$_4$; R=Me
i: Ar=4-CNC$_6$H$_4$; R=Me
j: Ar=Ph; R=Ph
k: Ar=4-MeC$_6$H$_4$; R=4-MeC$_6$H$_4$
l: Ar=4-ClC$_6$H$_4$; R=4-ClC$_6$H$_4$
m: Ar=Ph; R=4-ClC$_6$H$_4$
n: Ar=R=fluorenyl

Experimental procedures:

A mixture of **1** (1 g), Zn powder (5 g) and ZnCl$_2$ (1 g) was kept at room temperature for 3 h. The reaction mixture was combined with 3N HCl (5 mL) and toluene (10 mL) and filtered to remove Zn powder. The filtrate was extracted with toluene, and the toluene solution was washed with water and dried over MgSO$_4$. The toluene solution was evaporated to give **2**.

References: K. Tanaka, S. Kishigami, F. Toda, *J. Org. Chem.*, **55**, 2981 (1990).

Type of reaction: C–C bond formation
Reaction condition: solvent-free
Keywords: aldehyde, ketone, halide, Reformatsky reaction, Luche reaction, Zn, NH$_4$Cl, β-hydroxy ester

$$\underset{\textbf{1}}{Ar\text{-}CHO} \quad + \quad \underset{\textbf{2}}{BrCH_2COOEt} \quad \xrightarrow[NH_4Cl]{Zn} \quad \underset{\underset{\displaystyle OH}{|}}{\overset{\overset{\displaystyle H}{|}}{Ar\text{-}C\text{-}CH_2COOEt}}$$

3

a: Ar=Ph
b: Ar=4-BrC$_6$H$_4$
c: Ar=3,4-CH$_2$O$_2$C$_6$H$_3$
d: Ar=4-PhC$_6$H$_4$
e: Ar=2-Naphthyl

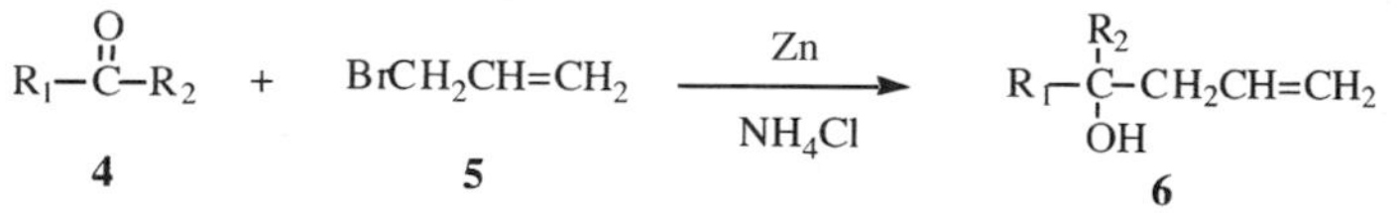

$$R_1-\overset{\overset{\displaystyle O}{\|}}{C}-R_2 \quad + \quad BrCH_2CH=CH_2 \quad \xrightarrow[NH_4Cl]{Zn} \quad R_1-\overset{\overset{\displaystyle R_2}{|}}{\underset{\underset{\displaystyle OH}{|}}{C}}-CH_2CH=CH_2$$

    4   **5**   **6**

a: R_1=Ph; R_2=H
b: R_1=2-Naphtyl; R_2=H
c: R_1=C_5H_{11}; R_2=H
d: R_1=*trans*-CH_3CH=CH; R_2=H
e: R_1=C_5H_{11}; R_2=CH_3
f: R_1, R_2=-$(CH_2)_5$-

Experimental procedures:

A mixture of aromatic ketone **1** (5.1 mmol), ethyl bromoacetate **2** (2.56 g, 15.3 mmol), Zn powder (5 g) and NH_4Cl (2 g) was thoroughly ground in an agate mortar and pestle, and the mixture was kept at room temperature for 23 h. The reaction product was mixed with aqueous NH_4Cl and extracted with ether. The ether solution was washed with water and dried over anhydrous $MgSO_4$. Evaporation of the solvent and volatile ketone in vacuo gave the product **3** in pure form.

References: K. Tanaka, S. Kishigami, F. Toda, *J. Org. Chem.*, **56**, 4333 (1991).

Type of reaction: C–C bond formation
Reaction condition: solvent-free
Keywords: alkyl 2-halocarboxylate, unsaturated carboxylic ester, radical addition, lactone

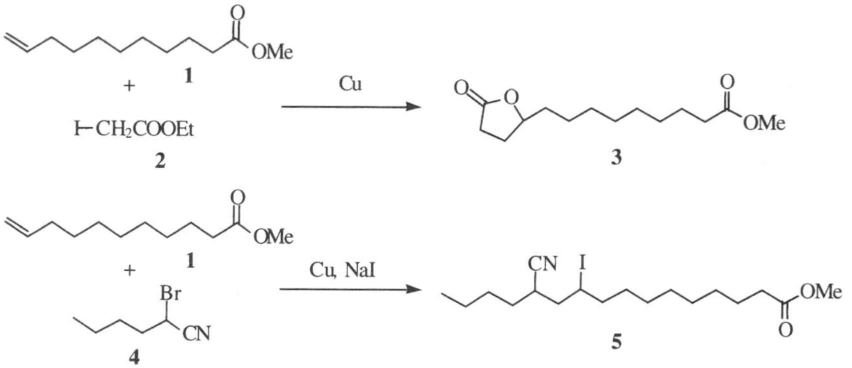

Experimental procedures:

A 50-mL, two-necked flask equiped with a magnetic stirrer, reflux condenser, and a gas inlet tube was charged with 1.3 equiv. of alkene, and 1.3 equiv. of copper powder (purum, Merck). The mixture was heated with constant stirring at 130 °C in an inert atmosphere. The reaction was monitored by thin-layer chromatography and was usually complete within 3–7 h. The product was isolated by distillation; alternatively, the reaction mixture was taken up into diethyl ether and after removal of copper salts by filtration and evaporation of the solvent on a rotary evaporator, the product was isolated by Kugelrohr distillation and recrystallization or flash chromatography.

References: J.O. Metzger, R. Mahler, *Angew. Chem. Int. Ed. Engl.,* **34**, 902 (1995).

Type of reaction: C–C bond formation
Reaction condition: solvent-free
Keywords: aldehyde, enantioselective addition, diethylzinc, secondary alcohol

a: R=Ph	e: R=p-ClC$_6$H$_4$
b: R=Tolyl	f: R=p-MeOC$_6$H$_4$
c: R=1-Naphthyl	g: R=c-C$_6$H$_{11}$
d: R=2-Naphthyl	h: R=PhCH$_2$CH$_2$

Experimental procedures:
To an ice-cooled two-necked flask containing (1S,2R)-DBNE **2** (13.7 mg, 0.05 mmol), neat Et$_2$Zn (370 mg, 3 mmol) was transferred into the flask through a cannula under an argon atmosphere. After the mixture was stirred for 10 min at 0 °C, aldehyde **1b** (120 mg, 1.0 mmol) was added slowly to the mixture, and the mixture was stirred for 2 h. After excess Et$_2$Zn was removed under a reduced pressure, the reaction was quenched by saturated aq. ammonium chloride. The mixture was extracted with ether and the organic layer was dried over magnesium sulfate. Concentration and purification on silica gel TLC gave (S)-**3b** (149 mg, 99%). Ee was determined to be 90% by HPLC analysis using a chiral stationary phase (Chiralcel OB-H).

References: I. Sato, T. Saito, K. Soai, *Chem. Commun.,* 2471 (2000).

Type of reaction: C–C bond formation
Reaction condition: solvent-free
Keywords: benzaldehyde, acetophenone, condensation, basic alumina, chalcone

R$_2$—⟨benzene⟩—CHO + H$_3$CCO—⟨benzene⟩—R$_1$ $\xrightarrow{\text{alumina}}$ R$_2$—⟨benzene⟩—CH=CH—CO—⟨benzene⟩—R$_1$

    1   **2**   **3**

    a: R$_1$=R$_2$=H
    b: R$_1$=OCH$_3$; R$_2$=H
    c: R$_1$=H; R$_2$=NO$_2$
    d: R$_1$=H; R$_2$=COOCH$_3$

⟨benzene⟩—CHO + R$_1$R$_2$C(CH$_2$)C=O $\xrightarrow{\text{alumina}}$ ⟨benzene⟩—CH=C(R$_1$)—C(=O)—C(R$_2$)=CH—⟨benzene⟩

  1a   **4**   **5**

    a: R$_1$, R$_2$=-CH$_2$-CH$_2$-
    b: R$_1$, R$_2$=-(CH$_2$)$_3$-

Experimental procedures:
Basic alumina (13 g) was added to a mixture of methyl 4-formylbenzoate **1d** (1.00 g, 6 mmol) and acetophenone **2** (0.48 g, 4 mmol) at room temperature. (When the reactants were solid, a minimum amount (2×3 mL) of dichloromethane was used to dissolve them prior to the addition of the alumina.) The reaction mixture was then agitated at room temperature for 2.5 h using a Fisher vortex mixer. The product was extracted into dichloromethane (5×15 mL). Removal of the solvent, under reduced pressure, yielded the solid product. Further purification (removal of traces of benzyl alcohol and aldehyde) was carried out by recrystallization from a petroleum ether-ether mixture to afford 1-phenyl-3-[4-(carbomethoxy)phenyl]-2-propen-1-one **3d** (4-carbomethoxychalcone), mp 119–120 °C (81%).

References: R. S. Varma, G. W. Kabalka, L. T. Evans, R. M. Pagni, *Synth. Commun.,* **15**, 279 (1985).

Type of reaction: C–C bond formation
Reaction condition: solvent-free
Keywords: aldehyde, ketone, Aldol condensation, chalcone

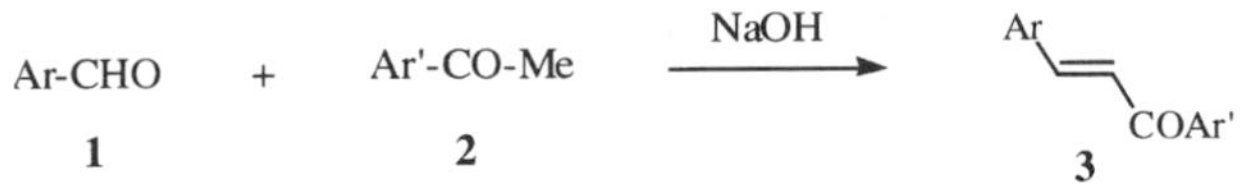

$$\text{Ar-CHO} \quad + \quad \text{Ar'-CO-Me} \quad \xrightarrow{\text{NaOH}} \quad$$

1   **2**   **3**

a: Ar=Ph; Ar'=Ph
b: Ar=4-MeC$_6$H$_4$; Ar'=Ph
c: Ar=4-MeC$_6$H$_4$; Ar'=4-MeC$_6$H$_4$
d: Ar=4-ClC$_6$H$_4$; Ar'=Ph
e: Ar=4-ClC$_6$H$_4$; Ar'=4-MeOC$_6$H$_4$
f: Ar=4-ClC$_6$H$_4$; Ar'=4-BrC$_6$H$_4$
g: Ar=3,4-Me$_2$C$_6$H$_3$; Ar'=4-BrC$_6$H$_4$

Experimental procedures:

When a slurry mixture of *p*-methylbenzaldehyde **1b** (1.5 g, 12.5 mmol), aceto-phenone **2b** (1.5 g, 12.5 mmol) and NaOH (0.5 g, 12.5 mmol) was ground by pestle and mortar at room temperature for 5 min, the mixture turned to a pale yellow solid. The solid was combined with water and filtered to give *p*-methyl-chalcone **3b** (2.7 g) in 97% yield.

References: F. Toda, K. Tanaka, K. Hamai, *J. Chem. Soc., Perkin Trans. 1*, 3207 (1990).

Type of reaction: C–C bond formation
Reaction condition: solvent-free
Keywords: active methylene compound, aldehyde, Knoevenagel condensation, ZnCl$_2$, styrene

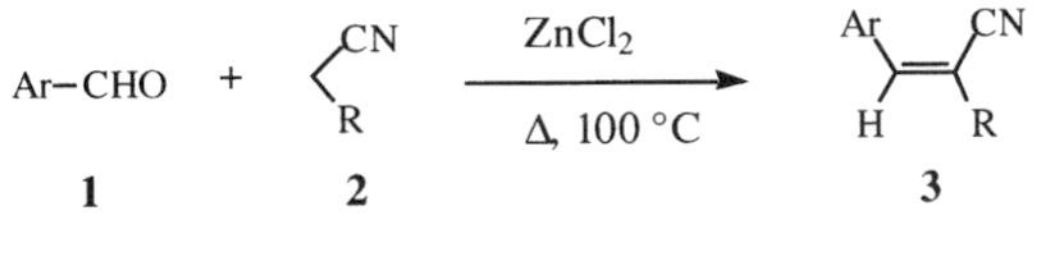

$$\text{Ar-CHO} \quad + \quad \begin{matrix} \text{CN} \\ \text{R} \end{matrix} \quad \xrightarrow[\Delta,\ 100\,°C]{\text{ZnCl}_2} \quad$$

1   **2**   **3**

a: Ar=Ph; R=CN
b: Ar=4-MeOC$_6$H$_4$; R=CN
c: Ar=4-ClC$_6$H$_4$; R=CN
d: Ar=3-MeO-4-HOC$_6$H$_3$; R=CN
e: Ar=PhCH=CH; R=CN
f: Ar=Ph; R=CONH$_2$
g: Ar=4-MeOC$_6$H$_4$; R=CONH$_2$
h: Ar=4-ClC$_6$H$_4$; R=CONH$_2$

i: Ar=3-MeO-4-HOC$_6$H$_3$; R=CONH$_2$
j: Ar=PhCH=CH; R=CONH$_2$
k: Ar=Ph; R=COOEt
l: Ar=4-MeOC$_6$H$_4$; R=COOEt
m: Ar=4-ClC$_6$H$_4$; R=COOEt
n: Ar=3-MeO-4-HOC$_6$H$_3$; R=COOEt
o: Ar=PhCH=CH; R=COOEt

Experimental procedures:

To a mixture of 0.01 mol of aldehyde **1** and 0.01 mol of active methylene com-pound **2**, zinc chloride (0.001 mol) was added and kept at 100 °C with constant stirring for the specified time. Then the mixture was cooled to room temperature and treated with a solution of 1% aqueous alcohol to obtain the product **3** in good purity. It was filtered and dried. Recrystallization is not necessary.

References: P. S. Rao, R. V. Venkataratnam, *Tetrahedron Lett.,* **32**, 5821 (1991).

Type of reaction: C–C bond formation
Reaction condition: solid-state
Keywords: lithium ester enolate, 3,3-dimethyl butanoate, aromatic aldehyde, Michael addition, β-hydroxy ester

a: R=2-MeOC$_6$H$_4$
b: R=4-ClC$_6$H$_4$
c: R=4-NO$_2$C$_6$H$_4$
d: R=3-NO$_2$C$_6$H$_4$
e: R=2-NO$_2$C$_6$H$_4$
f: R=5-NO$_2$-2-thienyl

Experimental procedures:
The freshly ground lithium enolate **1** (1.2 equiv.) was mixed with *o*-anisaldehyde **2a** (1 equiv.) in argon atmosphere at room temperature. The reaction was allowed to continue at room temperature under vacuum for three days, quenched with aqueous NH$_4$Cl and the mixture extracted with three portions of diethyl ether. The combined organic extract was washed with water and dried with anhydrous Na$_2$SO$_4$. The solvent was removed using a rotary evaporator at reduced pressure to yield the crude product. The crude product was found to contain mainly *anti* aldol product (*syn/anti* ratio 8:92). Further purification was carried out using preparative TLC with methanol-benzene (5:95 in volume) as eluent. The purified product thus isolated was a colorless solid (mp 64–65 °C, yield 70%) with the same *syn/anti* ratio as that of the crude product.

References: Y. Wei, R. Bakthavatchalam, *Tetrahedron Lett.,* **32**, 1535 (1991).

Type of reaction: C–C bond formation
Reaction condition: solvent-free
Keywords: cyclohexanone, benzaldehyde, Aldol condensation, RuCl$_3$, α,α'-bis-benzylidene cycloalkanone

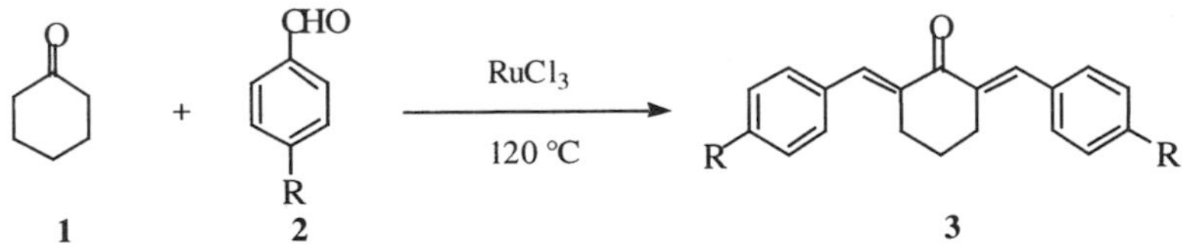

a: R=H
b: R=Cl
c: R=NO$_2$
d: R=Me
e: R=MeO
f: R=PhCH=CH

Experimental procedures:

Cyclohexanone **1** (5 mmol), aldehyde **2** (10.1 mmol) and anhydrous RuCl$_3$ (0.1 mmol) were placed in a glass tube and sealed. The sealed tube was placed in an oil bath and heated at 120 °C for 4–24 h. After cooling to room temperature, the sealed tube was opened. The reaction mixture was purified by either of the following procedures: (a) The reaction mixture was powdered, poured into cold ethanol (25 mL), stirred for 5 min and filtered. The crystalline product was further washed subsequently with water, 10% aqueous sodium bicarbonate, water, cold ethanol and dried. (b) The reaction mixture was added to 10% aqueous sodium bicarbonate (10 mL). The product was extracted with CHCl$_3$ (3×20 mL). The organic solution was washed with water, dried with anhydrous sodium sulfate and evaporated. Recrystallization of the product with ethanol or chromatography on a short column of silica-gel afforded the crystalline products (92–96%).

References: N. Iranpoor, F. Kazemi, *Tetrahedron,* **54**, 9475 (1998).

Type of reaction: C–C bond formation
Reaction condition: solvent-free
Keywords: diethyl adipate, diethyl pimelate, Dieckmann condensation, cyclic β-keto ester

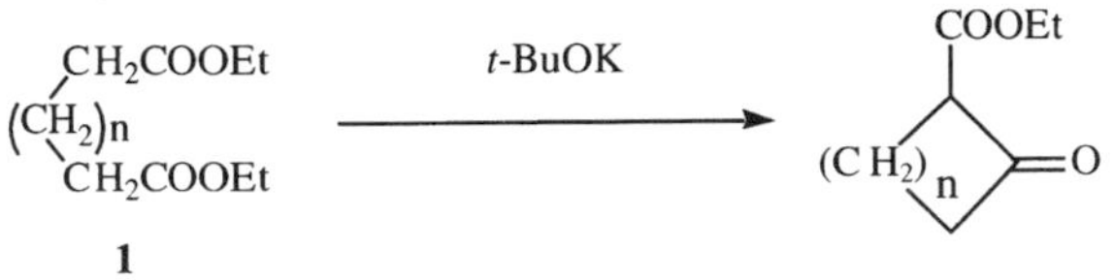

a: n=2
b: n=3

Experimental procedures:
When **1a** (10.2 g, 50.4 mmol) and *t*BuOK powder (8.44 g, 75.2 mmol) were mixed using a mortar and pestle for 10 min, the reaction mixture solidified. The solidified mixture was kept in a desiccator for 60 min. The reaction mixture was neutralized by addition of *p*-TsOH·H$_2$O and was distilled under 20 mmHg to give **2a** (5.8 g, 82% yield).

References: F. Toda, T. Suzuki, S. Higa, *J. Chem. Soc., Perkin Trans. 1*, 3521 (1998).

Type of reaction: C–C bond formation
Reaction condition: solvent-free
Keywords: benzaldehyde, pyrrole, alumina, corrole, oligopyrromethene

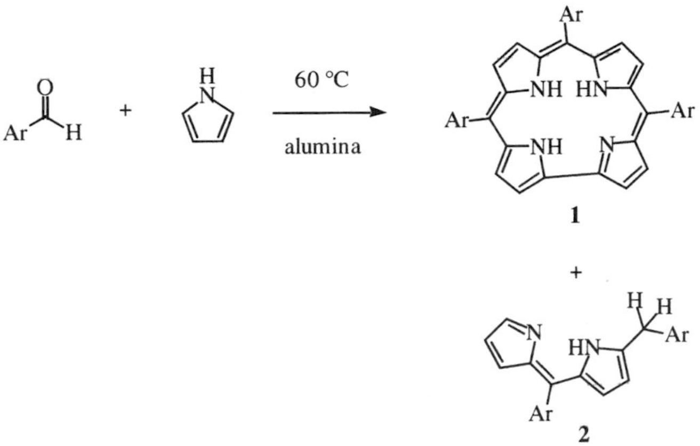

Experimental procedures:
The reagents (15 mmol of each) were first dissolved in CH$_2$Cl$_2$, mixed with alumina, and heated in an open vessel to 60 °C. The reaction starts only after the solvent evaporates, and a 4 h reaction time is sufficient. Column chromatographic separation gave 380 mg (12.7% yield) of the crude product **1** along with **2** (<1% yield).

References: Z. Gross, N. Galili, L. Simkhovich, I. Saltsman, M. Botoshansky, D. Blaser, R. Boese, I. Goldberg, *Org. Lett.,* **1**, 599 (1999).

Type of reaction: C–C bond formation
Reaction condition: solvent-free
Keywords: cyclohexanone, diethyl succinate, Stobbe condensation, cyclohexylidenesuccinic acid, cyclohexenylsuccinic acid

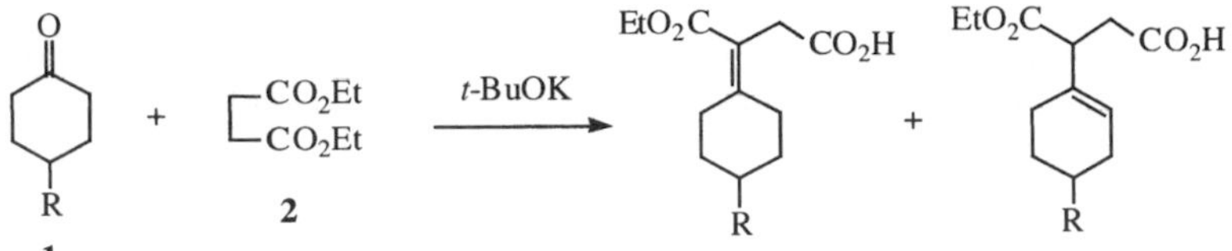

a: R=H
b: R=Me

Experimental procedures:
To a mixture of **1a** (1.0 g, 10.2 mmol) and **2** (1.78 g, 10.2 mmol) was added powdered *t*-BuOK (1.37 g, 12.3 mmol) in a mortar which was well ground with a pestle at room temperature for 10 min. The reaction was exposed to the air. The reaction mixture was neutralized with diluted HCl and then the crystals formed were isolated by filtration to give β-carbethoxy-β-cyclohexylidenepropionic acid **3a** (colorless plates, 1.73 g) in 75% yield after recrystallization from acetone.

References: K. Tanaka, T. Sugino, F. Toda, *Green Chem.,* **2**, 303 (2000).

Type of reaction: C–C bond formation
Reaction condition: solvent-free
Keywords: ninhydrin, dimedone, ball-milling, condensation

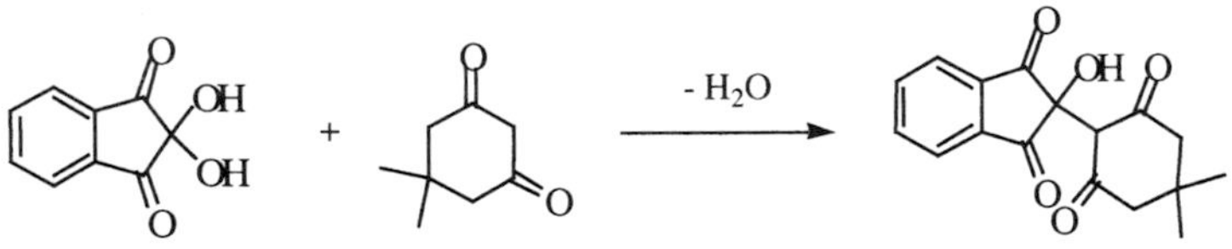

Experimental procedures:
Preparation of **3**: Ninhydrin **1** (178 mg, 1.00 mmol) and dimedone **2** (140 mg, 1.00 mmol) were ball-milled for 1 h at room temperature. Pure **3** (300 mg, 100%) was obtained (mp 193–195 °C, decomp.).

References: G. Kaupp, M. R. Naimi-Jamal, J. Schmeyers, *Chem. Eur. J.,* **8**, 594 (2002).

Type of reaction: C–C bond formation

Reaction condition: solvent-free

Keywords: veratraldehyde, 4-phenylcyclohexanone, 1-indanone, aldol reaction, a,β-unsaturated ketone

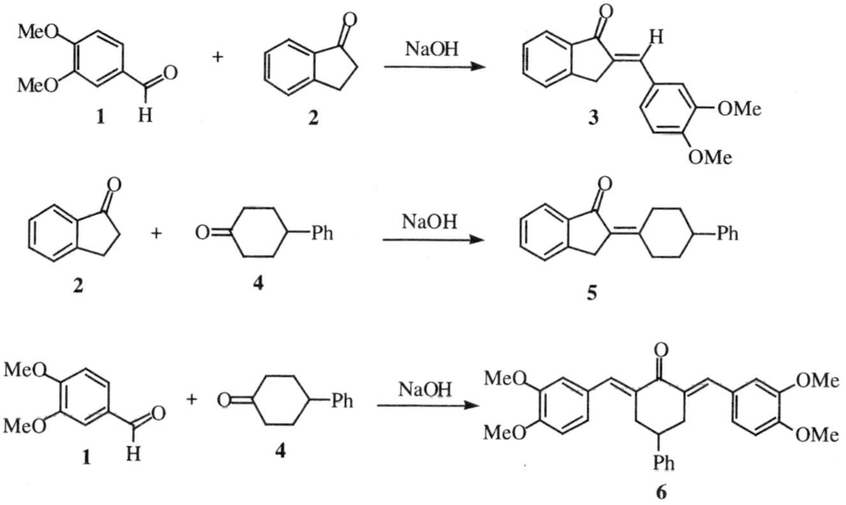

Experimental procedures:

Powdered ketone and aldehyde or ketone and ketone were ground with powdered sodium hydroxide. Powdered reagents were ground intermittently, over 10 min, in a mortar and pestle or in a vibrating ball mill for a total of 2 min over a ten-minute period. In many cases reaction was observed to proceed on mixing of powdered reagents but was accelerated by grinding. The sticky solid or viscous liquids so obtained were allowed to stand (unprotected from atmospheric oxygen or water) overnight whereupon the reaction mixtures solidified. TLC of the solids after quenching with 1 M aqueous HCl indicated the presence of a single major product and the solids were worked up by quenching with dilute aqueous HCl followed by filtration of the resultant suspension. The crude product so obtained was washed with water and recrystallized from an appropriate solvent.

References: C.L. Raston, J.L. Scott, *Green Chem.,* **2**, 49 (2000).

Type of reaction: C–C bond formation

Reaction condition: solvent-free

Keywords: Meldrum's acid, *o*-vanillin, Knoevenagel condensation, 3-carboxy-coumarin

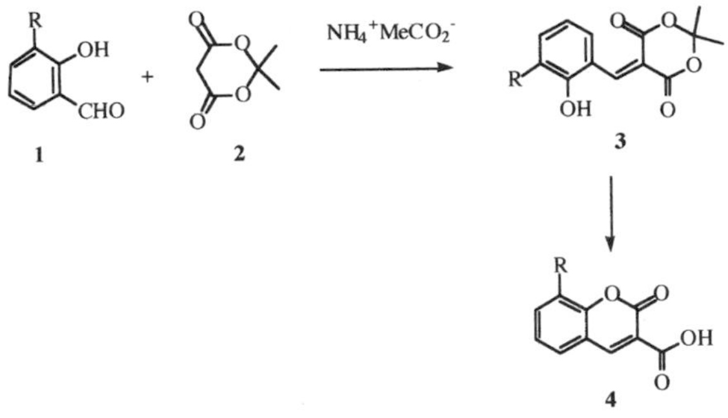

Experimental procedures:

Powdered reagents (1:1 molar ratio) were gently ground together in a mortar with a pestle and a catalytic amount of $NH_4^+MeCO_2^-$ (0.05–0.15 mol equiv. based on the benzaldehyde derivative) added and the reagents thoroughly mixed by grinding. The resulting sticky mass was allowed to stand, with occasional grinding, overnight or until no starting material was detectable by TLC analysis. The ground mixture underwent a series of color and consistency changes from white powders to sticky bright yellow material to off-white or beige solid. After regrinding, the product was suspended in water to dissolve the catalysts, filtered off, washed with water and dried on air.

References: J.L. Scott, C.L. Ratoson, *Green Chem.*, **2**, 245 (2000).

Type of reaction: C–C bond formation

Reaction condition: solvent-free

Keywords: epoxyaldehyde, ethyl nitroacetate, alumina surface, 3-(ethoxycarbo-nyl)-4-hydroxy-5-(1-hydroxyalkyl)-2-isoxazoline-2-oxide

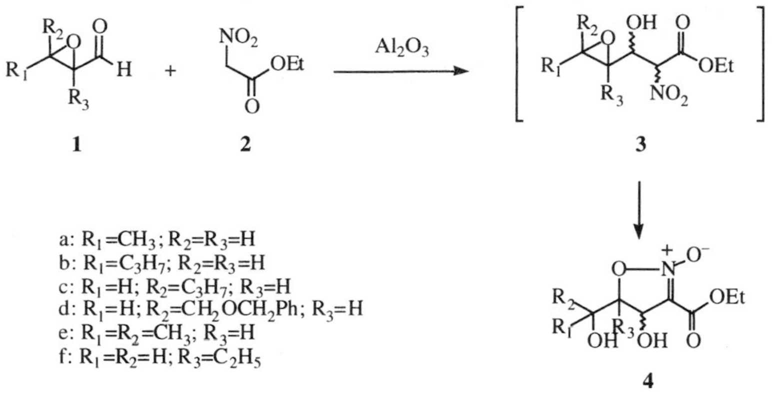

a: R_1=CH$_3$; R_2=R_3=H
b: R_1=C$_3$H$_7$; R_2=R_3=H
c: R_1=H; R_2=C$_3$H$_7$; R_3=H
d: R_1=H; R_2=CH$_2$OCH$_2$Ph; R_3=H
e: R_1=R_2=CH$_3$; R_3=H
f: R_1=R_2=H; R_3=C$_2$H$_5$

Experimental procedures:

Reactions are performed simply by mixing equimolar amounts of starting materials **1** and **2** and adding to the mixture, cooled at 0 °C and under vigorous stirring, sufficient commercial chromatographic alumina to absorb it completely. After standing for 2–20 h at room temperature with occasional stirring, products are isolated in fair to good yields by washing with dichloromethane, filtration of organic extracts, and evaporation of the solvent under reduced pressure. The separation of diastereomers is accomplished by flash chromatography on silica gel using diethyl ether as eluent.

References: G. Rosini, R. Galarini, E. Marotta, P. Righi, *J. Org. Chem.*, **55**, 781 (1990).

Type of reaction: C–C bond formation
Reaction condition: solvent-free
Keywords: aldehyde, urea, β-dicarbonyl compound, Biginelli reaction, lanthanide triflate, dihydropyrimidinone

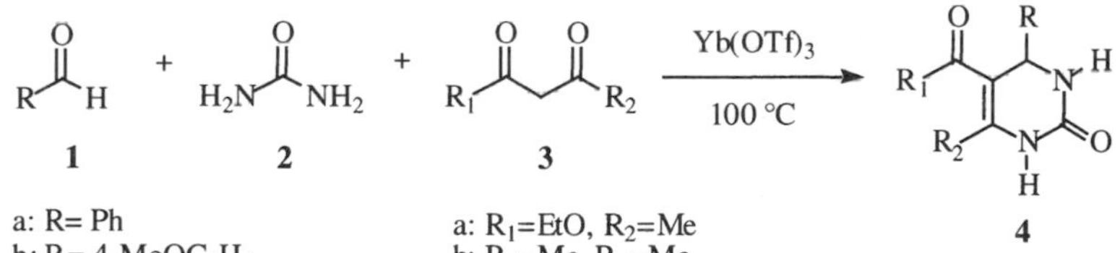

a: R= Ph
b: R= 4-MeOC$_6$H$_4$
c: R= 4-NO$_2$C$_6$H$_4$
d: R= 4-ClC$_6$H$_4$
e: R= 4-FC$_6$H$_4$
f: R= 2,4-(Cl)$_2$C$_6$H$_3$
g: R= 4-CF$_3$C$_6$H$_4$
h: R= 2-BrC$_6$H$_4$
i: R= Ph-CH=CH
j: R= n-Bu
k: R= i-Pr

a: R_1=EtO, R_2=Me
b: R_1=Me, R_2=Me
c: R_1=Ph, R_2=CF$_3$
d: R_1=2-thienyl, R_2=CF$_3$
e: R_1=MeO, R_2=Me

Experimental procedures:
Aldehyde (2.5 mmol), β-dicarbonyl compound (2.5 mmol), urea (3.7 mmol), and Yb(OTf)$_3$ (0.125 mmol, 5 mol%) were heated at 100 °C under stirring for 20 min. Then water was added, and the product was extracted with ethyl acetate. After the organic layer was dried (Na$_2$SO$_4$) and evaporated, the residue was recrystallized by ethyl acetate and hexane to produce **4**. The catalyst remaining in the aqueous phase can be recovered by removing the water through heating and then drying under vacuum at 100 °C for 2 h.

References: Y. Ma, C. Qian, L. Wang, M. Yang, *J. Org. Chem.*, **65**, 3864 (2000).

Type of reaction: C–C bond formation
Reaction condition: solvent-free
Keywords: ester, Claisen reaction, Cannizzaro reaction, β-keto ester

a: R=H
b: R=Me
c: R=Ph
d: R=*n*-Pr
e: R=*i*-Pr

Experimental procedures:
After a mixture of ethyl acetate (**1a**) (3.52 g, 40 mmol) and powdered *t*-BuOK (3.68 g, 28 mmol) was kept at 80 °C for 20 min, the reaction mixture was neutralized by addition of dil. HCl and extracted with ether. The oil left after evaporation of the solvent from the dried ether solution was distilled in vacuo by Kugelrohr apparatus to give **2a** (1.9 g, 73% yield).

References: K. Yoshizawa, S. Toyota, F. Toda, *Tetrahedron Lett.*, **42**, 7983 (2001).

Type of reaction: C–C bond formation
Reaction condition: solvent-free
Keywords: active methylene compound, unsaturated alkyl halide, alumina, potassium *tert*-butoxide, 4,4-bis-functionalized 1,6-diene, 1,6-diyne

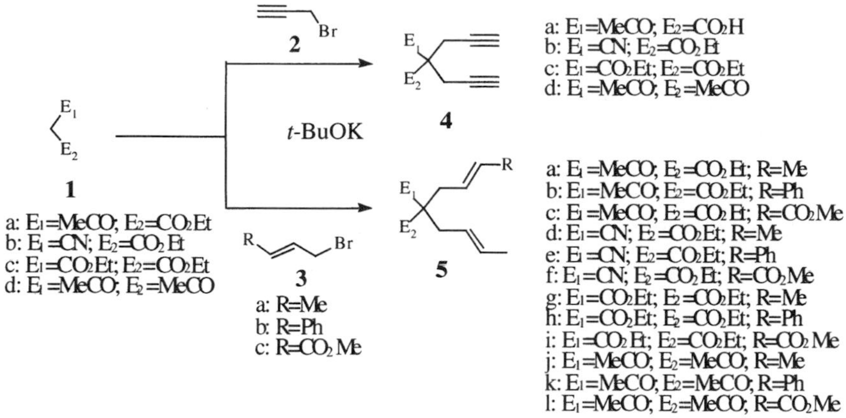

a: E_1=MeCO; E_2=CO$_2$H
b: E_1=CN; E_2=CO$_2$Et
c: E_1=CO$_2$Et; E_2=CO$_2$Et
d: E_1=MeCO; E_2=MeCO

a: E_1=MeCO; E_2=CO$_2$Et; R=Me
b: E_1=MeCO; E_2=CO$_2$Et; R=Ph
c: E_1=MeCO; E_2=CO$_2$Et; R=CO$_2$Me
d: E_1=CN; E_2=CO$_2$Et; R=Me
e: E_1=CN; E_2=CO$_2$Et; R=Ph
f: E_1=CN; E_2=CO$_2$Et; R=CO$_2$Me
g: E_1=CO$_2$Et; E_2=CO$_2$Et; R=Me
h: E_1=CO$_2$Et; E_2=CO$_2$Et; R=Ph
i: E_1=CO$_2$Et; E_2=CO$_2$Et; R=CO$_2$Me
j: E_1=MeCO; E_2=MeCO; R=Me
k: E_1=MeCO; E_2=MeCO; R=Ph
l: E_1=MeCO; E_2=MeCO; R=CO$_2$Me

1

a: E_1=MeCO; E_2=CO$_2$Et
b: E_1=CN; E_2=CO$_2$Et
c: E_1=CO$_2$Et; E_2=CO$_2$Et
d: E_1=MeCO; E_2=MeCO

a: R=Me
b: R=Ph
c: R=CO$_2$Me

Experimental procedures:

To a solution of sodium ethoxide in dry ethanol (or potassium *tert*-butoxide in dry *tert*-butanol as the case may be) (15 mmol), neutral alumina (15 times the weight of the active methylene substrate, activated by heating at 180 °C under vacuum for 2 h followed by cooling and strage under argon) was added with stirring followed by evaporation of the solvent under reduced pressure to obtain an easy flowing powder. The substrate (5 mmol) was added dropwise to the supported reagent under nitrogen with vigorous stirring. Stirring was continued for 10 min at room temperature. Then the reaction mixture was cooled in ice and the alkyl halide (11 mmol) was added dropwise under stirring condition. The reaction mixture was then allowed to attain room temperature and left at room temperature with intermittent stirring (to ensure complete mixing) until completion of the reaction (monitored with TLC). The product was extracted from the solid mass by filtration chromatography over a short plug of neutral alumina using dichloromethane as solvent. Evaporation of solvent under reduced pressure furnished the crude product, which was further purified by column chromatography over neutral alumina or short path distillation.

References: S. Bhar, S. K. Chaudhuri, S. G. Sahu, C. Panja, *Tetrahedron,* **57**, 9011 (2001).

Type of reaction: C–C bond formation
Reaction condition: solvent-free
Keywords: phenol, β-keto ester, Pechmann reaction, Knoevenagel reaction, coumarin

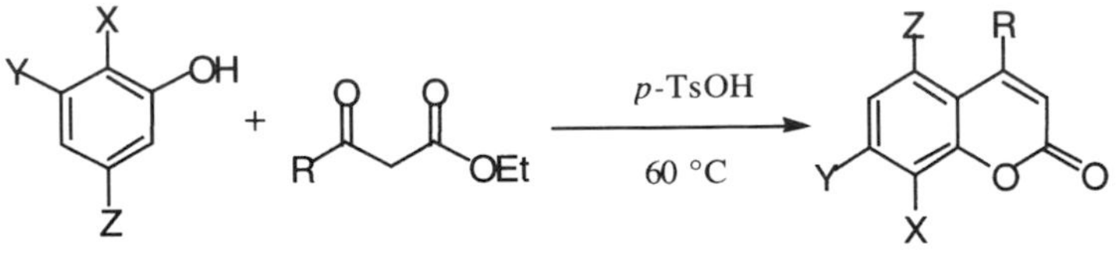

1

a: X=H; Y=OH;Z=H
b: X=OH;Y=OH;Z=H
c: X=H;Y=OH;Z=OH
d: X=H;Y=Me;Z=OH

2

a: R=Me
b: R=Ph
c: R=CH$_2$CO$_2$Et
d: R=CH$_2$Cl

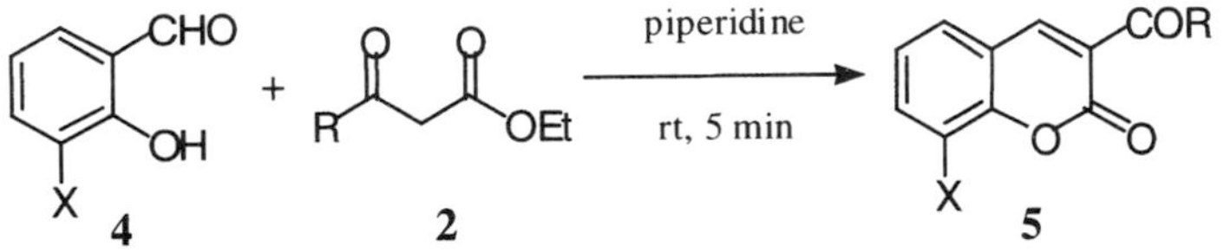

4

a: X=H
b: X=OMe

2

a: R=Me
b: R=Ph
c: R=CH$_2$CO$_2$Et
e: R=OEt

Experimental procedures:

To an equivalent mixture of resorcinol **1a** (1.1 g, 10.0 mmol) and ethyl acetoacetate **2a** (1.3 g, 10.0 mmol) was added TsOH (0.3 g, 1.5 mmol) in a mortar and ground well with a pestle at room temperature. The mixture was heated at 60 °C for 10 min under atmosphere. After cooling, water was added to the reaction mixture and the crystalline products were collected by filtration to give 7-hydroxy-4-methylcoumarin **3a** (1.73 g) in 98% yield. The crude crystals thus obtained were recrystallized from EtOH to give pure **3a** as colorless prisms (mp 185–187 °C).

References: T. Sugino and K. Tanaka, *Chem. Lett.,* 110 (2001).

Type of reaction: C–C bond formation
Reaction condition: solvent-free
Keywords: benzaldehyde, resorcinol, cyclocondensation, *p*-TsOH, calix[4]resorcinarene

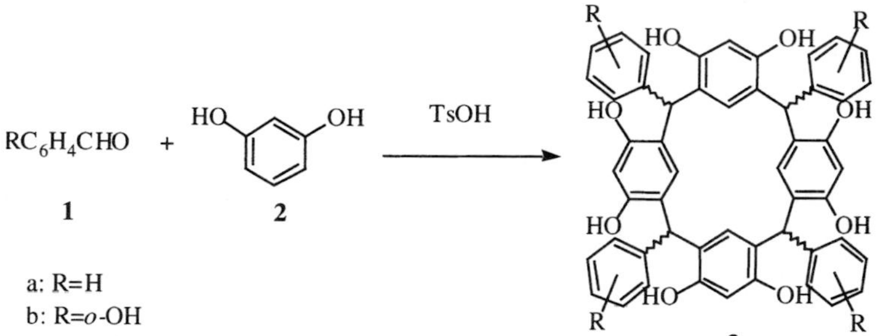

RC$_6$H$_4$CHO   +   [structure 2: HO, OH resorcinol]   →(TsOH)→   [structure 3]

1   **2**   **3**

a: R=H
b: R=o-OH
c: R=p-O(CH$_2$)$_3$CH$_3$
d: R=p-O(CH$_2$)$_7$CH$_3$
e: R=p-O(CH$_2$)$_4$Br
f: R=p-NO$_2$

Experimental procedures:

A 1:1 mixture of the starting aldehyde and resorcinol (0.5 to 1.0 g scale), along with a catalytic amount of p-toluenesulfonic acid (ca. 5%) were added together in a mortar and ground with a pestle vigorously. Within seconds a viscous paste forms which hardens on further grinding. The paste was left to stand for up to 1 h, during which time it solidified to yield a red solid. The solid was reground, washed with water to remove any acid, filtered and the pure product recrystallized with hot methanol.

References: B. A. Roberts, G. W. V. Cave, C. L. Raston, J. L. Scott, *Green Chem.*, **3**, 280 (2001).

Type of reaction: C–C bond formation
Reaction condition: solvent-free
Keywords: β-dicarbonyl compound, aldehyde, urea, Biginelli reaction, Yb(III)-resin, dihydropyrimidinone

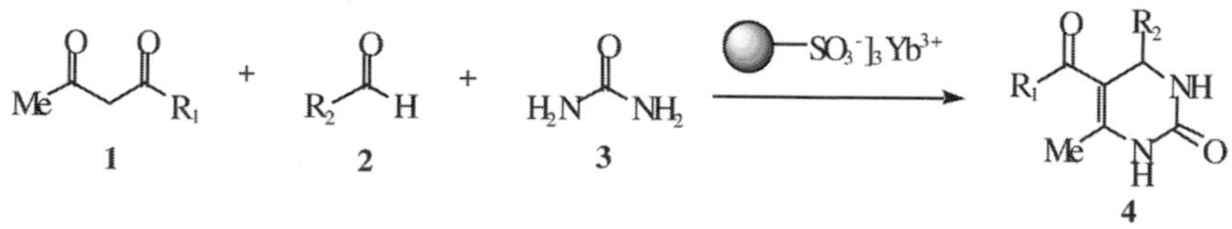

[reaction scheme]

Me–CO–CH$_2$–CO–R$_1$ +   R$_2$–CHO   +   H$_2$N–CO–NH$_2$ →(●—SO$_3^-$]$_3$ Yb^{3+})→   [structure 4]

1   **2**   **3**   **4**

a: R$_1$=OEt; R$_2$=C$_6$H$_5$
b: R$_1$=OMe; R$_2$=C$_6$H$_5$
c: R$_1$=Oi-Pr; R$_2$=C$_6$H$_5$
d: R$_1$=OBn; R$_2$=C$_6$H$_5$
e: R$_1$=Me; R$_2$=C$_6$H$_5$
f: R$_1$=OEt; R$_2$=4-FC$_6$H$_4$
g: R$_1$=OMe; R$_2$=4-FC$_6$H$_4$
h: R$_1$=Oi-Pr; R$_2$=4-FC$_6$H$_4$
i: R$_1$=OBn; R$_2$=4-FC$_6$H$_4$
j: R$_1$=Me; R$_2$=4-FC$_6$H$_4$

k: R$_1$=OEt; R$_2$=4-MeOC$_6$H$_4$
l: R$_1$=OMe; R$_2$=4-MeOC$_6$H$_4$
m: R$_1$=Oi-Pr; R$_2$=4-MeOC$_6$H$_4$
n: R$_1$=OBn; R$_2$=4-MeOC$_6$H$_4$
o: R$_1$=Me; R$_2$=4-MeOC$_6$H$_4$
p: R$_1$=OEt; R$_2$=3-MeOC$_6$H$_4$
q: R$_1$=OMe; R$_2$=3-MeOC$_6$H$_4$
r: R$_1$=Oi-Pr; R$_2$=3-MeOC$_6$H$_4$
s: R$_1$=OBn; R$_2$=3-MeOC$_6$H$_4$
t: R$_1$=Me; R$_2$=3-MeOC$_6$H$_4$

Experimental procedures:
A screw-capped vial, containing a magnetic stirring bar, was charged first with 160 mg of Yb(II)-resin then with urea **3** (1.5 mmol), aldehyde **2** (0.5 mmol), and β-dicarbonyl compound (0.5 mmol) **1** and heated at 120 °C for 5 min. Then 170 mg of Yb(III)-resin were added. The reaction mixture was heated at 120 °C under gentle stirring for 48 h. After cooling to 60 °C, methanol (1 mL) was added. The suspension was stirred for an additional 30 min then the resin was filtered off and washed thoroughly with EtOAc. Amberlyst 15 (400 mg) and Ambersep 900 OH (400 mg) were added to the combined filtrates. The suspension was shaken for 2 h then the resins were filtered off and washed thoroughly with methanol. The combined filtrates were concentrated to give dihydropyrimidinones **4**.

References: A. Dondoni, A. Massi, *Tetrahedron Lett., **42**,* 7975 (2001).

Type of reaction: C–C bond formation
Reaction condition: solvent-free
Keywords: nitroalkane, olefine, Michael addition, Amberlyst A-27, nitro compound

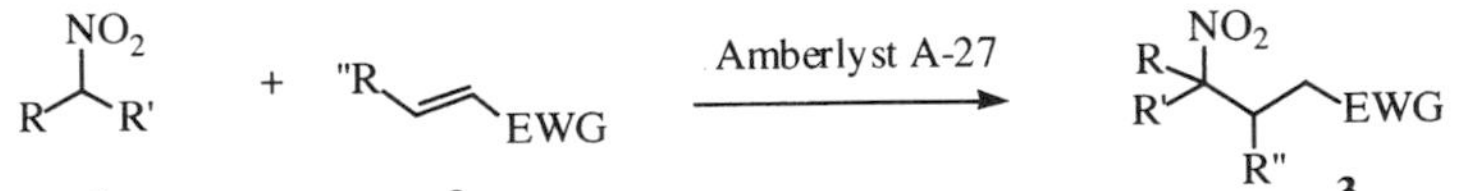

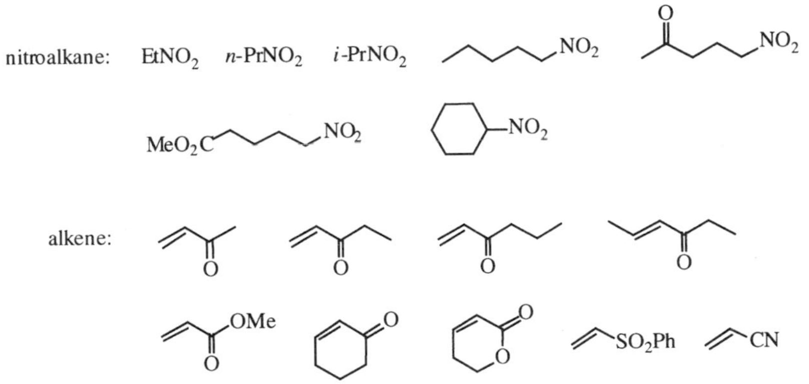

Experimental procedures:
A 100-mL two-necked flask equipped with a mechanical stirrer was charged with the nitro compound **1** (0.05 mol) and cooled with an ice-water bath. After 5 min the alkene **2** (0.05 mol) was added, and the mixture was stirred for 10 min. Am-

berlyst A-27 (810 g) was added, and after being stirred for 15 min, the mixture was left at room temperature for the appropriate time. The Amberlyst was washed with Et_2O (4×40 mL), the filtered extract was evaporated, and the crude nitroderivative **3** was purified by flash chromatography (cyclohexane-EtOAc).

References: R. Ballini, P. Marziali, A. Mozzicafreddo, *J. Org. Chem.*, **61**, 3209 (1996).

Type of reaction: C–C bond formation
Reaction condition: solvent-free
Keywords: 2-phenylcyclohexanone, a,β-unsaturated ketone, Michael addition, phase transfer catalys, 1,5-dicarbonyl compound

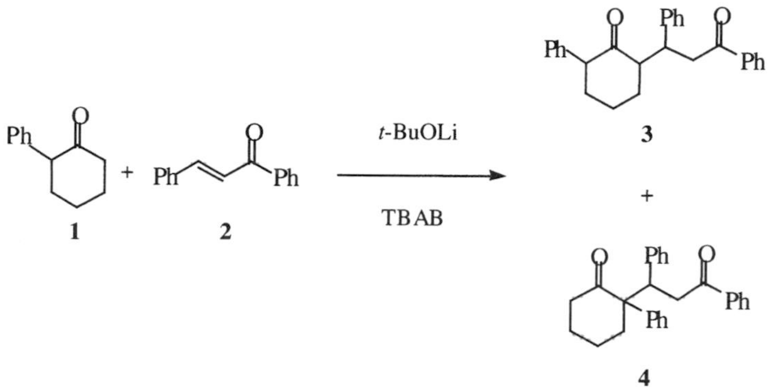

Experimental procedures:
A mixture of 2-phenylcyclohexanone **1** (1.5 mmol), base (6% mol) and catalyst (6% mol) was stirred for 5 min. The Michael acceptor **2** (1.5 mmol) was then added and the reaction mixture was kept at 60 °C for 24 h. The crude mixture was extracted with dichloromethane (20 mL) and filtered. Removal of solvent and column chromatography yielded the pure products.

References: E. D. Barra, A. de la Hoz, S. Merino, P. S. Verdu, *Tetrahedron Lett.*, **38**, 2359 (1997).

Type of reaction: C–C bond formation
Reaction condition: solvent-free
Keywords: 1,3-dicarbonyl compound, a,β-unsaturated ketone, Michael reaction, $FeCl_3 \cdot H_2O$, 1,5-dicarbonyl compound

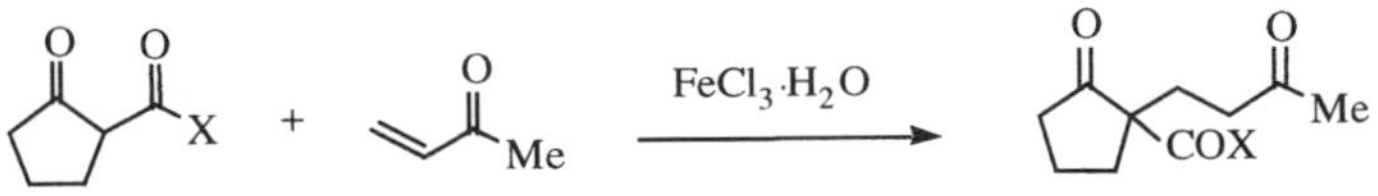

1   **2**   **3**

a: X=OEt
b: X=O*i*-Bu
c: X=Me

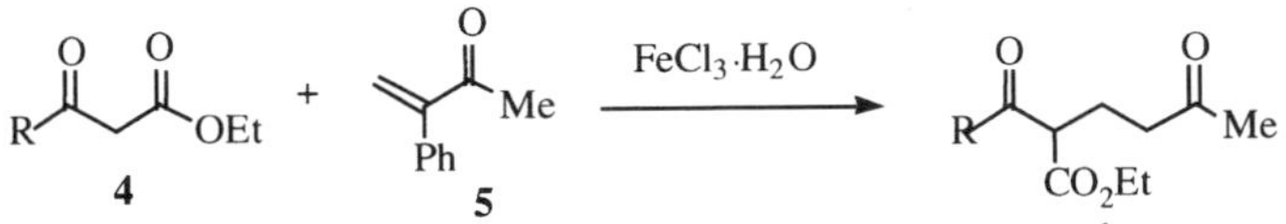

4   **5**   **6**

a: R=Me
b: R=Ph

Experimental procedures:

A mixture of the oxo ester **1a** (875 mg, 5.60 mmol), the enone **2** (0.500 mL, 6.00 mmol) and $FeCl_3 \cdot H_2O$ (15 mg, 0.055 mmol) was stirred overnight at room temperature, after which it was chromatographed on silica gel (hexane-MTB, 1:5; R_f 0.41) to afford **3a** as a colorless oil (1.23 g, 5.44 mmol, 97%).

References: J. Christoffers, *J. Chem. Soc., Perkin Trans. 1*, 3141 (1997).

Type of reaction: C–C bond formation
Reaction condition: solvent-free
Keywords: ketene silyl enol ether, α,β-unsaturated ketone, Michael addition, indium trichloride, 1,5-dicarbonyl compound

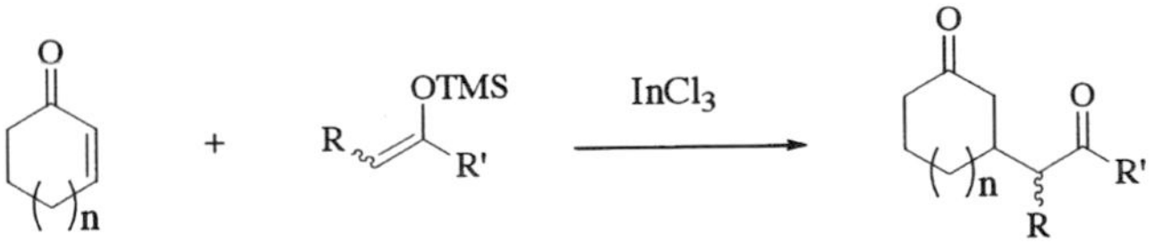

1   **2**   **3**

a: n=1
b: n=0

a: R=H; R'=Ph
b: R=Me; R'=Ph
c: R=H; R'=OEt

Experimental procedures:

2-Cyclohexene-1-one **1a** (48 mg, 48.4 µL, 0.5 mmol) and indium trichloride (22.1 mg, 0.1 mmol, 20 mol%) were stirred at room temperature for 15 min and then silyl enol ether **2** (192.3 mg, 1 mmol) was added. The resulting mixture was stirred at room temperature for 0.5 h and then 3 mL distilled water was added. The suspension was stirred at room temperature for 0.5 h, extracted with ethyl acetate and purified in the usual manner. The corresponding product was obtained in 67% yield (72 mg) after silica gel chromatography.

References: T. Loh, L. Wei, *Tetrahedron,* **54**, 7615 (1998).

Type of reaction: C–C bond formation
Reaction condition: solvent-free
Keywords: caclohexanone derivative, *a,β*-unsaturated carbonyl compound, phase transfer reaction, Michael addition, 1,5-dicarbonyl compound

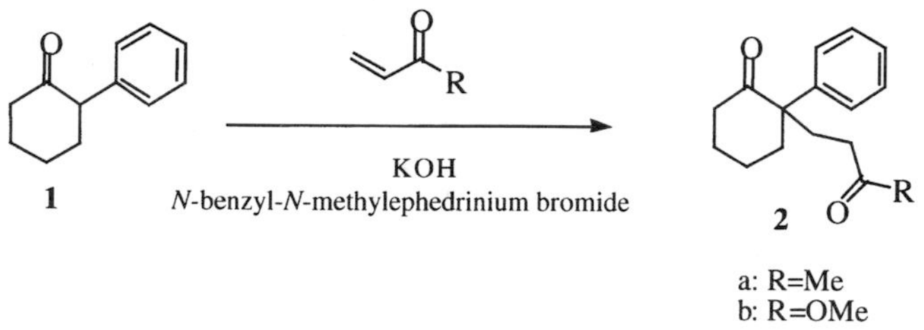

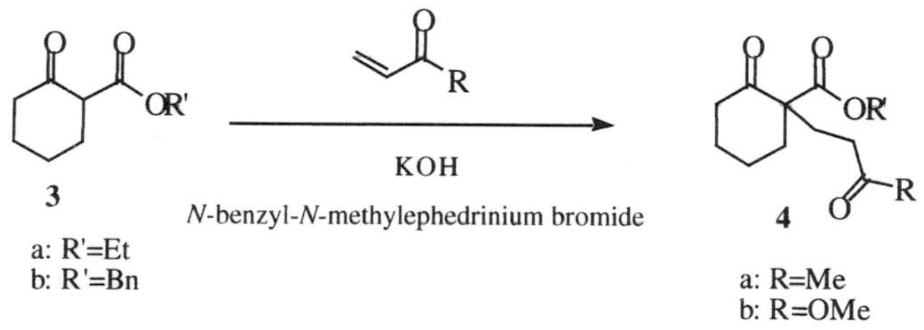

Experimental procedures:
A mixture of the Michael donor (1.5 mmol) and catalytic quantities (6 mol%) of base (KOH) and the catalyst (–)-*N*-benzyl-*N*-methylephedrinium bromide was stirred for 5 min. The Michael acceptor (1.5 mmol) was then added and the mixture was stirred at 20 °C for 24 h. The crude mixture was extracted with dichloromethane (20 mL) and filtered. The solvent was removed under reduced pressure and purified by column chromatography (silica gel 230–400 mesh) to give the pure products.

References: E. Diez-Barra, A. de la Hoz, S. Merino, A. Rodriguez, P. Sanchez-Verdu, *Tetrahedron,* **54**, 1835 (1998).

Type of reaction: C–C bond formation
Reaction condition: solvent-free
Keywords: nitroalkane, aldehyde, nitroaldol reaction, alumina supported chromium(VI) oxide, α-nitro ketone

a: R=Me; R_1=H; R_2=Ph(CH$_2$)$_2$
b: R=Me; R_1=H; R_2=CH$_3$(CH$_2$)$_8$
c: R=Et; R_1=H; R_2=CH$_3$(CH$_2$)$_3$
d: R=Me; R_1=Me; R_2=CH$_3$(CH$_2$)$_7$
e: R=Br; R_1=H; R_2=CH$_3$(CH$_2$)$_8$
f: R=Me; R_1=H; R_2=c-C$_6$H$_{11}$
g: R=Me; R_1=H; R_2=CH$_3$(CH$_2$)$_9$
h: R=PhCH$_2$; R_1=H; R_2=Ph

i: R=Br; R_1=H; R_2=Ph
j: R=MeO$_2$C(CH$_2$)$_3$; R_1=H; R_2=CH$_3$(CH$_2$)$_7$
k: R=CH$_3$CO(CH$_2$)$_2$; R_1=H; R_2=CH$_3$(CH$_2$)$_8$
l: R=PhCH$_2$; R_1=Me; R_2=CH$_3$
m: R=CH$_3$(CH$_2$)$_3$; R_1=H; R_2=Ph(CH$_2$)$_2$
n: R=Me; R_1=H; R_2=Me
o: R=Et; R_1=H; R_2=CH$_3$

Experimental procedures:
The compound **3** can be easily prepared, in one pot, through a solvent-free procedure by nitroaldol reaction of nitroalkane **1** (2.2 mmol) and aldehyde **2** (2.2 mmol, freshly distilled), on activated neutral alumina (0.6 g, the alumina was added to a mechanically stirred solution of **1** and **2**, at 0 °C, then at room temperature for 20 h). Then, in situ addition (0 °C) of wet-alumina supported chromium(VI) oxide (0.88 g (8.8 mmol) of CrO$_3$ and 2.64 g of wet alumina). After standing for additional 20 h, the product was extracted with diethyl ether and passed through a bed of alumina. Evaporation of the organic solvent and flash chromatographic purification afforded the pure α-nitro ketone **3** in good yields (68–86%).

References: R. Ballini, G. Bosica, M. Parrini, *Tetrahedron Lett.,* **39**, 7963 (1998).

Type of reaction: C–C bond formation
Reaction condition: solvent-free
Keywords: β-dicarbonyl compound, aldehyde, urea, Biginelli reaction, montmorillonite KSF, dihydropyrimidinone

R-CHO + [2] + H$_2$N—CO—NH$_2$ $\xrightarrow[130\,°C]{KSF}$ [4]

1   2   3

a: R=C$_6$H$_5$; R'=Me; R"=OEt
b: R=4-ClC$_6$H$_4$; R'=Me; R"=OEt
c: R=4-MeOC$_6$H$_4$; R'=Me; R"=OEt
d: R=4-HOC$_6$H$_4$; R'=Me; R"=OEt
e: R=C$_6$H$_5$CH=CH; R'=Me; R"=OEt
f: R=C$_6$H$_5$; R'=C$_6$H$_5$; R"=OEt
g: R=C$_6$H$_5$; R'=Me; R"=Me
h: R=C$_6$H$_5$; R'=Me; R"=C$_6$H$_5$
i: R=C$_4$H$_9$; R'=Me; R"=OEt

Experimental procedures:
Aldehyde **1** (10 mmol), β-dicarbonyl compound **2** (10 mmol), urea **3** (0.9 g, 15 mmol) and montmorillonite KSF (0.5 g) were heated at 130 °C under stirring for 48 h. Hot methanol (100 mL) was added and the mixture was filtered to remove the catalyst. Products **4** crystallized after several hours and were recovered by filtration.

References: F. Bigi, S. Carloni, B. Frullanti, R. Maggi, G. Sartori, *Tetrahedron Lett.,* **40**, 3465 (1999).

Type of reaction: C–C bond formation
Reaction condition: solvent-free
Keywords: ethyl 2-oxo-cyclohexanecarboxylate, ethyl acrylate, Michael addition, trifluoromethanesulfonic acid

[structure 1a] COOEt + [structure 2a] COOEt $\xrightarrow{TfOH}$ [structure 3a]

1a   2a   3a

Experimental procedures:
TfOH (0.6 mmol) was added dropwise to a mixture of ethyl 2-oxocyclohexanecarboxylate **1a** (2.0 mmol) and ethyl acrylate **2a** (2.4 mmol) at 0 °C, and the resultant yellow mixture was allowed to stand at room temperature for 5 h. When the reaction was finished, the mixture turned brown. The cooled mixture was diluted with CH$_2$Cl$_2$ and neutralized by the addition of the minimum amount of

Et_3N. Concentration and purification by silica gel column chromatography gave the desired Michael adduct **3a** in 92% yield.

References: H. Kotsuki, K. Arimura, T. Ohishi, R. Maruzasa, *J. Org. Chem.*, **64**, 3770 (1999).

Type of reaction: C–C bond formation
Reaction condition: solvent-free
Keywords: a,β-unsaturated ketone, cyclohexanone, acyclic ketone, Robinson annulation reaction, decalenone, cyclohexenone

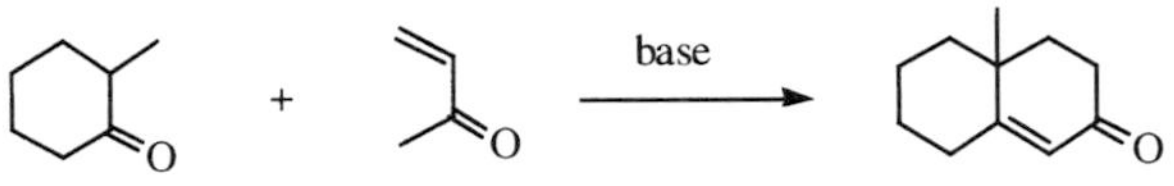

a: R_1=Ph; R_2=H
b: R_1=Ph; R_2=Me
c: R_1=Ph; R_2=Et
d: R_1=Ph; R_2=Ph
e: R_1=4-MeC_6H_4; R_2=Ph
f: R_1=4-BrC_6H_4; R_2=Ph
g: R_1=Me; R_2=H
h: R_1=Me; R_2=CO_2Me

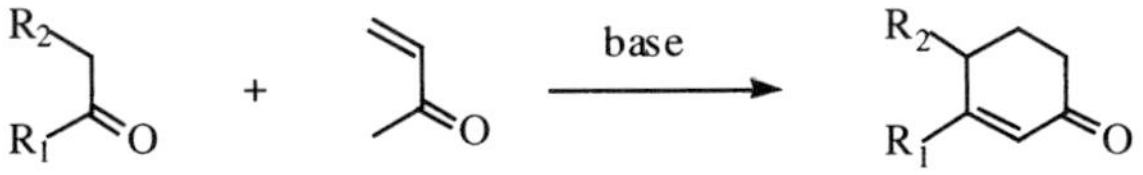

Experimental procedures:
2-Methylcyclohexanone **1** (2.21 g, 20 mmol), methyl vinyl ketone **2** (4.50 g, 40 mmol), and sodium methoxide (1.2 g, 24 mmol) were well mixed with agate mortar and pestle and the mixture was kept at room temperature for 3 h. The reaction product was combined with 3M HCl (20 mL), extracted with ether (4×20 mL), and the ether solution was washed with water and dried over $MgSO_4$. The dried ether solution was evaporated. Distillation of the residue in vacuo (150–170 °C/25 mmHg) gave **3** as colorless oil (1.05 g, 25% yield).

References: H. Miyamoto, S. Kanetaka, K. Tanaka, K. Yoshizawa, S. Toyota, F. Toda, *Chem. Lett.*, 888 (2000).

Type of reaction: C–C bond formation
Reaction condition: solvent-free
Keywords: chalcone, *N*-acetylaminomalonate, Michael reaction, phase transfer reaction

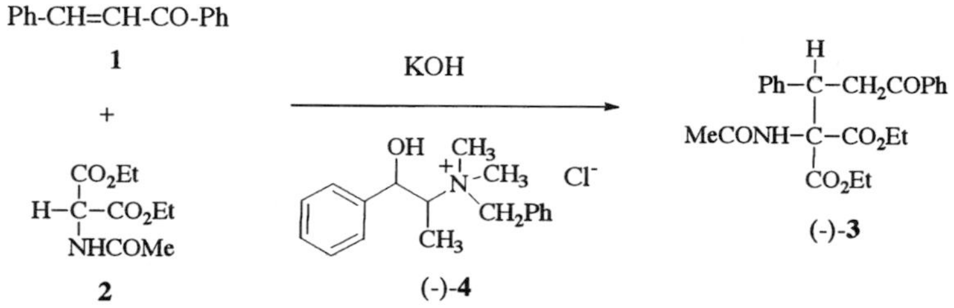

Ph-CH=CH-CO-Ph

1

+

CO$_2$Et
H–C–CO$_2$Et
NHCOMe

2

KOH
$\longrightarrow$

OH CH$_3$
| +|
Ph—CH—CH—N—CH$_3$ Cl$^-$
|
CH$_2$Ph
CH$_3$

(-)-4

H
|
Ph–C–CH$_2$COPh
|
MeCONH–C–CO$_2$Et
|
CO$_2$Et

(-)-3

Experimental procedures:
Chalcone **1** and malonate **2** in stoichiometric amounts (5 mmol) were vigorously shaken with a mechanical stirrer during 1 h at the required temperature (generally 60 °C) in the presence of catalytic quantities (6% mol) of KOH and chiral ammonium salt **4**. Then, the reaction mixture was cooled to room temperature and extracted with methylene chloride (50 mL). Product **3** was purified by chromatography on silica gel. Yield: 57%, 68% ee.

References: A. Loupy, J. Sansoulet, A. Zaparucha, C. Merienne, *Tetrahedron Lett.*, **30**, 333 (1989).

Type of reaction: C–C bond formation
Reaction condition: solvent-free
Keywords: 2-formylcycloalkanone, methyl vinyl ketone, proline, Robinson annulation, asymmetric annulation, spiro compound

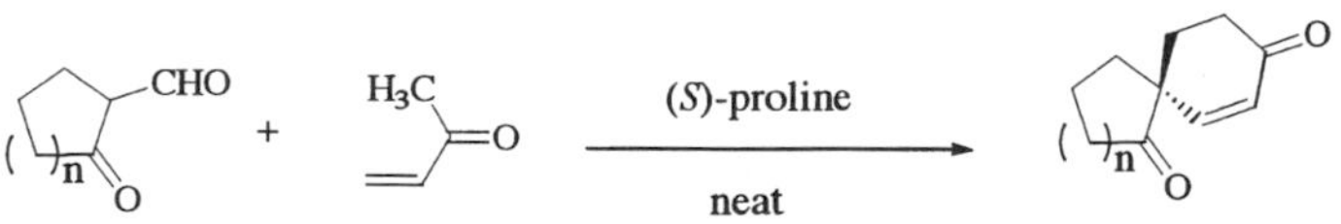

CHO
(   )$_n$
O

1

a: n=1
b: n=2
c: n=3
d: n=8

+

H$_3$C
=O

(*S*)-proline
$\longrightarrow$
neat

(   )$_n$
O

O

2

a: n=1, (*S*)-
b: n=2, (*R*)-
c: n=3, (*S*)-
d: n=8, (±)-

Experimental procedures:
A mixture of finely ground proline (0.01 mol) and 2-formylcycloalkanone (0.01 mol) was stirred at room temperature for 2 h under a nitrogen atmosphere. Freshly distilled MVK (0.012 mol) was then added dropwise over a 30-min period. The resultant brown viscous mass was stirred with CH_2Cl_2 (150 mL) and the organic extract was washed with water (2×50 mL), brine, then dried and the solvent was removed. The residue was purified by flash column chromatography (silica gel) using chloroform as eluent.

References: D. Rajagopal, R. Narayanan, S. Swaminathan, *Tetrahedron Lett.*, **42**, 4887 (2001).

Type of reaction: C–C bond formation
Reaction condition: solvent-free
Keywords: benzaldehyde, acetophenone, montmorillonite KSF, clay, *trans*-chalcone

a: R=H; R'=H
b: R=H; R'=4-Me
c: R=4-Cl; R'=4-Me
d: R=4-MeO; R'=4-Cl
e: R=2-Cl; R'=4-Cl

f: R=4-CN; R'=H
g: R=4-NO$_2$; R'=H
h: R=4-Ph; R'=H
i: R=4-CN; R'=4-Me
j: R=4-CN; R'=4-Cl

Experimental procedures:
Montmorillonite KSF (1.0 g), the selected benzaldehyde **1** (0.010 mol) and the selected acetophenone **2** (0.010 mol) were placed in a small autoclave and heated at 130 °C for 18 h. After cooling to room temperature, 95% ethanol (25 mL) was added, the mixture was filtered, the catalyst washed with hot 95% ethanol (25 mL) and the products crystallized from the same solvent. All the products gave melting points and spectral data consistent with the reported data.

References: R. Ballini, G. Bosica, R. Maggi, M. Ricciutelli, P. Righi, G. Sartori, R. Sartorio, *Green Chem.*, **3**, 178 (2001).

Type of reaction: C–C bond formation
Reaction condition: solvent-free
Keywords: ketone, aldehyde, dicyanomethane, benzyltriethylammonium chloride, Knoevenagel condensation

a: $R_1=C_6H_5$; $R_2=H$; $R_3=CN$
b: $R_1=C_6H_5$; $R_2=H$; $R_3=CONH_2$
c: $R_1=4\text{-}NO_2C_6H_4$; $R_2=H$; $R_3=COOEt$
d: $R_1=3\text{-MeO, }4\text{-}OHC_6H_3$; $R_2=H$; $R_3=CN$
e: $R_1=3\text{-MeO, }4\text{-}OHC_6H_3$; $R_2=H$; $R_3=CONH_2$
f: $R_1=$furyl; $R_2=H$; $R_3=COOEt$
g: $R_1=C_6H_5CH=CH$; $R_2=H$; $R_3=CN$
h: $R_1=C_6H_5CH=CH$; $R_2=H$; $R_3=COOEt$
i: $R_1=i\text{-Pr}$; $R_2=H$; $R_3=CN$
j: $R_1R_2=\text{-}(CH_2)_4\text{-}CH(CH_3)\text{-}$; $R_3=CN$
k: $R_1R_2=\text{-}CH(CH_3)\text{-}(CH_2)_3\text{-}CH(CH_3)\text{-}$; $R_3=CN$
l: $R_1=C_6H_5$; $R_2=CH_3$; $R_3=CN$
m: $R_1=2\text{-}HOC_6H_4$; $R_2=H$; $R_3=COOEt$

Experimental procedures:

To a mixture of the carbonyl compound **1a–m** (10 mmol) and active methylene compound **2** (10 mmol) was added benzyltriethylammonium chloride (2 mmol) at room temperature. After being stirred for 5 min, the resulting mixture was heated at 80 °C in a preheated oil bath for 1 h (monitored by TLC, EtOAc-hexane, 1:9, v/v). It was then stirred and allowed to cool to room temperature when it solidified. On completion, the reaction mixture was poured into water and extracted with Et$_2$O (2×25 mL), dried over Na$_2$SO$_4$ and concentrated in vacuo. The crude product thus obtained was purified by recrystallization or column chromatography to afford products **3a–m** in 40–95% yields.

References: D. S. Bose, A. V. Narsaiah, *J. Chem. Res. (S)*, 36 (2001).

Type of reaction: C–C bond formation
Reaction condition: solid-state
Keywords: inclusion complex, stereoselective Wittig-Horner reaction, carbethoxymethylene cyclohexane

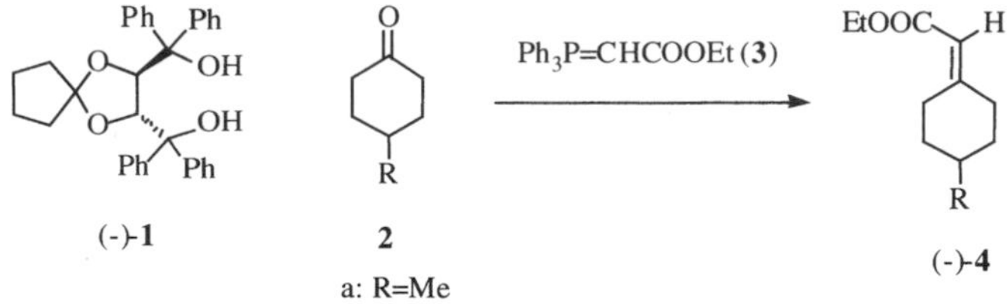

Ph   Ph
$\overbrace{\hspace{2cm}}$
(-)-1   2   (-)-4

a: R=Me
b: R=Et

Ph₃P=CHCOOEt (3)

EtOOC   H

Experimental procedures:

When a mixture of finely powdered 1:1 inclusion compound of (–)-**1** and 4-methyl-cyclohexanone **2a** (1.5 g) and (carbethoxymethyl)triphenylphosphorane **3** (2.59 g) was kept at 70 °C, the Wittig-Horner reaction was completed within 4 h. To the reaction mixture was added ether-petroleum (1:1), and the precipitated solid (triphenylphosphine oxide and excess **3**) was removed by filtration. The crude product left after evaporation of the solvent of the filtrate was distilled in vacuo to give (–)-4-methyl-1-(carbethoxymethylene)cyclohexane **4** of 42.3% ee in 73% yield.

References: F. Toda, H. Akehi, *J. Org. Chem.*, **55**, 3446 (1990).

Type of reaction: C–C bond formation
Reaction condition: solid-state
Keywords: chalcone, benzylidene aniline, trimethyloxosulfonium iodide, ylide reaction, cyclopropane, aziridine

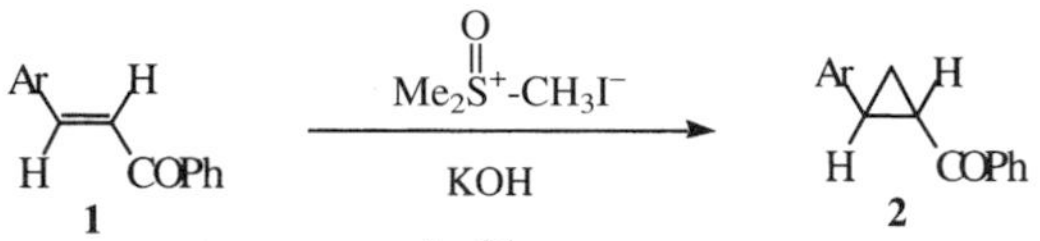

$$Me_2S^+\text{-}CH_3I^-$$

KOH

a: Ar=Ph
b: Ar=*m*-ClC₆H₄
c: Ar=*p*-PhC₆H₄
d: Ar=2-furyl
e: Ar=2-thienyl
f: Ar=*p*-MeOC₆H₄

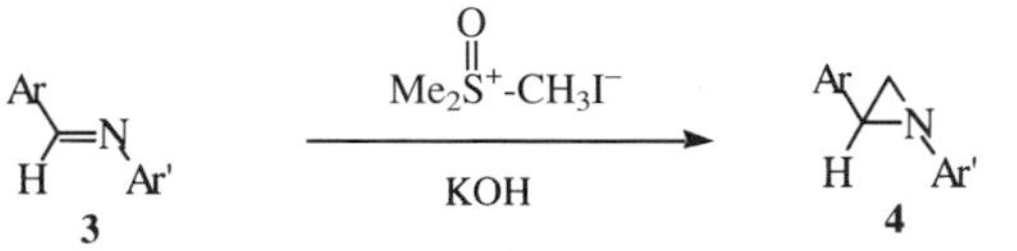

$$Me_2S^+\text{-}CH_3I^-$$

KOH

a: Ar=Ph; Ar'=Ph
b: Ar=Ph; Ar'=*p*-MeOC₆H₄
c: Ar=*p*-ClC₆H₄; Ar'=*p*-MeOC₆H₄
d: Ar=*p*-MeC₆H₄; Ar'=*p*-MeC₆H₄

Experimental procedures:
A mixture of powdered N-benzylideneaniline **3a** (0.5 g), trimethyloxosulfonium iodide (2 g) and KOH (1.8 g) was kept at 50 °C for 3 h, after which it was washed with water. The residual crude product was taken up in diethyl ether and the ethereal solution was worked up to give aziridine **4a** as pale yellow crystals (1.2 g, 56%), mp 38–39 °C.

References: F. Toda, N. Imai, *Chem. Commun.,* 2673 (1994).

Type of reaction: C–C bond formation
Reaction condition: solvent-free
Keywords: formylferrocene, triphenylbenzylphosphonium chloride, Wittig reaction, ferrocenylethene derivative

$$R_1\text{-CHO} \quad + \quad R_2CH_2P^+Ph_3X^- \quad \xrightarrow{\text{NaOH}} \quad R_1\text{-CH=CH-}R_2$$

$$\textbf{1} \qquad\qquad \textbf{2} \qquad\qquad\qquad\qquad \textbf{3}$$

a: R_1=Fc; R_2=C_6H_5; X=Cl
b: R_1=C_6H_5; R_2=Fc; X=I
c: R_1=Fc; R_2=p-ClC_6H_4; X=Br
d: R_1=Fc; R_2=p-BrC_6H_4; X=Br
e: R_1=Fc; R_2=m-$NO_2C_6H_4$; X=Br
f: R_1=Fc; R_2=C_6H_5CO; X=Br
g: R_1=Fc; R_2=p-BrC_6H_4CO; X=Br
h: R_1=Fc; R_2=H; X=Br
i: R_1=p-FcC_6H_4; R_2=C_6H_5; X=Cl
j: R_1=p-$MeOC_6H_4$; R_2=Fc; X=I
k: R_1=Fc; R_2=Fc; X=I

Experimental procedures:
A mixture of formylferrocene **1a** (1 mmol, 0.214 g), triphenylbenzylphosphonium chloride **2a** (1.1 mmol, 0.427 g) and NaOH (1.5 mmol, 0.06 g) was thoroughly ground with a pestle in an open mortar at room temperature under atmosphere. The reaction mixture was ground for 5 min until the reaction was complete by TLC monitoring, then extracted in dichloromethane (3×20 mL). The extracts were combined and dried with anhydrous $NaSO_4$. After filtration, the solvent was removed under vacuum to give crude product. The residue was chromatographed on silica gel using petroleum ether as eluent. The product from the first band was a yellow oily liquid (0.055 g), which is Z-ferrocenyl-2-phenylethylene **3a**, mp 122–124 °C.

References: W. Liu, Q. Xu, Y. Ma, Y. Liang, N. Dong, D. Guan, *J. Organomet. Chem.,* **625**, 128 (2001).

Type of reaction: C–C bond formation
Reaction condition: solvent-free
Keywords: imidate, dimethyl aminomalonate, aldehyde, 1,3-dipolar cycloaddition, 4,5-dihydrooxazol

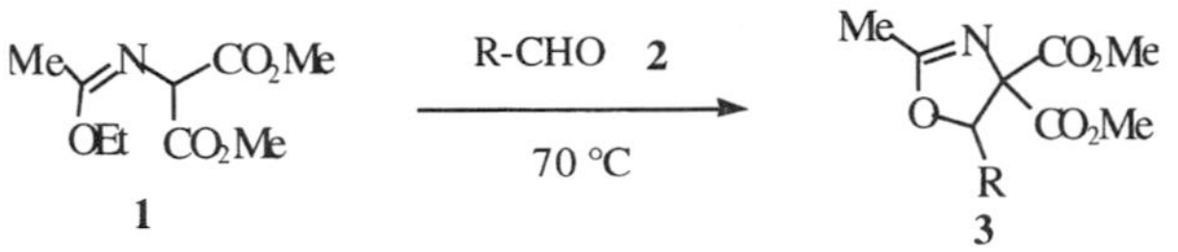

a: R=Ph
b: R=2-furyl
c: R=Me$_2$CH
d: R=2-OHC$_6$H$_4$
e: R=2-pyridyl
f: R=PhCH=CH

Experimental procedures:
A mixture of dimethyl 2-(1-ethoxyethylidene)aminomalonate **1** (1.0 g, 4.6 mmol) and freshly distilled aldehyde **2** (4.6 mmol) was heated to 70 °C during the appropriate time (monitored by TLC) under magnetic stirring. The reaction mixture was allowed to cool down. After removal of ethanol in vacuo, the crude residue was purified by chromatography on silica gel (60F 254, Merck) with appropriate eluent. Solvent evaporation gave the desired compounds which crystallized on standing.

References: J.M. Lerestif, L. Toupet, S. Simbandhit, F. Tonnard, J.P. Bazureau, J. Hamelin, *Tetrahedron*, **53**, 6351 (1997).

Type of reaction: C–C bond formation
Reaction condition: solvent-free
Keywords: benzylidenemethylamine, acetylacetone, montmorillonite K10, enamino ketone, β-oxo alkene, cyclohexene

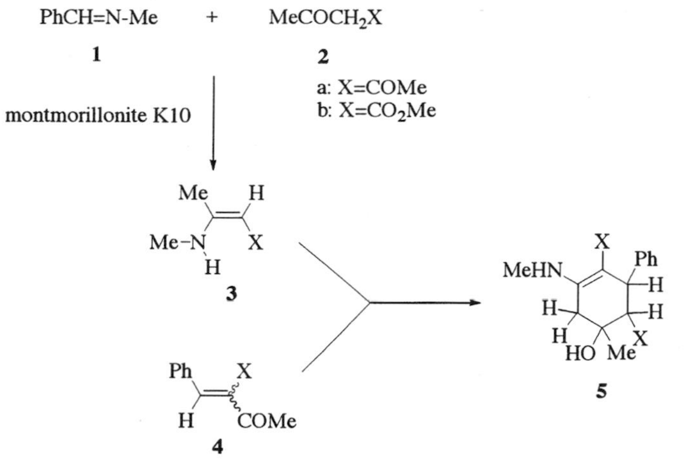

PhCH=N-Me   +   MeCOCH₂X

1

2

a: X=COMe
b: X=CO₂Me

montmorillonite K10

Me   H

Me–N   X
      H

3

Ph   X

H   COMe

4

X
MeHN   Ph
       –H
H–   –H
H
HO Me   X

5

Experimental procedures:

An equimolar mixture (10 mmol) of benzylidenemethylamine **1** (1.19 g) and acetylacetone **2a** (1 g) was adsorbed onto montmorillonite K10 (5 g) and allowed to stand at room temperature for 3 days. The mixture was extracted with CH₂Cl₂, the clay separated by filtration and the solvent evaporated under reduced pressure. Pure compound **3** could be isolated by short-path distillation (81% yield). The equimolar mixture of enamino ketone **3** and alkene **4a** (5 mmol) was allowed to stand at room temperature for a suitable time. Washing with suitable solvent afforded the pure solid product **5**.

References: S. A. Ayoubi, L. Toupet, F. Texier-Boullet, J. Hamelin, *Synthesis,* 1112 (1999).

Type of reaction: C–C bond formation
Reaction condition: solid-state
Keywords: enamine, 1,4-diphenyl-2-buten-1,4-dione, cascade reaction, solid-solid reaction, pyrroles

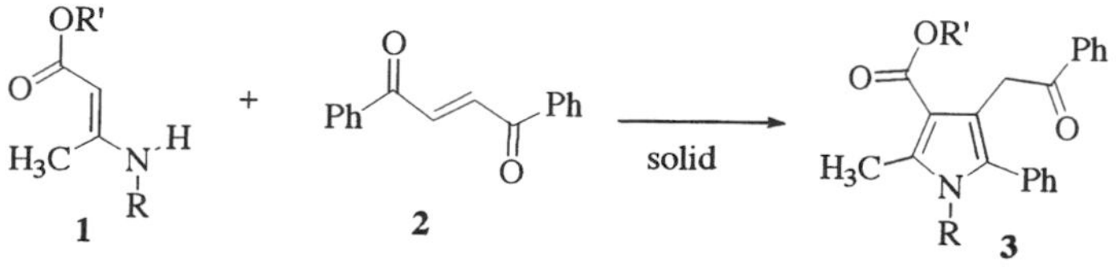

OR'

O

H₃C   N·H
      R

1

+

O

Ph   Ph

O

2

solid

OR'   Ph
O
H₃C   N   Ph
      R

3

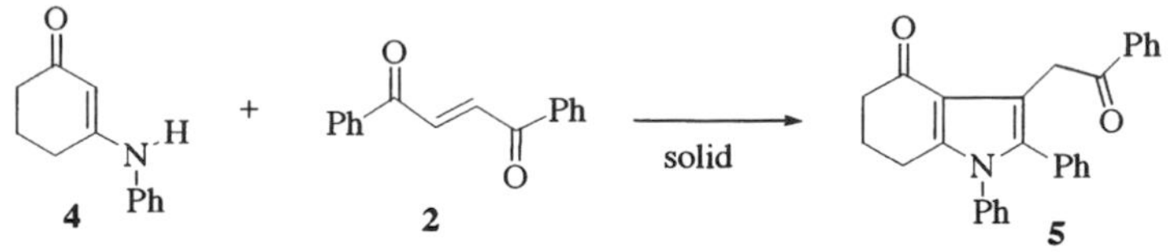

a: R=H; R'=CH$_3$
b: R=CH$_3$; R'=CH$_3$
c: R=CH$_3$; R'=C$_2$H$_5$
d: R=CH$_2$Ph; R'=C$_2$H$_5$

Experimental procedures:

A heatable/coolable ball-mill (Retsch MM2000) with 10-mL mill beakers made from stainless steel and two stainless steel balls (6.5 g) was run at 20–30 Hz for 3 h in order to achieve quantitative conversions. Milling of **1a–d** or **4** (2.00 mmol) and **2** (2.00 mmol; mp 111 °C) gave dust-dry powders that were heated to 80 °C for removal of the water of reaction. Compound **3c** was obtained as a 2:3 mixture with its precursor and was obtained in pure form after heating to 150 °C for 5 min in order to complete the elimination and removal of water.

References: G. Kaupp, J. Schmeyers, A. Atfeh, *Angew. Chem. Int. Ed. Engl.*, **38**, 2896 (1999).

Type of reaction: C–C bond formation
Reaction condition: solid-state
Keywords: 1,1-diaryl-2-propyn-1-ol, 2-naphthol, silica gel, TsOH, naphthopyran

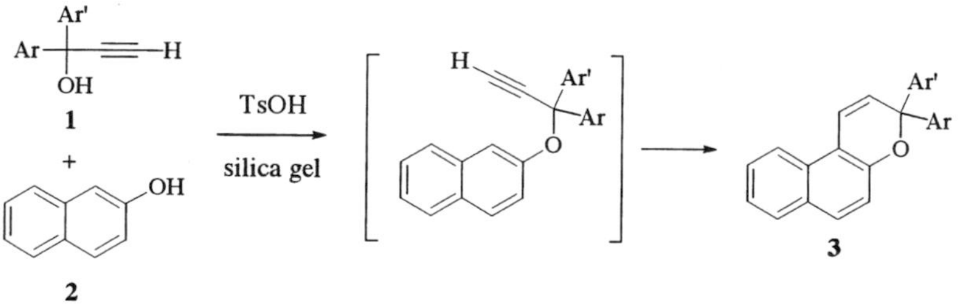

a: Ar=Ph; Ar'=Ph
b: Ar=Ph; Ar'=4-MeOC$_6$H$_4$
c: Ar=4-MeC$_6$H$_4$; Ar'=4-MeC$_6$H$_4$
d: Ar=Ph; Ar'=2,4-Me$_2$C$_6$H$_3$
e: Ar=4-ClC$_6$H$_4$; Ar'=4-ClC$_6$H$_4$
f: Ar, Ar'=Fluorenyl

Experimental procedures:
A mixture of **1a** (0.5 g, 2.4 mmol), **2** (0.34 g, 2.4 mmol), *p*-TsOH (0.046 g, 0.24 mmol), and silica gel (1.0 g) was ground for 10 min at room temperature using a mortar and pestle, and the mixture was kept for 1 h. The reaction mixture was chromatographed on silica gel using toluene as eluent to give **3a** (0.45 g, 56% yield, mp 160–162 °C) as colorless needles.

References: K. Tanaka, H. Aoki, H. Hosomi, S. Ohba, *Org. Lett.*, **2**, 2133 (2000).

Type of reaction: C–C bond formation
Reaction condition: solvent-free
Keywords: 2-naphthol, aldehyde, 1-phenylethylamine, asymmetric aminoalkylation aminoalkylnaphthol

2

a: R= Ph
b: R= 4-MeC$_6$H$_4$
c: R= 4-MeOC$_6$H$_4$
d: R= 4-ClC$_6$H$_4$
e: R= 4-NO$_2$C$_6$H$_4$
f: R= 1-naphthyl
g: R= 2-naphthyl
h: R= 1-hexyl
i: R= cyclohexyl
j: R= i-Pr

(R,R)-3

Experimental procedures:
A mixture of 2-naphthol **1** (0.72 g, 5.0 mmol), benzaldehyde **2a** (0.64 g, 6.00 mmol), and (*R*)-(+)-1-phenylethylamine (0.64 g, 5.25 mmol) was stirred at 60 °C for 8 h under nitrogen atmosphere. Following the progress of the reaction by TLC and ^{1}H NMR, it was seen that the formation of the product occurs during the first 4 h but the initial dr of (*R,R*)-**3a** (2.6 at 2 h) increases over time (99 at 8 h) with the formation of a solid and crystalline reaction mixture. The reaction mixture was dispersed at room temperature with EtOH (5 mL). The white crystals separated were collected and washed with EtOH (3×3 mL). The crystal-

line white residue, purified by crystallization from EtOAc-hexane, gives the pure
(*R,R*)-**3a** (1.64 g, 4.65 mmol, 93% yield).

References: C. Cimarelli, A. Mazzanti, G. Palmieri, E. Volpini, *J. Org. Chem.*, **66**, 4759
(2001).

Type of reaction: C–C bond formation
Reaction condition: solvent-free
Keywords: *N*-diphenylphosphinoylimine, diethylzinc, enantioselective addition,
N-diphenylphosphinoylamine

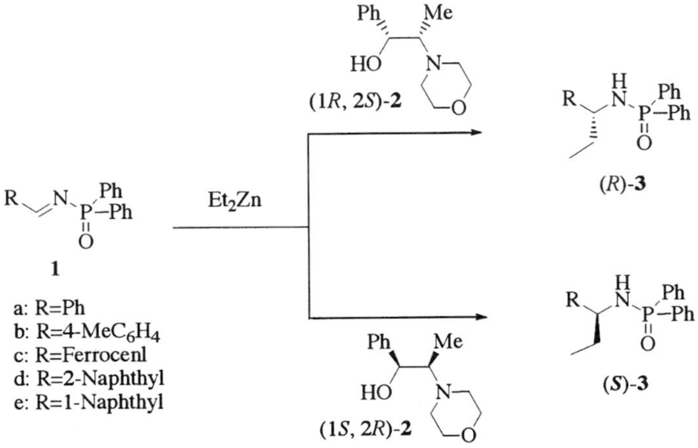

Experimental procedures:
To an ice-cooled two-necked flask containing *N*-diphenylphosphinoylimine **1c**
(0.21 g, 0.5 mmol) and (1*R*,2*S*)-2-morpholino-1-phenylpropan-1-ol **2** (0.11 g,
0.5 mmol), neat Et$_2$Zn (0.49 g, 4 mmol) was transferred through a cannula under
an argon atmosphere. After the mixture was stirred at 0 °C for 2 h, the completion of the reaction was confirmed by TLC analysis. After additional stirring for
2.5 h, excess Et$_2$Zn was removed under reduced pressure and saturated aq. ammonium chloride was added to the residue. The mixture was extracted with dichloromethane and the combined organic layer was dried over anhydrous sodium
sulfate. Concentration of the organic layer and purification of the residue on silica gel TLC gave (*R*)-**3c** (0.13 g, 58%). The ee was determined to be 97% by
HPLC analysis using a chiral stationary phase (Chiralpak AS).

References: I. Sato, R. Kodaka, K. Soai, *J. Chem. Soc., Perkin Trans. 1*, 2912 (2001).

Type of reaction: C–C bond formation
Reaction condition: solvent-free
Keywords: iodobenzene, alkylboronic acid, Suzuki coupling, palladium-doped KF/Al$_2$O$_3$, alkylbenzene

Experimental procedures:
To a mixture of KF/Al$_2$O$_3$ (0.950 g, 40 wt %) and palladium black (0.050 g, 0.470 mmol, 99.9+% as a submicron powder) was added *p*-methylphenylboronic acid (0.150 g, 1.1 mmol) contained in a clean, dry, round-bottomed flask. The solid mixture was stirred at room temperature in the open air until homogeneous. Iodobenzene (0.209 g, 1.02 mmol) was then added with stirring. The mixture was stirred at room temperature for an additional 1520 min to ensure efficient mixing. A condenser was put in place and the flask placed in a preheated oil bath (100 °C). Stirring was continued during the entire reaction period. After the allotted time period, the oil bath was removed and the reaction allowed to cool to room temperature. A small quantity of hexanes was then added and the slurry stirred at room temperature for an additional 20–30 min to ensure product removal from the surface. The mixture was vacuum filtered through a sintered glass funnel using Celite as a filter aid. The mixture was separated via flash chromatography to yield 4-methylbiphenyl in 89% yield.

References: R. S. Varma, K. P. Naicker, *Org. Lett.,* **1**, 189 (1999).

Type of reaction: C–C bond formation
Reaction condition: solid state
Keywords: cumulene derivative, [2+2]cycloaddition, [4]radialene tetracarboxylic acid

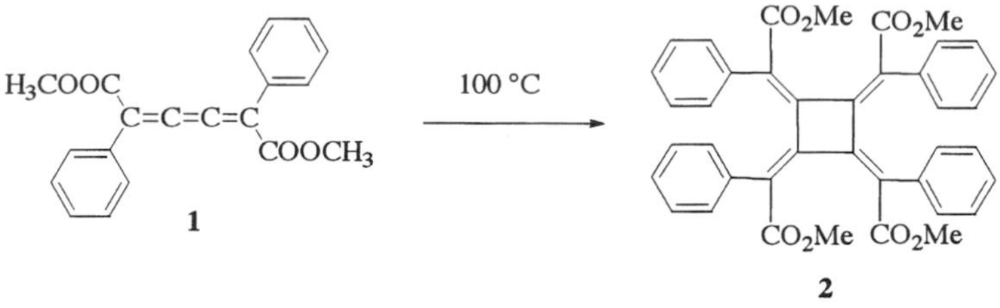

Experimental procedures:
The crystals of the ester *E*-**1** underwent complete reaction within 5 days without any observable melting during conversion. Aside from a small amount of polymeric material (<5%), only a bright yellow substance was formed, which, after column chromatographic separation of the polymeric material, could be isolated in crystalline form in 75% yield.

References: F. W. Nader, C. Wacker, *Angew. Chem. Int. Ed. Engl.*, **24**, 852 (1985).

Type of reaction: C–C bond formation
Reaction condition: solid state
Keywords: 1,3-di-*tert*-butyl-5-vinylidenecyclopentadiene, [6+2]cycloaddition

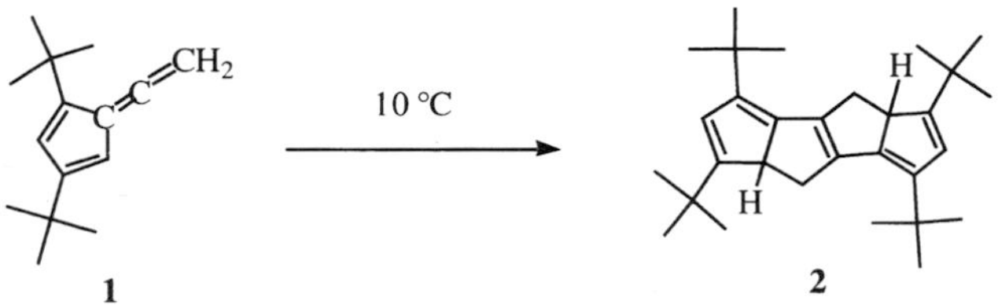

Experimental procedures:
1 dimerizes in a solid-state reaction at ca. 10 °C within approx. 14 days to give 1,3,5,7-tetra-*tert*-butyl-4,4a,8,8a-tetrahydrodicyclopenta[a,e]pentalene **2** (from *n*-hexane, pale yellow crystals, decomp. above 114 °C; yield 35%).

References: B. Stowasser, K. Hafner, *Angew. Chem. Int. Ed. Engl.*, **25**, 466 (1986).

Type of reaction: C–C bond formation
Reaction condition: solid-state
Keywords: potassium crotonate, dimerization, hex-1-ene-3,4-dicarboxylate

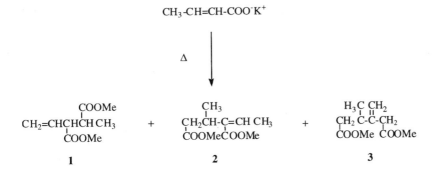

Experimental procedures:
A sample (ca. 1 g) of K-crotonate was placed in a Pyrex tube (6 mm i.d. (×200 mm) and heated in a metal bath maintained at a constant temperature. The tube was evacuated through the reaction. After a given time the reaction tube was taken out of the bath and allowed to cool to room temperature. The heat-treated salt was weighed and dissolved in aqueous hydrochloric acid solution and extracted with ether. The ether extract was treated with diazomethane. The methyl ester derivatives of the products were subjected to analysis by gas chromatography using diethyl maleate as internal standard.

Isolation and spectral analysis of the dimer: K-crotonate (100 g) was heated at 320 °C for 4 h and subsequently treated as described above. The methyl esters (42 g), bp 100–135 °C at 6 mmHg, were fractionally distilled to give three components: dimethyl hex-1-ene-3,4-dicarboxylate **1**, bp 97 °C at 8 mmHg; dimethyl 4-methyl-pent-2-ene-3,5-dicarboxylate **2**, bp 118 °C at 5 mmHg; dimethyl 2-methylene-3-methylbutyl-1,4-dicarboxylate **3**, bp 129 °C at 8 mmHg.

References: K. Naruchi, M. Miura, *J. Chem. Soc., Perkin Trans. 2*, 113 (1987).

Type of reaction: C–C bond formation
Reaction condition: solid-state
Keywords: (alkoxyphenyl)propiolic acid, Diels-Alder reaction, anhydride

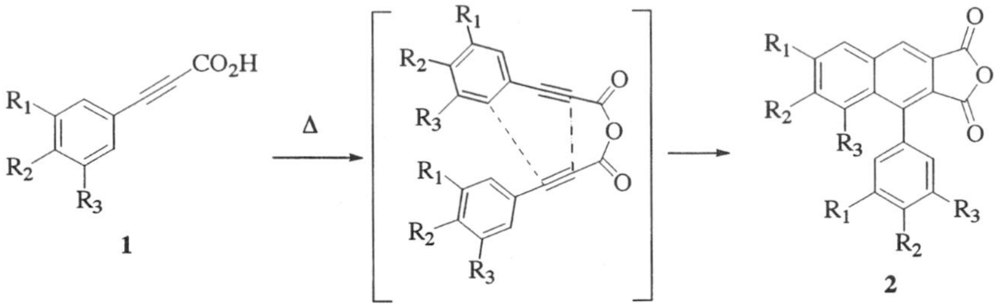

a: R_1, R_2=-OCH$_2$O-; R_3=H
b: R_1=R_2=CH$_3$O; R_3=H
c: R_1=R_2=R_3=CH$_3$O

Experimental procedures:

The solid-state reactions were carried out by taking approximately 100–500 mg of the powdered phenylpropiolic acids **2a–c** in round-bottomed flasks. The flasks were lightly plugged with cotton and kept in oil baths at appropriate temperatures between 70–110 °C. Within 27 days, the samples became discolored and some sublimed acid and phenylacetylene could be seen coated at the top of the flask. The mixtures were stirred occasionally, and after 46 weeks, the reaction mixtures were separated by column chromatography. The following reaction details were noted for acids **2a–c** (acid, percent conversion to Diels-Alder anhydride, percent unreacted plus sublimed acid, time of reaction, temperature of reaction): **2a**, 25%, 32%, 5 weeks, 95 °C; **2b**, 30%, 25%, 6 weeks, 90 °C; **2c**, 20%, 30%, 3 weeks, 90 °C.

References: G. R. Desiraju, K. V. R. Kishan, *J. Am. Chem. Soc.,* **111**, 4838 (1989).

Type of reaction: C–C bond formation
Reaction condition: solid state
Keywords: fullerenes, topochemistry, anthracene

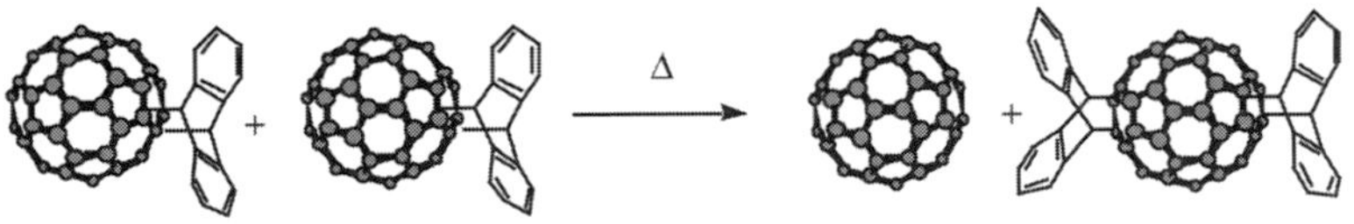

Experimental procedures:

Heating a deoxygenated sample of crystalline **2** at 180 °C for 10 min afforded 48% each of the fullerene **1** and the antipodal bisadduct **3** (96% conversion; no other components detected). The bisadduct **3** was isolated and its structure confirmed by NMR spectroscopy and FAB mass spectrometry, as well as by single crystal structure analysis.

References: B. Krautler, T. Muller, J. Maynollo, K. Gruber, C. Kratky, P. Ochsenbein, D. Schwarzenbach, H.-B. Burgi, *Angew. Chem. Int. Ed. Engl.*, **35**, 1204 (1996).

Type of reaction: C–C bond formation
Reaction condition: solid-state
Keywords: [60]fullerene, zinc, ethyl bromoacetate, 1-ethoxycarbonylmethyl-1,2-dihydro[60]fullerene

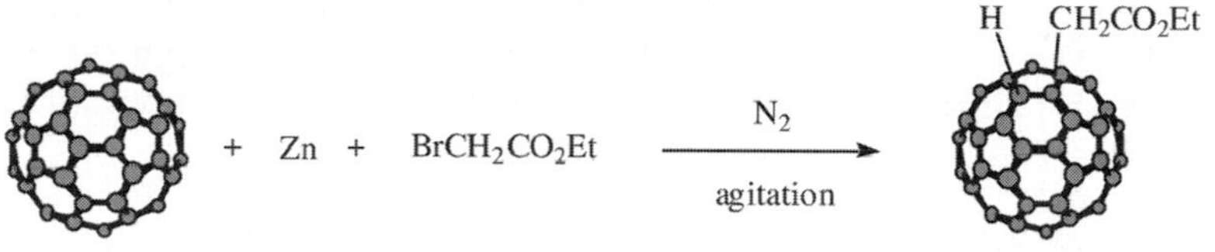

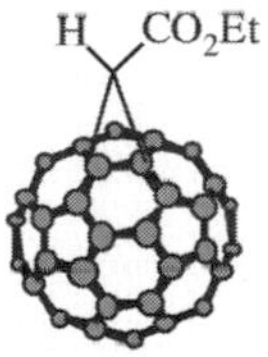

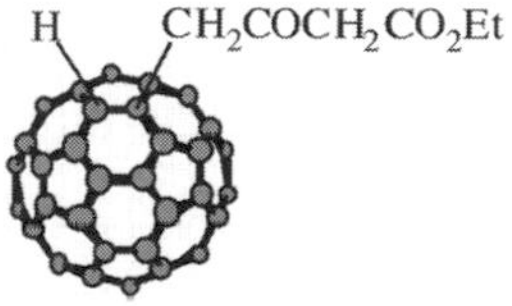

Experimental procedures:

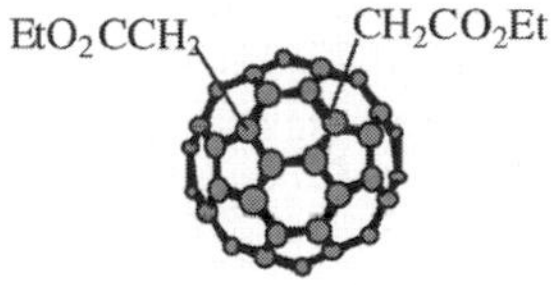

In the nitrogen bag, 50.2 mg of [60]fullerene, ethyl bromoacetate (5 equiv.), zinc dust (20 equiv.) and the stainless-steel ball were placed in the capsule. The above mixture was vigorously agitated for 20 min at room temperature, quenched with 0.5 mL of CF_3CO_2H in 20 mL of *o*-dichlorobenzene, and carefully separated by silica gel chromatography with hexane-toluene as the eluent to give the expected adduct **1** (17.2%) (62.5% based on consumed C60) together with **2** (0.8%), **3** (3.9%), **4** (1.8%) and unreacted C60 (72.5%).

References: G. Wang, Y. Murata, K. Komatsu, T.S.M. Wan, *Chem. Commun.*, 2059 (1996).

Type of reaction: C–C bond formation
Reaction condition: solid-state
Keywords: C60, KCN, [2+2]cycloaddition, C120

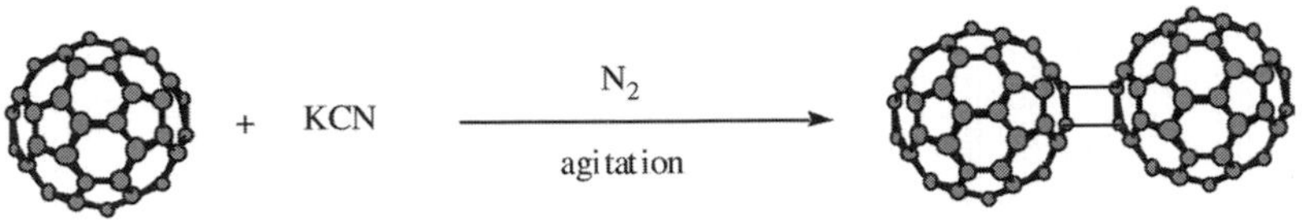

1

Experimental procedures:
A mixture of C60 and 20 molar equiv. of KCN powder was vigorously vibrated for 30 min under nitrogen according to our previous procedure. Analysis by high-performance liquid chromatography of the reaction mixture dissolved in *o*-dichlorobenzene (ODCB) on a Cosmosil Buckyprep column with toluene as the eluent showed only one major product besides unchanged C60. Separation by flash chromatography on silica gel, eluted with hexane-toluene and then with toluene-ODCB, gave 70% of recovered C60 and 18% of C120 (**1**).

References: G. Wang, K. Komatsu, Y. Murata, M. Shiro, *Nature,* **387**, 583 (1997).

Type of reaction: C–C bond formation
Reaction condition: solid state
Keywords: *trans*-diallene, [2+2]conrotatory cyclization, dimethylenecyclobutene

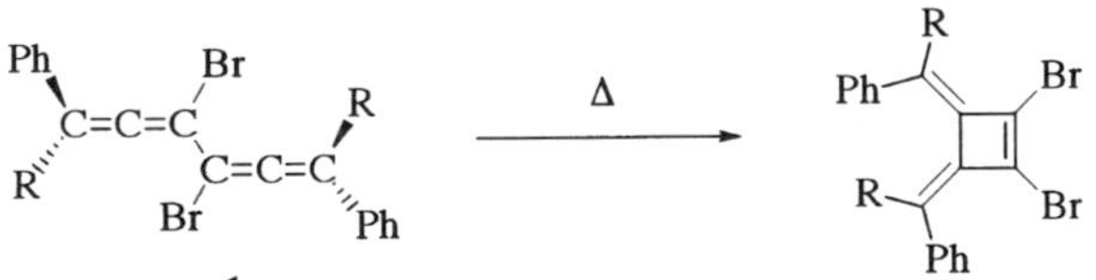

meso-**1**

in, out-**2**

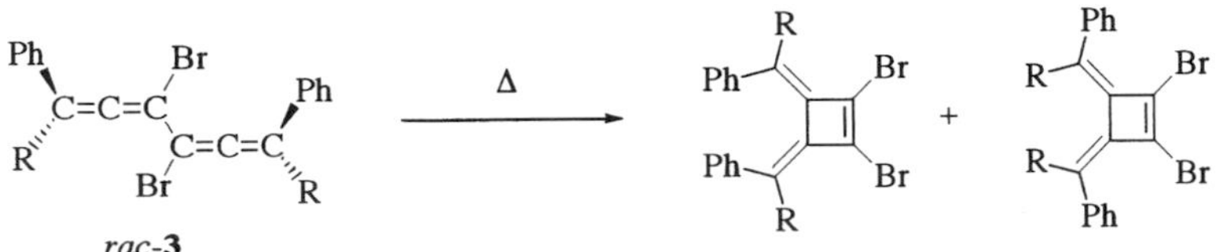

rac-**3**

in, in-**4**

out, out-**5**

Experimental procedures:
Compound **1** (0.2 g) was heated in the crystalline state at 135 °C for 1.5 h to give **2** (0.2 g, 100% yield) as colorless needles after recrystallization from AcOEt (mp 214–215 °C).

Compound **3** (0.4 g) was heated in the crystalline state at 125 °C for 1.5 h to give a 1:1 mixture of **4** and **5** (0.4 g, 100% yield). Fractional recrystallization of the mixture gave pure **4** as colorless needles (mp 180–183 °C) and pure **5** as colorless needles (mp 215–218 °C).

References: F. Toda, K. Tanaka, T. Tamashima, M. Kato, *Angew. Chem. Int. Ed. Engl.*, **37**, 2724 (1998).

Type of reaction: C–C bond formation
Reaction condition: solvent-free
Keywords: ketene silyl acetale, alkine, [2+2]cycloaddition, cyclobutane

$X=O, S$
$n= 1, 2, 3$

$R_1 = CO_2Et; R_2 = H$
$R_1 = R_2 = CO_2Me$
$R_1 = COMe; R_2 = H$

Experimental procedures:
To the ketene silyl (thio)acetal (1 equiv.) was added at room temperature neat ethyl propynoate, dimethyl acetylenedicarboxylate or ethynyl methyl ketone (1 equiv.). The mixture was stirred for 5 h at room temperature. The volatile products were removed at room temperature in vacuo (0.1 Torr) for 30 min. Purification of the crude reaction mixture by silica gel column chromatography (ethyl acetate-hexane, 5:95) afforded cycloadducts.

References: M. Miesch, F. Wendling, *Eur. J. Org. Chem.*, 3381 (2000).

Type of reaction: C–C bond formation
Reaction condition: solid-state
Keywords: 1,4-dithiin, anthracene, Diels-Alder reaction, single-crystal-to-single-crystal transformation

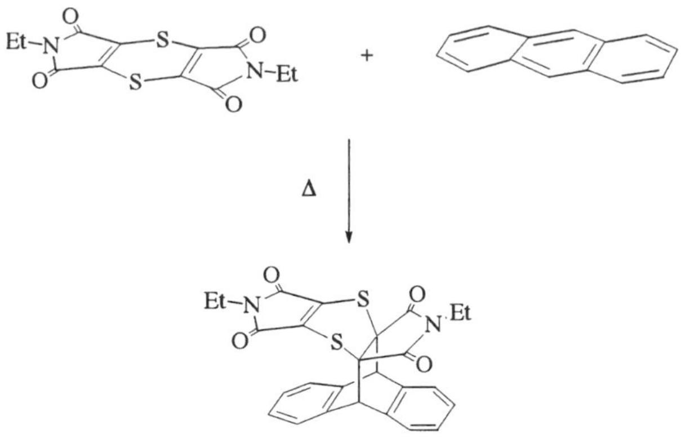

Experimental procedures:

The charge-transfer crystal of 1,4-dithiin and anthracene was grown in dichloromethane solution (50 mM) by slow evaporation of solvent at low temperature (–4 °C). After one week, a brown single crystal was isolated. The solid-state reaction of the anthracene-1,4-dithiin CT crystals was carried out at four temperatures (viz. 50, 60, 70 and 80 °C), and the conversion was monitered by ^{1}H NMR spectroscopy.

References: J. H. Kim, S. M. Hubig, S. V. Lindeman, J. K. Kochi, *J. Am. Chem. Soc.,* **123**, 87 (2001); J. H. Kim, S. V. Lindeman, J. K. Kochi, *ibid,* **123**, 4951 (2001).

Type of reaction: C–C bond formation
Reaction condition: solid-state
Keywords: allene, thermal cycloaddition, cyclobutane, anthrocyclobutene

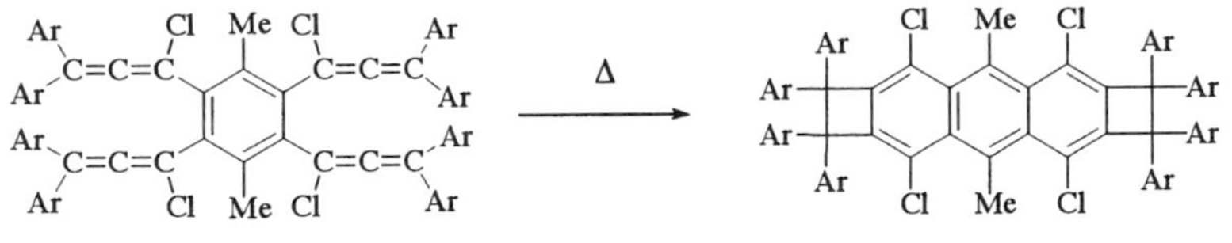

a: Ar=Ph
b: Ar=4-MeC$_6$H$_4$
c: Ar=4-FC$_6$H$_4$
d: Ar=4-ClC$_6$H$_4$

Experimental procedures:
Compound **1a** (0.5 g) was heated in the crystalline state at 180 °C on a hot plate for 30 min to give **2a** (0.5 g, 100% yield) as yellow prisms (mp >300 °C).

References: K. Tanaka, N. Takamoto, Y. Tezuka, M. Kato, F. Toda, *Tetrahedron,* **57**, 3761 (2001).

Type of reaction: C–C bond formation
Reaction condition: solid-state
Keywords: *N*-vinylpyrrolidene, acid catalysis, linear dimerization, gas-solid reaction

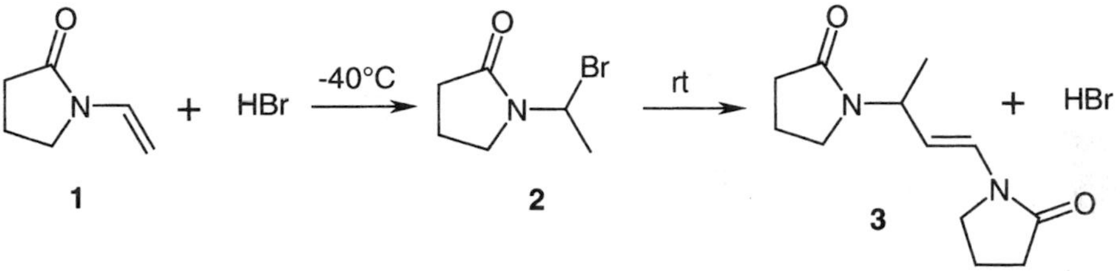

Experimental procedure:
N-Vinylpyrrolidone **1** (1.00 g, 9.0 mmol) was spread on cylindric glass rings (Raschig coils) in a 500-mL flask and crystallized by cooling to –40 °C in a vacuum. HBr gas (1 bar) was applied for 2 h at that temperature. Excess gas was pumped off to a recipient at –196 °C for further use and the solid reaction product **2** was left at room temperature with repeated evacuation for removal of the liberated HBr. The racemic product **3** was recrystallized from acetone to yield 650 mg (65%) of the pure compound (mp 70–73 °C) that was spectroscopically characterized.

Reference: G. Kaupp, D. Matthies, *Chem. Ber.,* **120**, 1897 (1987).

Type of reaction: C–C bond formation
Reaction condition: solid-state
Keywords: 1,1-diarylethylene, HCl, catalysis, linear dimerization, head-to-tail, waste free, gas-solid reaction, 1,1,3,3-tetraaryl-1-butene

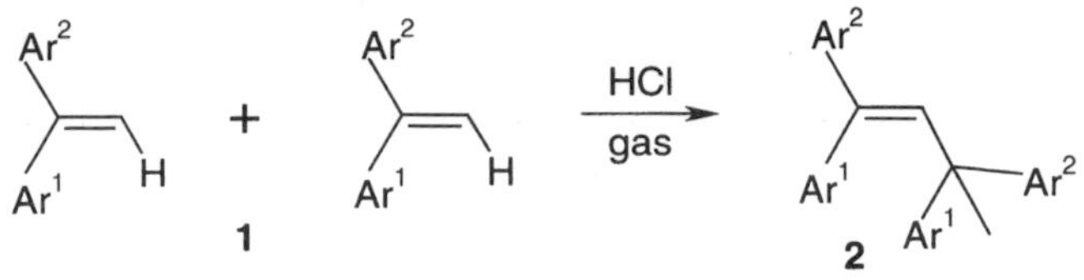

a: Ar^1=Ar^2=Ph (-50°C; 100%)
b: Ar^1=Ph, Ar^2=*p*-Tol (-25°C; 100%; *E/Z*=33:67)
c: Ar^1=Ar^2=*p*-Tol (-25°C; 100 %)
d: Ar^1=Ph, Ar^2=*p*-Anis (-25°C; 100%; *E/Z*=15:85)
e: Ar^1=*p*-Tol; Ar^2=*p*-Anis (-25°C; 100%; *E/Z*=52:48)

Experimental procedures:

The 1,1-diarylethylene **1** (1.00 g) was crystallized in a 100-mL flask by cooling to the appropriate temperature in a vacuum and HCl gas (1 bar) was added and left at that temperature overnight. The pure products **2** were obtained after removal of the HCl gas and thawing up to room temperature.

Reference: G. Kaupp, A. Kuse, *Mol. Cryst. Liq. Cryst.,* **313**, 361 (1998).

Type of reaction: C–C bond formation
Reaction condition: solid-state
Keywords: arylidene malodinitril, dimedone, Michael addition, cyclization, pyrane

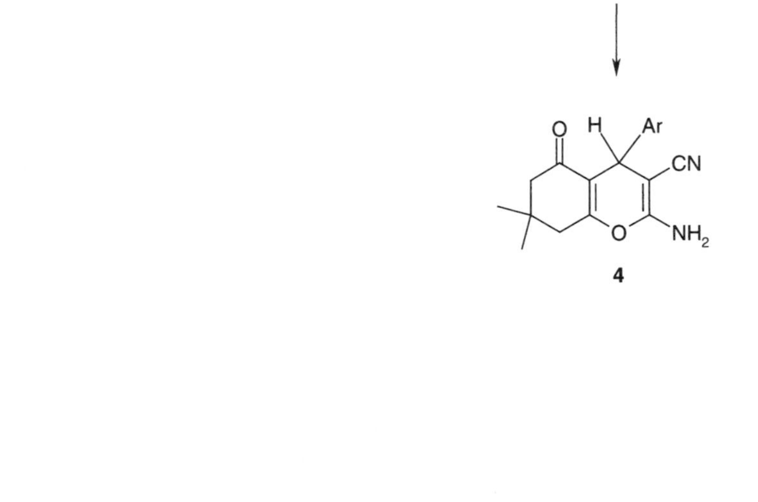

Experimental procedure:
A mixture of dimedone **1** (2.00 mmol) and **2** (2.00 mmol) was ball-milled at 70 °C for 1 h to give an 80% conversion to **4**. Completion of the reaction was achieved by heating the yellow powder to 100 °C for 2 h. The yield was 620 mg (100%) of **4**, mp 222–223 °C.

Reference: G. Kaupp, M.R. Naimi-Jamal, J. Schmeyers, submitted to *Tetrahedron*.

Type of reaction: C–C bond formation
Reaction condition: solvent-free
Keywords: 2-naphthol, methyleneimine, aminomethylation, solid-solid reaction, 1-(N-arylaminomethyl)-2-naphthol

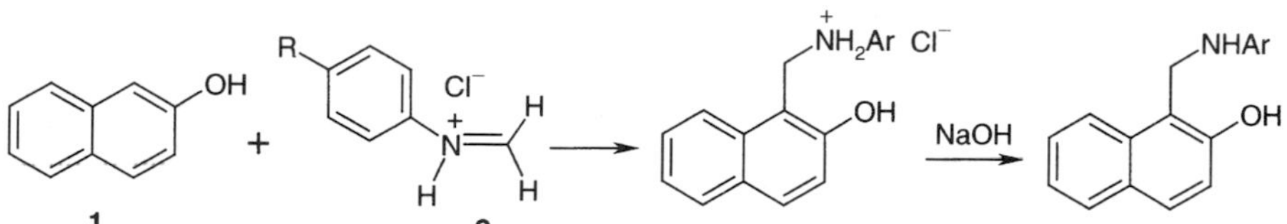

a: R=H   (50%)
b: R=CH$_3$ (65%)
c: R=OCH$_3$ (45%)

Experimental procedures:
The arylmethylene imine hydrochloride **2** was sampled under dry argon, precisely weighed (ca. 2 mmol) and ball-milled with the equivalent of 2-naphthol **1** for 1 h. The solid-solid character was finally lost during the reaction as the material became sticky. The free base **3** was obtained after neutralization of **3**·HCl in CH$_2$Cl$_2$ by washing with NaHCO$_3$ solution and crystallization from ethanol.

Reference: G. Kaupp, J. Schmeyers, J. Boy, *J. Prakt. Chem.*, **342**, 269 (2000).

Type of reaction: C–C bond formation
Reaction condition: solid-state
Keywords: Viehe salt, acetophenone, solid-solid reaction, cinnamamide

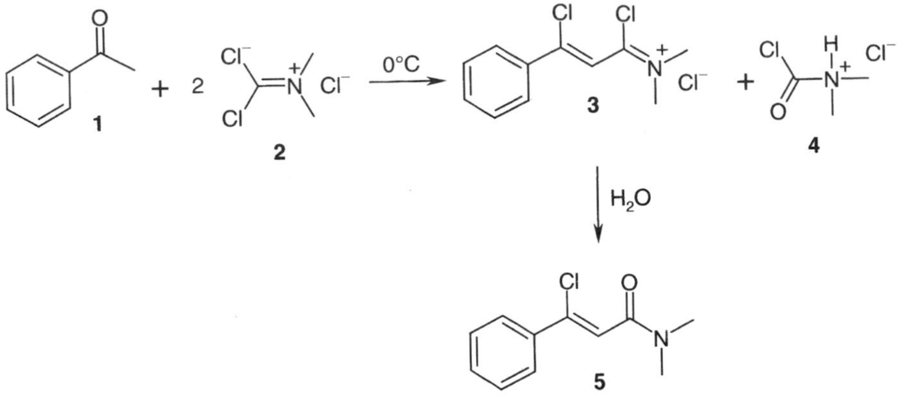

Experimental procedure:

Acetophenone **1** (600 mg, 5.00 mmol) was cooled to 0 °C in a 10-mL milling beaker and Viehe salt **2** (815 mg, 5.00 mmol) was added unter N_2. Ball-milling at 0 °C was performed for 2 h. A solid mixture of product **3** and couple product **4** was quantitatively obtained. Compound **5** was obtained by treatment with water, filtration and drying.

Reference: G. Kaupp, J. Boy, J. Schmeyers, *J. Prakt. Chem.,* **340**, 346 (1998).

Type of reaction: C–C bond formation
Reaction condition: solvent-free
Keywords: lactam, base catalysis, complex reaction, spiroaminal, aminoalkylimin

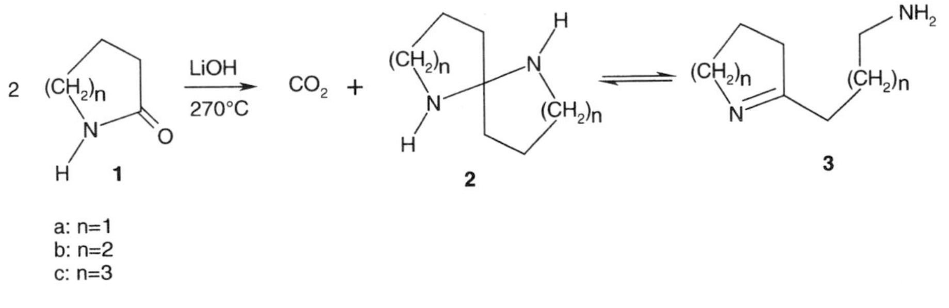

a: n=1
b: n=2
c: n=3

Experimental procedures:

Valerolactam **1b** (10 g, 100 mmol) and LiOH · H_2O (4.25 g, 100 mmol) were heated to 200 °C to remove water, then under N_2 to 270 °C for 4 h under reflux. The product **2b** and residual **1** were distilled off while the temperature was in-

creased up to 320 °C. The raw material of two runs was separated by fraction-
ation at a heatable 80-cm spinning band column under vacuum. Pure **2b** (7.6 g,
54%; bp14 84 °C) was obtained as a useful precursor for diaziridine synthesis.

Pyrrolidon **1a** (8.51 g, 100 mmol) or caprolactam **1c** (11.30 g, 100 mmol) and
LiOH·H₂O (4.25 g, 100 mmol) were dissolved in water (30 mL) and evaporated
to dryness in order to get a homogeneous mixture that was heated under N₂ to
270 °C for 4 h under reflux, then distilled over and pure **3a** (88%, bp₁₄ 83 °C) or
3c (55%, bp₁ 94 °C) was obtained by spinning band column distillation.

Reference: S. N. Denisenko, E. Pasch, G. Kaupp, *Angew. Chem.*, **101**, 1397 (1989); *Angew. Chem. Int. Ed. Engl.*, **28**, 1381 (1989).

Type of reaction: C–C bond formation
Reaction condition: solvent-free or solid-state
Keywords: cyanoacetamide, malodinitril, methyl cyanoacetate, aromatic aldehyde, Knoevenagel condensation, solid-solid reaction, base catalysis, melt reaction, uncatalyzed, cinnamamide, cinnamonitril, methyl cinnamate

a: R=OH
b: R=Cl
c: R=NMe₂
d: R=H

Experimental procedures:
3 (*catalyzed*): Cyanoacetamide **2** (2.00 mmol) and aldehyde **1** (2.00 mmol) were
ball-milled for 10 min. The powder was transferred to a 250-mL flask that was
evacuated and filled with trimethylamine gas and left for 24 h for the completion
of the reaction. The gaseous base catalyst and the water of reaction were evacuated and a quantitative yield of pure **3a–c** obtained.

3 (*uncatalyzed*): Equimolar mixtures of **1** and **2** were heated to 170 °C (**1d** and **2** to 100 °C) for 1 h to obtain **3a** (100%), **3b** (100%), **3c** (87%), or **3d** (92%). **3c**, d were recrystallized from ethanol.

5 (*uncatalyzed*): Malodinitril **4** (2.00 mmol) and aldehyde **1** (2.00 mmol) were heated to 150 °C for 1 h to obtain **5a** (100%), **5b** (100%), or **5c** (100%). **5d** was quantitatively obtained as a yellow powder by ball-milling of an initially liquid 1:1 mixture of **1d** and **4** (3.00 mmol) for 1 h.

7 (*uncatalyzed*): Methyl cyanoacetate (**6**) (2.00 mmol) and aldehyde **1a** and **1c** (2.00 mmol) were heated to 170 °C for 1 h to obtain pure **7a** and **7c** in quantitative yield.

Reference: G. Kaupp, M. R. Naimi-Jamal, *J. Schmeyers, submitted to Tetrahedron.*

Type of reaction: C–C bond formation
Reaction condition: solid-state
Keywords: aromatic aldehyde, barbituric acid, Knoevenagel condensation, uncatalyzed, large scale

a: R=NMe$_2$, R'=H, R''= Et, X=S (rt, 1 h)
b: R=OH, R'=H, R''= Me, X=O (rt, 1 h)
c: R=OH, R'=OMe, R''= H, X=O (50 °C, 1 h)
d: R=OH, R'= R''= H, X=O (from rt to 50 °C, 3 h)
e: R=R'=H, R''=Me, X=O (rt, 1 h; initially partly liquid, becomes solid)

Experimental procedures:
Large scale: A stoichiometric mixture of **1b** and **2b** (200 g per batch) was milled in a water cooled (14 °C) horizontal ball-mill (2 L Simoloyer®), Zoz GmbH) with 2 kg steel balls (5 mm diameter) for 1 h at 1000 rpm. The product **3b** was milled out at 600–1000 rpm. 100% yield (recovery) was obtained in the second batch etc. The purity of **3b** was checked by IR, ^{1}H, ^{13}C NMR and mp (297 °C) after drying in a vacuum at 80 °C.

Similarly, **1d** and **2d** reacted with quantitative yield in the 2-L ball-mill that was not water cooled while care was taken that the interior temperature did not rise above 50 °C (1 h per 200 g batch). Mp 299–301 °C.

Small scale: p-Dimethylaminobenzaldehyde **1a** (2.00 mmol) and the thiobarbituric acid **2a** (2.00 mmol) or **1c** (2.00 mmol) and **2c** (2.00 mmol) were ball-milled at room temperature or 50 °C for 1 h or 3 h, to give quantitative yields of **3a** (mp 207–209 °C) or **3c** (mp 313 °C) after drying at 80 °C in a vacuum.

Liquid-solid: Benzaldehyde **1e** (212 mg, 2.00 mmol) and **2b** (2.00 mmol, poorly soluble in **1e**) were ball-milled at room temperature for 1 h. The solid product was dried at 80 °C in a vacuum to give **3e** (100%). **3e** (mp 260–261 °C) was recrystallized from ethanol.

Reference: G. Kaupp, M. R. Naimi-Jamal, J. Schmeyers, submitted to *Tetrahedron.*

Type of reaction: C–C bond formation
Reaction condition: solvent-free
Keywords: aromatic aldehyde, Meldrum's acid, Knoevenagel condensation, uncatalyzed

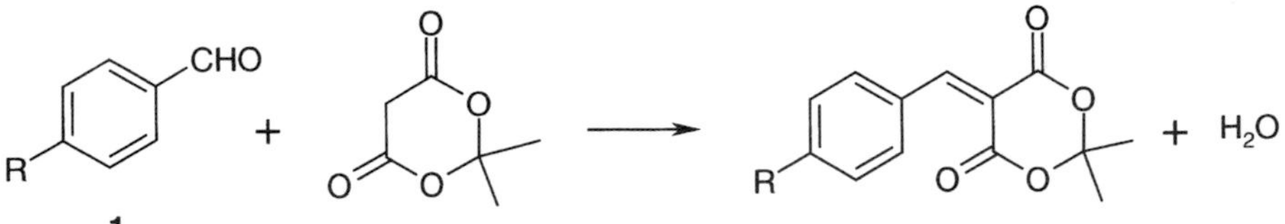

a: R=OH
b: R=NMe$_2$

Experimental procedures:
Meldrum acid **2** (288 mg, 2.00 mmol) and aldehyde **1** (2.00 mmol) were co-ground and heated to 50 °C when **3a, b** crystallized after intermediate melting. The quantitatively obtained pure product **3a, b** was dried at 80 °C in a vacuum. Mp 193 °C and 162 °C.

Reference: G. Kaupp, M. R. Naimi-Jamal, J. Schmeyers, submitted to *Tetrahedron.*

Type of reaction: C–C bond formation
Reaction condition: solvent-free
Keywords: dimedone, aromatic aldehyde, Knoevenagel condensation, Michael addition, uncatalyzed

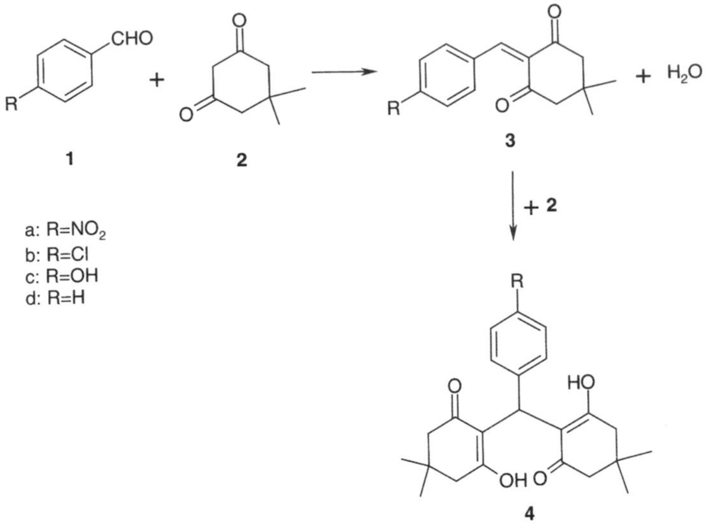
CHO
+
R
O
1
2
3
+ H_2O
+ 2
R
HO
OH O
4
a: R=NO_2
b: R=Cl
c: R=OH
d: R=H

Experimental procedures:
Dimedone **2** (4.00 mmol) and aldehyde **1** (2.00 mmol) were co-ground in a mortar and heated to 80 °C for 3 h (**1a, 1b**) or for 1 h (**1c**). The product **4** crystallized quantitatively from an intermediate melt and was obtained in pure form. Mp 189 °C, 139 °C and 189 °C.

Compound **4d** was quantitatively obtained from an initially liquid 2:1 mixture of **2** and **1d** at 100 °C for 90 min and drying at 80 °C in a vacuum. Equimolar mixtures of **2** and **1c** or **1d** did not provide compound **3c, d** under the reaction conditions but provided **4c, d** and unreacted **1c, d.**

Reference: G. Kaupp, M.R. Naimi-Jamal, J. Schmeyers, submitted to *Tetrahedron.*

Type of reaction: C–C bond formation
Reaction condition: solid-state
Keywords: ninhydrin, dimedone, aminocrotonate, , cyclization, cascade reaction, waste-free, solid-solid reaction

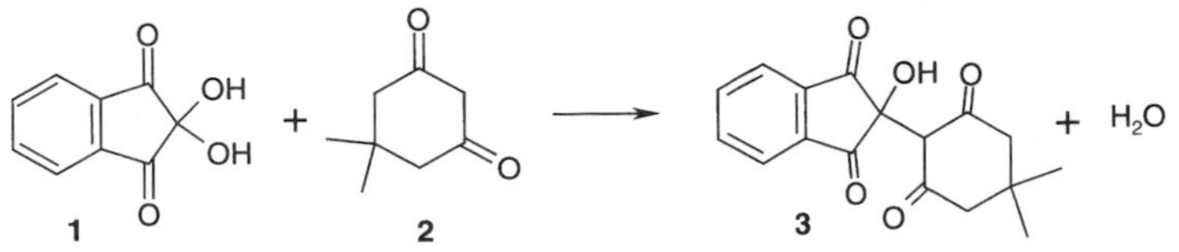

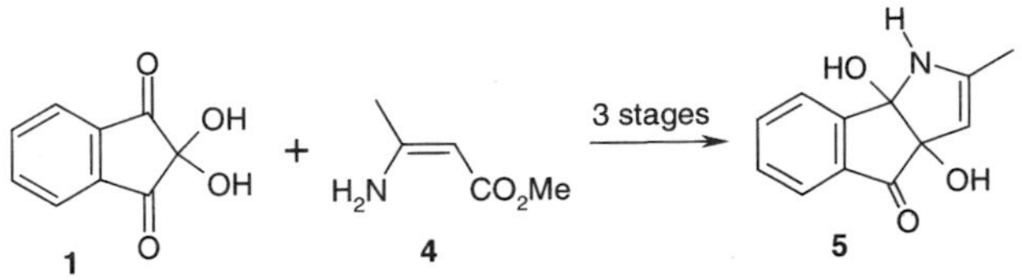

Experimental procedures:

Preparation of **3**: Ninhydrin **1** (178 mg, 1.00 mmol) and dimedone **2** (140 mg, 1.00 mmol) were ball-milled for 1 h at room temperature. Pure **3** (300 mg, 100%) was obtained. Mp 193–195 °C (decomp.).

   Preparation of 5: Ninhydrin **1** (356 mg, 2.00 mmol) and 3-aminocrotonate **4** (230 mg, 2.00 mmol) were ball-milled for 1 h at room temperature. Pure **5** (545 mg, 99%) was obtained after drying in a vacuum at 80 °C. Mp 201–202 °C.

Reference: G. Kaupp, M.R. Naimi-Jamal, J. Schmeyers, *Chem. Eur. J.,* **8**, 594 (2002).

3.2 Solvent-Free C–C Bond Formation under Microwave Irradiation

Type of reaction: C–C bond formation
Reaction condition: solvent-free
Keywords: barbituric acid, aromatic aldehyde, montmorillonite KSF, microwave irradiation, condensation

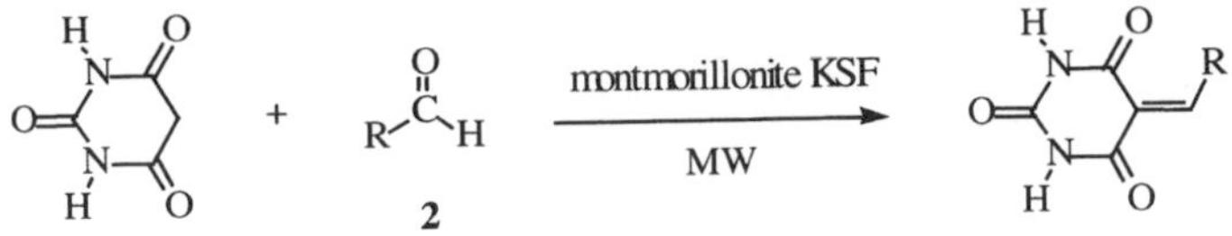

a: R=3,4-OCH$_2$OC$_6$H$_3$
b: R=3,4,5-(MeO)$_3$C$_6$H$_2$
c: R=4-Me$_2$NC$_6$H$_4$
d: R=4-ClC$_6$H$_4$
e: R=2-thienyl
f: R=2-furyl

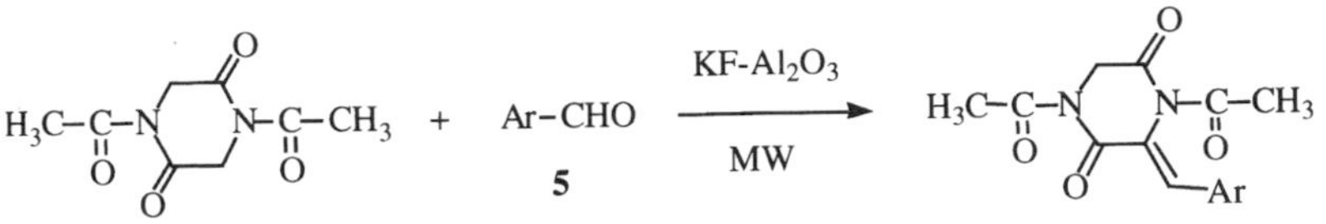

a: Ar=Ph
b: Ar=p-MeC$_6$H$_4$
c: Ar=p-Me$_2$NC$_6$H$_4$
d: Ar=p-MeOC$_6$H$_4$
e: Ar=4-pyridyl
f: Ar=p-NO$_2$C$_6$H$_4$

Experimental procedures:

Barbituric acid **1** (5 mol), aldehyde **2** (6 mmol) and montmorillonite KSF (1 g) were ground with solid aldehyde or were mixed with liquid aldehyde. The mixture placed in an open Pyrex Erlenmeyer flask (25 mL) was irradiated with a microwave oven. After cooling the product was extracted with DMF (15 mL) and the solvent was evaporated in vacuo. The resulting solid was ground and washed with ether (30 mL), filtered and dried.

References: D. Villemin, B. Labiad, *Synth. Commun.*, **20**, 3333 (1990); D. Villemin, A.B. Alloum, *Synth. Commun.*, **20**, 3325 (1990).

Type of reaction: C–C bond formation
Reaction condition: solvent-free
Keywords: creatinine, aldehyde, microwave irradiation, condensation, Z-arylidene creatinine

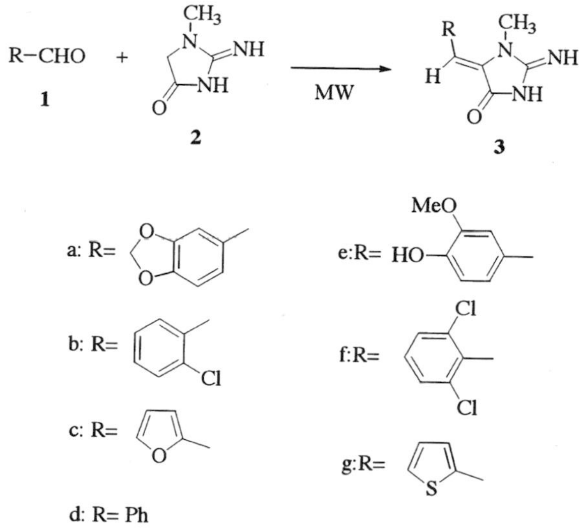

$$R-CHO \quad + \quad \text{(creatinine 2)} \quad \xrightarrow{MW} \quad \text{(product 3)}$$

1   **2**   **3**

a: R= (benzodioxole methyl)

e: R= HO, MeO substituted phenyl

b: R= (2-chloro methylphenyl)

f: R= 2,6-dichlorophenyl (Cl)

c: R= (furyl methyl)

g: R= (thienyl methyl) S

d: R= Ph

Experimental procedures:

Creatinine **2** (1 mmol) and solid aldehyde **1** (1.5 mmol) were crushed with a mortar and pestle. The mixture was irradiated in an open Pyrex tube 8 mm diameter with forcussed microwaves (40 W or 60 W) in resonance cavity TE01 at 2450 MHz with a universal generator MES 73-800, previously described. The mixture was washed with water and then with ether to remove excess aldehyde. Recrystallization from alcohol yielded the condensation product.

References: D. Villemin, B. Martin, *Synth. Commun.*, **25**, 3135 (1995).

Type of reaction: C–C bond formation
Reaction condition: solvent-free
Keywords: sulfone, aldehyde, KF-alumina, microwave irradiation, condensation

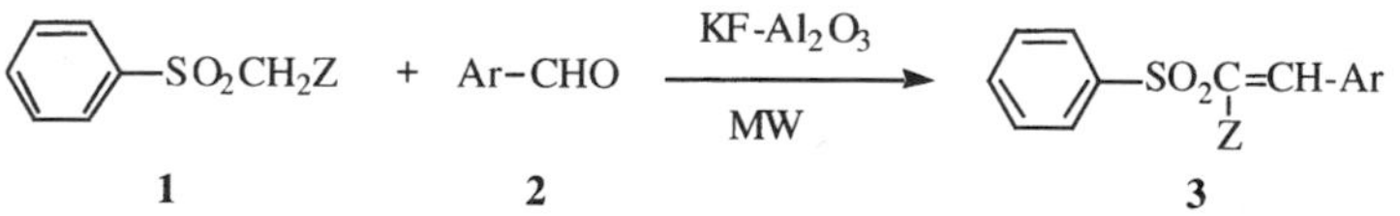

$$\text{C}_6\text{H}_5-SO_2CH_2Z \quad + \quad Ar-CHO \quad \xrightarrow[MW]{KF-Al_2O_3} \quad \text{C}_6\text{H}_5-SO_2\underset{Z}{C}=CH-Ar$$

1   **2**   **3**

a: $Z=COOC_2H_5$ a: Ar=Ph
b: $Z=CN$ b: $Ar=3,4-(Me)_2C_6H_3$
c: $Z=COPh$ c: $Ar=4-MeOC_6H_4$
                 d: $Ar=3-NO_2C_6H_4$

Experimental procedures:
To a solution of sulfone **1** (5 mmol) in the minimum of dichloromethane, aldehyde **2** (5 mmol) and potassium fluoride on alumina (3 g) were added. The solvent was evaporated in vacuo with a rotary evaporator. The resulting solid was
placed in a open Pyrex Erlenmeyer flask (25 mL) and was irradiated in a microwave oven (280 W, 5 min).

The product formed was extracted with CH_2Cl_2 (4×30 mL) and filtered on
Celite. After evaporation of solvent, the product was purified by preparative thin
layer chromatography (cyclohexane-ethyl acetate, 9:1) on silica and crystallized.

References: D. Villemin, A. B. Alloum, *Synth. Commun.*, **21**, 63 (1991).

Type of reaction: C–C bond formation
Reaction condition: solvent-free
Keywords: 5-nitrofuraldehyde, methylene compound, clay, microwave irradia-
tion, 5-nitrofurfurylidene derivative

a: X, Y=-CO-NH-CO-NH-CO-
b: X=CN; Y=COOEt
c: X=COMe; Y=COMe
d: X=CN; Y=CONH$_2$
e: X=CN; Y=CN
f: X=COOEt; Y=COOEt
g: X=CN; Y=CSNH$_2$

Experimental procedures:
Equimolar quantities of the aldehyde **1** (4 mmol) and the active methylene compound **2** were stirred in CH_2Cl_2 (30 mL) in the presence of K10, $ZnCl_2$ (4 g).
The solvent was evaporated in vacuo and the obtained solid was subjected to microwave irradiation in an open Erlenmeyer flask (25 mL). The condensation
product was extracted into acetonitrile. The extract was filtered and the solvent
was evaporated in vacuo. The colored solid was washed with ethanol-water (1:1)
and recrystallized from ethanol.

References: D. Villemin, B. Martin, *J. Chem. Res. (S)*, 146 (1994).

Type of reaction: C–C bond formation
Reaction condition: solvent-free
Keywords: nitromethane, benzylidene methylcyanoacetate, Michael addition, microwave irradiation

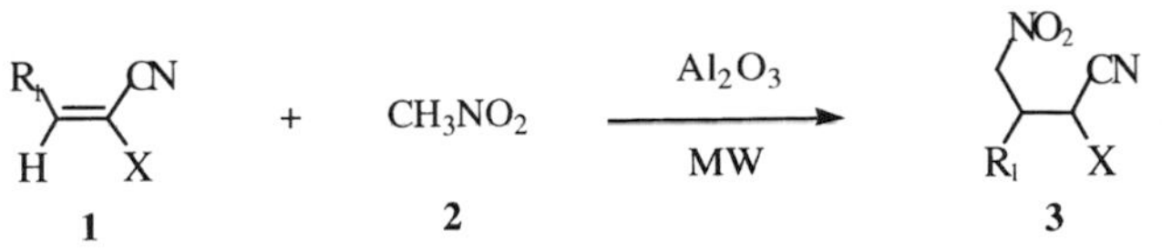

a: $R_1=C_6H_5$; $X=CO_2Me$
b: $R_1=4\text{-}ClC_6H_4$; $X=CO_2Me$
c: $R_1=2\text{-}BrC_6H_4$; $X=CO_2Me$
d: $R_1=2\text{-}FC_6H_4$; $X=CO_2Me$
e: $R_1=3\text{-}FC_6H_4$; $X=CO_2Me$
f: $R_1=i\text{-}Bu$; $X=CO_2Et$
g: $R_1=C_6H_5$; $X=CN$
h: $R_1=C_6H_5$; $X=CONH_2$

Experimental procedures:
The mixture of 5 mmol of benzylidene methylcyanoacetate **1a** and 15 mmol of nitromethane adsorbed over alumina was placed in a quartz tube and introduced into Synthewave 402 microwave reactor with a temperature monitored at 90 °C obtained after 3 and maintained 9 min. The mixture was cooled down and extracted with CH_2Cl_2. After filtration of Al_2O_3 and removal of the solvent the pure product **3a** (two diastereomers A/B, 50:50) was isolated by crystallization from cold ether.

References: D. Michaud, F. T. Boullet, J. Hamelin, *Tetrahedron Lett.,* **38**, 7563 (1997).

Type of reaction: C–C bond formation
Reaction condition: solvent-free
Keywords: nitroalkane, aromatic aldehyde, Henry reaction, microwave irradiation, nitroalkene

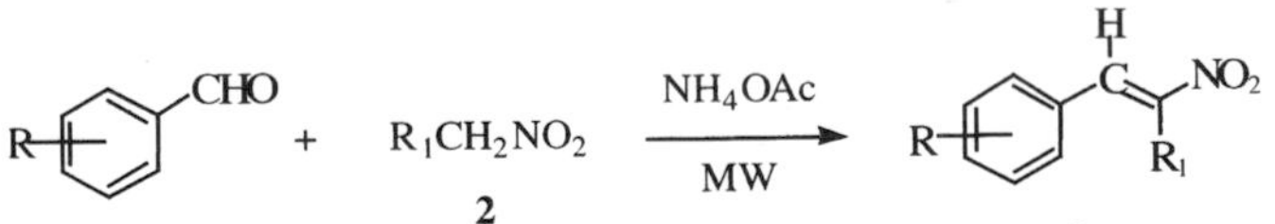

a: R=R$_1$=H
b: R=4-OH; R$_1$=H
c: R=3-MeO-4-OH; R$_1$=H
d: R=3,4-(MeO)$_2$; R$_1$=H
e: R=H; R$_1$=Me
f: R=4-OH; R$_1$=Me
g: R=3-MeO-4-OH; R$_1$=Me
h: R=3,4-(MeO)$_2$; R$_1$=Me
i: R=4-MeO; R$_1$=Me
j: R=1-naphthyl; R$_1$=H
k: R=2- naphthyl; R$_1$=H

Experimental procedures:

A mixture of benzaldehyde **1a** (0.106 g, 1 mmol) and nitromethane **2a** (0.061 g, 1 mmol) is placed in a glass tube containing ammonium acetate (0.019 g, 0.25 mmol) and is exposed to pulsed microwave irradiation in an alumina bath for 2 min using an unmodified microwave oven operating at its 40% power. The reaction mixture is cooled to room temperature ($\sim$ 1 min) and is irradiated again for 2 min. After two such successive irradiations (2 min) and cooling ($\sim$ 1 min) sequences, another 1 mmol of nitromethane is added to the reaction mixture that is further irradiated for two similar successions. After completion of the reaction (followed by TLC), the reaction mixture is passed through a short silica gel column using hexane-AcOEt (9:1, v/v) as an eluent. The evaporation of the solvent on a rotary evaporator affords β-nitrostyrene **3a**.

References: R.S. Varma, R. Dahiya, S. Kumar, *Tetrahedron Lett.*, **38**, 5131 (1997).

Type of reaction: C–C bond formation
Reaction condition: solvent-free
Keywords: aromatic aldehyde, active methylene compound, Knoevenagel condensation, microwave irradiation, benzylidene compound

a: Ar=Ph; R=CN
b: Ar=2-OHC$_6$H$_4$; R=CN
c: Ar=4-ClC$_6$H$_4$; R=CN
d: Ar=4-NO$_2$C$_6$H$_4$; R=CN
e: Ar=PhCH=CH; R=CN
f: Ar=2-furyl; R=CN
g: Ar=2-thienyl; R=CN

h: Ar=Ph; R=COOEt
i: Ar=2-OHC$_6$H$_4$; R=COOEt
j: Ar=4-ClC$_6$H$_4$; R=COOEt
k: Ar=4-NO$_2$C$_6$H$_4$; R=COOEt
l: Ar=PhCH=CH; R=COOEt
m: Ar=2-furyl; R=COOEt
n: Ar=2-thienyl; R=COOEt

Experimental procedures:

A mixture of aldehyde **1** (0.01 mole), active methylene compound **2** (0.01 mole) and LiCl or MgCl$_2$ (0.001 mol) was subjected to microwave irradiation in a Pyrex test tube at output of about 600 W for a given time, then extracted with 50 mL of ethyl acetate and washed with water, dried over Na$_2$SO$_4$ and concentrated in vacuo to afford pure products **3**.

References: G. Sabitha, B. V. S. Reddy, R. S. Satheesh, J. S. Yadav, *Chem. Lett.*, 773 (1998).

Type of reaction: C–C bond formation
Reaction condition: solvent-free
Keywords: α-tosyloxyketone, salicylaldehyde, KF-alumina, microwave irradiation, 2-aroylbenzo[*b*]furan

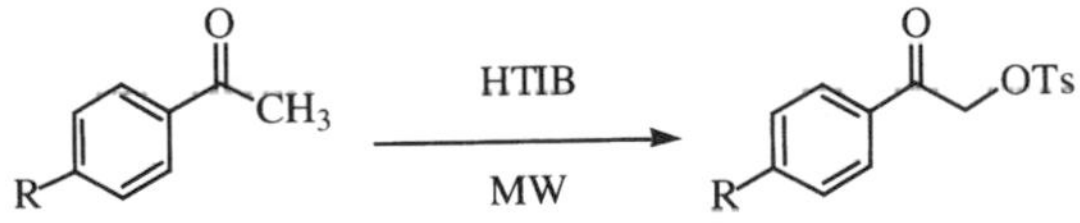

a: R=H; b: R=Cl; c: R=Me; d: R=OMe

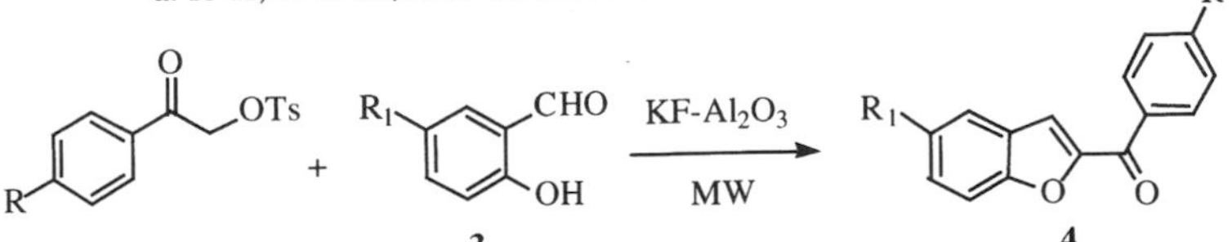

a: R=R$_1$=H
b: R=Cl; R$_1$=H
c: R=Me; R$_1$=H
d: R=OMe; R$_1$=H
e: R=H; R$_1$=Cl
f: R=R$_1$=Cl
g: R=Me; R$_1$=Cl
h: R=OMe; R$_1$=Cl

Experimental procedures:
General procedure for the synthesis of α-tosyloxyketones **2** (a–d):

A mixture of aryl methyl ketone **1** (1 mmol) and [hydroxy(tosyloxy)iodo]ben-zene (1.2 mmol) taken in a glass tube, was placed in an alumina bath inside the MW oven and irradiated for 30 s at 50% power level. After completion of the re-action, as determined by TLC examination, the crude product was washed with hexane to afford pure tosyloxymethyl aryl ketone **2**.

Salicylaldehyde **3** (0.122 mg, 1 mmol), KF-alumina (0.620 g, 2 mmol of KF) and α-tosyloxyketone **2** (1 mmol) were placed in a glass tube and mixed thor-oughly on a vortex mixer. The glass tube was then placed in an alumina bath inside the microwave oven and irradiated (intermittently at 1.5-min intervals; 130 °C) for a specified time. On completion of the reaction, followed by TLC ex-amination (hexane-EtOAc, 9:1), the product was extracted into methylene chlo-ride (3×10 mL). The solvent was then removed under reduced pressure and the residue was crystallized from ethanol to afford a nearly quantitative yield of 2-ar-oylbenzo[*b*]furans **4a–h**.

References: R.S. Varma, D. Kumar, P.J. Liesen, *J. Chem. Soc., Perkin Trans. 1*, 4093 (1998).

Type of reaction: C–C bond formation
Reaction condition: solvent-free
Keywords: aromatic aldehyde, nitromethane, nitroethane, Henry reaction, acti-vated SiO_2, microwave irradiation, nitroalcohol

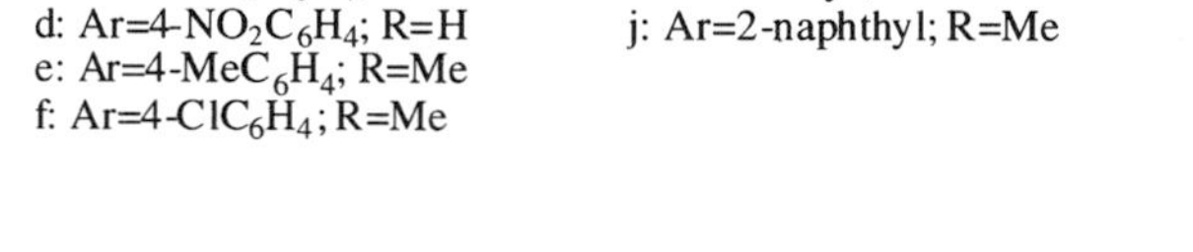

$$ArCHO + R{-}CH_2NO_2 \xrightarrow[\text{MW}]{SiO_2} \underset{OHH}{\overset{H\ NO_2}{Ar{-}\underset{|}{\overset{|}{C}}{-}\underset{|}{\overset{|}{C}}{-}R}}$$

1   **2**   **3**

a: Ar=Ph; R=Me
b: Ar=Ph; R=H
c: Ar=4-$NO_2C_6H_4$; R=Me
d: Ar=4-$NO_2C_6H_4$; R=H
e: Ar=4-MeC_6H_4; R=Me
f: Ar=4-ClC_6H_4; R=Me

g: Ar=4-BrC_6H_4; R=Me
h: Ar=2-thienyl; R=Me
i: Ar=2-furyl; R=Me
j: Ar=2-naphthyl; R=Me

Experimental procedures:
Nitroalcohols **3a–j** were isolated in good yields when a mixture of aromatic alde-hydes **1** (10 mmol) and nitro compound **2** (10 mmol) adsorbed on SiO_2 (finer than 200 mesh, 5 g) and taken in a open Pyrex test tube was subjected to micro-wave irradiation in a domestic microwave oven (BPL, BMO 700T) at an out-put of about 600 W. After cooling the reaction mass to room temperature, the prod-ucts were isolated by extracting with dichloromethane and evaporation of the sol-

vent in vacuo. The catalyst recovered during workup, could be effectively reused, after activation.

References: H. M. S. Kumar, B. V. S. Reddy, J. S. Yadav, *Chem. Lett.*, 637 (1998).

Type of reaction: C–C bond formation
Reaction condition: solvent-free
Keywords: *β*-enamino ester, 3-dimethylamino acrylate, microwave irradiation

1

X=NH, O, S

2

4

a: R=Me$_2$CH
b: R=Me$_3$C
c: R=Et(Me)CH
d: R=CH$_3$CH$_2$CH$_2$
e: R=(EtO)$_2$CHCH$_2$

Experimental procedures:
Ethyl 3-dimethylamino-acrylate (**2**) were easily prepared in good yields without solvent from a mixture of the respective compounds (**1**) and *N,N*-dimethylformamide-diethylacetal (DM-DEA) at 90 °C during 15 min under focused microwave irradiation.

Compound **2** (80 mmol) was mixed with an excess of amine **3** (1.5 mL) in the microwave reactor (ϕ=4 cm). Then, the mixture was immediately submitted to focused microwave irradiation at the suitable temperature during 30 min. Extraction of the reaction mixture from the microwave reactor with 15 mL of methylene chloride, elimination of solvent and excess of **3** in vacuo followed by the analysis of the crude reaction mixture by ^{1}H NMR spectroscopy indicate the formation of the desired compound **4**. The compounds **4** were purified by recrystallization.

References: Z. Dahmani, M. Rahmouni, R. Brugidou, J. P. Bazureau, J. Hamelin, *Tetrahedron Lett.*, **39**, 8453 (1998).

Type of reaction: C–C bond formation
Reaction condition: solvent-free
Keywords: hydroxybenzaldehyde, ethyl acetate, Knoevenagel condensation, microwave irradiation, coumarin

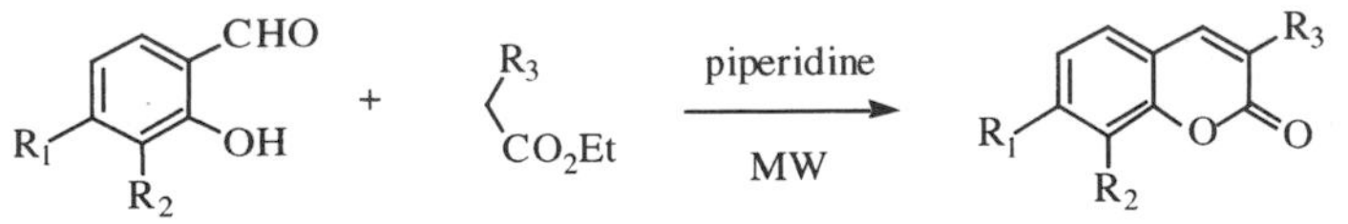

a: $R_1=R_2=H$; $R_3=CO_2Et$
b: $R_1=R_2=H$; $R_3=COMe$
c: $R_1=R_2=H$; $R_3=CN$
d: $R_1=R_2=H$; $R_3=p\text{-}NO_2C_6H_4$
e: $R_1=H$; $R_2=MeO$; $R_3=CO_2Et$
f: $R_1=H$; $R_2=MeO$; $R_3=COMe$

g: $R_1=H$; $R_2=MeO$; $R_3=CN$
h: $R_1=H$; $R_2=MeO$; $R_3=p\text{-}NO_2C_6H_4$
i: $R_1=Et_2N$; $R_2=H$; $R_3=CO_2Et$
j: $R_1=Et_2N$; $R_2=H$; $R_3=COMe$
k: $R_1=Et_2N$; $R_2=H$; $R_3=CN$
l: $R_1=Et_2N$; $R_2=H$; $R_3=p\text{-}NO_2C_6H_4$

Experimental procedures:

A mixture of a hydroxybenzaldehyde **1** (100 mmol), carbonyl compound **2** (110 mmol) and piperidine (0.20 g, 2.4 mmol) was irradiated with microwaves for the time indicated in Table 1. At the end of exposure to microwaves, the reaction mixture was cooled to room temperature, and the crude product recrystallized from an appropriate solvent to obtain the coumarins **3**.

References: D. Bogdal, J. *Chem. Res. (S)*, 468 (1998).

Type of reaction: C–C bond formation
Reaction condition: solvent-free
Keywords: phenol, 1,3-dihydroxybenzene, α,β-unsaturated carboxylic compound, ethyl acetoacetate, microwave irradiation, coumarin

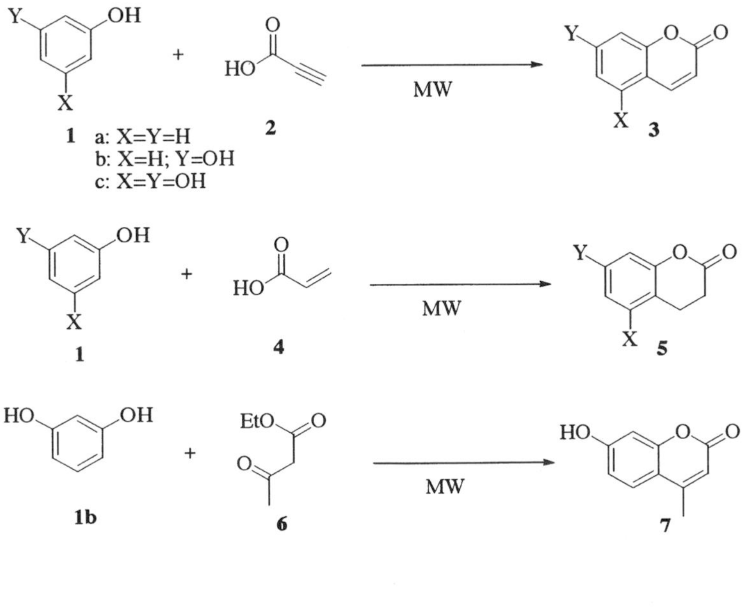

1 a: X=Y=H
     b: X=H; Y=OH
     c: X=Y=OH

Experimental procedures:

A mixture of phenol **1** (10 mmol) and α,β-unsaturated carboxylic derivative **2**, **4** or ethyl acetoacetate **6** (15–20 mmol) supported on 1 g of solid acid support (by dissolving the mixture in 5 mL of diethyl ether followed by evaporation of the solvent) was exposed to microwave irradiation in a focused microwave reactor (Prolabo MX350). For isolation of the compounds, the solid support was removed by extraction with ethanol (for reactants **2** and **6**) or acetonitrile (reactions of **4**). After solvent evaporation, the products were purified by column chromatography on silica gel (hexane-ethyl acetate, 3:1).

References: A. de la Hoz, A. Moreno, E. Vazquez, *Synlett*, 608 (1999).

Type of reaction: C–C bond formation
Reaction condition: solvent-free
Keywords: 1,4-cyclohexanedione, aromatic aldehyde, KF-Al$_2$O$_3$, microwave irradiation, hydroquinone

a: R=Ph
b: R=4-MeOC$_6$H$_4$
c: R=4-MeC$_6$H$_4$
d: R=4-BrC$_6$H$_4$
e: R=4-NO$_2$C$_6$H$_4$
f: R=4-ClC$_6$H$_4$
g: R=2-napthyl
h: R=9-anthryl

Experimental procedures:

A mixture of benzaldehyde **2a** (1.6 g, 10 mmol), cyclohexa-1,4-dione **1** (1.12 g, 10 mmol) and 37% w/w KF-Al$_2$O$_3$ (3 wt. equiv. of aldehyde) was placed in a Pyrex test tube and subjected to microwave irradiation at an output of 600 W. After completion of the reaction (3 min) as indicated by TLC, the reaction mass was cooled to room temperature, directly charged on a silica gel column (100–200 mesh) and eluted (ethyl acetate-hexane, 3:7) to afford 2-benzylhydroquinone **3a** as a white crystalline solid (1.7 g, 85%).

References: H. M. S. Kumar, B. V. S. Reddy, E. J. Reddy, J. S. Yadav, *Green Chem.*, **1**, 141 (1999).

Type of reaction: C–C bond formation
Reaction condition: solvent-free
Keywords: 2-hydroxybenzaldehyde, 2-hydroxyacetophenone, Meldrum's acid, kaolinitic clay, microwave irradiation, 3-carboxycoumarin

a: R_1=H; R_2=H; R_3=H
b: R_1=5-Cl; R_2=H; R_3=H
c: R_1=4-MeO; R_2=H; R_3=H
d: R_1=4-MeO; R_2=H; R_3=Me
e: R_1=H; R_2=Me; R_3=H
f: R_1=4-OH; R_2=Me; R_3=H
g: R_1=5-Cl; R_2=Me; R_3=H

Experimental procedures:
A mixture of the carbonyl compound **1** (5 mmol) and Meldrum's acid **2** (5 mmol) was placed in a beaker containing EPZG/EPZ10 (100 mg) or natural clay (100 mg, with a drop of conc. H_2SO_4) and exposed to pulsed microwave irradiation for 2 s using an unmodified microwave oven (Kelvinator, T-37 model, India, 760 W, 2450 MHz) operating at 100% power. The reaction mixture was then removed from the microwave oven and allowed to cool to room temperature (1 min) and irradiated again for 2 s. After completion of the reaction (TLC), the reaction mixture was filtered and the catalyst was washed with diethyl ether (4×10 mL). The solvent was removed under reduced pressure to obtain crude 3-carboxycoumarin **3** which was purified by column chromatography using ethyl acetate-light petroleum (1:9) as eluent.

References: B.P. Bandgar, L.S. Uppalla, D.S. Kurule, *Green Chem.*, **1**, 243 (1999).

Type of reaction: C–C bond formation
Reaction condition: solvent-free
Keywords: methyl phenyl ketone, 2-phenyl indole, formylation, Vilsmeier reagent, silica gel, microwave irradiation

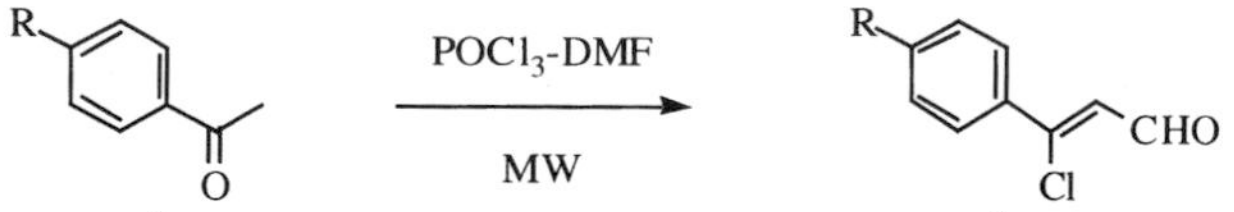

1

R=Cl, Br, OMe

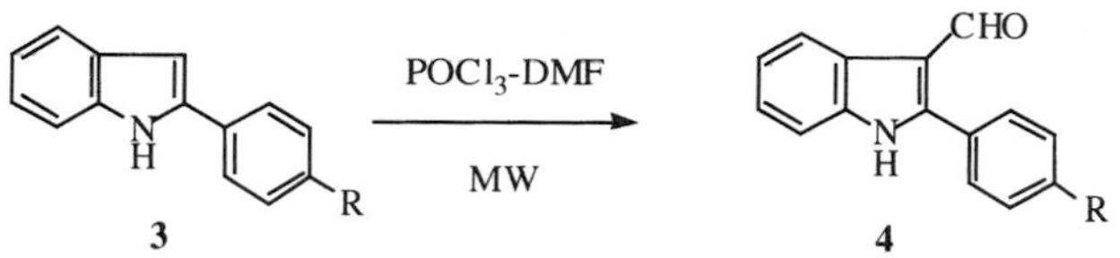

3

R=H, Me, Cl

Experimental procedures:

To the Vilsmeier reagent (4 mmol or 6 mmol) at 0–5 °C, substrate (2 mmol) was added in portions. After the addition was complete, the reaction vessel was kept at room temperature for 5 min and silica gel (1.5–2 or 3–4 g) was added and properly mixed with the help of a glass rod, till free flowing powder was obtained. This powder is then irradiated in a microwave oven for the appropriate time. After irradiation, cold saturated aqueous NaOAc solution (10 mL) was added and stirred. Finally, the product was extracted with dichloromethane (2×15 mL). After removal of the solvent under vacuum, the product was obtained by recrystallization from suitable solvent.

References: S. Paul, M. Gupta, R. Gupta, *Synlett*, 1115 (2000).

Type of reaction: C–C bond formation
Reaction condition: solvent-free
Keywords: aldehyde, ketone, active methylene compound, Knoevenagel condensation, basic alumina, microwave irradiation, olefin

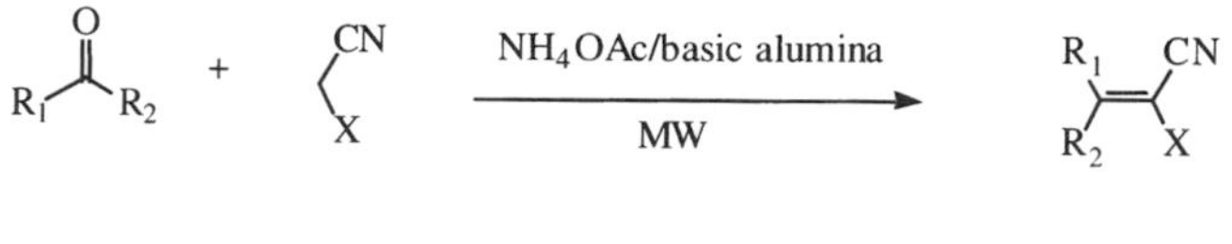

a: R_1=Ph; R_2=H; X=CN
b: R_1=3-$NO_2C_6H_4$; R_2=H; X=CN
c: R_1=4-$NO_2C_6H_4$; R_2=H; X=CN
d: R_1=2-furyl; R_2=H; X=CN
e: R_1=PhCH=CH; R_2=H; X=CN
f: R_1=4-$CH_3OC_6H_4$; R_2=H; X=CN
g: R_1=4-$CH_3C_6H_4$; R_2=H; X=CN
h: R_1=Ph; R_2=CH_3; X=CN
i: R_1=i-Pr; R_2=H; X=CN
j: R_1=CH_3; R_2=CH_3; X=CN

k: R_1=Ph; R_2=H; X=COOMe
l: R_1=3-$NO_2C_6H_4$; R_2=H; X=COOMe
m: R_1=4-$NO_2C_6H_4$; R_2=H; X=COOMe
n: R_1=2-furyl; R_2=H; X=COOMe
o: R_1=PhCH=CH; R_2=H; X=COOMe
p: R_1=4-$CH_3OC_6H_4$; R_2=H; X=COOMe

Experimental procedures:
Aldehyde or ketone (3 mmol), ammonium acetate (3 mmol, 231 mg), basic alumina (3 g) and active methylene compound **2** (3 mmol) were mixed thoroughly in a mortar. The reaction mixture was placed in a beaker and irradiated under the conditions shown in Table. The progress of reaction was monitored by TLC using petroleum ether-CH_2Cl_2 (30:70) as eluent. The mixture was extracted into CH_2Cl_2 then filtered and washed with water. The solvent was removed under reduced pressure by rotatory evaporator. Further purification by column chromatography on silica gel gave the desired product.

References: S. Balalaie, N. Nemati, *Synth. Commun.*, **30**, 869 (2000).

Type of reaction: C–C bond formation
Reaction condition: solvent-free
Keywords: Meldrum's acid, piperonal, methylene compound, Knoevenagel reaction, $Zr(O_3POK)_2$, microwave irradiation

a: R_1, R_2= -CO_2-CH(CH$_3$)$_2$-O_2C-
b: R_1=R_2=CN
c: R_1=CN; R_2=COPh
d: R_1, R_2= -CONH-CO-NHCO-

Experimental procedures:
In a typical experiment 200 mg of $Zr(O_3POK)_2$ were added to a solution of piperonal **1** (75 mg, 5 mmol) and Meldrum's acid **2a** (72 mg, 5 mmol) in dichloromethane (2 mL) placed in a 100-mL round-bottomed flask. The solvent was removed under reduced pressure. The round-bottomed flask was placed in a microwave oven (Whirlpool, M430, Jet 900) and irradiated (450 W) for a total of 14 min and in 2-min intervals. After cooling, the crude product was purified by sublimation (10 mTorr, isomantle at a temperature between 180 and 220 °C) to give 5-(3,4-methylenedioxyphenyl)methylene-2,2-dimethyl[1,3]dioxane-4,6-dione **3a** (0.14 g, 100%), mp 179 °C.

References: D. Fildes, V. Caignaert, D. Villemin, P.-A. Jaffres, *Green Chem.*, **3**, 52 (2001).

Type of reaction: C–C bond formation
Reaction condition: solvent-free
Keywords: aromatic hydroxyaldehyde, ethyl malonate, methyl malonate, Knoevenagel reaction, microwave irradiation, 3-substituted coumarin

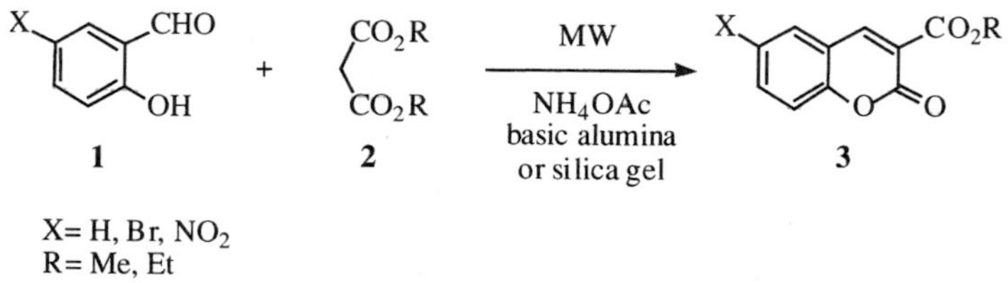

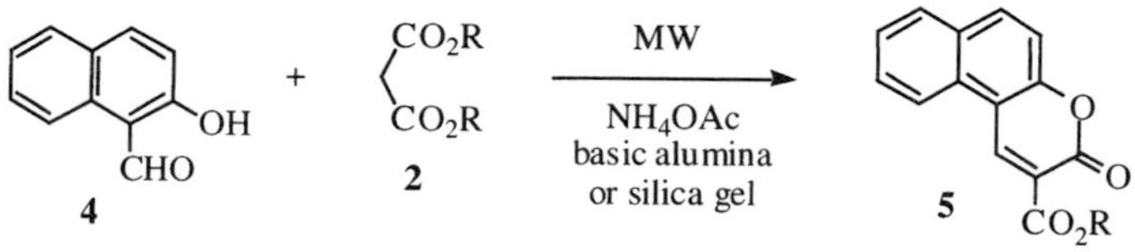

X = H, Br, NO$_2$
R = Me, Et

Experimental procedures:
2-Hydroxyaldehyde **1** or **4** (3 mmol), ethyl or methyl malonate **2** (3 mmol), ammonium acetate (231 mg, 3 mmol) and basic alumina or silica gel (3 g) were mixed thoroughly in a mortar. The reaction mixture was placed in a beaker and irradiated. The progress of reaction was monitored by TLC using (petroleum ether-CH$_2$Cl$_2$, 30:70). The mixture was extracted into methylene chloride (3×30 mL) then filtered and washed with water, the organic phase was removed under reduced pressure by rotary evaporator. Further purification by column chromatography on silica gel gave the desired product. Crystallization was carried out in EtOH.

References: S. Balalaie, N. Nemati, *Heterocycl. Commun.*, **7**, 67 (2001).

Type of reaction: C–C bond formation
Reaction condition: solvent-free
Keywords: malonic acid, aromatic aldehyde, condensation, microwave irradiation, benzylidenemalonic acid

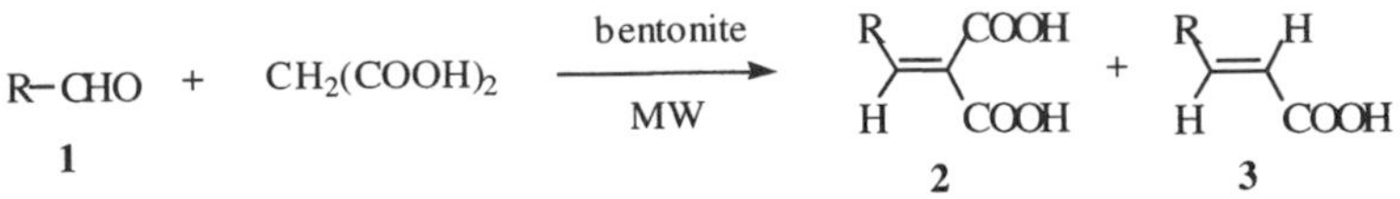

a: R=Ph
b: R=o-MeC$_6$H$_4$
c: R=p-MeC$_6$H$_4$
d: R=p-MeOC$_6$H$_4$
e: R=o-ClC$_6$H$_4$
f: R=α-naphthyl
g: R=2-furyl

Experimental procedures:

In a screw-capped vial were placed malonic acid (0.52 g, 5.0 mmol, 3 equiv.), p-tolualdehyde **1c** (0.20 g, 1.67 mmol) and bentonite (0.72 g). The tube was capped and the contents of the tube were thoroughly mixed with a vortex mixer and then irradiated in the microwave oven for 5 min at a power of 1050 W. After the reaction the mixture was cooled to room temperature and washed successively with hexane (3×10 mL) and cold water (3×10 mL). The resulting mixture was immersed in ethyl acetate (2×10 mL) for 5 min. After removal of bentonite by filtration under vacuum, the mixture was evaporated under reduced pressure to give 2-(4-methylbenzylidene)malonic acid **2c** as a white solid (0.29 g, 86%). This solid was recrystallized from hot, distilled water, mp 205–208 °C.

References: A. Loupy, S. Song, S. Sohn, Y. Lee, T. Kwon, *J. Chem. Soc., Perkin Trans. 1*, 1220 (2001).

Type of reaction: C–C bond formation
Reaction condition: solvent-free
Keywords: 2-methylquinoline, benzoyl chloride, silica gel, microwave irradiation, 2-ketomethylquinolines

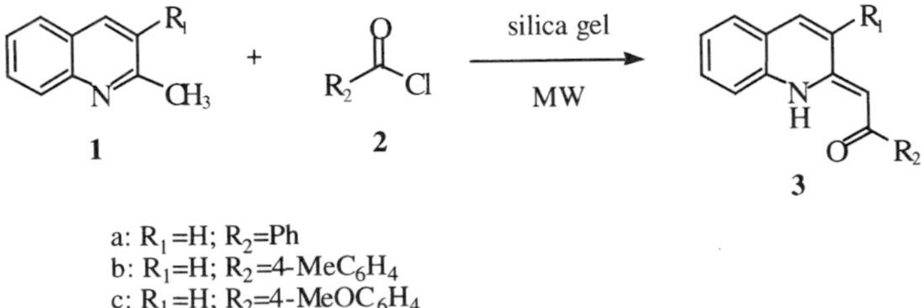

a: R$_1$=H; R$_2$=Ph
b: R$_1$=H; R$_2$=4-MeC$_6$H$_4$
c: R$_1$=H; R$_2$=4-MeOC$_6$H$_4$
d: R$_1$=H; R$_2$=4-ClC$_6$H$_4$
e: R$_1$=H; R$_2$=2-ClC$_6$H$_4$
f: R$_1$=H; R$_2$=4-BrC$_6$H$_4$
g: R$_1$=H; R$_2$=2-pyridyl
h: R$_1$=H; R$_2$=1-naphtyl
i: R$_1$=Me; R$_2$=Ph

Experimental procedures:

To an equimolar (1 mmol) mixture of 2-methylquinoline **1a** (143 mg) and benzoyl chloride **2a** (140 mg) placed in an open glass container, silica gel (silica gel 60, 230–240 mesh, Merck) (300 mg) was added and the reaction mixture was irradiated in a microwave oven at 400 W power for 4 min. Upon completion of the reaction, as followed by TLC examination, the product is extracted into dichloromethane (3×10 mL). The solvent was evaporated and the resulting crude material was purified on a silica gel plate (eluent; CCl_4-Et_2O) affording the 1-phenyl-2-(quinol-2-yl)-ethane-1-one **3a** in 91% yield.

References: H. L. Khouzani, M. M. Safari, A. Minaeifar, *Tetrahedron Lett.,* **42**, 4363 (2001).

Type of reaction: C–C bond formation
Reaction condition: solvent-free
Keywords: 2,4-dinitrotoluene, benzaldehyde, pyrrolidine, microwave irradiation, nitrostilbene

a: R_1=R_2=H
b: R_1=R_2=OMe
c: R_1=H; R_2=OMe
d: R_1=H; R_2=OH
e: R_1=NO_2; R_2=H

Experimental procedures:

Synthesis of stilbene **3e** is representative for the general procedure followed. A mixture of 2,4-dinitrotoluene **1** (0.93 g, 5 mmol) and aldehyde **2e** (0.63 g, 5 mmol) was irradiated under microwave of 800 W in a domestic-type microwave oven in the presence of pyrrolidine (0.106 g, 1.5 mmol) for 30 s. The resultant mixture was washed with ethanol. The solid product was filtered and dried. Yield: 80%; mp 180 °C.

References: S. Saravanan, P. C. Srinivasan, *Synth. Commun.,* **31**, 823 (2001).

Type of reaction: C–C bond formation
Reaction condition: solvent-free
Keywords: aromatic aldehyde, Meldrum's acid, dimedone, NH$_4$OAc, microwave irradiation, 2,5-dioxo-1,2,3,4,5,6,7,8-octahydroquinoline

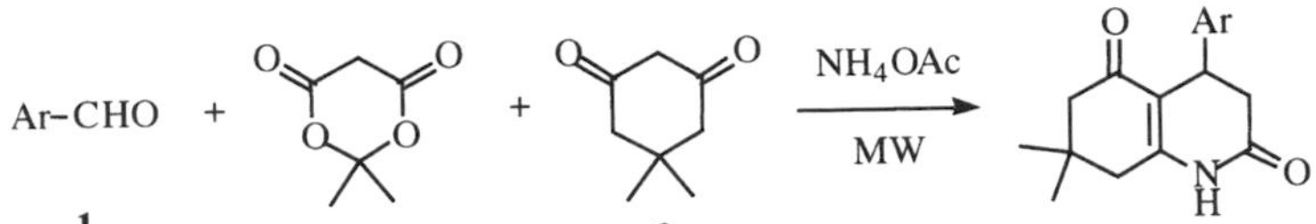

a: Ar=C$_6$H$_5$
b: Ar=4-ClC$_6$H$_4$
c: Ar=4-MeOC$_6$H$_4$
d: Ar=3,4-(MeO)$_2$C$_6$H$_3$
e: Ar=2-ClC$_6$H$_4$
f: Ar=4-BrC$_6$H$_4$
g: Ar=3,4-OCH$_2$OC$_6$H$_3$

Experimental procedures:
A mixture of the appropriate aromatic aldehyde **1** (5 mmol), Meldrum's acid **2** (0.72 g, 5 mmol), dimedone **3** (0.7 g, 5 mmol) and ammonium acetate (0.6 g, 8 mmol) was irradiated by microwave for 3–5 min. The mixture was cooled to room temperature. A little ethanol (2 mL) was added. The solid that precipitated was collected by filtration. Further purification was accomplished by recrystallization from ethanol.

References: S. Tu, Q. Wei, H. Ma, D. Shi, Y. Gao, G. Cui, *Synth. Commun.,* **31**, 2657 (2001).

Type of reaction: C–C bond formation
Reaction condition: solvent-free
Keywords: 2-ethoxycarbonylcyclohexanone, lithium bromide, microwave irradiation, 2-alkylcyclohexanone

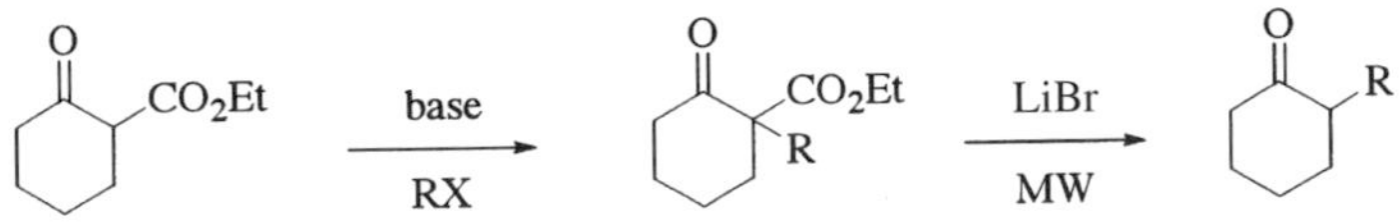

a: R=Et
b: R=Bu
c: R=C$_6$H$_{13}$

Experimental procedures:
Alkylation of **1**. To a mixture of the ester **1** (5 g, 25 mmol) and 6% of Aliquat 336 (715 mg, 1.5 mmol) was added *t*-BuOK (3.3 g, 25 mmol, 1 equiv.) under magnetic stirring over 15 min; alkyl bromide (25 mmol, 1 equiv.) was then added slowly. The flask was left under the experimental conditions indicated in Table 1. Finally, the mixture was diluted with ethyl acetate (50 ml) and filtered on Florisil (10 g). The crude products were analysed by GC and characterised by MS and ^{1}HNMR. 2-Alkylcyclohexanones. A mixture of lithium bromide (1.73 g, 20 mmol), tetrabutylammonium bromide (323 mg, 1 mmol), water (360 mm^3, 0.36 mL, 20 mmol) and ethyl 1-alkyl-2-oxocyclohexanecarboxylate **2** (10 mmol) was placed in a Pyrex tube. The tube was then introduced into a Maxidigest MX 350 Prolabo microwave reactor fitted with a rotational system. Microwave irradiation was carried out for a suitable power and time (Table 2). An approximate final temperature was measured by introducing a digital thermometer at the end of irradiation. The mixture was cooled to ambient temperature. After elution with ethyl acetate (50 mL) and subsequent filtration on Florisil, organic products were analyzed by GC and finally purified by chromatography on silica gel (pentane-ethyl acetate, 95:5).

References: J. P. Barnier, A. Loupy, P. Pigeon, M. Ramdani, P. Jacquault, *J. Chem. Soc., Perkin Trans. 1*, 397 (1993).

Type of reaction: C–C bond formation
Reaction condition: solvent-free
Keywords: active methylene compound, chalcone, Michael reaction, alumina, microwave irradiation

a: X=NO$_2$; Y=H
b: X=CN; Y=H
c: X=Y=COOEt
d: X=Y=COMe

Experimental procedures:
Nitromethane **2a** (1.2 mmol), alumina (1 g, basic activated at 200 °C) and benzalacetophenone **1** (1.2 mmol) were mixed together without solvent in an Erlenmeyer flask and placed in a commercial microwave oven (operating at 2450 Hz frequency) and irradiated for 18 min. The reaction mixture was allowed to reach room temperature and extracted with chloroform. Removal of solvent and the re-

sidue on purification by thin layer chromatography using chloroform-petroleum ether (60–80), 3:2 as eluent gave the corresponding Michael adduct in 90% yield, mp 98–99 °C without the formation of any side products.

References: A. Boruah, M. Baruah, D. Prajapati, J.S. Sandhu, *Chem. Lett.*, 965 (1996).

Type of reaction: C–C bond formation
Reaction condition: solvent-free
Keywords: 1,3-dicarbonyl compound, α,β-unsaturated carbonyl compound, Michael addition, $EuCl_3$, microwave irradiation

a: $R_1=R_2=R_3=Me$; $R_4=H$
b: $R_1=Me$; $R_2=OEt$; $R_3=Me$; $R_4=H$
c: $R_1=Me$; $R_2=OEt$; $R_3,R_4=-(CH_2)_3-$
d: $R_1=C_6H_5$; $R_2=Me$; $R_3=Me$; $R_4=H$
e: $R_1=C_6H_5$; $R_2=Me$; $R_3,R_4=-(CH_2)_3-$
f: $R_1=C_6H_5$; $R_2=OEt$; $R_3=Me$; $R_4=H$
g: $R_1=C_6H_5$; $R_2=OEt$; $R_3,R_4=-(CH_2)_3-$

Experimental procedures:
A mixture of 1,3-dicarbonyl compound **1** (2 mmol), Michael acceptor **2** (2.4 mmol) and Eu^{3+} catalyst is submitted to microwave irradiation in a kitchen oven under the conditions reported in the Table. Then the crude mixture is directly poured onto the top of silica gel chromatographic column and elution with petroleum ether-ethyl acetate mixtures affords pure adducts **3**.

References: A. Soriente, A. Spinella, M.D. Rosa, M. Giordano, A. Scettri, *Tetrahedron Lett.*, **38**, 289 (1997).

Type of reaction: C–C bond formation
Reaction condition: solvent-free
Keywords: 1,3-dicarbonyl compound, methyl vinyl ketone, chalcone, Michael addition, $BiCl_3$, $CdCl_2$, microwave irradiation

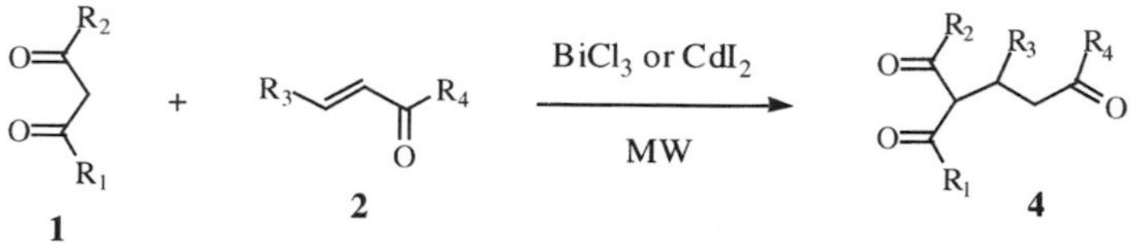

a: R_1=Me; R_2=Me; R_3=H; R_4=Me
b: R_1=Me; R_2=OEt; R_3=H; R_4=Me
c: R_1=OEt; R_2=OEt; R_3=H; R_4=Me
d: R_1=Me; R_2=Me; R_3=Ph; R_4=Ph
e: R_1=Me; R_2=OEt; R_3=Ph; R_4=Ph
f: R_1=MOEt; R_2=OEt; R_3=Ph; R_4=Ph

Experimental procedures:

Acetylacetone **1a** (10 mmol), methyl vinyl ketone **2a** (10 mmol) and bismuth trichloride (0.32 g, 10% mol) were mixed together without solvent in an Erlenmeyer flask and placed in a commercial microwave oven (operating at 2450 MHz frequency) and irradiated for 15 min. The reaction mixture was allowed to reach room temperature and extracted with chloroform. Removal of solvent and the residue on purification by passing through a short column of silica gel using chloroform as eluent, affords the Michael adduct **4a** in 90% yield without the formation of any side products. Similarly cadmium iodide (10% mol) was used in place of bismuth trichloride and the corresponding Michael adduct was isolated in 85% yields.

References: B. Baruah, A. Boruah, D. Prajapati, J.S. Sandhu, *Tetrahedron Lett.,* **38**, 1449 (1997).

Type of reaction: C–C bond formation
Reaction condition: solvent-free
Keywords: 1,3-dicarbonyl compound, cycloalkenone, Michael addition, alumina, microwave irradiation

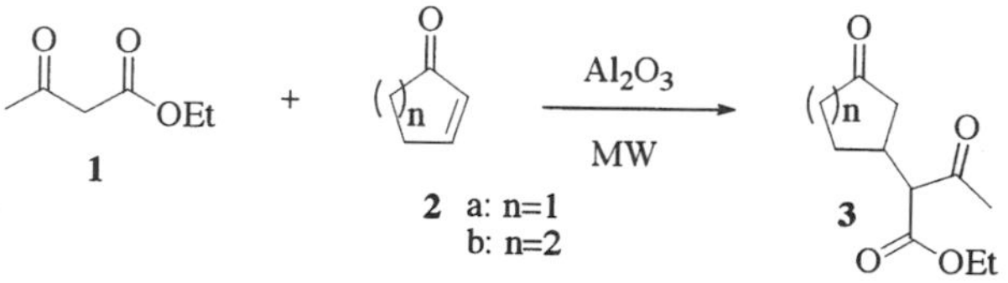

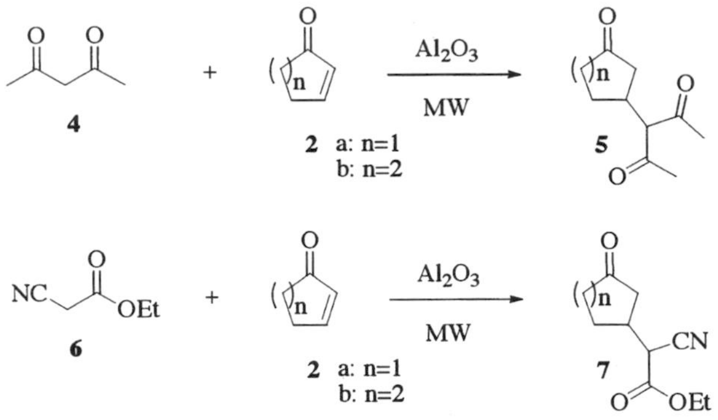

4

2 a: n=1
   b: n=2

5

6

2 a: n=1
   b: n=2

7

Experimental procedures:

A mixture of Michael donor (1 mmol) and Michael acceptor (1 mmol) was added to alumina (neutral, Brockmann activity; grade **1** for column chromatography, SRL, India; 1 g, activated at 180 °C for 4 h at 0.05 mmHg) in a small Pyrex round-bottomed flask and shaken well. The flask was then placed in a porcelain basin containing alumina and irradiated with microwave at 200 W in a domestic microwave oven (BPL, model OBMO-700T) for a certain period of time as required for completion. After being allowed to cool, the solid mass was then taken in a column with a short plug of silica gel and eluted with CH_2Cl_2. Evaporation of solvent furnished the practically pure product which was further purified by column chromatography over silica gel.

References: B.C. Ranu, M. Saha, S. Bhar, *Synth. Commun.,* **27**, 621 (1997).

Type of reaction: C–C bond formation
Reaction condition: solvent-free
Keywords: aldehyde, 1-alkene, hydroacylation, $RhCl(PPh_3)_3$, microwave irradiation

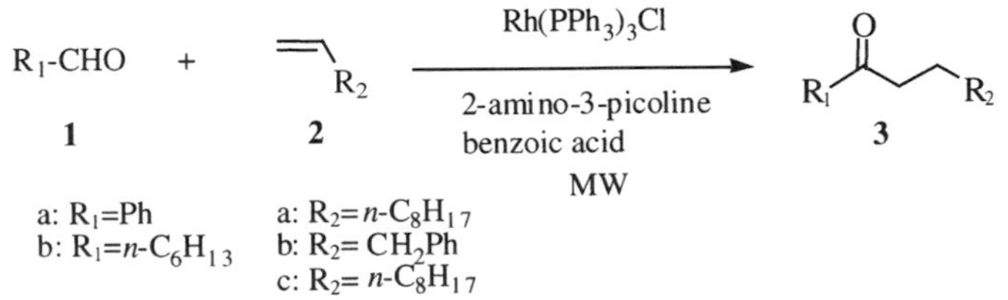

1   **2**   3

a: R_1=Ph   a: R_2=n-C_8H_{17}
b: R_1=n-C_6H_{13} b: R_2= CH_2Ph
                c: R_2= n-C_8H_{17}

Experimental procedures:
Aldehyde **1** (0.2 mmol), alkene **2** (1.0 mmol), 2-amino-3-picoline (0.08 mmol), benzoic acid (0.02 mmol) and RhCl(PPh$_3$)$_3$ (0.01 mmol) were mixed in the absence of any organic solvent and then submitted for 10 min to microwave irradiation inside a domestic microwave oven (Samsung, RE-431H, 700 W) or in a monomode Synthewave 402 Prolabo where power is enslaved to temperature at 140 °C. After the reaction, the product **3** was purified by column chromatography 38–85%.

References: C.H. Lo, J.H. Chung, D.Y. Lee, A. Loupy, S. Chatti, *Tetrahedron Lett.,* **42**, 4803 (2001).

Type of reaction: C–C bond formation
Reaction condition: solvent-free
Keywords: 1,3-dipole, keten acetal, cycloaddition, microwave irradiation

a: Ar=Ph
b: Ar=*p*-CF$_3$C$_6$H$_4$

Experimental procedures:
A mixture of the ketene acetal **1** (1 equiv.) and 1,3-dipole **2** (1 equiv.) was placed in an open vessel in the microwave oven and irradiated at the power specified for the indicated time. The crude reaction mixture was purified by flash chromatography (silica gel) using light petroleum (40–60 °C)-ethyl acetate (7:1) as eluent, unless otherwise stated, to give the corresponding cycloadduct **3**.

References: A. Diaz-Ortiz, E. Diez-Barra, A. de la Hoz, P. Prieto, A. Moreno, *J. Chem. Soc., Perkin Trans. 1,* 3595 (1994).

Type of reaction: C–C bond formation
Reaction condition: solvent-free
Keywords: nitron, dipolarophile, 1,3-dipolar cycloaddition, microwave irradiation, tetrahydroisoxazol

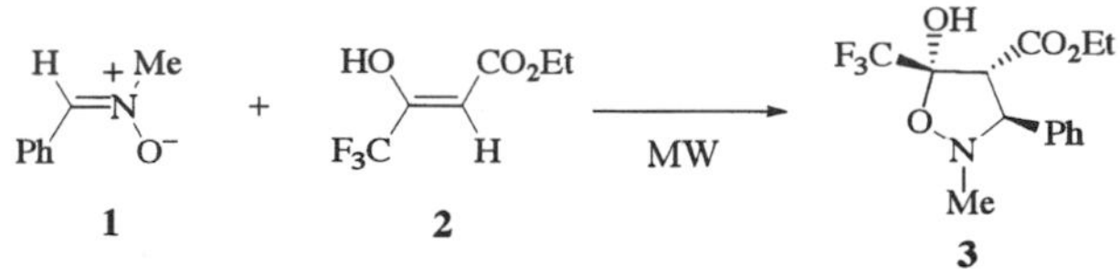

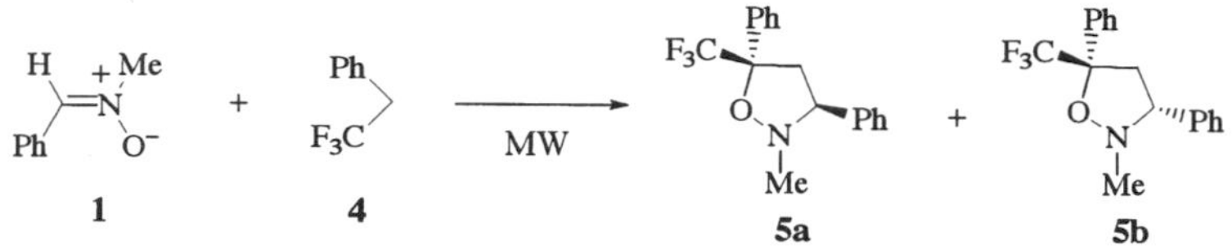

Experimental procedures:

All reactions were performed in a cylindrical Pyrex vessel using 10 mmol of nitrone **1** and 20 mmol of dipolarophiles **2** or **4**. The mixtures were introduced into the monomode reactor (Maxidigest MX 350 Prolabo) at the powers and times indicated in Table 1. Temperatures were recorded throughout the reaction using an IR detector connected to the reactor. At the end of the reaction, after cooling down and extraction with CH_2Cl_2, products **3**, **5a** and **5b** were analyzed by GC methods using an internal standard and authentic samples.

References: A. Loupy, A. Petit, D. Delpon, *J. Fluor. Chem.,* **75**, 215 (1995).

Type of reaction: C–C bond formation
Reaction condition: solvent-free
Keywords: vinylpyrazole, dimethyl acetylenedicarboxylate, Diels-Alder cycloaddition, microwave irradiation

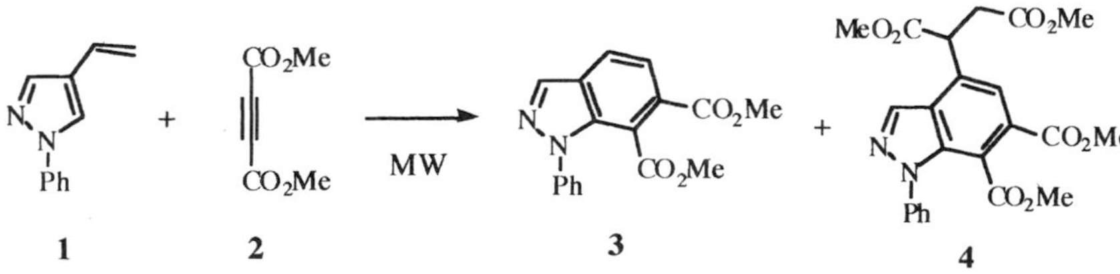

Experimental procedures:

A Teflon vessel was charged with 1-phenyl-4-vinylpyrazole **1** (170 mg, 1 mmol) and dimethyl acetylenedicarboxylate **2** (426 mg, 3 mmol) and then closed and the reaction mixture irradiated in a domestic oven at 780 W for 6 min (final temperature 130 °C). Flash chromatography (hexane-ethyl acetate, 2:1) afforded the products **3** (31 mg, 10%) and **4** (281 mg, 62%).

References: A. Ortiz, J.R. Carrillo, E. Barra, A. Hoz, M.J. Escalonilla, A. Moreno, F. Langa, *Tetrahedron,* **52**, 9237 (1996).

Type of reaction: C–C bond formation
Reaction condition: solvent-free
Keywords: nitroalkene, 2-azadiene, pyrazolylimine, Diels-Alder reaction, microwave irradiation, pyridine derivative

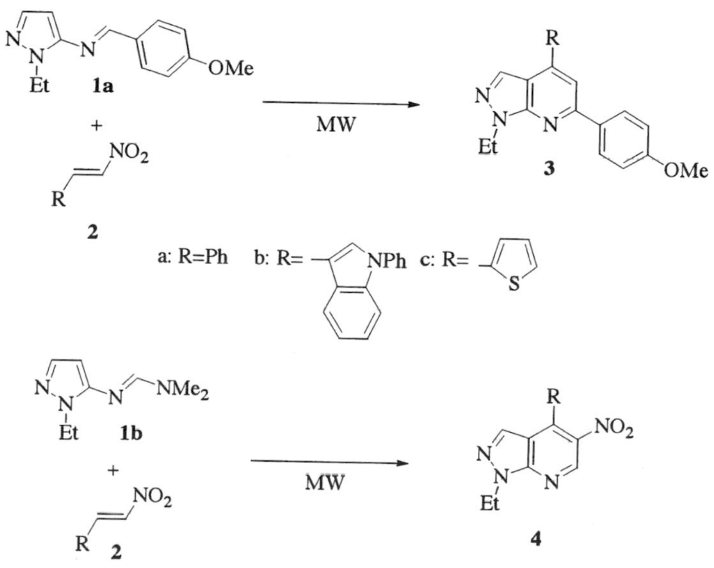

Experimental procedures:
A mixture of the 2-azadiene **1** (1 mmol) and the nitroalkene **2** (2 mmol) was irradiated at atmospheric pressure in a focused microwave reactor Prolabo MX350 with measurement and control of power. The crude reaction mixture was purified by flash chromatography on silica gel (Merck type 60, 230–400 mesh).

References: A. Diaz-Ortiz, J.R. Carrillo, M.J. Gomez-Escaonilla, A. de la Hoz, A. Moreno, P. Prieto, *Synlett,* 1069 (1998).

Type of reaction: C–C bond formation
Reaction condition: solvent-free
Keywords: phenyl acetaldehyde, cyclic secondary amine, salicylaldehyde, microwave irradiation, 2-amino-isoflav-3-ene

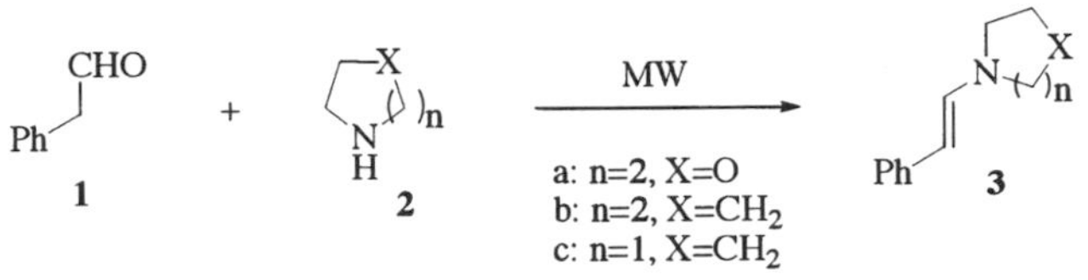

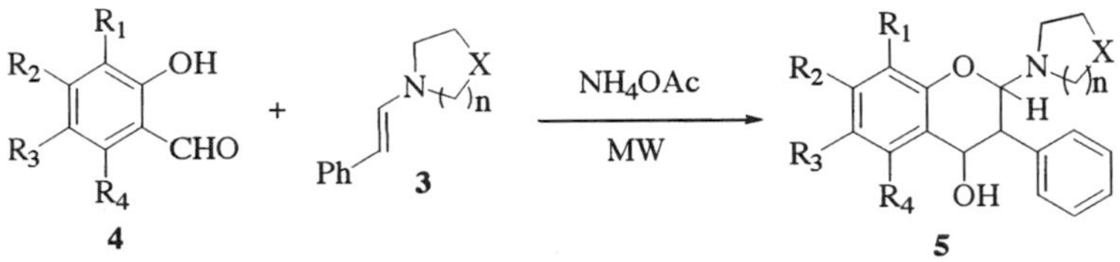

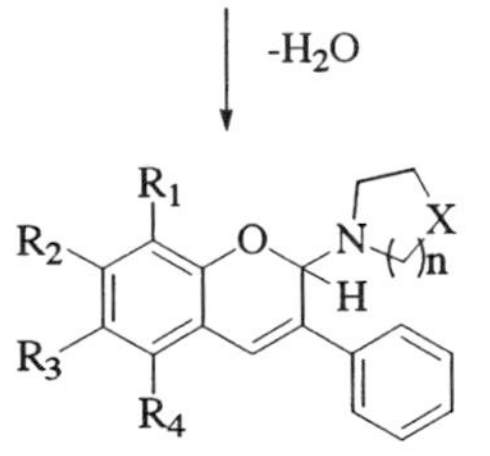

a: n=2, X=O
b: n=2, X=CH$_2$
c: n=1, X=CH$_2$

4	R$_1$	R$_2$	R$_3$	R$_4$
a	H	H	H	H
b	H	H	Cl	H
c	H	H	NO$_2$	H
d	OMe	H	H	H
e	H	OMe	H	OMe

Experimental procedures:
A mixture of phenyl acetaldehyde **1** (0.6 g, 5 mmol) and morpholine **2a** (0.48 g, 5.5 mmol) was placed in a small beaker and irradiated in an unmodified household microwave oven at its full power (900 W) for 2 min. Salicylaldehyde **4a** (0.61 g, 5 mmol) and ammonium acetate (0.02 g, 0.25 mmol) were then added to the same reaction vessel, and the reaction mixture was further irradiated in the microwave oven at its 50% power for 5 min using a pulse technique. Upon completion of the reaction, followed by TLC, the reaction mixture was passed through a bed of basic alumina using hexane-ether (9 : 1, v/v) as an eluent to afford pure 2-aminomorpholinoisoflav-3-enes **6a** (yield 80%, mp 103–105 °C).

References: R. S. Varma, R. Dahiya, *J. Org. Chem.*, **63**, 8038 (1998).

Type of reaction: C–C bond formation

Reaction condition: solvent-free

Keywords: 1,3-azadiene, microwave irradiation, *N-tert*-butyldimethylsilyl azetidinone

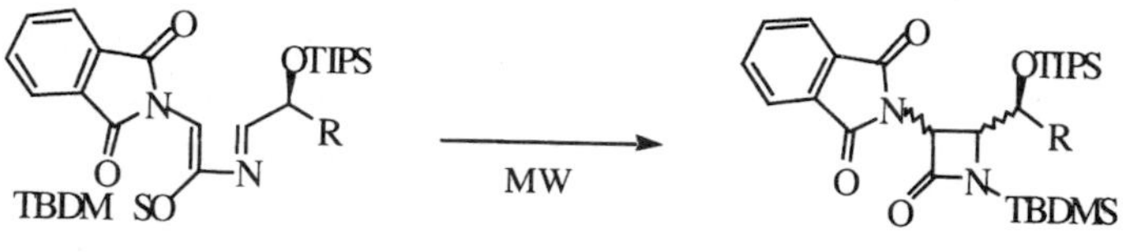

1   2

a: R=Me
b: R=*i*-Pr
c: R=*t*-Bu

Experimental procedures:

The azadiene **1a** (45 g) was placed in an Erlenmeyer flask and irradiated for 3 min at 650 W to give the azetidinone **2a** (32.4 g, 72%).

References: G. Martelli, G. Spunta, M. Panunzio, *Tetrahedron Lett.,* **39**, 6257 (1998).

Type of reaction: C–C bond formation

Reaction condition: solvent-free

Keywords: phenol, cinnamic acid, phenylpropiolic acid, K10-clay, microwave irradiation, 4-phenylcoumarin, 3,4-dihydro-4-phenylcoumarin

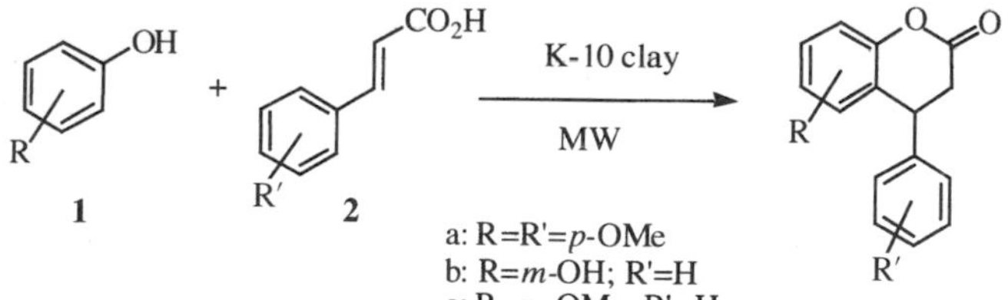

1   2

a: R=R'=*p*-OMe
b: R=*m*-OH; R'=H
c: R=*m*-OMe; R'=H
d: R=H; R'=*m,p*-OCH$_2$O-
e: R=R'=H

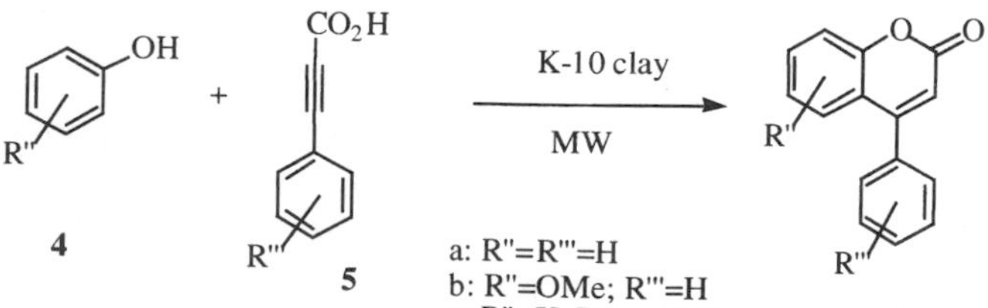

3

4   5

a: R"=R'''=H
b: R"=OMe; R'''=H
c: R"=H; R'''=*p*-OMe

6

Experimental procedures:
To activated montmorillonite K-10 clay (2 g) in a 100-mL Erlenmeyer flask was added a mixture of freshly distilled phenol (0.50 g, 5.3 mmol) and recrystallized *p*-methoxycinnamic acid (0.94 g, 5.3 mmol) dissolved in CH_2Cl_2 (5 mL) along with one drop of concentrated H_2SO_4. The solvent was evaporated and the resultant free-flowing solid placed on a silica bath and subjected to microwave irradiation at 640 W for 10 min. Dichloromethane (20 mL) was added, the reaction mixture filtered and the filtrate washed with saturated $NaHCO_3$ solution, brine and dried over Na_2SO_4. Evaporation of the solvent in vacuo yielded the product, 1.05 g (82%).

References: J. Singh, J. Kaur, S. Nayyar, G.L. Kad, *J. Chem. Res. (S)*, 280 (1998).

Type of reaction: C–C bond formation
Reaction condition: solvent-free
Keywords: arylidenemalanonitrile, ketone, microwave irradiation, 2-amino-3-cyanopyridine

3a: R=H; R_1=Ph; R_2=H
3b: R=OMe; R_1=Ph; R_2=H
3c: R=H; R_1+R_2=-(CH$_2$)$_4$-
3d: R=OMe; R_1+R_2=-(CH$_2$)$_4$-

5a: R=R_3=R_4=H; n=2
5b: R=OMe; R_3=R_4=H; n=2
5c: R=H; R_3=OEt; R_4=OMe; n=1
5d: R=OMe; R_3=OEt; R_4=OMe; n=1

Experimental procedures:
A mixture of arylidenemalanonitrile **1** (3 mmol), ketone **2** or **4** (3 mmol), ammonium acetate (24 mmol) and *o*-dichlorobenzene (0.5 mL) was placed in a borosil beaker (100 mL) and mixed thoroughly with the help of a glass rod. The mixture was then subjected to microwave irradiation for an optimized time at a power output of 275 W. After completion of the reaction (monitored by TLC), ethanol

(4 mL) was added to the reaction mixture and kept at room temperature for 5–10 min, the crystalline product obtained was filtered, washed with ethanol and recrystallized from the appropriate solvent (benzene, light petroleum (bp 40–60 °C)-ethanol or tetrahydrofuran).

References: S. Paul, R. Gupta, A. Loupy, *J. Chem. Res. (S),* 330 (1998).

Type of reaction: C–C bond formation
Reaction condition: solvent-free
Keywords: *a,β*-unsaturated carbonyl compound, amine, nitroethane, microwave irradiation, pyrrole

a: R_1=Ph; R_2=H; R_3=H; R_4=PhCH$_2$
b: R_1=Ph; R_2=H; R_3=Ph; R_4=PhCH$_2$
c: R_1=Ph; R_2=H; R_3=Me; R_4=PhCH$_2$
d: R_1=H; R_2=H; R_3=Me; R_4=PhCH$_2$
e: R_1=Ph; R_2=H; R_3=Me; R_4=cyclohexyl
f: R_1=Ph; R_2=H; R_3=H; R_4=Me$_2$CH
g: R_1=2-furyl; R_2=H; R_3=Me; R_4=Me$_2$CH
h: R_1=n-Pr; R_2=Et; R_3=H; R_4=PhCH$_2$
i: R_1=Ph; R_2=H; R_3=H; R_4=n-Pr
j: R_1=Ph; R_2=H; R_3=Me; R_4=n-Bu

Experimental procedures:
A mixture of *a,β*-unsaturated aldehyde or ketone **1** (1 mmol), an amine **2** (1 mmol) and nitroethane **3** (3 mmol) was uniformly adsorbed on the surface of silica gel (HF 254, 2 g) by stirring for 5 min in a Pyrex round-bottomed flask at room temperature under moisture guard of anhydrous CaCl$_2$. The flask was then placed on a bed of silica gel in a porcelain basin and irradiated by microwave in a domestic microwave oven (manufactured by BPL-Sanyo, India; monomode, multipower; power source: 230 V, 50 Hz; microwave frequency: 2450 MHz) at 120 W for a certain period of time as required to complete the reaction. The reaction mixture was eluted with ether and the ether extract was evaporated to leave the crude product which was purified by column chromatography over silica gel.

References: B.C. Ranu, A. Hajra, U. Jana, *Synlett,* 75 (2000).

Type of reaction: C–C bond formation
Reaction condition: solvent-free
Keywords: N-allylaniline, 3-aza-Cope rearrangement, Zn^{2+} montmorillonite, microwave irradiation, 2-methylindoline

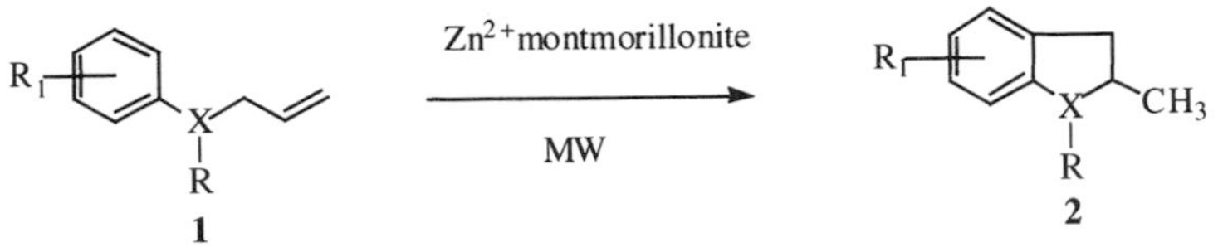

R=H, alkyl, benzyl
R_1=H, OMe, CN, NO_2, Me, Cl, F
X=NR, S

Experimental procedures:
N-allylaniline (1.33 g, 10 mmol) was admixed with Zn^{2+} montmorillonite and subjected to microwave irradiation at 650 W (BPL, BMO-700T, operating at 2450 MHz) for 3 min. After cooling down to room temperature, the reaction mass was directly charged on a silica gel column (100–200 mesh) and eluted with ethyl acetate-n-hexane (3:7) to afford 2-methylindoline (1.06 g, 80% yield) as pale yellow liquid.

References: J.S. Yadav, B.V.S. Reddy, M.A. Rasheed, H.M.S. Kumar, *Synlett*, 487 (2000).

Type of reaction: C–C bond formation
Reaction condition: solvent-free
Keywords: aromatic aldehyde, ethyl 2-cyano-2-(1-ethoxyethylidene)amino-ethanoate, dipolar cycloaddition, microwave irradiation, 4-cyano-2-oxazoline-4-carboxylate

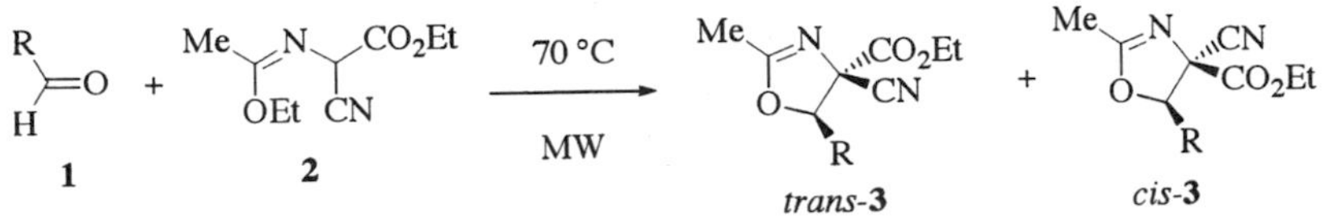

a: R=Ph
b: R=2-pyridyl
c: R=4-MeOC_6H_4

d: R=2-BrC_6H_4
e: R=2-HOC_6H_4
f: R=3-HOC_6H_4

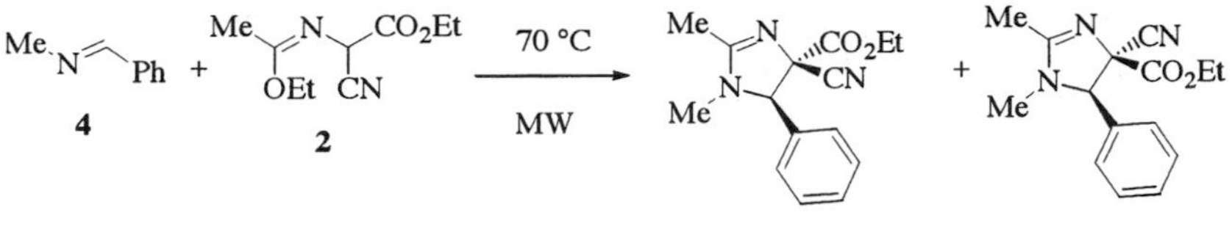

$$trans\text{-}\mathbf{5} \qquad cis\text{-}\mathbf{5}$$

Experimental procedures:

A mixture of ethyl 2-cyano-2-(1-ethoxyethylidene)aminoethanoate **2** (1 g, 5 mmol) and freshly distilled aromatic aldehyde **1** (5 mmol) was placed in a cylindrical quartz tube ($\phi=4$ cm). The tube was then introduced into a Synthewave R 402 Prolabo microwave reactor (2.45 GHz, adjusted power within the range 0–300 W and a wave guide (single mode T01) fitted with a stirring device and an IR temperature detector). Microwave irradiation was carried out at 70 °C for the appropriate reaction time. The mixture was allowed to cool down. After addition of methylene chloride (15 mL) in the reactor and removal of solvent in vacuo, the crude residue was purified by chromatography on silica gel (60F 254, Merck) with the appropriate eluent. Solvent evaporation gave the desired compound **3** as a viscous oil which crystallized on standing and was characterized by ^{1}H, ^{13}C NMR and HRMS analysis.

References: J. Fraga-Dubreuil, J.R. Cherouvrier, J.P. Bazureau, *Green Chem.*, **2**, 226 (2000).

Type of reaction: C–C bond formation
Reaction condition: solvent-free
Keywords: aldimine, ethyl-α-mercaptoacetate, ethyl-α-cyanoacetate, basic alumina, microwave irradiation, β-lactams

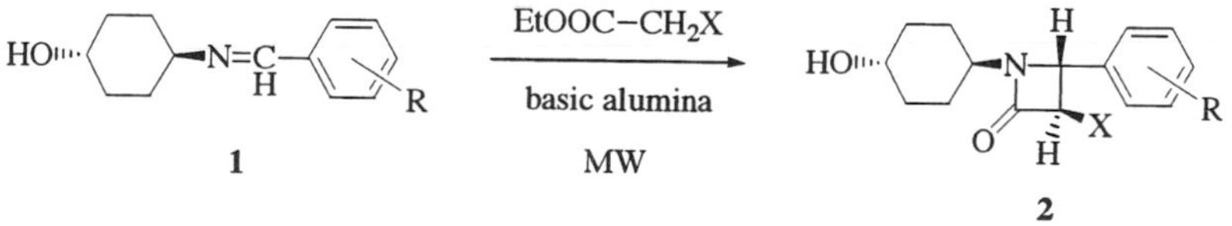

X=SH	a: R=H b: R=4-OH c: R=4-OMe d: R=2-OH e: R=3-NO$_2$ f: R=4-Cl
X=CN	a: R=H b: R=4-OH c: R=4-OMe d: R=2-OH e: R=3-NO$_2$ f: R=4-Cl

Experimental procedures:
To a solution of aldimine **1** (0.01 mol) in acetone, added ethyl-*a*-mercapto/*a*-cyanoacetate (0.01 mol) followed by basic alumina (20 g) with constant stirring. The reaction mixture taken in a beaker, was thoroughly mixed and the adsorbed material was dried in air. The adsorbed reactant in the beaker was placed in an alumina bath and subjected to microwave irradiation for 1–2 min. On completion of the reaction as followed by TLC examination, the mixture was cooled to room temperature and the product was extracted into acetone (3×15 mL). Recovering of solvent under reduced pressure yielded the product, which was purified by recrystallization from the mixture of ethanol-acetone.

References: M. Kidwai, R. Venkataramanan, S. Kohli, *Synth. Commun.*, **30**, 989 (2000).

Type of reaction: C–C bond formation
Reaction condition: solvent-free
Keywords: benzaldehyde, pyrrole, microwave irradiation, mesotetraphenylporphirin

Experimental procedures:
A 25-mL Erlenmeyer flask, a standard Pyrex watch glass and a 1000-W microwave are used to carry out the microwave synthesis. 0.43 mL of benzaldehyde **1** and 0.3 mL of pyrrole **2** are mixed in the flask. Once the reactants are thoroughly mixed 0.63 g of silica gel is added, the flask stoppered, and the reagents mixed well until the silica gel is evenly and completely covered with the reactant mixture. The flask containing the reaction mixture is then placed in the microwave oven, covered with the watch glass and heated for 10 min in five 2-min intervals. Once the reaction is complete, it is allowed to cool to room temperature and ca. 15 mL of ethyl acetate added. The solution is filtered to remove the silica gel and then the ethyl acetate is removed using a rotary evaporator. Prior to

chromatographic separation, the crude reaction mixture is extracted into 1 mL of CH$_2$Cl$_2$. It is this fraction that is used in subsequent purification.

References: M. G. Warner, G. L. Succaw, J. E. Hutchison, *Green Chem.*, **3**, 267 (2001).

Type of reaction: C–C bond formation
Reaction condition: solvent-free
Keywords: *β*-naphthol, benzaldehyde, piperidine, Mannich reaction, acidic alumina, aminoalkylation, microwave irradiation

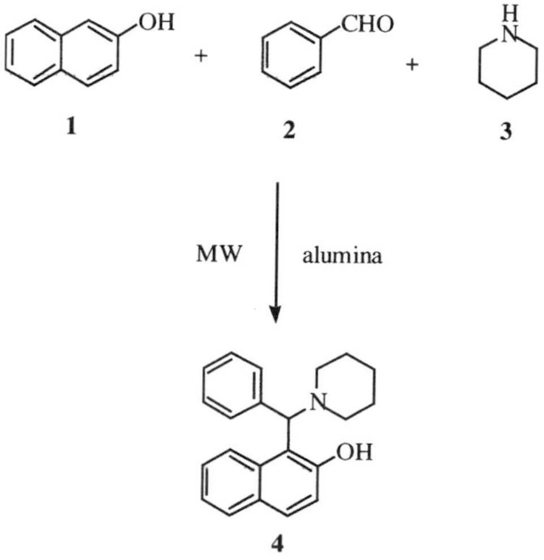

Experimental procedures:
A mixture of 1 mmol *β*-naphthol, 1 mmol aldehyde, and 3 mmol amine were mixed and ground with 2 g acidic alumina in a mortar. The mixture was transferred to a 25-mL beaker and irradiated in a microwave oven for the required time. The progress of the reaction was monitored with TLC. After cooling, the mixture was extracted with ethyl acetate. Evaporation of the solvent under reduced pressure gave the crude product, which was purified by flash column chromatography (silica gel; eluent: petrol ether-ethyl acetate, 8 : 1).

References: A. Sharifi, M. Mirzaei, M. R. Naimi-Jamal, *Monatsh. Chem.*, **132**, 875 (2001).

Type of reaction: C–C bond formation
Reaction condition: solvent-free
Keywords: benzaldehyde, alkyl propiolate, primary amine, cyclocondensation, microwave irradiation, 1,4-dihydropyridine

$$\text{Ar}\overset{O}{\underset{H}{\bigvee}} + H-C\equiv C-CO_2R_1 + R_2NH_2 \xrightarrow{MW}$$

1   2   3   4

a: Ar=4-MeOC$_6$H$_4$; R$_1$=Me; R$_2$=CH$_2$Ph
b: Ar=4-MeOC$_6$H$_4$; R$_1$=Et; R$_2$=CH$_2$Ph
c: Ar=4-MeOC$_6$H$_4$; R$_1$=Me; R$_2$=Me
d: Ar=Ph; R$_1$=Me; R$_2$=CH$_2$Ph
e: Ar=Ph; R$_1$=Et; R$_2$=CH$_2$Ph
f: Ar=4-BrC$_6$H$_4$; R$_1$=Me; R$_2$=CH$_2$Ph
g: Ar=4-MeOC$_6$H$_4$; R$_1$=Et; R$_2$=n-Bu
h: Ar=4-MeOC$_6$H$_4$; R$_1$=Et; R$_2$=cyclohexyl

Experimental procedures:
The benzaldehyde derivative **1** (4 mmol), alkyl propiolate **2** (8 mmol), primary amine **3** (4 mmol), and 2 g silica gel were mixed thoroughly in a mortar. Then the reaction mixture was transferred to a beaker and irradiated with microwaves for 4 min. The progress of reaction was monitored by TLC. The mixture was extracted with 3×30 mL CHCl$_3$, filtered, and the solvent was removed by a rotary evaporator under reduced pressure. Further purification by recrystallization gave the desired pure products **4**.

References: S. Balalaie, E. Kowsari, *Monatsh. Chem.,* **132**, 1551 (2001).

Type of reaction: C–C bond formation
Reaction condition: solvent-free
Keywords: epoxyisophorone, diethyl malonate, epoxide ring-opening, KF-alumina, microwave irradiation, isophorone

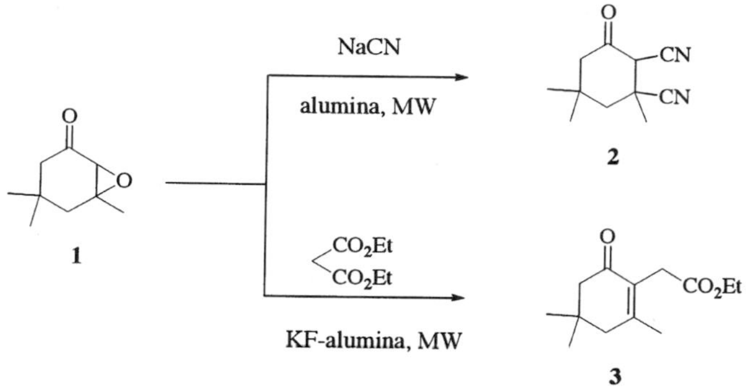

Experimental procedures:
Neutral alumina (20 g) was stirred for 15 min with an aqueous solution of KF
(15 g KF/150 mL H$_2$O). The mixture was concentrated to dryness under vacuum
and kept at 100 °C for 30 min. The reagent (15 mmol) was intimately blended
with KF/Al$_2$O$_3$ (4×reagent mass), epoxyisophorone (1.54 g, 10 mmol) was
quickly added to the support and immediately heated in an oil bath or exposed to
microwave irradiation. When the reagent was solid, it was dissolved into 5 mL
CH$_2$Cl$_2$ and mixed with the solid support, and the solvent was evaporated to dryness. At the end of the reaction, the products were eluted from the solid support
by washing with 3×10 mL methylene chloride.

References: B. Rissafi, N. Rachiqi, A. E. Louzi, A. Loupy, A. Retit, S. F. Tetouani, *Tetrahedron*, **57**, 2761 (2001).

Type of reaction: C–C bond formation
Reaction condition: solvent-free
Keywords: 5-amino pyrimidin-4(3*H*)-one, cyanomethyl phenyl ketone, benzalde-
hyde, microwave irradiation, dihydropyrido[2,3-*d*]pyrimidin-4(3*H*)-one

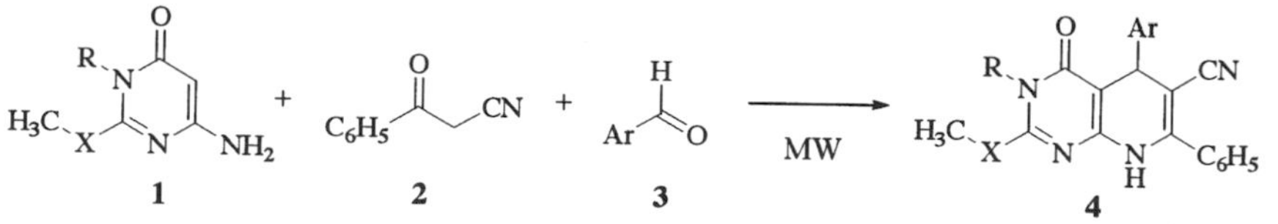

X=O, S
R=H, CH$_3$
Ar=C$_6$H$_5$, 4-CH$_3$OC$_6$H$_4$, 4-ClC$_6$H$_4$

Experimental procedures:
Equimolar amounts of starting compounds **1**, **2** and **3** were placed into Pyrex-glass open vessels and irradiated in a domestic microwave oven for 15–20 min (at 600 W). When the irradiation was stopped, the solid was treated with ethanol and filtered to give the products **4a–g** in 70–75% yields.

References: J. Quiroga, C. Cisneros, B. Insuasty, R. Abonia, M. Nogueras, A. Sanchez, *Tetrahedron Lett.,* **42**, 5625 (2001).

Type of reaction: C–C bond formation
Reaction condition: solvent-free
Keywords: chalcone, Michael reaction, microwave irradiation, flavanone

a: $R_1=R_2=R_3=H$
b: $R_1=Me$; $R_2=R_3=H$
c: $R_1=Cl$; $R_2=R_3=H$
d: $R_1=R_3=H$; $R_2=OMe$
e: $R_1=Cl$; $R_2=H$; $R_3=OMe$

Experimental procedures:
A mixture of substituted 2′-hydroxychalcone (1 mmol), the corresponding support (5 g) and the additive (5 mg) in a Pyrex tube was subjected to irradiation at 45 W for 20 min in a Maxidigest 350 Prolabo monomode, focused microwave (2.45 GHz) reactor with continuous rotation. The cooled mixture was washed with methylene chloride (2×10 mL) and concentrated in vacuo. The residue was purified by two subsequent column chromatographies (silica gel) by using hexane-acetone (4:1, v/v) and toluene or toluene-ethyl methyl ketone (30:1, v/v) as eluent.

References: T. Patonay, R. S. Varma, A. Vass, A. Levai, J. Dudas, *Tetrahedron Lett.,* **42**, 1403 (2001).

Type of reaction: C–C bond formation
Reaction condition: solvent-free
Keywords: aniline, microwave irradiation, 4-hydroxyquinoline

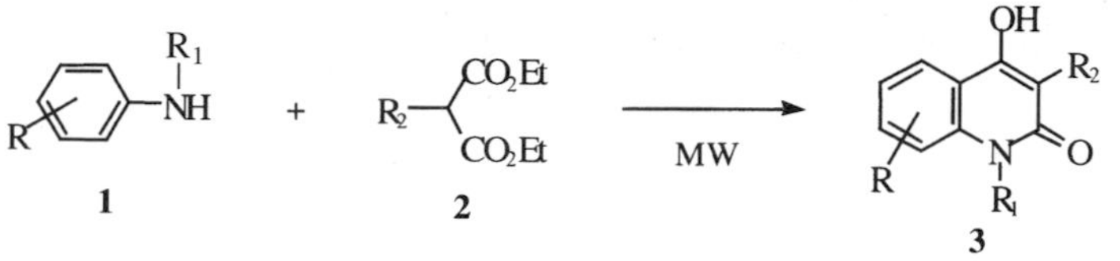

a: R=R$_1$=H; R$_2$=Ph
b: R=2-Cl; R$_1$=H; R$_2$=Ph
c: R=2-OMe; R$_1$=H; R$_2$=Ph
d: R=2-CF$_3$; R$_1$=H; R$_2$=Ph
e: R=3-Cl; R$_1$=H; R$_2$=Ph
f: R=3-OMe; R$_1$=H; R$_2$=Ph
g: R=3-F; R$_1$=H; R$_2$=Ph
h: R=3-CF$_{3;}$ R$_1$=H; R$_2$=Ph
i: R=4-Cl; R$_1$=H; R$_2$=Ph
j: R=4-OMe; R$_1$=H; R$_2$=Ph

k: R=4-CF$_3$; R$_1$=H; R$_2$=Ph
l: R=H; R$_1$=Me; R$_2$=Ph
m: R=H; R$_1$=n-C$_4$H$_9$; R$_2$=Ph
n: R=H; R$_1$=cyclohexyl; R$_2$=Ph
o: R,R$_1$=2-CH$_2$CH$_2$CH$_2$; R$_2$=Ph
p: R=3-Cl; R$_1$=H; R$_2$=4-OMeC$_6$H$_4$
q: R=3-Cl; R$_1$=H; R$_2$=4-MeC$_6$H$_4$
r: R=3-Cl; R$_1$=H; R$_2$=4-OPhC$_6$H$_4$
s: R=3-Cl; R$_1$=H; R$_2$=3-thienyl

Experimental procedures:

A mixture of 3-chloroaniline **1e** (2.10 mL, 20.0 mmol) and diethyl phenylmalonate **2e** (8.62 mL, 40.0 mmol) is stirred and irradiated in a microwave oven (type: Milestone Ethos 900) for 15 min at 500 W power under a gentle stream of nitrogen (in order to effectively remove the formed ethanol during the reaction). The mixture is allowed to attain room temperature and diluted with diethyl ether. The precipitate is collected by filtration and washed with diethyl ether to furnish pure 7-chloro-4-hydroxy-3-phenylquinolin-2(1*H*)-one **3e** (4.36 g, 81%) as white crystals.

References: J. H. M. Lange, P. C. Verveer, S. J. M. Osnabrug, G. M. Visser, *Tetrahedron Lett.*, **42**, 1367 (2001).

Type of reaction: C–C bond formation
Reaction condition: solvent-free
Keywords: *o*-ethynylphenol, secondary amine, paraformaldehyde, CuI, Mannich condensation, microwave irradiation, 2-(dialkylaminomethyl)benzo[*b*]furan

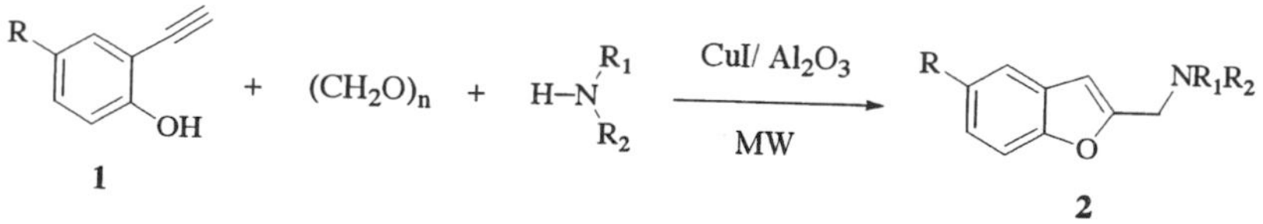

a: R=H
b: R=Me
c: R=OMe

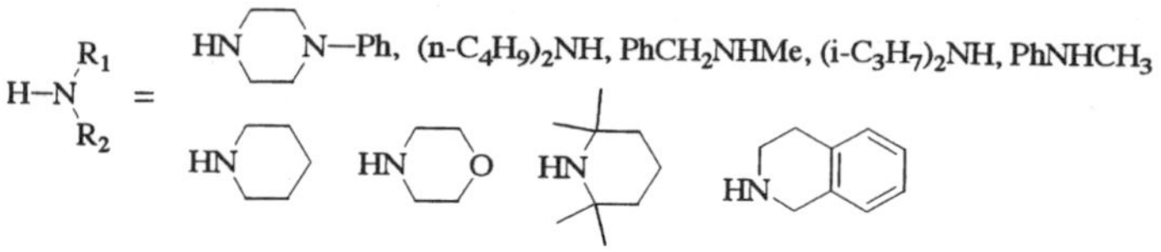

R₁ appears in structure:

$$H-N\begin{smallmatrix}R_1\\R_2\end{smallmatrix} = $$

HN⌒N—Ph, (n-C₄H₉)₂NH, PhCH₂NHMe, (i-C₃H₇)₂NH, PhNHCH₃

HN (cyclohexyl ring), HN O (morpholine), HN (2,2,6,6-tetramethylpiperidine), HN (tetrahydroisoquinoline/naphthalene)

Experimental procedures:

o-Ethynylphenol **1** (0.118 g, 1.00 mmol) and a secondary amine (1.00 mmol) were added to a mixture of cuprous iodide (0.572 g, 3.00 mmol), paraformalde-hyde (0.09 g, 3.00 mmol) and alumina (1.00 g) contained in a clean, dry, 10-mL round-bottomed flask. The mixture was stirred at room temperature to ensure efficient mixing. The flask was then fitted with a septum which had been punctured by an 18 gauge needle (to serve as a pressure relief valve), placed in the microwave oven and irradiated at 30% power for 5 min. After cooling, ether (4 mL) was added and the slurry stirred at room temperture to ensure product removal from the surface. The mixture was vacuum filtered using a sintered glass tunnel and the product purified by flash chromatography (hexane/EtOAc as eluent) to afford the desired 2-(dialkylaminomethyl)benzo[b]furan. Yields 38 ∼ 70%.

References: G. W. Kabalka, L. Wang, R. M. Pagni, *Tetrahedron Lett.,* **42**, 6049 (2001).

Type of reaction: C–C bond formation
Reaction condition: solvent-free
Keywords: aldehyde, β-ketonic ester, urea, Hantzsch synthesis, microwave irradiation, 1,4-dihydropyridine

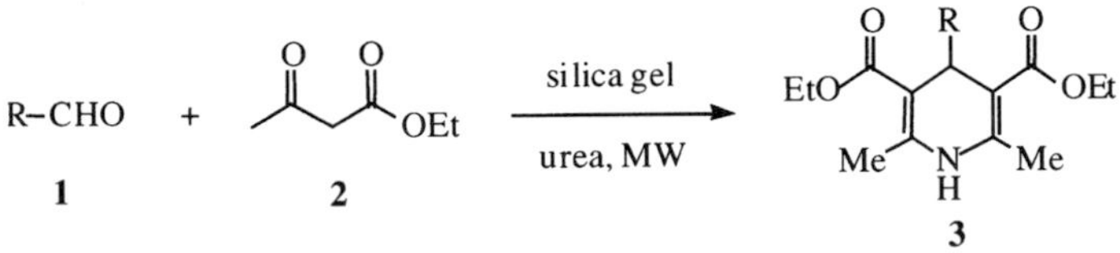

$$R-CHO + \underset{\textbf{2}}{\overset{O\quad O}{\text{Me}\diagdown\diagup\diagdown\text{OEt}}} \xrightarrow[\text{urea, MW}]{\text{silica gel}} \textbf{3}$$

1   **2**

a: R=Ph
b: R=α-naphthyl
c: R=2-thienyl
d: R=2-furyl
e: R=4-ClC₆H₄
f: R=4-Me₂NC₆H₄
g: R=2-MeOC₆H₄

h: R=4-NO₂C₆H₄
i: R=PhCH₂
j: R=i-Pr
k: R=n-hexyl
l: R=n-decyl
m: R=PhCH=CH
n: R=3-NO₂C₆H₄

Experimental procedures:

Aldehyde **1** (10 mmol), ethylacetoacetate **2** (20 mmol), and urea (10 mmol) were thoroughly admixed with silica gel (5 g, Aldrich 200–300 mesh) in solid state using pestle and mortar. The resulting powder was transferred to a 25-mL Erlen-meyer flask and subjected to microwave irradiation (BPL, BMO-700T) for 2–5 min at 650 W. After complete conversion, as shown by TLC, the reaction mass was charged directly on a small silica gel column (60–120 mesh), and eluted with ethyl acetate-hexane (3:7) to afford product **3**.

References: J. S. Yadav, B. V. S. Reddy, P. T. Reddy, *Synth. Commun.*, **31**, 425 (2001).

Type of reaction: C–C bond formation

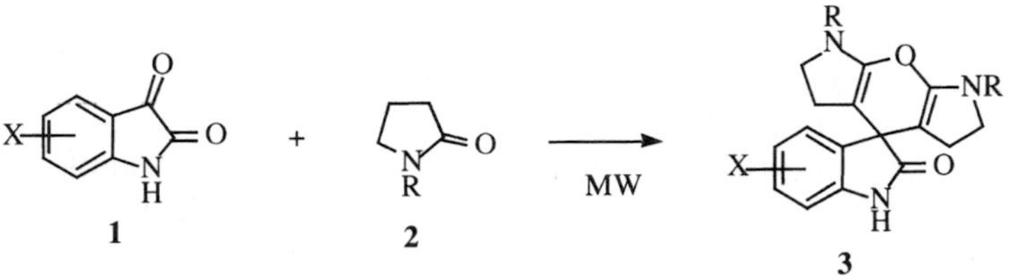

a: X=5-F; R=H
b: X=7-NO$_2$; R=H
c: X=5,7-Me$_2$; R=Me
d: X=5-NO$_2$; R=Me
e: X=7-NO$_2$; R=Me

Reaction condition: solvent-free
Keywords: indole-2,3-dione, 2-pyrrolidone, microwave irradiation, spiro[indole-dipyrrolopyridines]
Experimental procedures:
A mixture of appropriate indole-2,3-dione **1** (2 mmol) and 2-pyrrolidone **2** (5 mmol) was introduced in a microwave oven and irradiated at 840 W till the completion of the reaction. Progress of the reaction was checked by TLC. Extraction with acetic acid and removal of the solvent afforded the product which was analyzed by ^{1}H NMR and found to be of sufficient purity on TLC.

References: A. Dandia, H. Sachdeva, R. Singh, *Synth. Commun.*, **31**, 1879 (2001).

Type of reaction: C–C bond formation
Reaction condition: solvent-free
Keywords: bicarboxylic acid, decarboxylation, graphite, microwave irradiation, cycloalkanone

a: n=2; R=H
b: n=2; R=Me
c: n=3; R=H
d: n=4; R=H

Experimental procedures:
Conventional Heating: To a 2.19 g (15 mmol) sample of adipic acid **1c**, 5 g of graphite A were added, and the mixture was ground in a mortar prior to being placed in the reactor. The flask, surmounted by a Dean-Stark condenser equipped with an ice-cold finger, was immersed in a preheated tubular electric furnance and heated at 450 °C for 30 min. The ketone **2c** condensed mainly in the cold trap, and the remainder was extracted from graphite powder with ether. A 90% overall yield in **2c** was determined by GC analysis. To obtain a large sample of product, four successive runs were performed, and the combined organic phases were distilled to obtain 4.29 g (85% yield) of pure **2c** (bp 130–131 °C).

Microwave Heating: The same mixture as before was introduced in the quartz reactor of the microwave apparatus, surmounted by the same Dean-Stark condenser. Microwave irradiation was programmed using the computer for a sequential process in which the sample was exposed for periods of 2 min separated by periods of 2 min. This sequence was repeated 6 times with an incident power of 90 W for the first two irradiations and of 75 W for the others. After the mixture cooled, the same treatment as before gave a 90% yield in **2c**.

References: J. Marquie, A. Laporterie, J. Dubac, N. Roques, *Synlett*, 493 (2001).

Type of reaction: C–C bond formation
Reaction condition: solvent-free
Keywords: ketone, phosphorus ylide, Wittig reaction, microwave irradiation, olefin

a: R_1=Ph; R_2=Me
b: R_1=R_2=Ph
c: R_1=-$(CH_2)_6CH_3$; R_2=CH_3

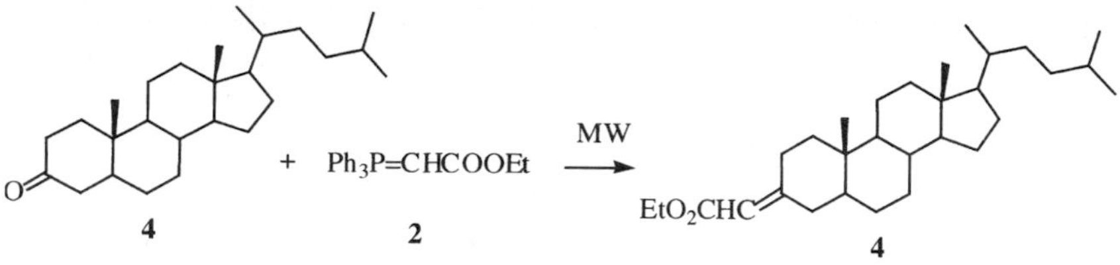

Experimental procedures:

Ketone (1 mmol) was mixed to carbethoxymethylenetriphenylphosphorane **2** (0.5 mmol) in a Ace glass tube (Aldrich). The reactor was put in a glass beaker (600 mL) filled with vermiculite beads (50 g) and subjected to microwave heating. The use of vermiculite beads was very important, not only for safety reasons but also for the thermal enhancement due to the absorption of microwave energy by the beads. In fact, without vermiculite the transformation yields were very low even at the highest microwave irradiation power (650 W). Using vermiculite, high temperatures were rapidly reached and the reactions were conducted more easily. The reaction mixtures obtained after irradiation were purified by chromatography on silica gel (Merck, 70–230 mesh; petroleum ether-diethyl ether mixture).

References: A. Spinella, T. Fortunati, A. Soriente, *Synlett*, 93 (1997).

Type of reaction: C–C bond formation
Reaction condition: solvent-free
Keywords: iodobenzene, *p*-methylphenylboronic acid, Suzuki coupling, KF/Al$_2$O$_3$, palladium powder, microwave irradiation, 4-methylbiphenyl

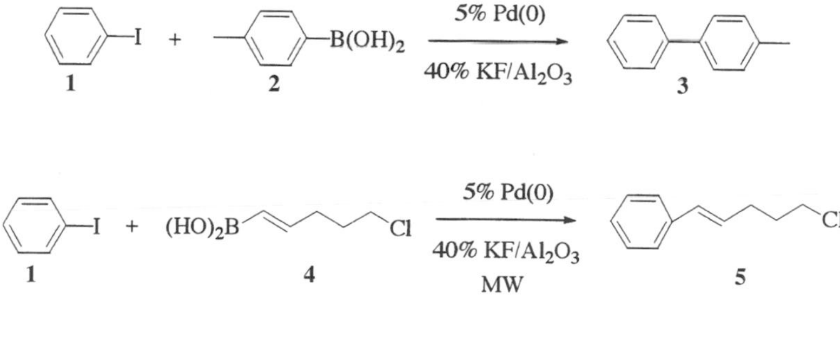

Experimental procedures:

To a mixture of KF/Al$_2$O$_3$ (0.950 g, 40% by weight) and palladium black (0.050 g, 0.470 mmol, 99.9+% as a submicron powder) contained in a clean, dry, round-bottomed flask was added *p*-methylphenylboronic acid **2** (0.150 g, 1.10 mmol). The solid mixture was stirred at room temperature in the open air until homogeneous. Iodobenzene **1** (0.209 g, 1.02 mmol) was then added with stirring. The mixture was stirred at room temperature for an additional 15–20 min to ensure efficient mixing. The flask was then fitted with a septum (punctured by an 18 gauge needle), placed in a microwave oven and irradiated at 100% power for 2 min. After cooling, a small quantity of hexane was added and the slurry stirred at room temperature for an additional 20–30 min to ensure product removal from the surface. The mixture was vacuum filtered through a sintered glass funnel and the product isolated via flash chromatography to yield 4-methylbiphenyl **3** (82%).

References: G.W. Kabalka, R.M. Pagni, L. Wang, V. Namboodiri, C.M. Hair, *Green Chem.*, **2**, 120 (2000).

Type of reaction: C–C bond formation
Reaction condition: solvent-free
Keywords: terminal alkyne, Glaser coupling, cupric chloride, microwave irradiation, butadiyne

$$R-C{\equiv}C-H \xrightarrow[\text{MW}]{\text{CuCl/KF/Al}_2\text{O}_3} R-C{\equiv}C-C{\equiv}C-R$$

$$\mathbf{1} \qquad\qquad\qquad\qquad \mathbf{2}$$

a: R=n-C_8H_{17} e: R=2-FC_6H_4
b: R=n-C_6H_{13} f: R=4-FC_6H_4
c: R=4-$CH_3C_6H_4$ g: R=Ph
d: R=2-ClC_6H_4

Experimental procedures:
Phenylacetylene **1g** (102 mg, 1.0 mmol) was added to a mixture of KF/Al_2O_3 (600 mg, 40% by weight) and cupric chloride (500 mg, 3.7 mmol) contained in a clean, dry 10-mL round-bottomed flask. The mixture was stirred at room temperature to ensure efficient mixing. The flask was then fitted with a septum (punctured by an 18 gauge needle), placed in the microwave and irradiated at 30% power for 2 min and then allowed to cool. The microwave irradiation was then continued for 6 min. After cooling, hexane (3 mL) was added and the slurry stirred at room temperature to ensure product removal from the surface. The mixture was filtered and the product was purified by flash chromatography to yield 76 mg of diphenylbutadiyne **2g** (75%).

References: G.W. Kabalka, L. Wang, R.M. Pagni, *Synlett*, 108 (2001).

Type of reaction: C–C bond formation

Reaction condition: solvent-free

Keywords: aryl iodide, terminal alkyne, Sonogashira coupling, KF-alumina, microwave irradiation, aryl alkyne

$$R_1\text{-}X \quad + \quad R_2\text{-}C{\equiv}C\text{-}H \quad \xrightarrow[\text{MW}]{\text{Pd-CuI-PPh}_3/\ \text{KF-Al}_2\text{O}_3} \quad R_2\text{-}C{\equiv}C\text{-}R_1$$

$$\textbf{1} \qquad\qquad \textbf{2} \qquad\qquad\qquad\qquad \textbf{3}$$

a: $R_1X=C_6H_5I$; $R_2=n\text{-}C_8H_{17}$

b: $R_1X=C_6H_5I$; $R_2=n\text{-}C_6H_{13}$

c: $R_1X=C_6H_5I$; $R_2=C_6H_5$

d: $R_1X=p\text{-}MeC_6H_4I$; $R_2=n\text{-}C_8H_{17}$

e: $R_1X=p\text{-}MeOC_6H_4I$; $R_2=n\text{-}C_8H_{17}$

f: $R_1X=o\text{-}FC_6H_4I$; $R_2=n\text{-}C_8H_{17}$

g: $R_1X=m\text{-}FC_6H_4I$; $R_2=n\text{-}C_8H_{17}$

h: $R_1X=p\text{-}MeCOC_6H_4I$; $R_2=C_6H_5$

i: $R_1X=p\text{-}NO_2C_6H_4I$; $R_2=C_6H_5$

j: $R_1X=o\text{-}Me_2NC_6H_4I$; $R_2=C_6H_5$

k: $R_1X=C_6H_5I$; $R_2=p\text{-}MeC_6H_4$

l: $R_1X=C_6H_5I$; $R_2=o\text{-}ClC_6H_4$

m: $R_1X=C_6H_5I$; $R_2=p\text{-}FC_6H_4$

n: $R_1X=C_6H_5I$; $R_2=o\text{-}FC_6H_4$

o: $R_1X=C_6H_5I$; $R_2=p\text{-}MeCOC_6H_4$

p: $R_1X=C_6H_5I$; $R_2=p\text{-}BrC_6H_4$

Experimental procedures:

Aryl iodide **1** (1.00 mmol) and terminal alkyne **2** (1.05 mmol) were added to a mixture of KF/Al$_2$O$_3$ (1.00 g, 40% by weight), palladium powder (0.040 g, 0.376 mmol, 99.9+% as a submicron powder), cuprous iodide (0.070 g, 0.368 mmol) and triphenylphosphine (0.180 g, 0.686 mmol) containing in a clean, dry, 25-mL round-bottomed flask. The mixture was then fitted with a septum (punctured by an 18 gauge needle to serve as a pressure release valve), placed in the microwave oven and irradiated at 100% power for 2.5 min. After cooling, hexane (5 mL) was added and the slurry stirred at room temperature to ensure product removal from the surface. The mixture was vacuum filtered using a sintered glass funnel and the product was purified by flash chromatography to yield the desired aryl alkyne **3**.

References: G. W. Kabalka, L. Wang, R. M. Pagni, *Tetrahedron,* **57**, 8017 (2001).

Type of reaction: C–C bond formation

Reaction condition: solvent-free

Keywords: 2-pyridone, benzylbromide, microwave irradiation, 3-benzyl-2-pyridone, 5-benzyl-2-pyridone, 3,5-dibenzyl-2-pyridone

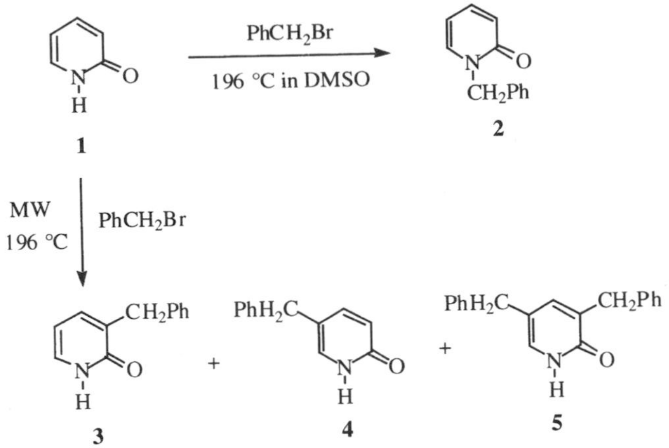

Experimental procedures:
In a Pyrex flask (25 mL), a mixture of 2-pyridone **1** (0.95 g, 10 mmol) and benzyl bromide (10 or 20 mmol) was introduced in the microwave oven and irradiated for 5 min at 196 °C. Extraction with dichloromethane (4×10 mL) and removal of the solvent afforded a mixture of 3-benzyl-2-pyridone **3**, 5-benzyl-2-pyridone **4** and 3,5-dibenzyl-2-pyridone **5** in 31, 44 and 25% yield, respectively. On the other hand, conventional heating of a mixture of 2-pyridone and benzyl bromide in a oil bath at 196 °C for 5 min gave only 1-benzyl-2-pyridone **2** in 100% yield.

References: I. Almena, A. D. Ortiz, E. D. Barra, A. de la Hoz, A. Loupy, *Chem. Lett.,* 333 (1996)

3.3   Solvent-Free C–C Bond Formation under Photoirradiation

Type of reaction: C–C bond formation
Reaction condition: solid-state
Keywords: 1-methyl-5,6-diphenyl-2-pyrazinone, [4+4]photodimerization

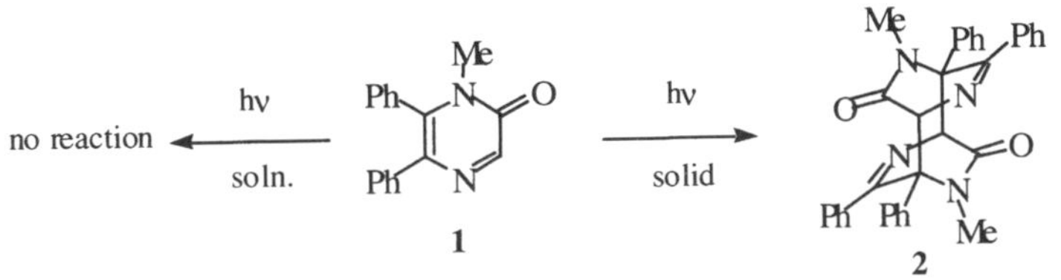

Experimental procedures:
Irradiation of **1** in the solid state with a high-pressure mercury lamp (400 W) at room temperature for 20 min gave the [4+4] *anti* dimer **2**, mp 148–150 °C, in 100% yield based on recovered starting pyrazinone **1**.

References: T. Nishio, N. Nakajima, Y. Omote, *Tetrahedron Lett.*, **21**, 2529 (1980).

Type of reaction: C–C bond formation
Reaction condition: solid-state
Keywords: coumarin, [2+2]photodimerization, cyclobutane

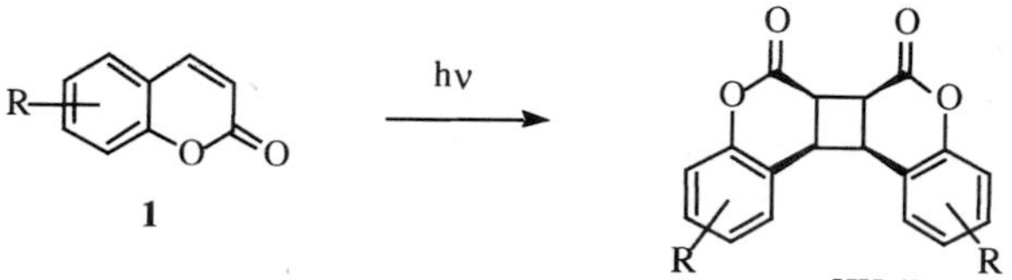

a: R=6-methoxy
b: R=7-methoxy
c: R=8-methoxy
d: R=6-acetoxy
e: R=7-acetoxy
f: R=4-methyl-7-acetoxy
g: R=6-chloro
h: R=7-chloro
i: R=4-methyl-6-chloro
j: R=4-methyl-7-chloro
k: R=7-methyl

Experimental procedures:
Powdered single crystals of coumarins kept in a petri dish were irradiated with a Hanovia 450-W medium-pressure mercury arc lamp from a distance of about 2 ft. Samples were turned around periodically to provide uniform exposure. Progress of the irradiation was monitored by variation in melting point and ^{1}H NMR and IR spectra. After complete conversion, the time of which was dependent on the nature of the coumarin, the dimer was separated from the monomer by TLC. Dimers were identified by their spectral properties.

References: K. Gnanaguru, N. Ramasubbu, K. Venkatesan, V. Ramamurthy, *J. Org. Chem.*, **50**, 2337 (1985).

Type of reaction: C–C bond formation
Reaction condition: solid-state
Keywords: α-cyano-4-[2-(2-pyridyl)ethenyl]cinnamate, [2+2]photodimerization, cyclobutane, [2.2]paracyclophane

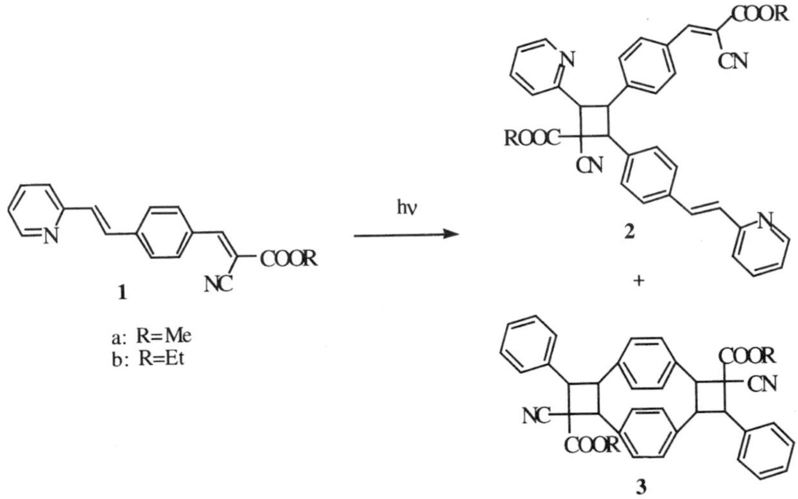

a: R=Me
b: R=Et

Experimental procedures:
Finely powdered monomer crystals were dispersed in distilled water containing a few drops of surfactant and irradiated, with vigorous stirring, by a 500-W super-high-pressure mercury lamp set outside of the flask. Dimers **2** were prepared by the irradiation of crystals **1** with λ>410 nm, and by successive purification by preparative TLC. [2.2]Paracyclophane **3a** was obtained by the photoirradiation of **1a** crystals with λ>300 nm at –40 °C, followed by successive washing with ethyl acetate.

References: C. Chung, A. Kunita, K. Hayashi, F. Nakamura, M. Hasegawa, *J. Am. Chem. Soc.,* **113**, 7316 (1991).

Type of reaction: C–C bond formation
Reaction condition: solid-state
Keywords: 2,3-di[(*E*)-styryl]pyrazine, [2+2]photodimerization

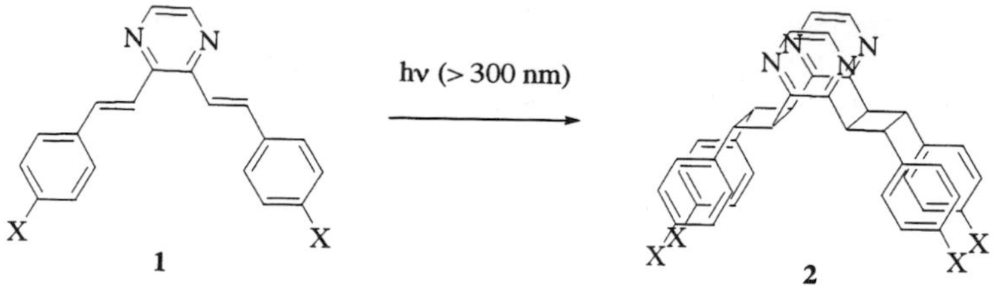

hν (> 300 nm)

1

a: X=Cl
b: X=CN

2

Experimental procedures:

Finely powdered crystals of **1** (100 mg) were dispersed in 100 mL of water containing a few drops of a surfactant (Nikkol TL-10FF) and irradiated from outside the flask with a 500-W super-high-pressure mercury lamp (Ushio USH 500 D) through an optical filter (Kenko UV-32 (cut off nm)) with vigorous stirring under a nitrogen atmosphere. The products were separated and purified by preparative TLC (dichloromethane).

References: A. Takeuchi, H. Komiya, T. Yukihiko, Y. Hashimoto, M. Hasegawa, Y. Iitaka, K. Siago, *Bull. Chem. Soc. Jpn.*, **66**, 2987 (1993).

Type of reaction: C–C bond formation
Reaction condition: solid-state
Keywords: thiocoumarin, [2+2]photodimerization, cyclobutane

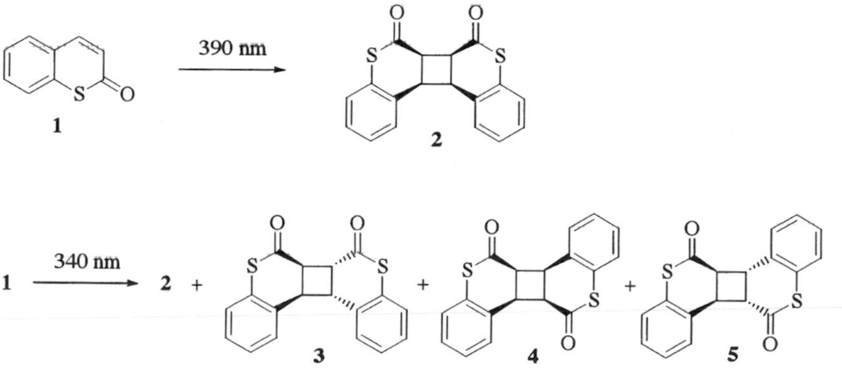

390 nm

1

2

340 nm

1 ⟶ **2** +

3   **4**   **5**

Experimental procedures:

A solution of 81 mg (0.5 mmol) of **1** in 10 mL of Et$_2$O (25 mL tapered flask) is slowly evaporated to produce a homogeneous solid film. The flask is then purged with Ar, fixed next to the immersion wall (filter A) of the lamp and turned around from time to time. After 140 h, chromatography (SiO$_2$/CH$_2$Cl$_2$) affords first 49 mg of **1** and then 25 mg (30 %) of **2**.

References: C.P. Klaus, C. Thiemann, J. Kopf, P. Margaretha, *Helv. Chim. Acta.*, **78**, 1079 (1995).

Type of reaction: C–C bond formation
Reaction condition: solid-state
Keywords: cinnamic acid derivative, [2+2]photodimerization, α-truxillic acid, cyclobutane

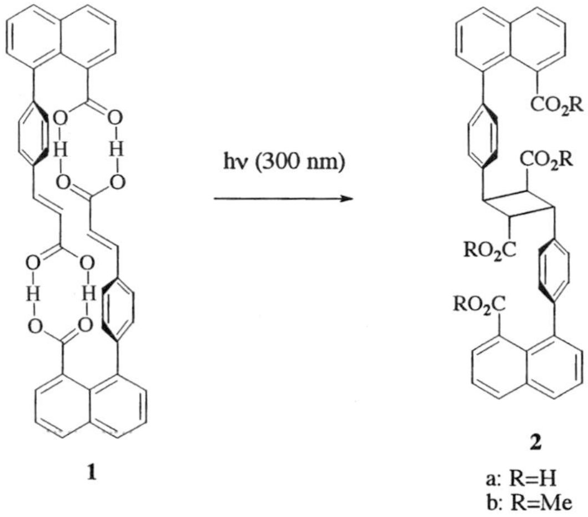

hv (300 nm)

2
a: R=H
b: R=Me

1

Experimental procedures:
Irradiation of a recrystallized sample (10–15 mg) of **1** through Pyrex in a Rayonet carousel photoreactor (300 nm) led to clean conversion of the starting material into a single cyclobutane product over the course of 20–50 h. The original colorless crystals visibly yellowed and lost definition around the sharp edges during irradiation but did not shatter or crumble. Treatment of the crude solid reaction product with etherial diazomethane permitted isolation of the cyclobutane product **2b** which was shown to have the α-truxillate-type stereochemistry.

References: K.S. Feldman, R.F. Campbell, *J. Org. Chem.*, **60**, 1924 (1995).

Type of reaction: C–C bond formation

Reaction condition: solid-state

Keywords: *trans*-cinnamic acid, double salt formation, [2+2]photodimerization, truxillic acid, truxinic acid, cyclobutane

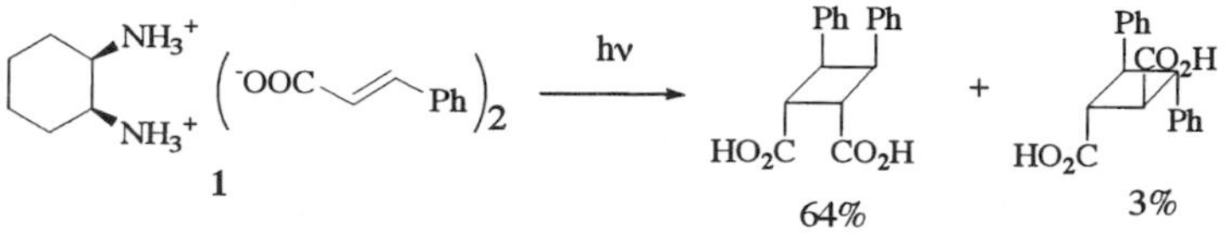

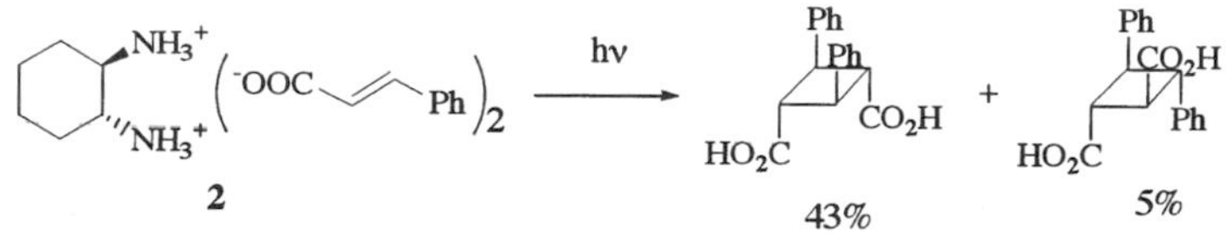

Experimental procedures:

A white powder (158 mg) obtained from photolysis of the double salt **2** was fractionally recrystallized by using 10 mL of methanol, resulting in the isolation of pure colorless plates of the 1:1 (from NMR) ε-truxillic acid *trans*-1,2-diaminocyclohexane salt; yield 57 mg, mp 146–150 °C. In addition, 2 mg of white crystals of the 1:1 δ-truxillic acid *trans*-1,2-diaminocyclohexane salt contaminated with a small amount of the above double salt was obtained. Free δ-truxillic acid was recovered by dissolving 3 mg of the double salt in 1 mL of water, followed by acidifying the solution with 1 M HCl to pH 1, as white needles (1.5 mg), mp 200–202 °C (lit. mp 192 °C).

References: Y. Ito, B. Borecka, J. Trotter, J. R. Scheffer, *Tetrahedron Lett.,* **36**, 6083 (1995).

Type of reaction: C–C bond formation

Reaction condition: solid-state

Keywords: acenaphthylene, tetracyanoethylene, CT complex, [2+2]photocycloaddition, cyclobutane

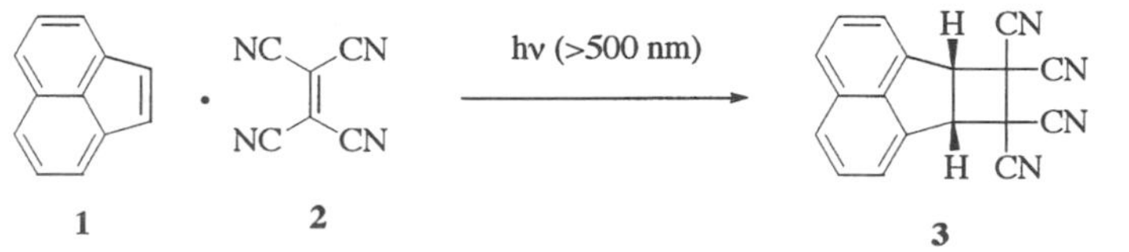

Experimental procedures:
A 40 mg sample of **1·2** placed between two disk glass filters was irradiated using a 400-W high-pressure Hg lamp in a thermostat. The products (**3** and dimers of **1**) and unreacted **1** were isolated by column chromatography. Compound **3** and the dimers of **1** were identified by comparison with authentic samples prepared by alternative methods.

References: N. Haga, H. Nakajima, H. Takayanagi, K. Tokumaru, *Chem. Commun.*, 1171 (1997); N. Haga, H. Nakajima, H. Takayanagi, K. Tokumaru, *J. Org. Chem.*, **63**, 5372 (1998).

Type of reaction: C–C bond formation
Reaction condition: solid-state
Keywords: fluoro-substituted styrylcoumarin, [2+2]photodimerization, cyclobutane

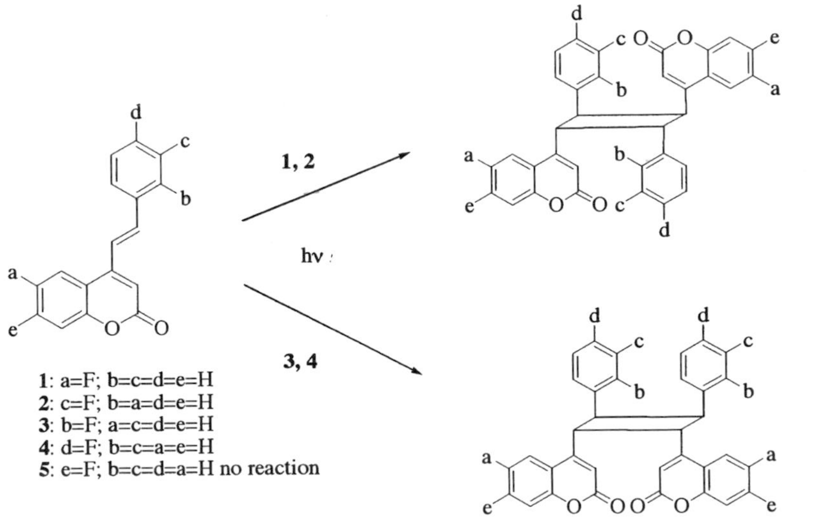

1: a=F; b=c=d=e=H
2: c=F; b=a=d=e=H
3: b=F; a=c=d=e=H
4: d=F; b=c=a=e=H
5: e=F; b=c=d=a=H no reaction

Experimental procedures:
The powder samples of **1** and **2** were irradiated with UV-light in a Rayonet photochemical reactor (λ_{max} 320 nm) at room temperature. The progress of the reaction was monitored by ^{1}H NMR spectroscopy and TLC. Corresponding photodimers were purified by column chromatography using 10% EtOAc in light petroleum.

References: K. Vishnumurthy, T. N. G. Row, K. Venkatesan, *J. Chem. Soc., Perkin Trans. 2*, 1475 (1996); K. Vishnumurthy, T. N. G. Row, K. Venkatesan, *J. Chem. Soc., Perkin Trans. 2*, 615 (1997).

Type of reaction: C–C bond formation
Reaction condition: solid-state
Keywords: vinylquinone, [2+2]photodimerization, 1,4-benzoquinone, cyclobutane

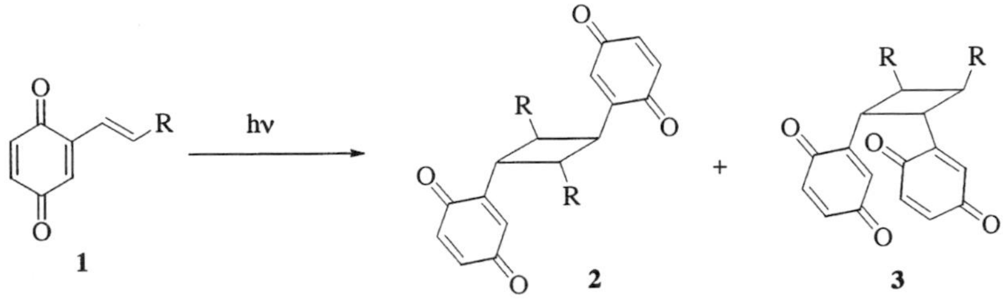

Experimental procedures:
The irradiation of 200–400 mg of the powdered derivatives **1** were carried out with a 1200-W high-pressure mercury lamp in a glass vessel at 20 °C for 5 h with continuous stirring. The melting range of the irradiated quinones was raised by 35–70 °C. For removal of unreacted starting materials and isolation of the generated photoproducts, the mixture was treated with a suitable solvent at room temperature. After recrystallization, the yellow cyclobutanes were obtained in analytical purity.

References: H. Irngartinger, B. Stadler, Eur. *J. Org. Chem.*, **605** (1998).

Type of reaction: C–C bond formation
Reaction condition: solid-state
Keywords: 1,4-dihydropyridine, [2+2]photodimerization, cyclobutane

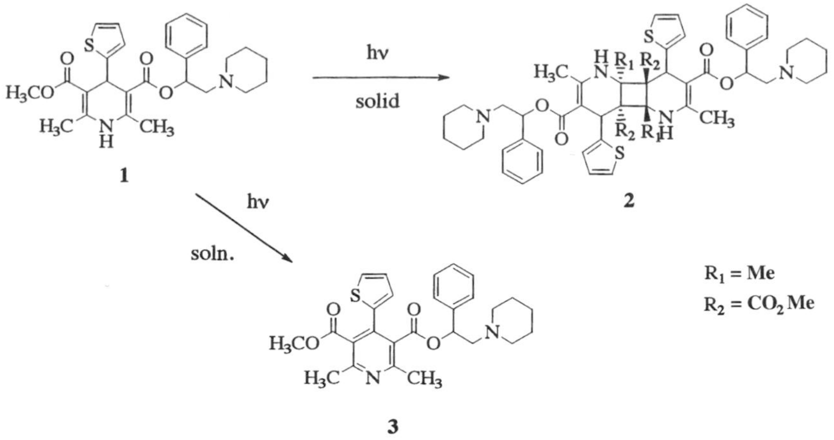

Experimental procedures:

Photolyses were performed at room temperature using a Nagano Science LT-120 irradiatior equipped with a Toshiba chemical lamp FIR-20S-BL/M (800 mW cm^{-2}). The crystalline samples were packed between two glass plates and placed in the irradiator. The time course of the reaction was checked by HPLC periodically. Irradiation of **1** for 3 h afforded the photoproduct **2** in 100% yield. The pyridine derivatives of **1** and **3** were identified by means of HPLC analyses using photodiode array detection by comparison of the retention time and the UV spectra to those of the authentic samples.

References: N. Marubayashi, T. Ogawa, T. Hamasaki, N. Hirayama, *J. Chem. Soc., Perkin Trans. 2*, 1309 (1997); N. Marubayashi, T. Ogawa, N. Hirayama, *Bull. Chem. Soc. Jpn., **71**, 321 (1998).*

Type of reaction: C–C bond formation
Reaction condition: solid-state
Keywords: 1,4-dihydropyridine, [2+2]photodimerization, cyclobutane

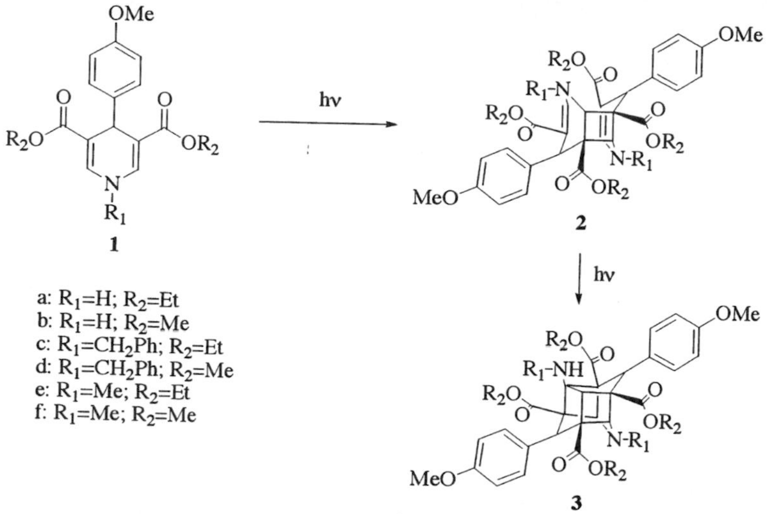

a: R_1=H; R_2=Et
b: R_1=H; R_2=Me
c: R_1=CH$_2$Ph; R_2=Et
d: R_1=CH$_2$Ph; R_2=Me
e: R_1=Me; R_2=Et
f: R_1=Me; R_2=Me

Experimental procedures:

1 g of crystalline 1,4-dihydropyridine **1** with a layer thickness of 1 mm was irradiated with an Ultra-Vitalux lamp from a distance of 60 cm at a measured temperature of 25 °C. After 3–4 d of irradiation (product formation monitored by TLC) the products **2** and **3**, were dissolved in boiling toluene and ethanol, respectively, from which they crystallized. The following yields are based on 1 g of **1**, corresponding to a 100% yield of **3** obtained by the direct irradiation of **1**.

References: A. Hilgeroth, U. Baumeister, F. W. Heinemann, *Eur. J. Org. Chem., 1213* (1998); A. Hilgeroth, *Chem. Lett.,* 1269 (1997).

Type of reaction: C–C bond formation
Reaction condition: solid-state
Keywords: 2,5-bis(2-alkoxycarbonylethenyl)-1,4-benzoquinone, [2+2]photo-dimerization, cyclobutane

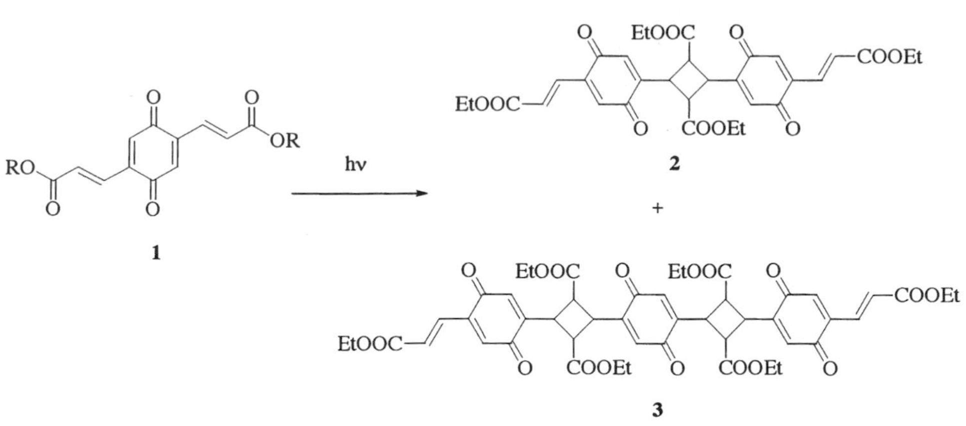

Experimental procedures:
The powdered quinones were dispersed homogenously in a dish made of aluminum foil and covered with the appropriate filter. All quinones were irradiated successively under dried nitrogen for 4 h at 20 °C.

References: H. Irngartiger, R. Herpich, *Eur. J. Org. Chem.*, 595 (1998).

Type of reaction: C–C bond formation
Reaction condition: solid-state
Keywords: *trans*-stilbene, cinnamic acid, [2+2]photodimerization, photopolymerization, cyclobutane

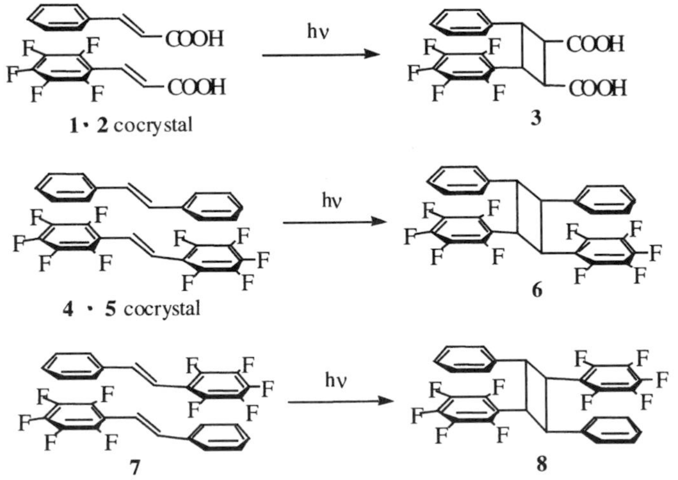

Experimental procedures:
trans-Cinnamic acid (**1**, 1.00 g, 6.75 mmol) and *trans*-2,3,4,5,6-pentafluorocin-namic acid (**2**, 1.61 g, 6.75 mmol) were dissolved in 8 mL of hot absolute etha-nol. The solution was allowed to cool slowly to room temperature, during which fine white crystals were formed. After crystallizing overnight, the cocrystals were collected by filtration and dried in vacuo to give cocrystal **1·2** (2.45 g, 94%). Crystals of **1·2** (20.0 mg) were placed in a quartz reaction tube and photolyzed for 7 h at ambient temperature. ^{1}H NMR spectroscopy revealed the cyclodimer **3** (87%).

References: G.W. Coates, A.R. Dunn, L.M. Henling, J.W. Ziller, E.B. Lobkovsky, R.H. Grubbs, *J. Am. Chem. Soc.,* **120**, 3641 (1998).

Type of reaction: C–C bond formation
Reaction condition: molten-state
Keywords: chalcone, [2+2]photodimerization, molten-state, cyclobutane

*anti-head-to-head-***2**

a: Ar=Ph; Ar'=Ph
b: Ar=Ph; Ar'=4-MeC$_6$H$_4$
c: Ar=Ph; Ar'=4-ClC$_6$H$_4$
d: Ar=4-MeOC$_6$H$_4$; Ar'=Ph
e: Ar=4-MeC$_6$H$_4$; Ar'=4-MeC$_6$H$_4$

Experimental procedures:
Chalcone **1a** (1 g, 4.8 mmol) was placed in between two Pyrex plates and the sample was melted by being heated at 60 °C on a hot-plate, and thereafter was ir-radiated with a 400-W high-pressure Hg lamp for 24 h. The oily crude product was crystallized by addition of a small amount of MeOH to give, after recrystal-lization from MeOH, compound **2a** as prisms (0.31 g, 31%), mp 123–125 °C.

References: F. Toda, K. Tanaka, M. Kato, *J. Chem. Soc., Perkin Trans. 1,* 1315 (1998).

Type of reaction: C–C bond formation
Reaction condition: solid-state
Keywords: 9-methylanthracene, C60, fullerene, Diels-Alder reaction, photoirra-diation

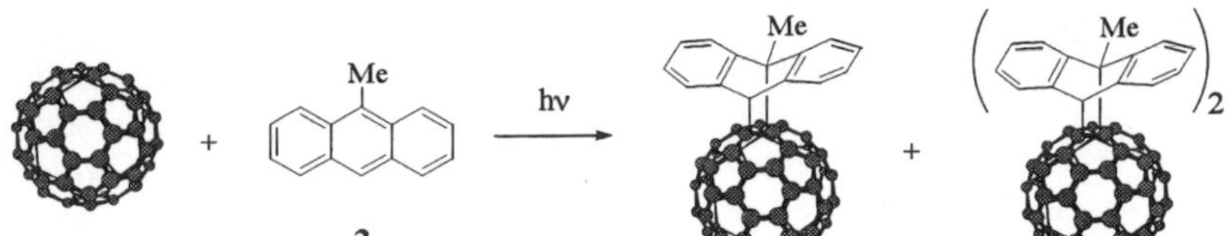

Experimental procedures:
A well ground mixture of C60 **1** and 9-methylanthracene **2** (1:1) was exposed to high-pressure mercury lamp through Pyrex vessel with stirring by using cooling jacket to keep the reaction temperature at 28 °C. The reaction mixture was then dissolved in C_6D_6/CS_2 and yields of mono- and bis-adducts were determined by ^1H NMR analysis using triptycene as an internal standard.

References: K. Mikami, S. Matsumoto, T. Tonoi, Y. Okubo, *Tetrahedron Lett.*, **39**, 3733 (1998).

Type of reaction: C–C bond formation
Reaction condition: solid-state
Keywords: diarylacetylene, dichlorobenzoquinone, donor-acceptor complex, photoirradiation

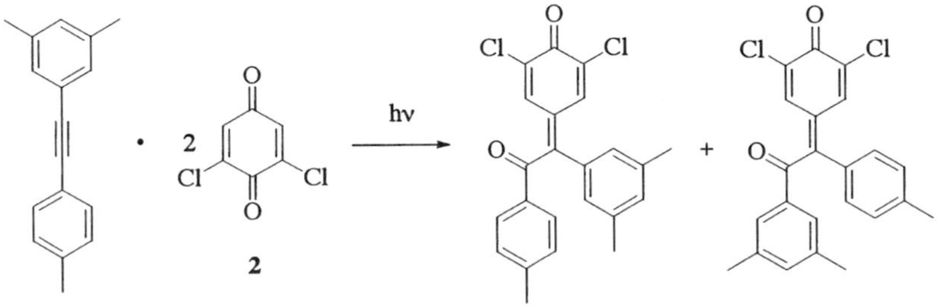

Experimental procedures:
To a thin Pyrex tube were added several orange crystals (7 mg, 0.012 mmol) of EDA complex (**1·2(2)**), and the tube was sealed under an argon atmosphere. The reaction tube was placed in a clear Dewar filled with acetone cooled to –60 °C and the sample irradiated with a medium-pressure mercury lamp fitted with an aqueous IR filter and a Corning cutoff filter to effect irradiation with visible light (λ > 410 nm). The tube was rotated each hour, and the temperature in the Dewar

was controlled by the periodic addition of dry ice. After 5 h of irradiation, the crystals extensively cracked. The crystals were dissolved in dichloromethane, and the immediate analysis by quantitative GC. The crude products from several experiments were combined and purified by thin-layer chromatography with hexane-ethyl acetate (10:1) as eluent.

References: E. Bosch, S. M. Hubig, S. V. Lindeman, J. K. Kochi, *J. Org. Chem.*, **63**, 592 (1998).

Type of reaction: C–C bond formation
Reaction condition: solid-state
Keywords: 2,3-bis(2-phenylethenyl)-4,5-dicyanopyrazine, [2+2]photocycloaddition, cyclobutane

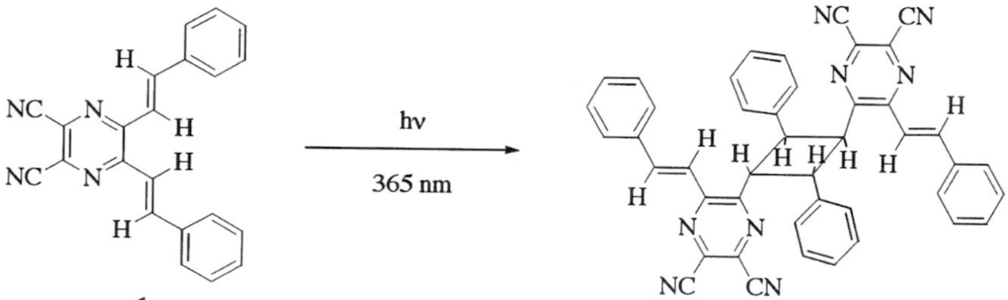

Experimental procedures:
Compound **1** was recrystallized in two different forms; one was yellow colored crystal **1a** (benzene), and the other was orange crystal **1b** (a mixture of THF and acetonitrile). Photoirradiation on **1a** in the single crystals at 366 nm for 15 h gave mainly the insoluble polymer and small amounts of **2**, but **1b** did not react at all.

References: J. H. Kim, M. Matsuoka, K. Fukunishi, *Chem. Lett.*, 143 (1999).

Type of reaction: C–C bond formation
Reaction condition: solid-state
Keywords: 4-methoxy-6-methyl-2-pyrone, maleinimide, [2+2]photocycloaddition, cyclobutane

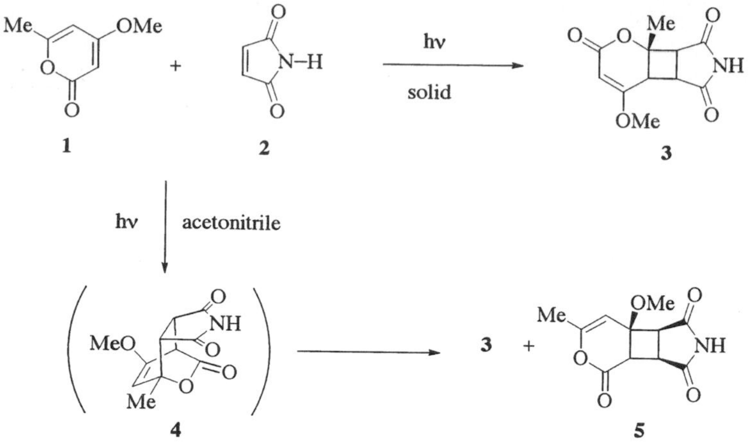

Experimental procedures:

Irradiation of a 1:1 complex crystal (mp 95–97 °C, plates) between **1** and 2, which was prepared by recrystallization of the equimolar substrates from acetonitrile, with a 400-W high-pressure mercury lamp through a Pyrex filter under nitrogen at room temperature gave [2+2]cycloadduct **3** in 54% yield as a sole product. On the other hand, direct photoirradiation to an acetonitrile solution of **1** and **2** gave another type of [2+2]cycloadduct 5 in 25% yield together with a small amount of 3.

References: T. Obata, T. Shimo, S. Yoshimoto, K. Somekawa, M. Kawaminami, *Chem. Lett.*, 181 (1999); T. Obata, T. Shimo, M. Yasutaka, T. Shimyozu, M. Kawaminami, R. Yoshida, K. Somekawa, *Tetrahedron*, **57**, 1531 (2001).

Type of reaction: C–C bond formation
Reaction condition: solid-state
Keywords: tetrabenzo[*ab,f,jk,o*][18]annulene, [2+2]photodimerization

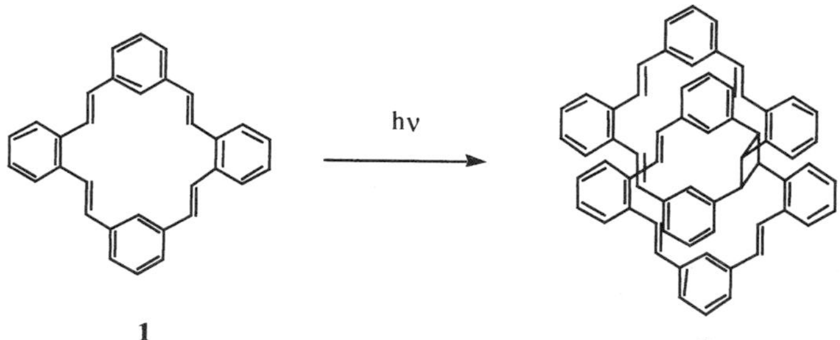

Experimental procedures:

A suspension of 16.0 mg (0.04 mmol) **1** in 500 mL water was circulated through a photoreactor equipped with a 150-W mercury lamp (Hanau TQ 150 Z 3) with a main emission between 300 and 400 nm. After 1 h the irradiation was stopped and the suspension filtered. The dry residue was dissolved in $CDCl_2$-$CDCl_2$. The ^{1}H NMR spectrum at 50 °C shows the signals of **1** and its dimer **2** in a ratio of 30:70 and a small amount of oligomers. Prolonged irradiation yielded higher amounts of oligomeric material.

References: R. Yu, A. Yakimansky, I.G. Voigt-Martin, M. Fetten, C. Schnorpfeil, D. Schollmeyer, H. Meier, *J. Chem. Soc., Perkin Trans. 2*, 1881 (1999).

Type of reaction: C–C bond formation
Reaction condition: solid-state
Keywords: styryldicyanopyrazine, topochemical photodimerization, cyclobutane

1a: R_1=R_2=H
1b: R_1=*t*-Bu; R_2=Et

Experimental procedures:

Upon photoirradiation of single crystals of **1a** and **1b** with 366 nm UV light for 15 h a photodimer of **1a** having an anti head-to-head form **3a** was obtained in excellent yield (90%), whereas **1b** gave **3b** in poor yield (17%) together with unreacted **1b** (83%). Neither of **2a** and **2b** were obtained in the solid state.

References: J.H. Kim, M. Matsuoka, K. Fukunishi, *J. Chem. Res. (S)*, 132 (1999).

Type of reaction: C–C bond formation
Reaction condition: solid-state
Keywords: tryptamine, 3-nitrocinnamic acid, [2+2]photocycloaddition, cyclobutane

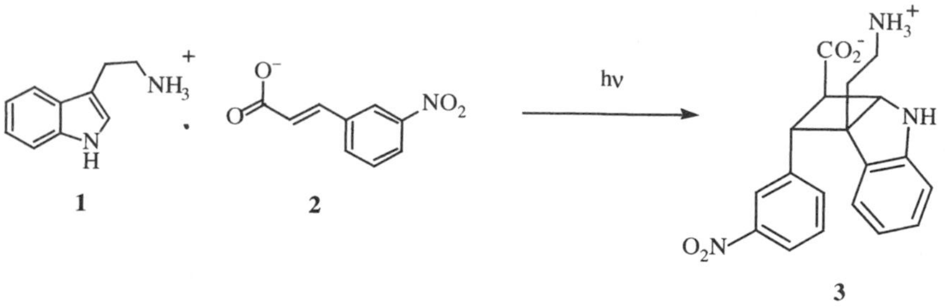

Experimental procedures:

A crystalline salt of **1** and **2** (20 mg) was crushed and spread between two Pyrex plates and this was irradiated with a 400-W high-pressure mercury lamp for 20 h under an argon atmosphere. The irradiation vessel was cooled from the outside by tap water. After the irradiation, the reaction mixture was analyzed by ^{1}H NMR. The only peaks observable were those for the starting materials (**1** and **2**) and a cross photoadduct (**3**) (47% conversion, 39% yield). In a separate experiment, 105 mg of the salt was irradiated under similar conditions. The product **3** could be isolated simply by fractional recrystallization from ethanol (15 mL). Polymeric products (6 mg) first separated and then 18 mg (17% yield) of nearly pure **3** separated as a pale brown powder: mp 215–225 °C.

References: Y. Ito, H. Fujita, *Chem. Lett.,* 288 (2000).

Type of reaction: C–C bond formation
Reaction condition: solid-state
Keywords: isothiocoumarin, [2+2]photodimerization, cyclobutane

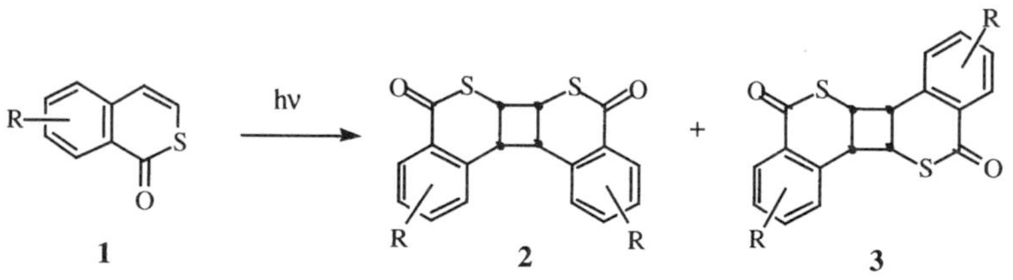

a: R=H
b: R=7-Me
c: R=5-CF$_3$
d: R=5,6-C$_4$H$_4$

Experimental procedures:
Irradiation (350 nm) of **1b** as homogeneous solid film leads to the selective (94%) formation of the *HH-cis-cisoid-cis* photocyclodimer **2b** and traces (6%) of the *HT-cis-cisoid-cis* dimer **3b** (monitoring by ^{1}H NMR). Dimer 2b was isolated and purified by chromatography. Similarly, irradiation of **1c** affords a 4:5 mixture of dimers **2c** and **3c** which were not separated.

References: M. A. Kinder, J. Kopf, P. Margaretha, *Tetrahedron,* **56**, 6763 (2000).

Type of reaction: C–C bond formation
Reaction condition: solid-state
Keywords: *trans*-cinnamamide, dicarboxylic acid, cocrystals, [2+2]photodimerization, cyclobutane

Experimental procedures:
Cocrystal **1** (51 mg) was irradiated for 20 h as described above. The slightly colored photolysate, which contained **2** with 29% yield (estimated by ^{1}H NMR and HPLC), was taken up in 2 mL of water and was stirred at 45 °C for 1 h. After cooling in a refrigerator, an insoluble pale yellow solid was collected by filtration. This was stirred into 2 mL of acetonitrile and an insoluble solid was filtered off to afford 5 mg (13% yield) of **2** as a white crystalline solid: mp 246–252 °C.

References: Y. Ito, H. Hosomi, S. Ohba, *Tetrahedron,* **56**, 6833 (2000).

Type of reaction: C–C bond formation
Reaction condition: solid-state
Keywords: *N*-[(*E*)-3,4-methylenedioxycinnamoyl]dopamine, [2+2]photodimerization, cyclobutane

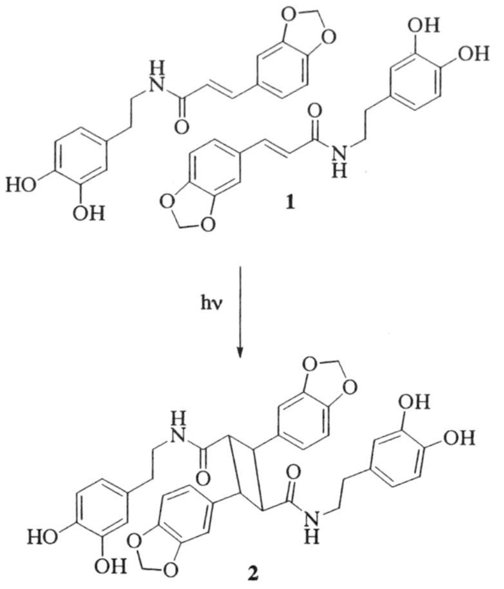

Experimental procedures:

Irradiation of crystalline **1** under an argon atmosphere with a 400-W high-pressure Hg lamp (Pyrex) for 20 h gave **2** (66% conversion, 100% yield).

References: Y. Ito, S. Horie, Y. Shindo, *Org. Lett.,* **3**, 2411 (2001).

Type of reaction: C–C bond formation
Reaction condition: solid-state
Keywords: 2-(dibenzylamino)ethyl 3-benzoylacrylate, [2+2]photocycloaddition, cyclobutane

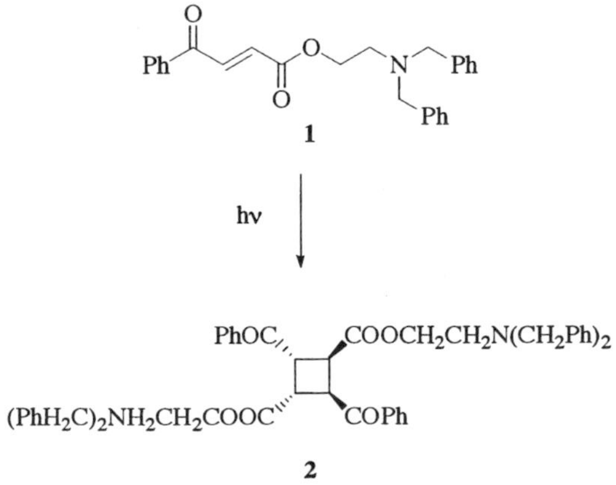

Experimental procedures:
The crystals of **1** (0.635 g, 1.6 mmol) were placed between Pyrex glass plates and irradiated for 75 min with a 400-W high-pressure mercury lamp. After the irradiation the color of the crystals had changed from pale yellow to white and bis[2-(dibenzylamino)ethyl]*c*-2,*t*-4-dibenzoylcyclobutane-*r*-1,*t*-3-dicarboxylate   **2** was obtained quantitatively: mp 143.0 °C (from a mixture of dichloromethane and hexane).

References: T. Hasegawa, K. Ikeda, Y. Yamazaki, *J. Chem. Soc., Perkin Trans. 1*, 3025 (2001).

Type of reaction: C–C bond formation
Reaction condition: solid-state
Keywords: cholest-4-en-3-one, photodimerization, cyclobutane

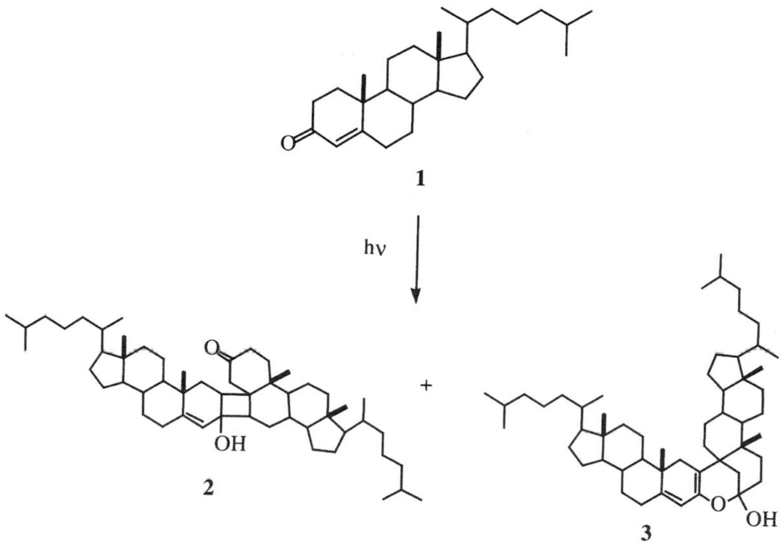

Experimental procedures:
Powdered **1** (200 mg), crystallized from Et₂O, was irradiated with a Hanovia 450-W lamp for 4 h to give a mixture of unreacted **1** (142 mg), **2** (18 mg) and **3** (14 mg) separated by flash chromatography on silica gel (hexane-Et₂O, 19:1).

References: M. DellaGreca, P. Monaco, L. Previtera, A. Zarrelli, A. Fiorentino, F. Giordano, C. Mattia, *J. Org. Chem.*, **66**, 2057 (2001).

Type of reaction: C–C bond formation
Reaction condition: solid-state
Keywords: enone, photolysis, [2+2]-cycloaddition, *head-to-tail-anti*, cyclobutane, spiro compound

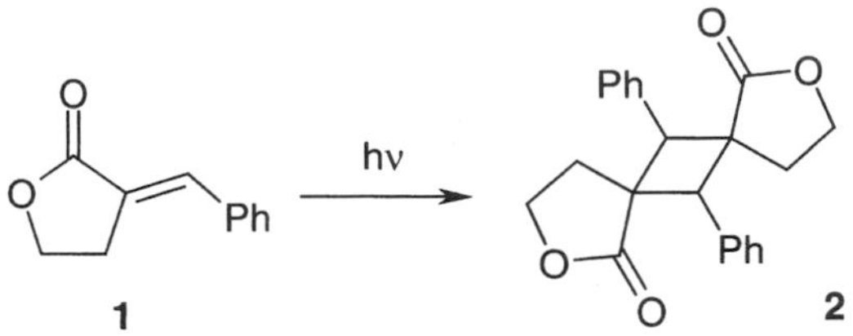

Experimental procedure:
a-Benzylidene-*γ*-butyrolactone **1** (1.0 g, mp 115–117 °C) was evenly spread on the inner wall of a mirrored Dewar vessel (diameter 14 cm, height 20 cm) with some dichloromethane. After heating to 80 °C for 1 h the crystalline film was irradiated from within for 5 h with a high-pressure Hg-lamp (Hanovia 450 W) through a 5% solution of benzophenone in benzene (5 mm; $\lambda > 380$ nm) and cooling with running water at 30–35 °C. The yield of *head-to-tail-anti*-dimer **2** was quantitative (mp 242 °C).

References: G. Kaupp, E. Jostkleigrewe, H.-J. Hermann, *Angew. Chem.*, **94**, 457 (1982); *Angew. Chem. Int. Ed. Engl.*, **21**, 435 (1982). Numerous early solid-state photodimerizations are listed in Houben-Weyl, Methoden der Organischen Chemie, Vol. IV/5a, *Chap. 4 and 5*, pp. 278–412, Thieme, Stuttgart (1975). Further examples are listed in *CRC Handbook of Organic Photochemistry and Photobiology* (Ed. W. M. Horspool), *Chap. 4*, pp. 50–63, CRC Press, Boca Raton (1995).

Type of reaction: C–C bond formation
Reaction condition: solid-state
Keywords: enone, photolysis, [2+2]-cycloaddition, cyclobutane, spiro compound

Experimental procedure:

2,5-Bisbenzylidene-cyclopentanone **1** (1.04 g, 2.0 mmol) was evenly spread on the inner wall of a mirrored Dewar vessel (diameter 14 cm, height 20 cm) with some dichloromethane. After heating to 80 °C for 1 h the crystalline film was ir-radiated from within for 3–5 h at 30–35 °C with a high-pressure Hg-lamp (Hano-via 450 W) through a benzophenone filter (5% in benzene; 5 mm; $\lambda > 380$ nm) and cooling with running water. ^{1}H NMR-analysis revealed 400 mg (38%) **3**, 275 mg (26%) **2** (structure clarified in Ref. 2), 85 mg (8%) **4** and 190 mg (18%) unreacted **1**. The separation of the products by preparative TLC at 200 g SiO$_2$ used dichloromethane. Recrystallization of the products was performed in 1,2-di-chloroethane, toluene, and methanol, respectively.

References: 1) G. Kaupp, I. Zimmermann, *Angew. Chem.,* **93**, 1107 (1981); *Angew. Chem. Int. Ed. Engl.,* **20**, 1018 (1981).
2) G. Kaupp, *J. Microscopy,* **174**, 15 (1994); G. Kaupp, *Mol. Cryst. Liq. Cryst.,* **252**, 259 (1994).
3) Numerous early solid-state photodimerizations are listed in Houben-Weyl, Methoden der Organischen Chemie, Vol. IV/5a, *Chap. 4* and 5, pp. 278–412, Thieme, Stuttgart (1975). Further examples are listed in *CRC Handbook of Organic Photochemistry and Photobiology* (Ed. W. M. Horspool), *Chap. 4*, pp. 50–63, CRC Press, Boca Raton (1995).

Type of reaction: C–C bond formation
Reaction condition: solid-state
Keywords: heterostilbene, [2+2]-cycloaddition, polar effect, photochemistry, cy-clobutane

(4 h, 70%)   (4 h, 95%)   (4 h, 69%)

(4 h, 85%)

(4 h, 29%)   (4 h, 75%)   (3 h, 64%)   (10 h, 0%)

Experimental procedures:

One to three mmol of the polar heterostilbene **1** was spread with some dichloro-methane in a sufficiently wide bottom-closed, top-stoppered glass cylinder with gas inlet and outlet to form a 10 cm high zone that matched with the burning zone of a 500-W high-pressure Hg-lamp behind a water cooled Solidex filter. After dry-ing in a vacuum, an even crystal film was obtained. This setup was immersed in a cold bath at 0 °C while the crystal film compartment was filled with argon. Irradia-tion times varied from 3 to 10 h. The indicated yields were obtained after recrystal-lization. There were no different stereoisomers of **2** except when the phenyl group was replaced by *p*-nitrophenyl- or *p*-cyanophenyl-groups or if the Het-group was replaced by methoxycarbonyl- or acetyl-groups in 1,2- bifunctional derivatives (MeO$_2$CCH=CHPhCH=CHCO$_2$Me, AcCH=CHPhCH=CHAc).

References: G. Kaupp, H. Frey, G. Behmann, *Chem. Ber.,* **121**, 2135 (1988).

Numerous early solid-state photodimerizations are listed in Houben-Weyl, Methoden der Organischen Chemie, Vol. IV/5a, *Chap. 4* and *5*, pp. 278–412, Thieme, Stuttgart (1975). Further examples are listed in *CRC Handbook of Or-ganic Photochemistry and Photobiology* (Ed. W.M. Horspool), *Chap. 4*, pp. 50–63, CRC Press, Boca Raton (1995).

Type of reaction: C–C bond formation
Reaction condition: solid-state
Keywords: anthracenophane, [4+4]-dimerization, [4+4]-cycloreversion, photo-chrome, topotaxy, photochemistry

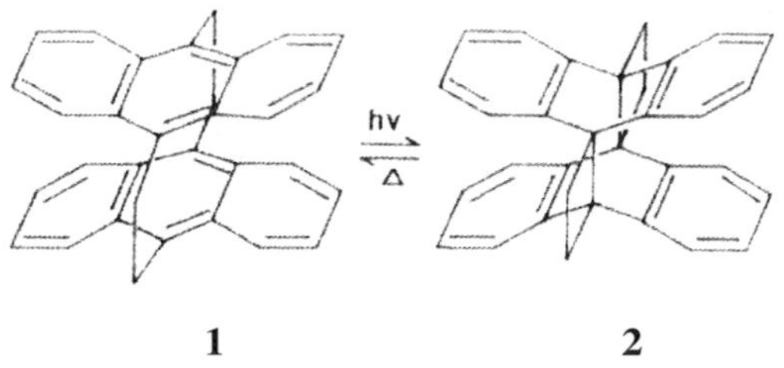

1   **2**

Experimental procedures:
The golden yellow crystals of **1** were exposed to glass-filtered daylight and thus formed the colorless "dimer" structure **2** with quantitative yield. If **2** was heated to 30 °C in the dark for 5 h compound **1** was quantitatively formed back in a topotactic manner without change of the crystal shape even on the nanoscopic scale as shown by atomic force microscopy (AFM). Numerous cycles were performed without loss using single crystals of **1/2** as well.

References: G. Kaupp, *Current Opinion in Solid State and Materials Science,* **6**, 131 (2002); *Adv. Photochem.,* **19**, 119–177 (1995); *Liebigs Ann. Chem.,* 844 (1973); D.A. Dougherty, C.S. Choi, G. Kaupp, A.B. Buda, J.M. Rudzinski, E. Osawa, *J. Chem. Soc. Perkin Trans. 2,* 1063 (1986).
Numerous earlier solid-state [4+4]-photodimerizations of solid (substituted) anthracenes are cited there and mechanistically interpreted on the basis of supermicroscopic AFM data.

Type of reaction: C–C bond formation
Reaction condition: solid-state
Keywords: 4,4′dimethylbenzophenone, photodimerization

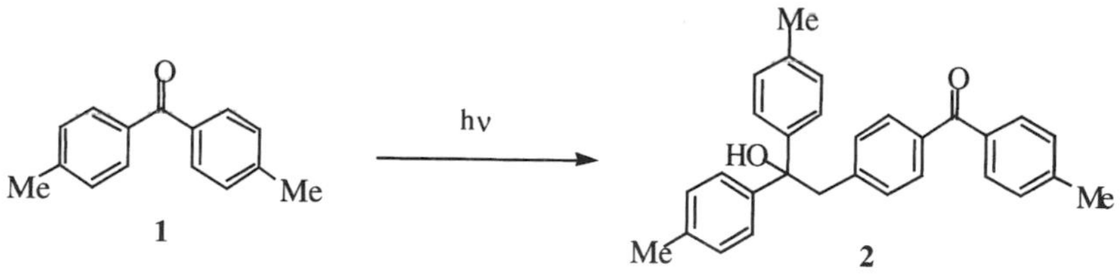

Experimental procedures:
Crystals of **1** were ground to a fine powder in a mortar. The powder (210 mg) was placed between two Pyrex disks (8 cm diameter) and irradiated under a nitrogen atmosphere for 10 h at 0°C (ice-water) by using a 400-W high-pressure mercury lamp as a light source. The separation between the lamp and the sample was approximately 5 cm. The reaction mixture was separated by preparative TLC to give 178 mg of the recovered **1** and 31 mg ($\sim$ 100% yield) of a dimeric product, 4-(2-hydroxy-2,2-di-*p*-tolylethyl)-4'-methylbenzophenone **2**. This was recrystallized from hexane-ethyl acetate (10:1, v/v) to give colorless crystals: mp 53–55°C.

References: Y. Ito, T. Matsuura, K. Tabata, M. Ben, *Tetrahedron*, **43**, 1307 (1987).

Type of reaction: C–C bond formation
Reaction condition: solid-state
Keywords: α-diketone, silica gel surface, Norrish type I reaction, photoreaction

a: $R_1=R_2=Ph$
b: $R_1=R_2=p$-MeC_6H_4
c: $R_1=Me$; $R_2=Ph$
d: $R_1=R_2=Me$
e: $R_1=R_2=p$-$MeOC_6H_4$

Table 1. Yields of photoproducts.

1	Yields (%)		
	2	3	4
a	Trace	59	16
b	Trace	41	23
c	60	60	0
d	30	0	–
e	0	0	0

Experimental procedures:
The α-diketone **1** (ca. 2.0 mmol) in dichloromethane (5 mL) was added to silica gel (5 g) in a 100-mL round-bottomed flask. The mixture was sonicated for 5 min after which the solvent was evaporated under reduced pressure. The coated silica gel, divided into six nearly equal portions, was then placed in Pyrex tubes (18×180 mm). In the experiments under oxygen-free conditions the tubes were degassed by three freeze-pump-thaw cycles and sealed. The tubes were rotated and irradiated for 48 h with a 100-W high-pressure mercury lamp. The irradiated silica gel was collected. Acetone (20 mL) was added to the silica gel to extract

the organic components and the mixture was sonicated for 10 min. The silica gel was filtered off and washed with acetone (10 mL). The combined filtrate and washings were then evaporated under reduced pressure. More than 93% of organic material based on the weight of the starting diketone used was recovered by this method. The residue was chromatographed on silica gel. Elution with a mixture of acetone-hexane (1:6, v/v) gave unchanged diketone and photoproducts (Table 1).

References: T. Hasegawa, M. Imada, Y. Imase, Y. Yamazaki, M. Yoshioka, *J. Chem. Soc., Perkin Trans. 1*, 1271 (1997).

Type of reaction: C–C bond formation
Reaction condition: solid-state
Keywords: 16-dehydroprogesterone, photodimerization

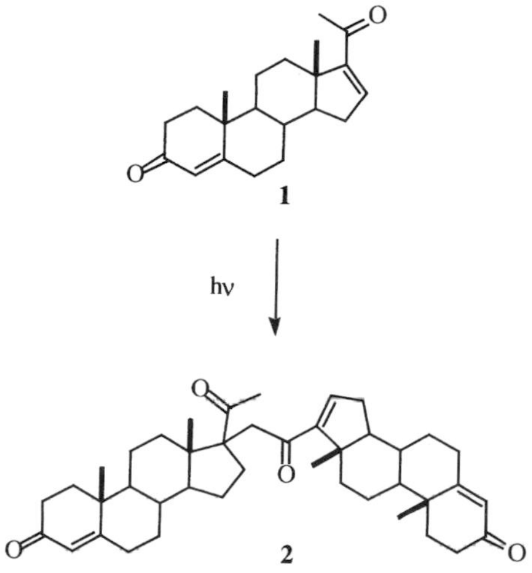

Experimental procedures:
An aqueous suspension (50 mL) of 16-dehydroprogesterone **1** (100 mg) in a Pyrex flask was irradiated with fluorescent lamp (Philips, TLD 30W/55) under magnetic stirring at 24 °C for 10 days. EtOAc extraction and flash chromatography (silica gel, EtOAc-hexane 3:2) followed by preparative TLC gave **2** (20 mg) as a colorless oil.

References: M. DellaGreca, P. Monaco, L. Previtera, A. Fiorentino, F. Giordano, C. Mattia, *J. Org. Chem.*, **64**, 8976 (1999).

Type of reaction: C–C bond formation
Reaction condition: solid-state
Keywords: tetracyanobenzene, benzyl cyanide, two-component crystal, photo-reaction

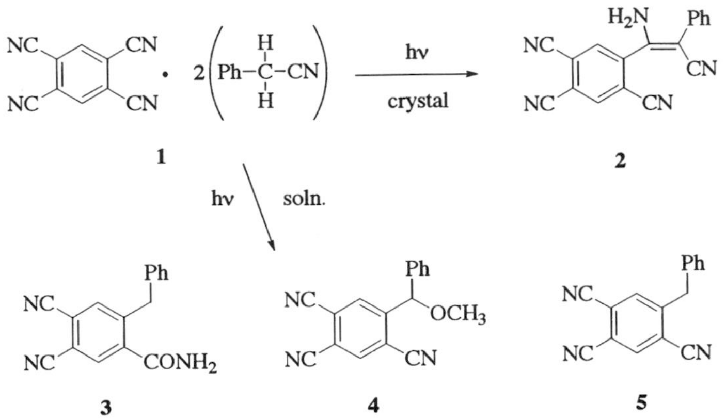

Experimental procedures:
The crystals of **1** (40 mg) were spread between two Pyrex plates and irradiated with a 400-W high-pressure mercury lamp for 20 h under argon. During the irradiation, the photolysis vessel was cooled from the outside by circulation of cold water (4 °C). After the irradiation, the orange photolysate was immediately evaporated in vacuo at 60 °C to remove unreacted BzCN. The residue, where **2** was present as a sole product in 65% yield on the basis of TCNB, was fractionally recrystallized with MeCN (0.6 mL) to furnish 3 mg of pure **2** as orange-yellow plates: mp >300 °C.

References: Y. Ito, S. Endo, S. Ohba, *J. Am. Chem. Soc.*, **119**, 5974 (1997); Y. Ito, H. Nakabayashi, S. Ohba, H. Hosomi, *Tetrahedron*, **56**, 7139 (2000).

Type of reaction: C–C bond formation
Reaction condition: solid-state
Keywords: *N,N*-dialkyl-α-oxoamide, Norrish type II reaction, photoirradiation, β-lactam

a: R_1 = Ph, R_2 = R_4 = H, R_3 = R_5 = Ph
b: R_1 = Me, R_2 = R_4 = H, R_3 = R_5 = Ph
c: R_1 = Ph, R_2 = R_3 = R_4 = R_5 = Me
d: R_1 = Ph, R_2 = R_4 = H, R_3, R_5 = -(CH$_2$)$_3$-
e: R_1 = R_2 = R_3 = R_4 = R_5 = Me

Experimental procedures:
The recrystallized α-oxoamide **1** (100–300 mg) was sandwiched between a pair of Pyrex plates (thickness 2 mm) and put into a polyethylene envelope. The envelope was sealed and placed in a cold medium (water, ice-water, or dry ice-methanol) and irradiated with a high-pressure mercury lamp (100 W) for 1–10 h. The lamp was used with a vessel for immersion type irradiation which contained dry air. The products were isolated by column chromatography on silica gel.

References: H. Aoyama, T. Hasegawa, Y. Omote, *J. Am. Chem. Soc.*, **101**, 5343 (1979).

Type of reaction: C–C bond formation
Reaction condition: solid-state
Keywords: cyclohexenone, photorearrangement, intramolecular allylic hydrogen abstraction

Experimental procedures:
The preparative solid-state photolyses were carried out either on the inner surface of an immersion well apparatus (method A) or by using Pyrex glass plate "sandwiches" sealed in polyethylene bags (method B). Naphthoquinol **1** (70 mg) was irradiated through a uranium glass filter sleeve (λ>330 nm) by method A at $-74\,^\circ$C under nitrogen for 2.1 h. Column chromatography (silica gel, 20% ethyl acetate-toluene) afforded 39 mg of starting material and 25 mg (81% based on unrecovered starting material) of a new photoproduct **3**.

References: W.K. Appel, Z.Q. Jiang, J.R. Scheffer, L. Walsh, *J. Am. Chem. Soc.*, **105**, 5354 (1983).

Type of reaction: C–C bond formation
Reaction condition: solid-state
Keywords: carbene, photorearrangement

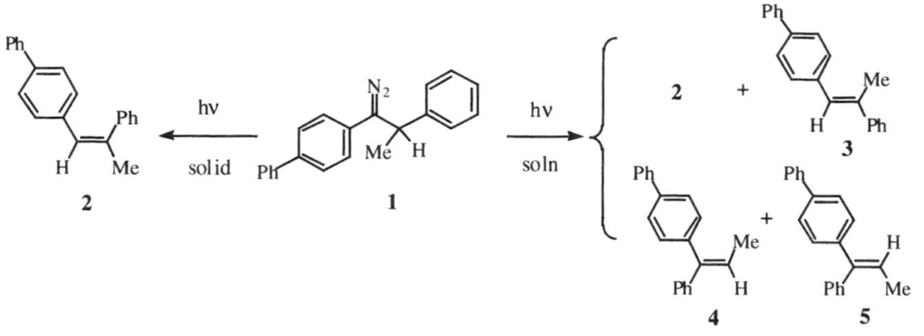

Experimental procedures:
Pure powdered samples of **1** irradiated at 0 °C with $\lambda > 380$ nm showed that compound **2** constitutes over 96% of the product mixture. Product **2** melts at 105 °C.

References: S. H. Shin, A. E. Keating, M. A. Garibay, *J. Am. Chem. Soc.*, **118**, 7626 (1996).

Type of reaction: C–C bond formation
Reaction condition: solid-state
Keywords: *N*-cinnamoyl-1-naphthamides, [2+2], [4+2]photodimerization, cyclobutane

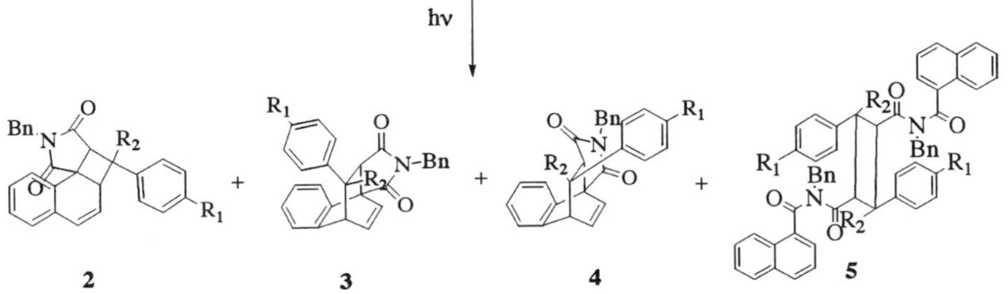

Experimental procedures:
Compound **1** (0.25 mmol) was ground to a powder and then sandwitched between two cover glasses (25×80 mm, thickness 1.5 mm) and dipped in a water bucket. Irradiation was carried out with a 400-W high-pressure mercury lamp for 24 h. Dimers **5** were separated from intramolecular cyclization products by preparative GPC runs. Isomeric products **2** and **3** were separated by preparative HPLC runs.

References: S. Kohmoto, T. Kobayashi T. Nishio, I. Iida, K. Kishikawa, M. Yamamoto, K. Yamada, *J. Chem. Soc., Perkin Trans. 1*, 529 (1996).

Type of reaction: C–C bond formation
Reaction condition: solid-state
Keywords: *cis*-9-decalyl aryl ketone, Norrish type II photochemistry, cyclopropanol, cyclobutanol, cyclopentanol

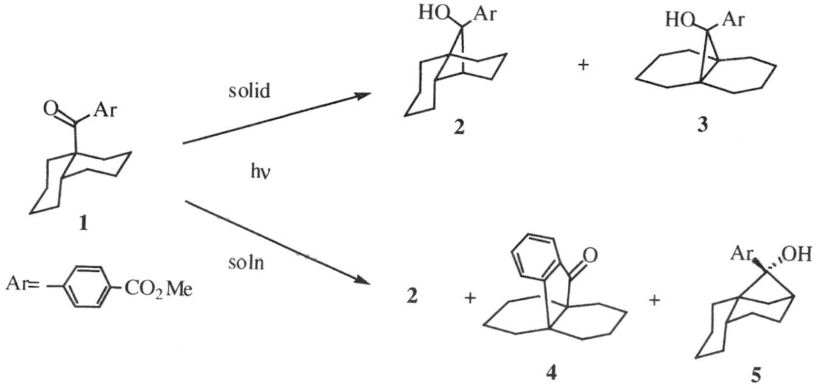

Experimental procedures:
Irradiation of CH_3CN solutions of *cis*-9-decalyl *p*-carbomethoxyphenyl ketone **1** (145 mg) through Pyrex afforded cyclobutanols **2** (47%) and **5** (47%) along with 6% of cyclopentanone **4**. In contrast, when crystals of ketone **1** were irradiated (Pyrex, –20 °C, 76 mg), a mixture of cyclobutanol **2** (81%) and cyclopropanol **3** (19%) was formed.

References: E. Cheung, T. Kang, J. R. Scheffer, J. Trotter, *Chem. Commun.*, 2309 (2000).

Type of reaction: C–C bond formation
Reaction condition: solid-state
Keywords: α-mesitylacetophenone, photocyclization, 2-indanol

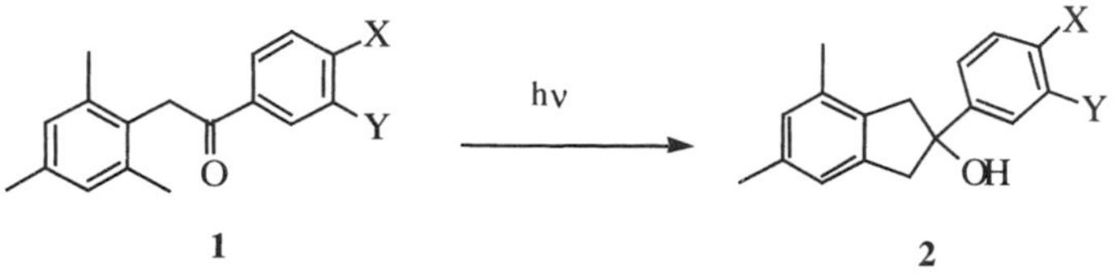

a: X=F; Y=H
b: X=CN; Y=H
c: X=COOH; Y=H
d: X=COOMe; Y=H
e: X=H; Y=F
f: X=H; Y=CN
g: X=H; Y=COOH
h: X=H; Y=COOMe
i: X=H; Y=Me

Experimental procedures:
5–10 mg samples were placed between two Pyrex microscope slides, and by sliding the top and bottom plates back and forth, the sample was distributed over the surface in a thin, even layer. The microscope plates were Scotchtaped together at the top and bottom ends, place in a polyethylene bag, and thoroughly degassed with nitrogen. The bag was then sealed under a positive pressure of nitrogen with a heat-sealing device and placed 5 cm from the immersion well. Workup of the solid-state photolysis samples consisted of washing the solid off the plate with ether into a suitable container and determining the conversion by gas chromatography.

References: E. Cheung, K. Rademacher, J.R. Scheffer, J. Trotter, *Tetrahedron,* **56**, 6739 (2000).

Type of reaction: C–C bond formation
Reaction condition: solid-state
Keywords: 1,1,3,3-tetramethyl diadamantyl-1,3-acetonedicarboxylate, decarbonylation, photoirradiation

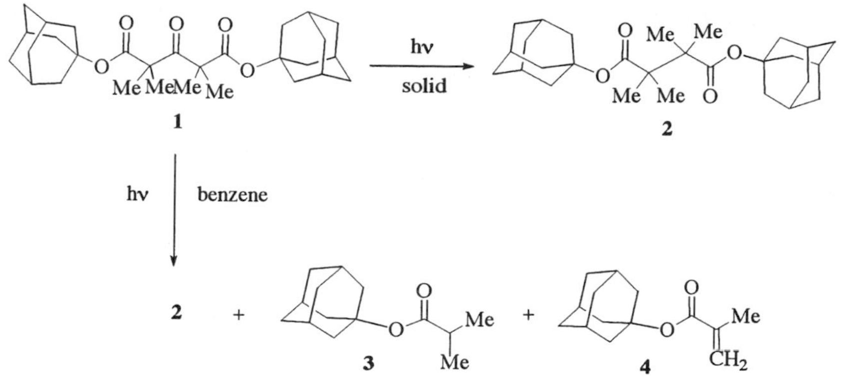

Experimental procedures:

Finely powdered solid samples of ketodiester **1** (5 mg) were placed evently between two microscope slides (which also acted as a $\lambda > 305$ nm filter) and placed at a similar distance from a medium-pressure Hg Hanovia lamp. After 4 h of irradiation, compound **2** was obtained in 87% yield.

References: Z. Yang, D. Ng, M. A. Carcia-Garibay, *J. Org. Chem.*, **66**, 4468 (2001); Z. Yang, M. A. Carcia-Garibay, *Org. Lett.*, **2**, 1963 (2000).

Type of reaction: C–C bond formation
Reaction condition: solid-state
Keywords: 1-phenylcyclopentyl ketone, Paterno-Buchi reaction, Norrish type I photoreaction, oxetane

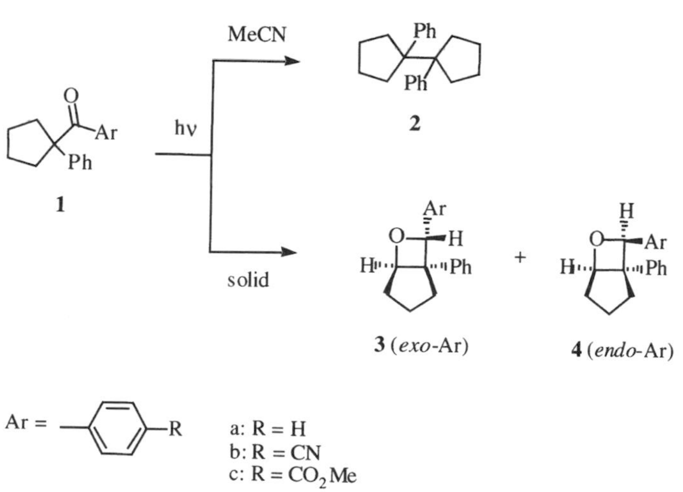

3 (*exo*-Ar)   4 (*endo*-Ar)

Ar = ⟨benzene⟩–R   a: R = H
b: R = CN
c: R = CO_2Me

Experimental procedures:
Solid-state photolysis was performed using a 450-W hanovia medium-pressure mercury lamp fitted with a Pyrex filter ($\lambda > 290$ nm). The crystalline **1a** was irradiated for 22.5 h at $-20\,°C$ to give a mixture of *exo*-isomer **3a** and *endo*-isomer **4a** in 63 and 12% yield, respectively.

References: T. Kang, J. R. Scheffer, *Org. Lett.*, **3**, 3361 (2001).

Type of reaction: C–C bond formation
Reaction condition: solid state
Keywords: *o*-alkyl aromatic aldehyde, photocyclization, benzocyclobutanol

a: X = Y = H
b: X = Br; Y = H
c: X = CN; Y = H
d: X = Y = Br
e: X = Y = CN
f: X = Br; Y = CN

Experimental procedures:
In a typical reaction, ca. 100 mg of the gently ground aldehyde was dispersed in a Pyrex container and purged with a stream of nitrogen gas for 15–10 min. The solid sample was subsequently irradiated, under a nitrogen gas atmosphere, in a Rayonet reactor fitted with 350 nm lamps for 24 h. After this period, the irradiated mixture was subjected to silica gel column chromatography. The combined fractions from the chromatography corresponding to the cyclobutanol were stripped off the solvent in vacuo at room temperature and characterized.

References: J. N. Moorthy, P. Mal, R. Natarajan, P. Venugopalan, *J. Org. Chem.*, **66**, 7013 (2001).

Type of reaction: C–C bond formation
Reaction condition: solid-state
Keywords: 9-anthryl-*N*-(naphthylcarbonyl)carboxamide, [4+4]photocycloaddition

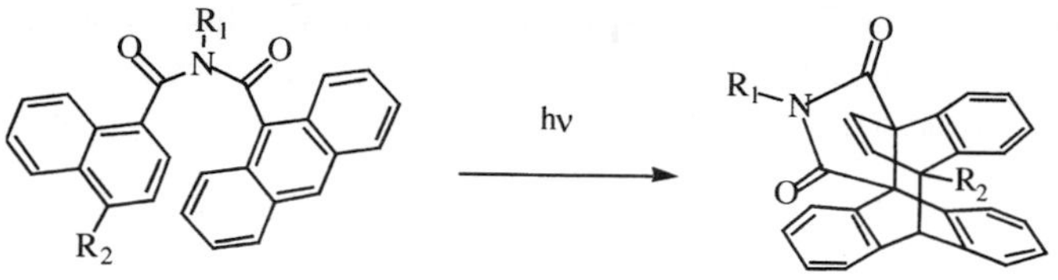

a: R_1=$C_6H_5CH_2$; R_2=H
b: R_1=o-$CH_3C_6H_4CH_2$; R_2=H
c: R_1=p-$CH_3C_6H_4CH_2$; R_2=H
d: R_1=o-$FC_6H_4CH_2$; R_2=H
e: R_1=o-$(CH_3)_3CC_6H_4$; R_2=H
f: R_1=n-C_3H_7; R_2=H
g: R_1=i-C_3H_7; R_2=CH_3

Experimental procedures:
Powdered single crystals of **1** (ca. 20 mg) were sandwiched between two Pyrex cover glasses and placed in a polyethylene bag, which was irradiated with a 400-W high-pressure mercury lamp for 6 h (3 h for each side of the sample) in an ice-water bath. Out of seven carboxamides examined, **1a**, **1e** and **1g** showed intramolecular photocycloaddition in solid state to give the [4+4] cycloadducts **2a**, **2e**, and **2g** in an almost quantitative yield after complete conversion, and **1f** gave **2f** in 9% yield.

References: S. Kohmoto, Y. Ono, H. Masu, K. Yamaguchi, K. Kishikawa, M. Yamamoto, *Org. Lett.*, **3**, 4153 (2001).

Type of reaction: C–C bond formation
Reaction condition: solid state
Keywords: sodium *trans*-2-butenoate, trimerization, γ-ray irradiation, tricarboxylic acid

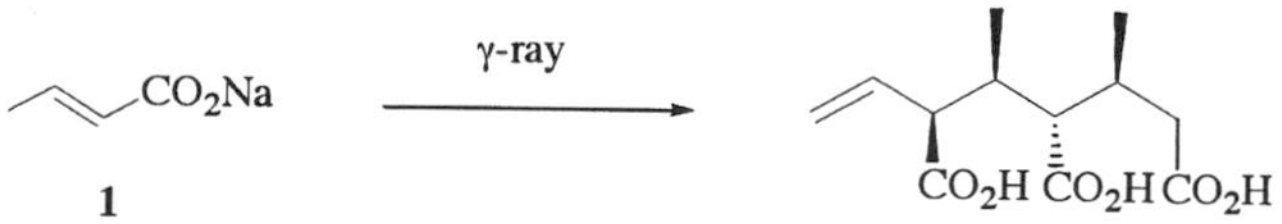

Experimental procedures:
1 (3.24 g, 29.9 mmol) was irradiated (15.6 Mrad) in a Gammacell 220 Irradiation Chamber (Atomic Energy of Canada, Ltd.), equipped with a ^{60}Co source (nominal activity as of June 1990, 0.049 Mrad h^{-1}). The resulting colorless powder

was dissolved in H_2O (20 mL) to give a homogeneous solution which was acidified with conc. HCl (3.5 mL). The colorless precipitate that formed was isolated by filtration and dried in vacuo (0.25 Torr) over anhydrous $CaSO_4$ to give **2** (colorless powder, 1.4 g, 55%).

References: G.C.D. Delgado, K.A. Wheeler, B.B. Snider, B.M. Foxman, *Angew. Chem. Int. Ed. Engl.*, **30**, 420 (1991).

Type of reaction: C–C bond formation
Reaction condition: solid-state
Keywords: ethyl 4-[2-(2-pyrazinyl)ethenyl]cinnamate, photopolymerization, cyclobutane

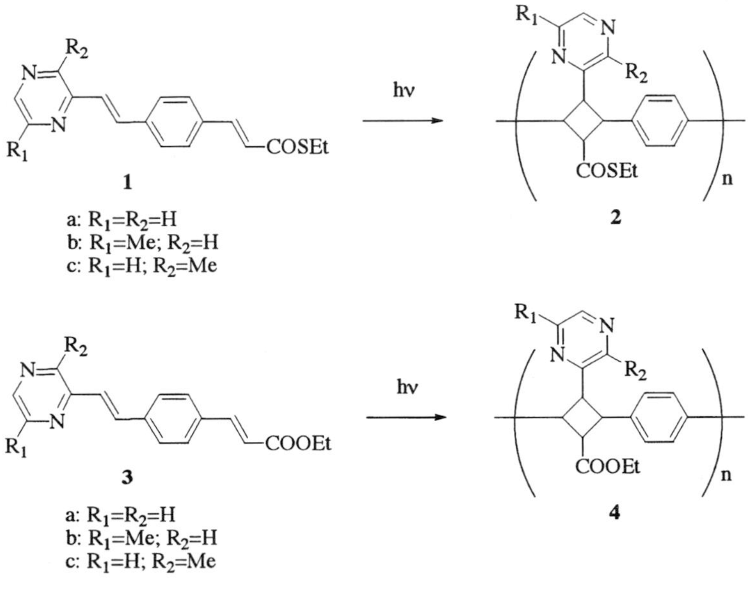
R2
N
N
R1
COSEt
1
a: R1=R2=H
b: R1=Me; R2=H
c: R1=H; R2=Me
hν
R1
N
N
R2
COSEt
n
2
R2
N
N
R1
COOEt
3
a: R1=R2=H
b: R1=Me; R2=H
c: R1=H; R2=Me
hν
R1
N
N
R2
COOEt
n
4

Experimental procedures:
Finely powdered crystals (100 mg) were dispersed in 300 mL of water containing a few drops of a surfactant (Nikkol TL-10FF) and irradiated with a 100-W high-pressure mercury lamp (Eikousha EHB WF-100), set inside of the flask, through a Pyrex glass filter. Vigorous stirring continued under a nitrogen atmosphere.

References: K. Siago, M. Sukegawa, Y. Maekawa, M. Hasegawa, *Bull. Chem. Soc. Jpn.*, **68**, 2355 (1995).

Type of reaction: C–C bond formation
Reaction condition: solid-state
Keywords: (Z,Z)-2,4-hexadienedicarboxylate, topochemical 1,4-polymerization, photoirradiation

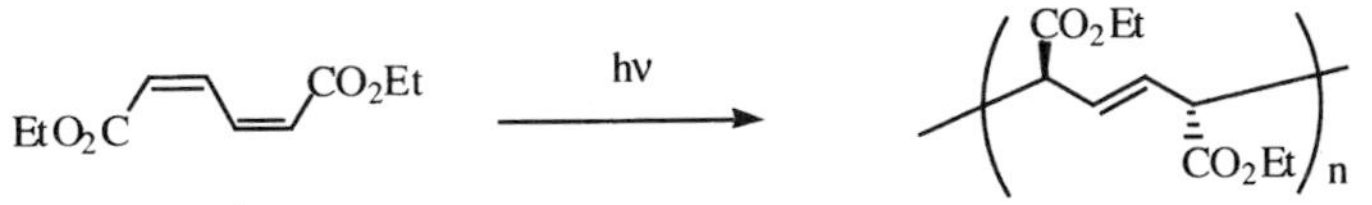

Experimental procedures:
Photopolymerization of the crystals was carried out in a sealed Pyrex ampule under irradiation of UV light using a high-pressure mercury lamp (Toshiba SHL-100-2, 100 W) at a distance of 10 cm. A typical polymerization procedure is described below. Monomer diethyl *cis,cis*-muconate **1** (200 mg, 1.0 mmol) was placed in an ample, which was then evacuated on a vacuum line. After irradiation, polymer was isolated by removal of the unreacted monomer with chloroform (20 mL) for 5–10 h at room temperature. Photopolymerization was also carried out by direct exposure to sunlight.

References: A. Matsumoto, K. Yokoi, S. Aoki, K. Tashiro, T. Kamae, M. Kobayashi, *Macromolecules*, **31**, 2129 (1998).

Type of reaction: C–C bond formation
Reaction condition: solid-state
Keywords: diacetylene, tetrathiafulvalene, γ-ray irradiation, polydiacetylene

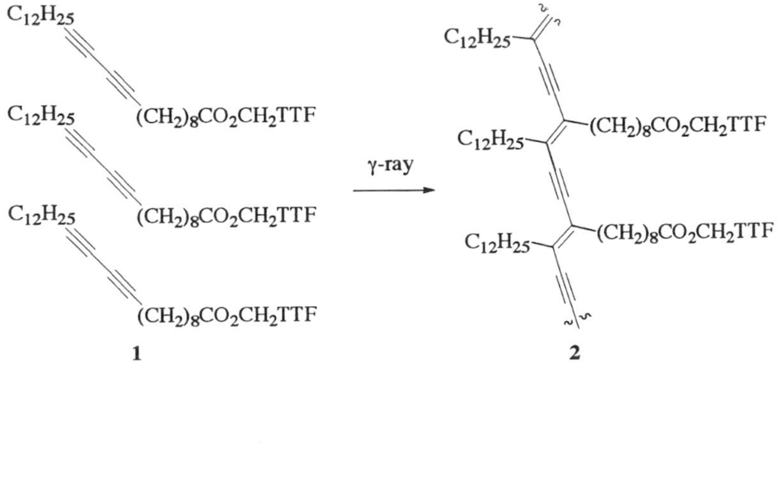

Experimental procedures:
The color of the burk crystals of monomer **1** changed to dark blue from light yellow after exposure to UV with a low-pressure mercury lamp (4 W) or ^{60}Co γ-ray irradiation.

References: S. Shimada, A. Masaki, K. Hayamizu, H. Matsuda, S. Okada, H. Nakanishi, *Chem. Commun.,* 1421 (1997).

Type of reaction: C–C bond formation
Reaction condition: solid-state
Keywords: 1,6-bis(2,5-dimethoxyphenyl)hexa-2,4-diyne, photopolymerization, diacetylene, polydiacetylene

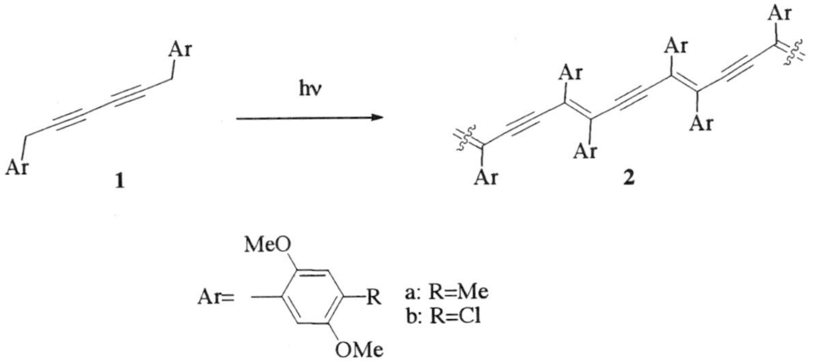

Experimental procedures:
A sample of 10.2 mg (24.4 mmol) of powdered **1b** was irradiated with UV light for 6 h. The resulting violet solid was then extracted with dichloromethane for 4 h to remove unreacted monomer. Drying in vacuo to constant mass gave 1.8 mg (18%) of **2b** as a light-violet solid, mp >320 °C.

References: H. Irngartiger, M. Skipinski, *Eur. J. Org. Chem.,* **917** (1999); H. Irngartiger, M. Skipinski, *Tetrahedron,* **56**, 6781 (2000).

Type of reaction: C–C bond formation
Reaction condition: solid-state
Keywords: 5-(2-methylthio-4-methylpyrimidin-5-yl)penta-2,4-diyn-1-ol, polymerization

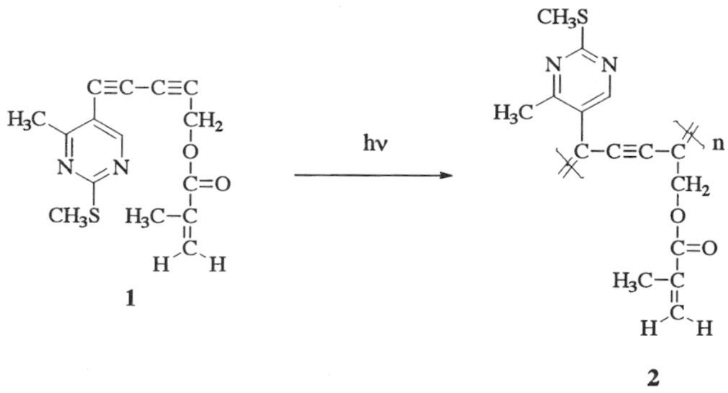

Experimental procedures:

The monomer **1** was polymerized by heating the crystals in a vacuum vessel below the melting point or by γ-ray or UV irradiation of the crystals at room temperature. ^{60}Co γ-ray irradiation with a dose rate of 0.1 Mrad h^{-1} or a high-pressure mercury lamp (200 W) without filter was used as the radiation sources for the polymerization; the conversion ratio was determined by extraction of residual monomer with ethanol. A comparison of the polymerization rates indicates that ^{60}C γ-ray irradiation is much more efficient than UV irradiation in inducing polymerization.

References: J.-H. Wang, Y.-Q. Shen, C.-X. Yu, J.-H. Si, *J. Chem. Soc., Perkin Trans. 1*, 1455 (2000).

Type of reaction: C–C bond formation
Reaction condition: solid-state
Keywords: chalcone, dibenzylidene acetone, inclusion crystal, [2+2]photo-cycloaddition, cyclobutane

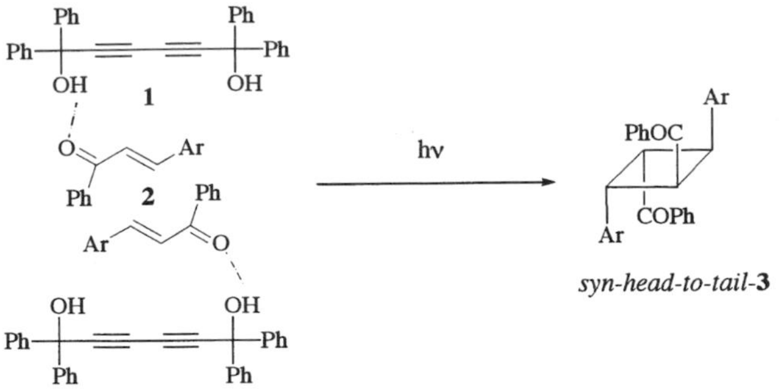

a: Ar=Ph
b: Ar=2-MeC$_6$H$_4$
c: Ar=2-MeOC$_6$H$_4$
d: Ar=1-Naphthyl
e: Ar=2-ClC$_6$H$_4$
f: Ar=2-Furyl
g: Ar=2-Thienyl

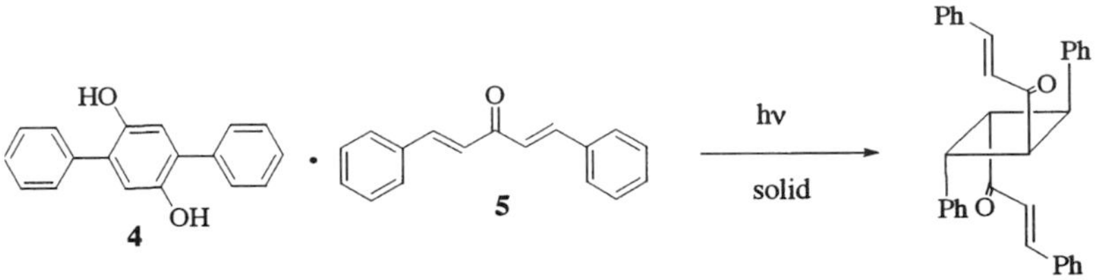

syn-head-to-tail-**3**

Experimental procedures:

Irradiation of powdered 1:2 inclusion complex of **1** and chalcone **2** by a high-pressure Hg-lamp at room temperature for 1 h gave a single photoaddition product *syn-head-to-tail* dimer **3a** in 82% yield.

References: K. Tanaka, F. Toda, *Chem. Commun.*, 593 (1983); M. Kaftory, K. Tanaka, F. Toda, *J. Org. Chem.*, **50**, 2154 (1985).

Type of reaction: C–C bond formation
Reaction condition: solid-state
Keywords: 2-pyridone, inclusion complex, [4+4]photocycloaddition

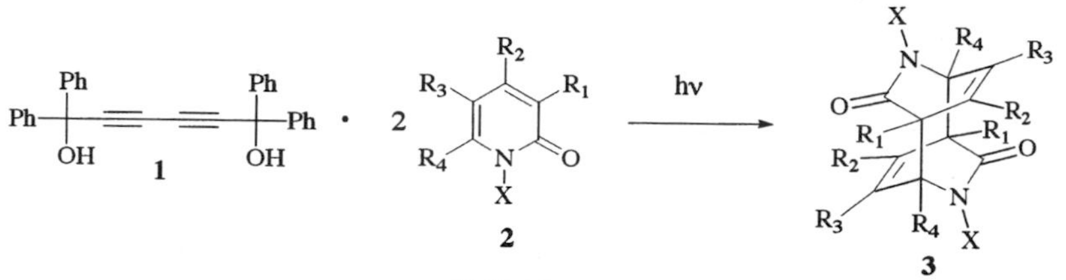

X=H, Me

a: $R_1=R_2=R_3=R_4=H$
b: $R_1=Me$; $R_2=R_3=R_4=H$
c: $R_1=H$; $R_2=Me$; $R_3=R_4=H$
d: $R_1=R_2=H$; $R_3=Me$; $R_4=H$
e: $R_1=R_2=R_3=H$; $R_4=Me$
f: $R_1=Me$; $R_2=H$; $R_3=Me$; $R_4=H$
g: $R_1=H$; $R_2=Me$; $R_3=H$; $R_4=Me$
h: $R_1=R_2=H$; $R_3=R_4=Me$
i: $R_1=H$; $R_2=R_3=R_4=Me$
j: $R_1=R_2=R_3=H$; $R_4=OMe$

Experimental procedures:

A powdered 1 : 2 inclusion complex of **1** and **2** was irradiated by 400-W high-pressure Hg-lamp for 6 h at room temperature. The reaction mixture was recrystallized from MeOH to give [4+4] dimer **3**.

References: K. Tanaka, F. Toda, *Nippon Kagakukaishi*, 141 (1984); M. Kuzuya, A. Noguchi, N. Yokota, T. Okuda, F. Toda, K. Tanaka, *Nippon Kagakukaishi*, 1746 (1986).

Type of reaction: C–C bond formation
Reaction condition: solid-state
Keywords: deoxycholic acid, acetophenone, host-guest complex, photoirradiation

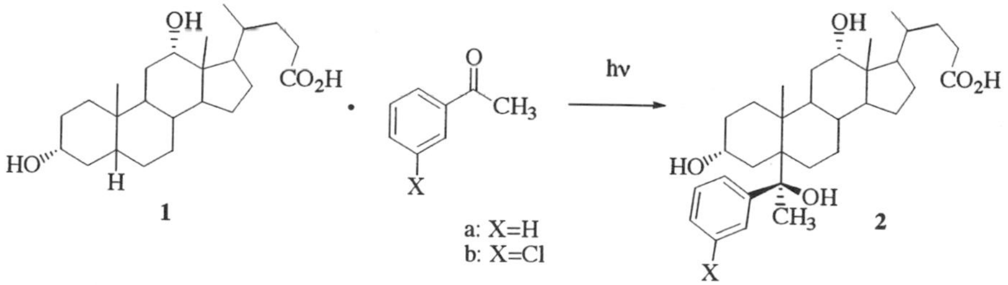

a: X=H
b: X=Cl

Experimental procedures:

In a typical experiment, 1–2 g of complex was irradiated at room temperature through Pyrex dishes, $\lambda > 300$ nm, for about 30 days. The crystals were generally in the form of powder. Single crystals preserved their integrity during irradiation. The products were separated by chromatography on silica gel 1:100 (eluted with $CH_2Cl_2/CH_3OH/AcOH$ in a ratio of 90.5:5.0:0.5) and by preparative TLC with the same eluent in a ratio of 90.5:9.0:0.5 by using UV detection and phosphomolybdic acid as coloring spray.

References: C.P. Tang, H.C. Chang, R. Popovitz-Biro, F. Frolow, M. Lahav, L. Leiserowitz, R.K. McMullan, *J. Am. Chem. Soc.*, **107**, 4058 (1985).

Type of reaction: C–C bond formation
Reaction condition: solid-state
Keywords: tropolone alkyl ether, inclusion crystal, [2+24]photocycloaddition

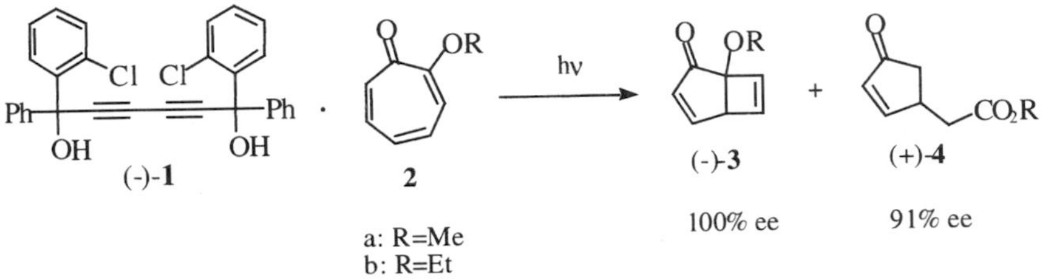

a: R=Me
b: R=Et

Experimental procedures:
Irradiation of powdered 1:1 inclusion complex of (–)-**1** and **2a** by a high-pressure Hg-lamp at room temperature for 72 h (50% conversion) gave (–)-**3a** (11% yield, $[a]_D$–168° *c* 0.2, CH_3OH, 100% ee) and (+)-**4a** (26% yield, $[a]_D$–89.5° (*c* 0.2, CH_3OH, 91% ee).

References: F. Toda, K. Tanaka, *Chem. Commun.*, 1429 (1986).

Type of reaction: C–C bond formation
Reaction condition: solid-state
Keywords: α-oxoamide, inclusion crystal, Norrish type II photoreaction, β-lactam

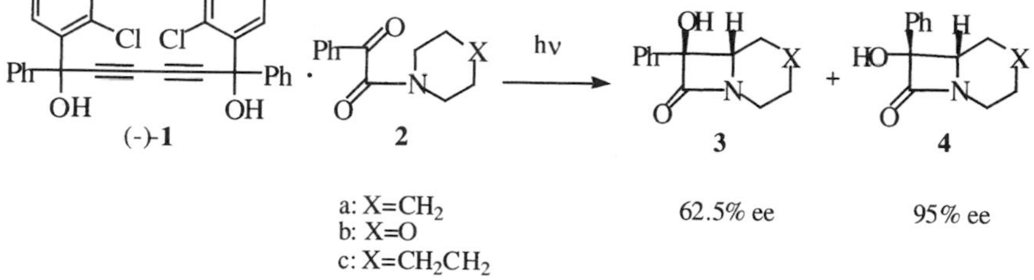

a: X=CH$_2$
b: X=O
c: X=CH$_2$CH$_2$

62.5% ee   95% ee

Experimental procedures:

Photoirradiation of a 1:1 inclusion complex of (–)-**1** and **2a** by high-pressure Hg-lamp for 100 h at room temperature gave optically active β-lactams (–)-**3a** of 62.5% ee and (–)-**4a** of 95% ee in 38 and 29% yields, respectively.

References: F. Toda, K. Tanaka, M. Yagi, *Tetrahedron*, **43**, 1495 (1987).

Type of reaction: C–C bond formation
Reaction condition: solid-state
Keywords: α-oxoamide, inclusion complex, Norrish type II photoreaction, β-lactam

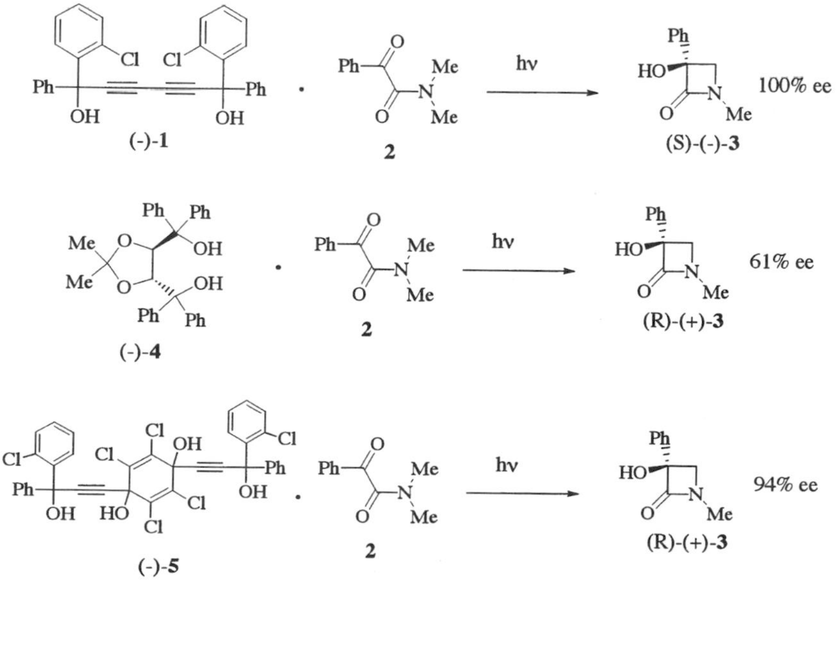

Experimental procedures:
The powdered 1:1 complex (0.85 g) of (–)-**1** and **2** was irradiated by a 400-W high-pressure Hg lamp at room temperature, for 27 h with occasional grinding with a pestle and mortar. The reaction mixture was chromatographed on silica gel with benzene-ethyl acetate as solvent to give (–)-**1** (1.0 g, 100%) and 100% ee (–)-**3**, $[a]_D$–99.7 $^\circ$ (*c* 0.34, CHCl$_3$) as colorless plates, mp 123–124 $^\circ$C.

References: M. Kaftory, M. Yagi, K. Tanaka, F. Toda, *J. Org. Chem.*, **53**, 4391 (1988); F. Toda, H. Miyamoto, K. Kanemoto, *Chem. Commun.,* 1719 (1995); F. Toda, H. Miyamoto, M. Inoue, S. Yasaka, I. Matijasic, *J. Org. Chem.*, **65**, 2728 (2000).

Type of reaction: C–C bond formation
Reaction condition: solid-state
Keywords: 2-pyridone, inclusion crystal, [2+2]photodimerization, β-lactam

a: R=H
b: R=Me
c: R=OMe
d: R=OEt

Experimental procedures:
A powdered 1:1 inclusion complex of (–)-**1** and 4-methoxy-1-methylpyridone **2c** was irradiated by 100-W high-pressure Hg-lamp at room temperature in the solid state. The crude reaction product was chromatographed on silica gel using CHCl$_3$ as a solvent to give a mixture of host compound and **3**, from which **3** was isolated by distillation as an oil ($[a]_D$–123° (*c* 0.026, CHCl$_3$), 100% ee).

References: F. Toda, K. Tanaka, *Tetrahedron Lett.,* **29**, 4299 (1988).

Type of reaction: C–C bond formation
Reaction condition: solid-state
Keywords: cycloocta-2,4,6-trien-1-one, cycloocta-2,4-dien-1-one, inclusion complex, [2+2]photocycloaddition, cyclobutane

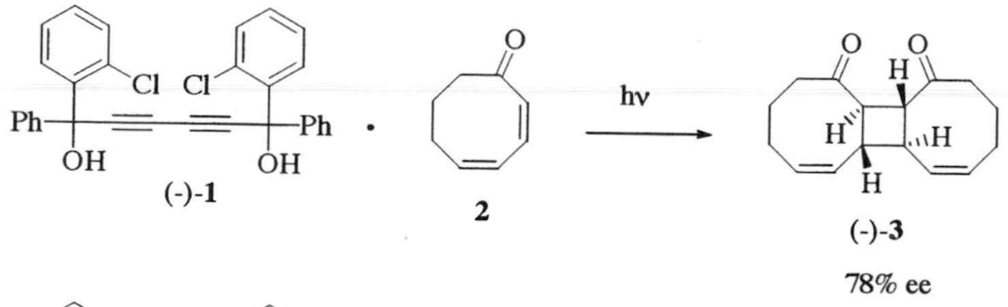

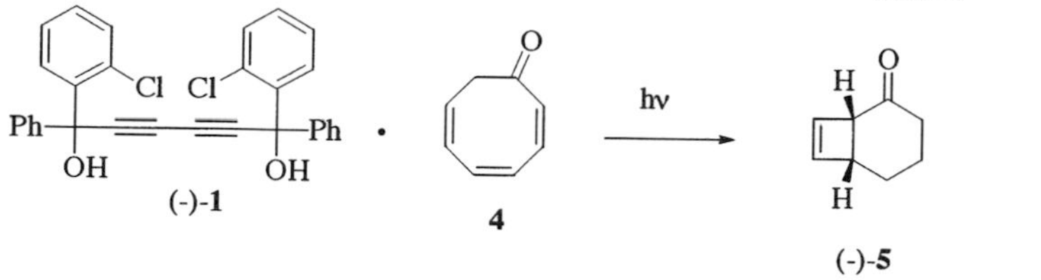

(-)-**3**

78% ee

(-)-**5**

Experimental procedures:

Irradiation of finely powdered 3:2 complex of (−)-**1** and **2** (2 g) in the solid state for 48 h gave a crude reaction product. Purification of the crude product by column chromatography on silica gel using CHCl$_3$ as solvent gave (−)-**3**, which on distillation gave pure (−)-**3** of 78% ee (0.16 g, 55%, bp 200 °C/1 mmHg, [a]$_D$− 34.7° (c 0.11, CHCl$_3$)).

Irradiation of finely powdered 1:2 complex of (−)-**1** and **4** (1.1 g) in the solid state for 168 h gave a crude reaction product. Purification of the crude product by column chromatography on silica gel using CHCl$_3$ as solvent gave (−)-**5**, which on distillation gave pure (−)-**5** (0.052 g, 14%, bp 200 °C/1 mmHg, [a]$_D$− 62.9° (c 0.12, CHCl$_3$)).

References: F. Toda, K. Tanaka, M. Oda, *Tetrahedron*, **29**, 653 (1988); T. Fujiwara, N. Nanba, K. Hamada, F. Toda, K. Tanaka, *J. Org. Chem.*, **55**, 4532 (1990).

Type of reaction: C–C bond formation
Reaction condition: solid-state
Keywords: β-cyclodextrin, benzaldehyde, asymmetric synthesis, photoirradiation, benzoin, 4-benzoylbenzaldehyde

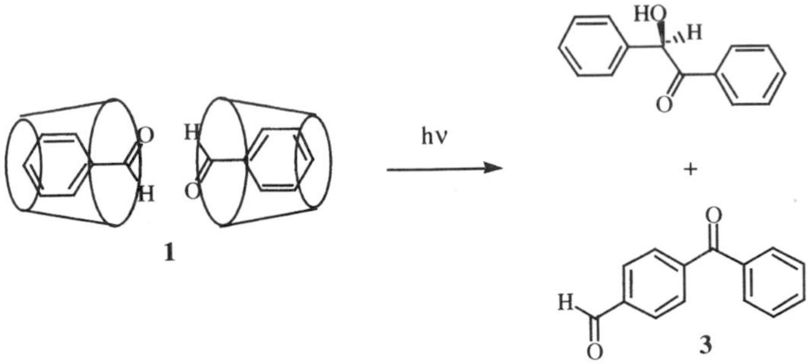

(R)-(-)-2

15% ee

Experimental procedures:
Photolyses of the solid cyclodextrin complexes **1** were carried out with a Hanovia 450-W medium-pressure Hg lamp for 3 h at room temperature in a quartz vessel under vacuum. The photolysis vessel was tumbled continuously during the irradiation to ensure homogeneous photolysis of the sample. Conversions were limited to less than 20%. After photolysis, the solid complexes were dissolved in excess water and extracted with diethyl ether and chromatographed with hexane-ethyl acetate (5:1) to isolate the products in pure form. Irradiation of solid *β*-cyclodextrin complexes of benzaldehyde resulted in an intramolecular reaction to give benzoin (*R*)-(–)-**2** and 4-benzoylbenzaldehyde **3** (7:3, 80%).

References: V.P. Rao, N.J. Turro, *Tetrahedron Lett.*, **30**, 4641 (1989).

Type of reaction: C–C bond formation
Reaction condition: solid-state
Keywords: *N*-methyl cyclohex-1-enylanilide, inclusion crystal, [2+2]potocycloaddition, 3,4-dihydroquinolin-2(1*H*)-one

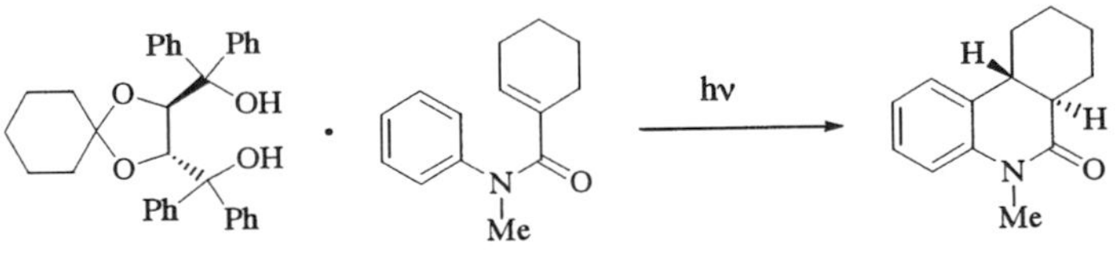

(-)-**1** **2** (-)-**3** 98% ee

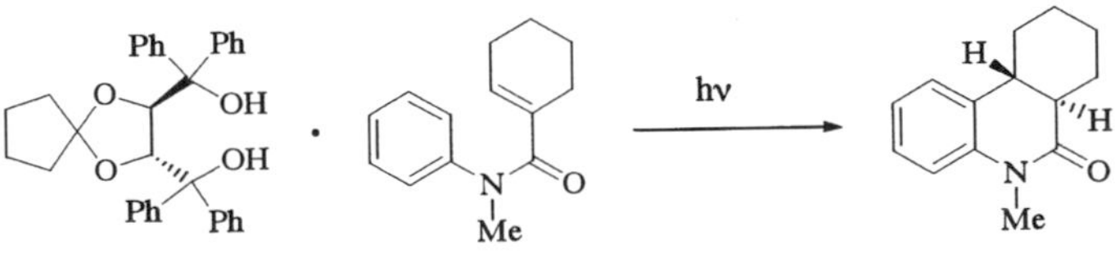

(-)-**4** **2** (+)-**3** 70% ee

Experimental procedures:
Photoirradiation of a suspension of powdered 1:1 complex of (–)-**1** and **2** (3.2 g) in water (150 mL) containing a small amount of sodium alkyl sulfate as a surfactant was carried out for 50 h using a 100-W high-pressure Hg lamp under stirring. The reaction product was filtered, air dried, and chromatographed on silica gel using CH_2Cl_2 as an eluent to give (–)-*trans*-**3** (0.67 g, 70% yield, mp 123–125 °C, $[\alpha]_D$–177° (*c* 0.5 CH_3OH), 98% ee).

References: K. Tanaka, O. Kakinoki, F. Toda, *Chem. Commun.*, 1053 (1992); S. Ohba, H. Hosomi, K. Tanaka, H. Miyamoto, F. Toda, *Bull. Chem. Soc. Jpn.*, **73**, 2075 (2000); H. Hosomi, S. Ohba, K. Tanaka, F. Toda, *J. Am. Chem. Soc.*, **122**, 1818 (2000).

Type of reaction: C–C bond formation
Reaction condition: solid-state
Keywords: coumarin, inclusion crystal, [2+2]photodimerization, cyclobutane

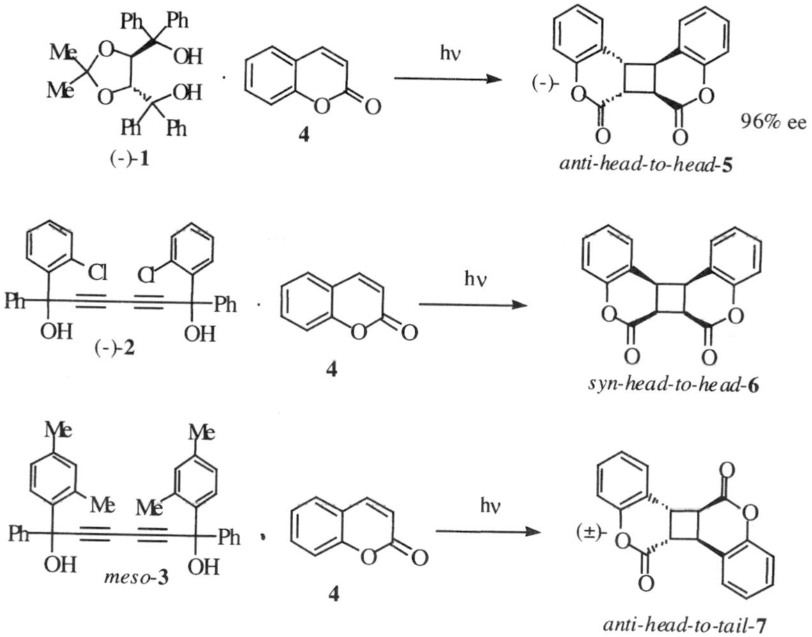

Experimental procedures:
Recrystallization of (–)-**1** (10.0 g, 21.5 mmol) and **4** (3.2 g, 21.9 mmol) from AcOEt (20 mL)-hexane (100 mL) gave the 1:1 inclusion crystal as colorless needles (5.7 g, 43%, mp 95–98 °C). Irradiation of the powdered 1:1 complex (1 g)

in water (100 mL) containing surfactant for 3 h gave a 2 : 1 complex of (–)-**1** and (–)-**5** of 96% ee as colorless needles (0.96 g, 96%, mp 228–232 °C).

References: K. Tanaka, F. Toda, *J. Chem. Soc., Perkin Trans. 1*, 943 (1992); K. Tanaka, F. Toda, E. Mochizuki, N. Yasui, Y. Kai, I. Miyahara, K. Hirotsu, *Angew. Chem. Int. Ed. Engl.*, **38**, 3523 (1999); K. Tanaka, F. Toda, E. Mochizuki, N. Yasui, Y. Kai, I. Miyahara, K. Hirotsu, *Tetrahedron*, **56**, 6853 (2000).

Type of reaction: C–C bond formation
Reaction condition: solid-state
Keywords: thiocoumarin, inclusion crystal, [2+2]photodimerization, cyclobutane

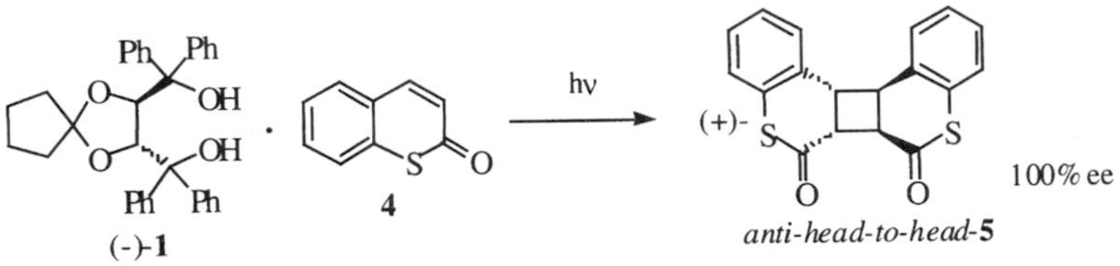

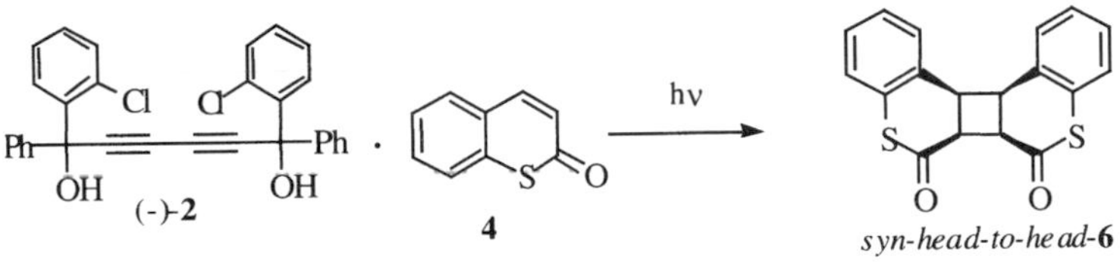

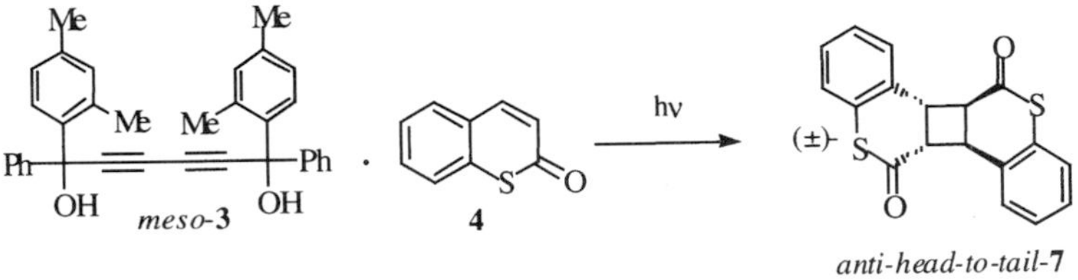

Experimental procedures:
When a mixture of thiocoumarin **4** and optically active host compound (*R,R*)-(–)-**1** in butyl ether was kept at room temperature for 12 h, a 1 : 1 inclusion complex (mp 106–108 °C) was obtained as colorless needles. The 1 : 1 complex gave *anti-head-to-head* dimer (+)-**5** (mp 254–255 °C) of 100% ee ([a]$_D$+182 °, *c* 0.02 in CHCl$_3$) in 73% yield upon photoirradiation in the solid state. The optical purity of **5** was determined by HPLC on the chiral stational phase Chiralpak AS with using hexane-EtOH (95 : 5) as an eluent.

References: K. Tanaka, F. Toda, *Mol. Cryst. Liq. Cryst.*, **313**, 179 (1998); K. Tanaka, F. Toda, E. Mochizuki, N. Yasui, Y. Kai, I. Miyahara, K. Hirotsu, *Tetrahedron,* **56**, 6853 (2000).

Type of reaction: C–C bond formation
Reaction condition: solid-state
Keywords: cyclohex-2-enone, inclusion crystal, [2+2]photodimerization, cyclo-butane

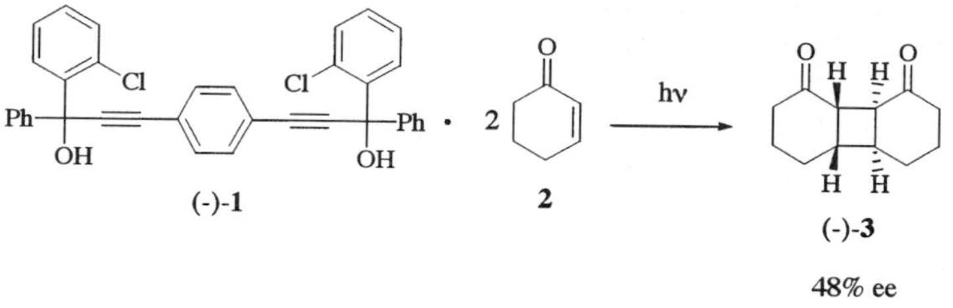

(-)-1   2   hv   (-)-3

48% ee

Experimental procedures:
A solution of (−)-**1** (5.0 g, 8.94 mmol) and **2** (1.72 g, 17.9 mmol) in ether-hexane (1:1, 10 mL) was kept at room temperature for 6 h to give a 1:2 complex of (−)-**1** and **2** as colorless prisms (6.1 g, 91%), mp 90–95 °C. A suspension of the powdered complex (4.2 g) in water (100 mL) containing a small amount of sodium alkylsulfate as surfactant was irradiated at room temperature for 24 h. The reaction mixture was filtered, dried, and distilled in vacuo to give (−)-**3** in 48% ee as an oil (0.8 g, 74.8%, [a]$_D$−61.0° (c 0.4, MeOH)).

References: K. Tanaka, O. Kakinoki, F. Toda, *J. Chem. Soc., Perkin Trans. 1*, 307 (1992); K. Tanaka, H. Mizutani, I. Miyahara, K. Hirotsu, F. Toda, *Cryst. Eng. Comm.*, 3 (1999).

Type of reaction: C–C bond formation
Reaction condition: solid-state
Keywords: *N*-(aryloylmethyl)-δ-valerolactam, inclusion complex, photocycliza-tion, azetidine

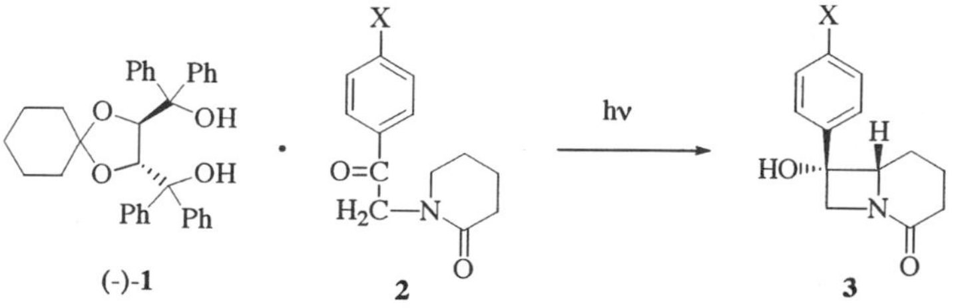

(-)-1   2   hv   3

a: X=H
b: X=Cl
c: X=Br
d: X=Me

98% ee

Experimental procedures:
A suspension of the powdered 1:1 inclusion compound of (−)-**1** and **2a** (3.21 g) in water (120 mL) containing sodium alkylsulfate as a surfactant was irradiated for 12 h while being stirred at room temperature. The reaction mixture was filtered to give crude crystals and an aqueous solution. The crude crystals were purified by column chromatography on silica gel followed by recrystallization from AcOEt to give (+)-**3a** of 98% ee (0.42 g, 59% yield) and recovered (−)-**1** (0.24 g, 25% yield).

References: F. Toda, K. Tanaka, O. Kakinoki, T. Kawakami, *J. Org. Chem.*, **58**, 3783 (1993); D. Hashizume, Y. Ohashi, K. Tanaka, F. Toda, *Bull. Chem. Soc. Jpn.*, **67**, 2383 (1994).

Type of reaction: C–C bond formation
Reaction condition: solid-state
Keywords: 2-[*N*-(2-propenyl)amino]cyclohex-2-enone, inclusion complex, photocyclization

(−)-**1**   **2**   (+)-**3**
                                99% ee

a: R=Me   a: X=H
b: R,R=-(CH$_2$)$_4$-   b: X=*m*-Cl
c: R,R=-(CH$_2$)$_5$-   c: X=*p*-Me

Experimental procedures:
A suspension of powdered 2:1 inclusion complex of (−)-**1a** and **2a** (1.2 g, 0.97 mmol) in water (120 mL) containing alkylsulfate (0.1 g) as a surfactant was irradiated under stirring for 17 h. The reaction product was filtered, dried and chromatographed on silica gel using AcOEt-hexane (1:1) as an eluent to give (+)-**3a** (0.16 g, 0.62 mmol, 64% yield, [a]$_D$+70° (*c* 1.0, CCl$_4$), mp 106–107 °C).

References: F. Toda, H. Miyamoto, K. Takeda, R. Matsugawa, N. Maruyama, *J. Org. Chem.*, **58**, 6208 (1993).

Type of reaction: C–C bond formation

Reaction condition: solid-state

Keywords: 3-oxo-2-cyclohexenecarboxamide, inclusion complex, photocyclization, spiro compound

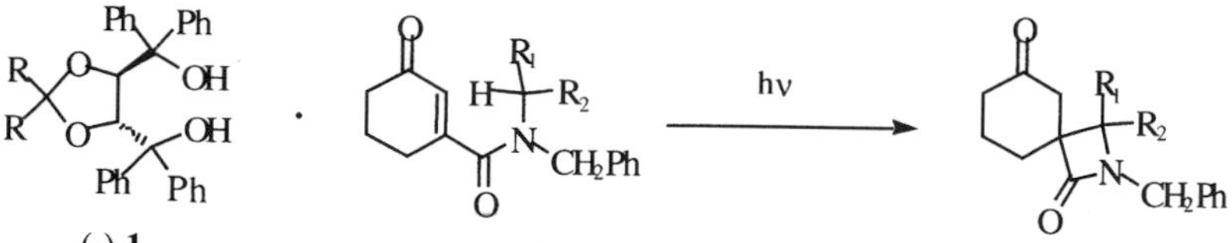

(-)-1

2

(-)-3

99.9% ee

a: R=Me
b: R,R=-(CH₂)₄-
c: R,R=-(CH₂)₅-

a: R₁=H; R₂=H
b: R₁=H; R₂=Me
c: R₁=H; R₂=Et
d: R₁=Me; R₂=Me
e: R₁=H; R₂=Ph
f: R₁=H; R₂=n-Pr

Experimental procedures:

A suspension of powdered 2:1 inclusion complex of (–)-**1b** and **2b** (6.72 g, 5.29 mmol) in water (120 mL) containing alkylsulfate (0.1 g) as a surfactant was irradiated under stirring for 4 h. The reaction product was filtered, dried and chromatographed on silica gel using AcOEt-hexane (1:1) as an eluent to give (–)-**3b** (1.18 g, 87% yield).

References: F. Toda, H. Miyamoto, K. Takeda, R. Matsugawa, N. Maruyama, *J. Org. Chem.*, **58**, 6208 (1993).

Type of reaction: C–C bond formation

Reaction condition: solid-state

Keywords: 4-(3-butenyl)cyclohexa-2,5-dien-1-one, inclusion complex, photocyclization

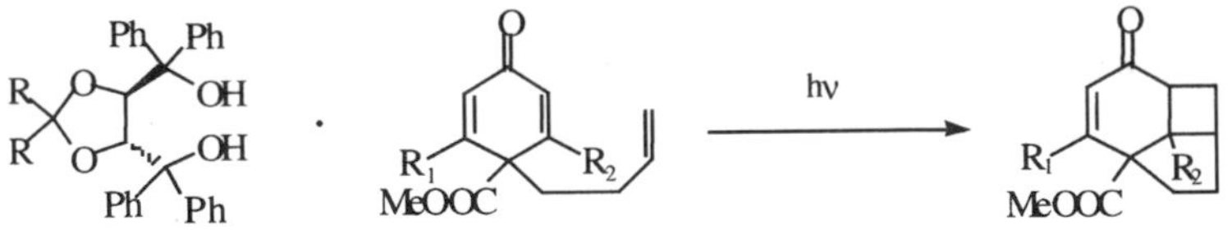

(-)-1

a: R=Me
b: R,R=-(CH₂)₄-
c: R,R=-(CH₂)₅-

2

a: R₁=R₂=OMe
b: R₁=OMe; R₂=H

(-)-3

73% ee
a: R₁=R₂=OMe
b: R₁=OMe; R₂=H

Experimental procedures:
A suspension of powdered 2:1 inclusion complex of (–)-**1a** and **2a** (0.36 g) in water (100 mL) containing alkylsulfate (0.1 g) as a surfactant was irradiated under stirring for 5 h. The reaction product was filtered, dried and chromatographed on silica gel using AcOEt-hexane (1:1) as an eluent to give (+)-**3b** of 73% ee (0.04 g, 50% yield, $[a]_D$+5.7° CH₃OH, mp 110–113 °C).

References: F. Toda, H. Miyamoto, K. Takeda, R. Matsugawa, N. Maruyama, *J. Org. Chem.*, **58**, 6208 (1993).

Type of reaction: C–C bond formation
Reaction condition: solid-state
Keywords: 2-arylthio-3-methylcyclohex-2-en-1-one, inclusion complex, photocyclization, thiolane
Experimental procedures:

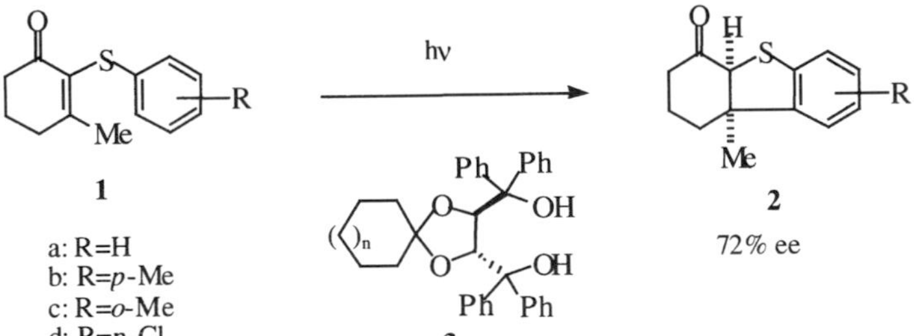

1

a: R=H
b: R=*p*-Me
c: R=*o*-Me
d: R=*p*-Cl
e: R=*o*-Cl
f: R=*p*-Br
g: R=*o*-Br

3

a: n=0
b: n=1

2

72% ee

When a benzene-hexane solution of (–)-**3b** (1.09 g, 2.15 mmol) and **1c** (0.5 g, 2.15 mmol) was left at room temperature for 24 h, a 1:1 inclusion compound of (–)-**3b** and **1c** was obtained as colorless crystals (1.51 g, 95%). Irradiation of the inclusion compound (1.22 g) with stirring for 30 h in water (200 mL) containing hexadecyltrimethylammonium bromide (0.1 g) gave a crude reaction product. This was filtered, air dried and distilled at 250 °C (2 mmHg) to give (+)-**2c** of 72% ee (0.33 g, 86% yield).

References: F. Toda, H. Miyamoto, S. Kikuchi, R. Kuroda, F. Nagami, *J. Am. Chem. Soc.,* **118**, 11315 (1996).

Type of reaction: C–C bond formation
Reaction condition: solid-state
Keywords: *N*-alkylfuran-2-carboxanilide, inclusion complex, photocyclization, lactam

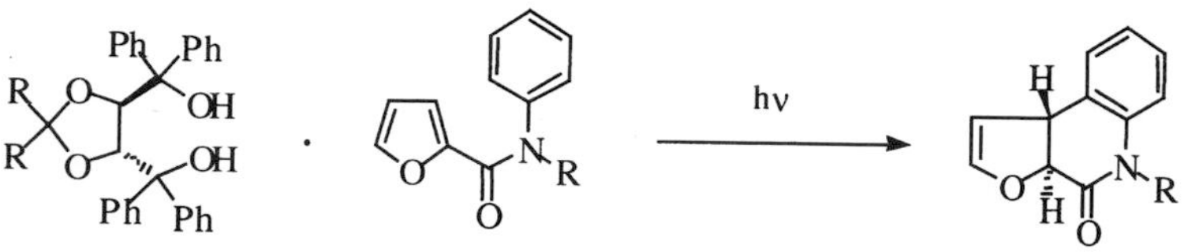

(-)-**1**   **2**   (-)-**3**

96% ee

a: R=Me
b: R,R=-(CH₂)₄-
c: R,R=-(CH₂)₅-

a: R=H
b: R=Me
c: R=CH₂CH=CH₂

Experimental procedures:
Photoirradiation of a suspension of powdered 1:1 complex of (–)-**1b** and **2c** (1.4 g) in water (80 mL) containing hexadecyltrimethylammonium bromide (0.2 g) as a surfactant was carried out at room temperature for 77 h under stirring. The reaction product was filtered, air-dried, and chromatographed on silica gel using toluene-AcOEt (10:1) as an eluent to afford (–)-*trans*-photocyclization product **3c** of 96% ee (0.22 g, 50% yield, mp 102–104 °C, $[a]_D$–289° (*c* 0.5, MeOH).

References: F. Toda, H. Miyamoto, K. Kanemoto, *J. Org. Chem.,* **61**, 6490 (1996).

Type of reaction: C–C bond formation
Reaction condition: solid-state
Keywords: *N*-phenyl enaminone, inclusion complex, photocyclization, dihydroindole derivative

(-)-1   2   3

97% ee

a: R =Me   a: R₁=H; R₂=Me
b: R,R=-(CH₂)₄-   b: R₁=Me; R₂=H
c: R,R=-(CH₂)₅-   c: R₁=Me; R₂=Me

Experimental procedures:

A suspension of powdered 1:1 inclusion complex of (–)-**1a** and **2a** (3.07 g, 4.41 mmol) in water (120 mL) containing hexadecyltrimethylammonium bromide (0.1 g) as a surfactant was irradiated under stirring for 16 h with 100-W high-pressure Hg-lamp. The reaction mixture was filtered, dried and chromatographed on silica gel using AcOEt-toluene (1:9) as an eluent to give (+)-**3a** of 97% ee after distillation at 180 °C/2 mmHg as colorless oil.

References: F. Toda, H. Miyamoto, T. Tamashima, M. Kondo, Y. Ohashi, *J. Org. Chem.*, **64**, 2690 (1999).

Type of reaction: C–C bond formation
Reaction condition: solid-state
Keywords: cyclohexenone, inclusion complex, photocyclization, β-lactam, spiro compound

(-)-1   2   3

97% ee

a: R =Me   a: R₁=PhCH₂; R₂=Ph; R₃=H
b: R,R=-(CH₂)₄-   b: R₁=PhCH₂; R₂=Me; R₃=H
c: R,R=-(CH₂)₅-   c: R₁=PhCH₂; R₂=H; R₃=H
                            d: R₁=i-Pr, R₂,R₃=-(CH₂)₅-

Experimental procedures:

A suspension of a powdered 1:1 inclusion compound of (–)-**1c** and **2a** (2.6 g, 3.2 mmol) in water (120 mL) containing hexadecyltrimethylammonium bromide (0.04 g) as a surfactant was irradiated with stirring for 8 h. The reaction product was filtered, dried, and chromatographed on silica gel using AcOEt-hexane (1:4) as the eluent to give (–)-**3a** in 97% ee as a colorless oil (0.27 g, 26% yield). $[a]_D$–48°(c 0.6, MeOH).

References: F. Toda, H. Miyamoto, M. Inoue, S. Yasaka, I. Matijasic, *J. Org. Chem.*, **65**, 2728 (2000).

Type of reaction: C–C bond formation
Reaction condition: solid-state
Keywords: 4-isopropyltropolone methyl ester, inclusion crystal, [2+2]photocycloaddition, bicyclo[3.2.0]hepta-3,6-dien-2-one, methyl-4-oxocyclopent-2-ene-1-acetate

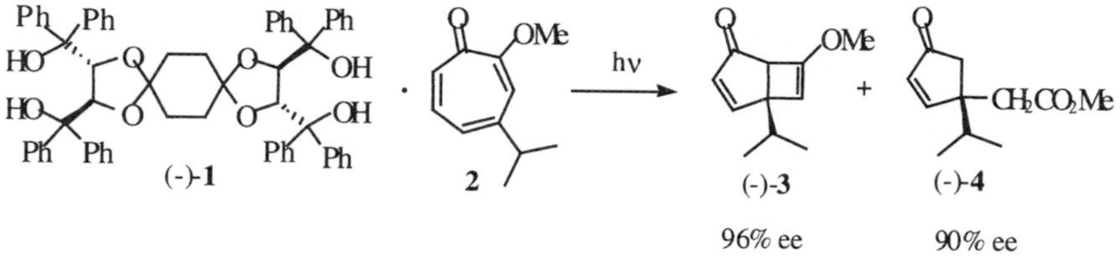

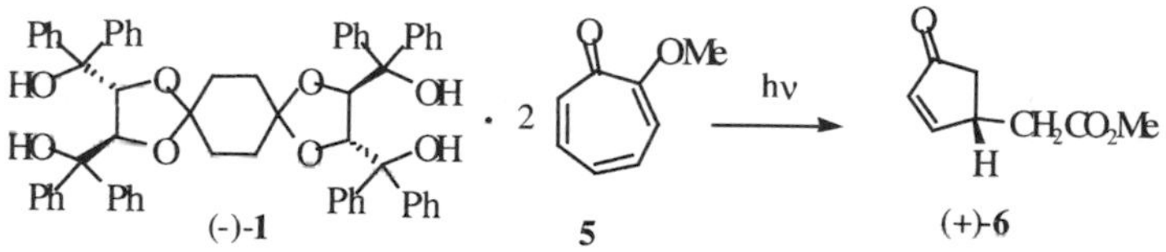

Experimental procedures:
A crystalline powder of the 2:1:1 complex of (–)-**1**, **2** and CHCl₃ (4.44 g) was irradiated by a 400-W high-pressure Hg-lamp (Pyrex filter) in the solid state for 70 h. Separation of the reaction mixture by silica gel column chromatography gave a mixture of (–)-5-isopropyl-7-methoxybicyclo[3.2.0]hepta-3,6-dien-2-one **3** ($[a]_D$–207° (c 0.23, CHCl₃), 96% ee) and (–)-methyl-1-isopropyl-4-oxocyclopent-2-ene-1-acetate **4** ($[a]_D$–94.6 °(c 0.10, CHCl₃), 90% ee). The optical purities were determined by HPLC (Chiralcel OD, Daicel).

References: K. Tanaka, R. Nagahiro, Z. Lipkowska, *Org. Lett.*, **3**, 1567 (2001).

Type of reaction: C–C bond formation
Reaction condition: solid-state
Keywords: [2+2]photodimerization, absolute asymmetric synthesis, cyclobutane

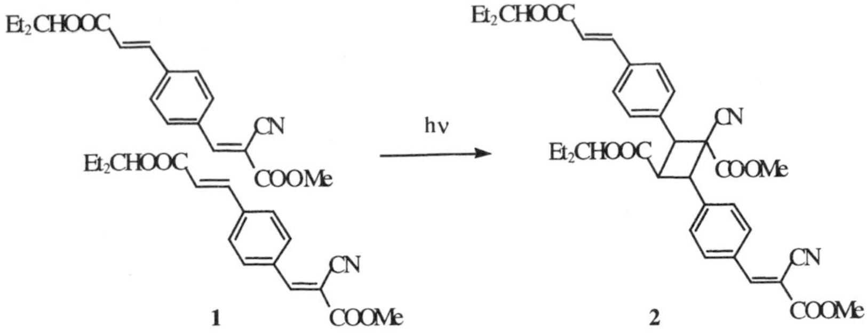

1 → 2

100% ee

Experimental procedures:

Achiral **1** crystallizes from EtOH-CH₂Cl₂ in a chiral structure of space group *P*21. The monomer was suspended in a mixture of methanol-water (10%) and ir- radiated under continuous stirring with a 450-W immersion high-pressure Hano- via Hg-lamp at –2 °C for periods of 20–60 h. Dimer **2**: mp (racemic dimer) 170 °C; $[a]_D$ 114° (100% ee).

References: L. Addadi, J. v. M. Lahav, *J. Am. Chem. Soc.*, **104**, 3422 (1982).

Type of reaction: C–C bond formation
Reaction condition: solid-state
Keywords: di-π-methane and Norrish type II photorearrangement, chiral crystal

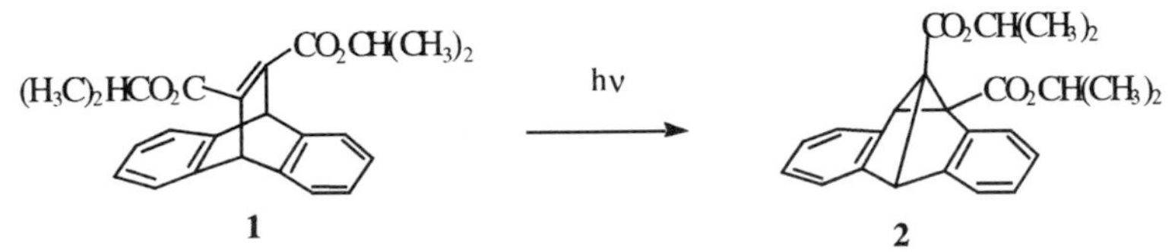

1
$P2_1 2_1 2_1$

2
100% ee

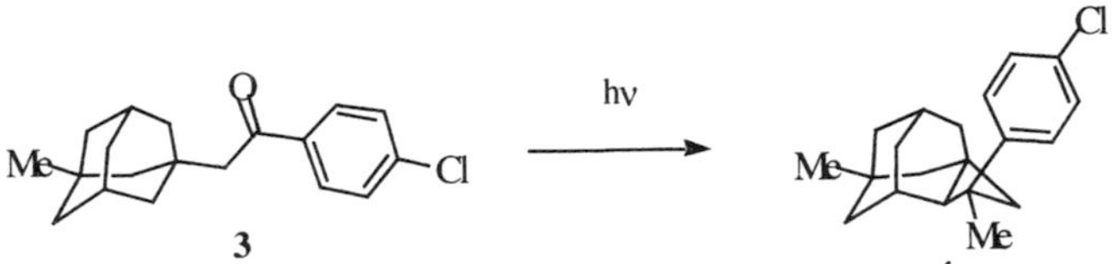

3
$P2_1 2_1 2_1$

4
80% ee

Experimental procedures:
Large (20–85 mg) single crystals of diester **1** were grown by slow evaporation and were photolyzed by the output from a Molectron UV-22 nitrogen laser (337 nm, 330 mW). The optical activity produced in each photolysis was determined by dissolving the sample in chloroform and measuring its $[a]_D$ values.

References: S. V. Evance, M. G. Garibay, N. Omkaram, J. R. Scheffer, J. Trotter, F. Wireko, *J. Am. Chem. Soc.*, **108**, 5648 (1986).

Type of reaction: C–C bond formation
Reaction condition: solid-state
Keywords: oxoamide, chrial crystal, Norrish type II photoreaction, β-lactam

1
chiral crystal

2
93% ee

a: R=H	e: R=*p*-Me	h: R=*o*-Me	k: R=3,5-Me$_2$
b: R=*m*-Me	f: R=*p*-Cl	i: R=*o*-Cl	l: R=3,4-Me$_2$
c: R=*m*-Cl	g: R=*p*-Br	j: R=*o*-Br	
d: R=*m*-Br			

Experimental procedures:
The powdered crystals of (+)-**1** (1 g) were irradiated with the Hg-lamp. The photoproduct (+)-**2**, after purification by column chromatography, received 93% enantiomeric excess, 74% chemical yield (0.74 g), $[a]_D$+123 °(c 0.5, CHCl$_3$), and mp 149–150 °C.

References: F. Toda, M. Yagi, S. Soda, *Chem. Commun.*, 1413 (1987); A. Sekine, K. Hori, Y. Ohashi, M. Yagi, F. Toda, *J. Am. Chem. Soc.*, **111**, 697 (1989); D. Hashizume, H. Kogo, A. Sekine, Y. Ohashi, H. Miyamoto, F. Toda, *J. Chem. Soc., Perkin Trans. 2*, 61 (1996).

Type of reaction: C–C bond formation
Reaction condition: solid-state
Keywords: 4-[2-(4-pyridyl)ethenyl]cinnamate, chiral crystal, [2+2]photodimerization, cyclobutane

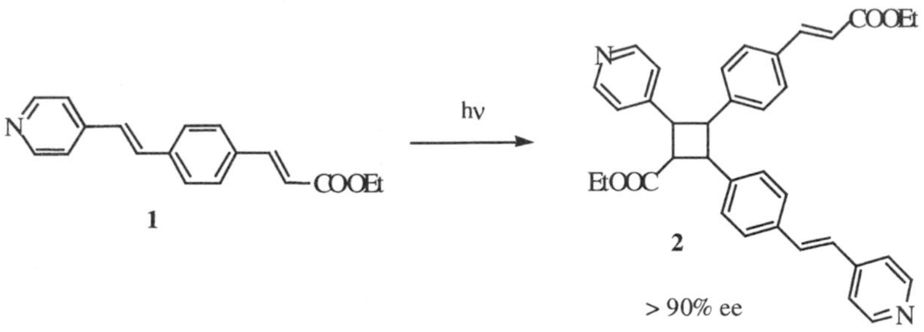

1   **2**

> 90% ee

Experimental procedures:

Finely powdered monomer crystals were dispersed in distilled water containing a few drops of surfactant and irradiated, with vigorous stirring, by a 500-W super-high-pressure mercury lamp set outside of the flask. Optically active **2** was obtained by the irradiation of crystals **1** with $\lambda > 365$ nm, and by successive purification by preparative TLC.

References: C. Chung, M. Hasegawa, *J. Am. Chem. Soc.*, **113**, 7311 (1991).

Type of reaction: C–C bond formation
Reaction condition: solid-state
Keywords: α-adamanthyl ketone, prolinol, Norrish type II photoreaction, fused cyclobutane

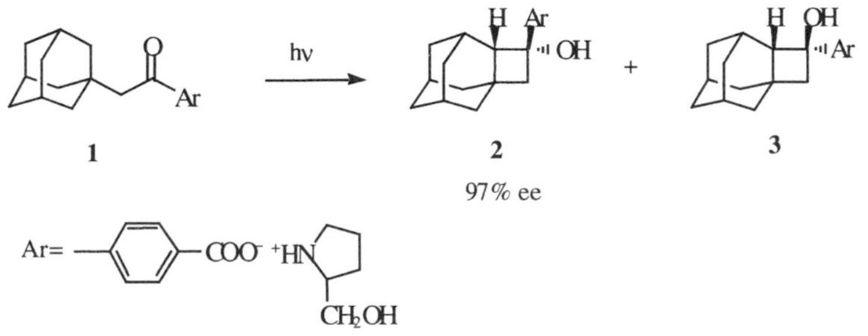

1   **2**   **3**

97% ee

$$Ar = -\!\!\!\!\!\bigcirc\!\!\!\!\!-COO^- \ ^+HN\!\!\!\searrow \quad CH_2OH$$

Experimental procedures:

The samples were prepared for solid state photolysis by crushing the crystals between two Pyrex plates and sliding the plates back and forth so as to distribute the crystals over the surface in a thin, even layer. The sample plates were then taped together at the top and bottom ends, placed in polyethylene bags, degassed with nitrogen and sealed under a positive pressure of nitrogen with a heat-sealing device. The bags were immersed in a cooling bath maintained at $-40\,^{\circ}$C by means of a cryomat and irradiated with the output from a 450-W Hanovia medium-pressure lamp. Conversions of 70% and higher were possible in the solid state runs without significant sample melting or loss of enantioselectivity. In a typical run, a "sandwich" containing 120 mg of salt required 24 h of photolysis.

References: R. Jones, J.R. Scheffer, J. Trotter, J. Yang, *Tetrahedron Lett.,* **33**, 5481 (1992).

Type of reaction: C–C bond formation
Reaction condition: solid-state
Keywords: di-π-methane photorearrangement, chiral crystal, fused cyclopropane

Experimental procedures:
Crystalline **1** (6 g, 26.4 mmol) was ground to a fine powder and poured into a Pyrex vessel mounted in a Rayonet reactor ($\lambda = 350 \pm 40$ nm) containing vigorously stirred H_2O (300 mL). A cooling finger was lowered into the mixture and the vessel was irradiated for 90 h. The turbid lightly orange mix was poured into a separatory funnel and the remaining orange and yellow solid/melt was washed from the vessel with CH_2Cl_2. Extractive workup afforded a dark reddish orange oil from which **1** (2.93 g, 49%), **2** (1.36 g, 23%) and **3** (1.68 g, 28%) were chromatographically isolated.

References: A.L. Roughton, M. Muneer, M. Demuth, *J. Am. Chem. Soc.,* **115**, 2085 (1993).

Type of reaction: C–C bond formation
Reaction condition: solid-state
Keywords: 9,10-dihydro-9,10-ethenoanthracene derivative, di-π-methane photorearrangement, chiral crystal, dibenzosemibullvalene

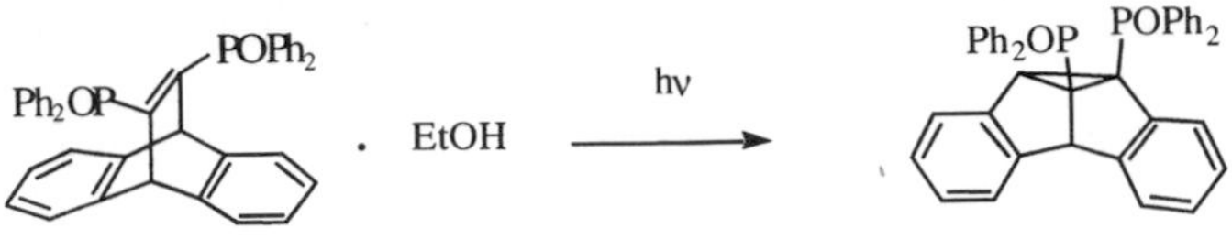

1

$P2_12_12_1$

2

89% ee

Experimental procedures:
Single crystals of the ethanol complex of **1** were irradiated through Pyrex and the chiral photoproduct **2** was tested for optical activity by polarimetry and chiral HPLC. This indicated an enantiomeric excess of 89% (84% conversion at room temperature).

References: T. Y. Fu, Z. Liu, J. R. Scheffer, J. Trotter, *J. Am. Chem. Soc.*, **115**, 12202 (1993).

Type of reaction: C–C bond formation
Reaction condition: solid-state
Keywords: *N*-(thiobenzoyl)methacrylanilide, chiral crystal, photoirradiation, *β*-lactam

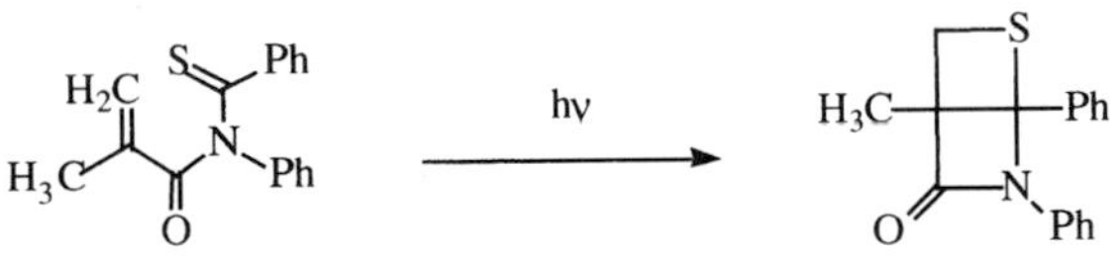

1

2

40 % ee

Experimental procedures:
When powdered crystals of the monothioimide **1**, recrystallized from hexane, were irradiated under nitrogen through a Pyrex filter with the UV light from a 500-W high-pressure mercury lamp at 0 °C for 12 h, a transformation occurred and the photoproduct was purified by column chromatography on silica gel; optically active *β*-lactam (+)-**2** was obtained in 75% yield, $[a]_D+23°$ (*c* 0.1, CHCl$_3$), 10% ee. The solid-state photoreaction proceeded even at –45 °C, and optically active (+)-**2** which showed a higher ee value was formed, $[a]_D+93°$ (*c* 0.1, CHCl$_3$), 40% ee (conv. 30%, yield 70%).

References: M. Sakamoto, N. Hokari, M. Takahashi, T. Fujita, S. Watanabe, I. Iida, T. Nishio, *J. Am. Chem. Soc.*, **115**, 818 (1993).

Type of reaction: C–C bond formation

Reaction condition: solid-state

Keywords: 3,4-bis(diphenylmethylene)-*N*-methylsuccimide, [4+2]photodimerization cyclization, 2-methyl-4,4,9-triphenyl-3a,4-dihydro-benzo[f]isoindole-1,3-dione

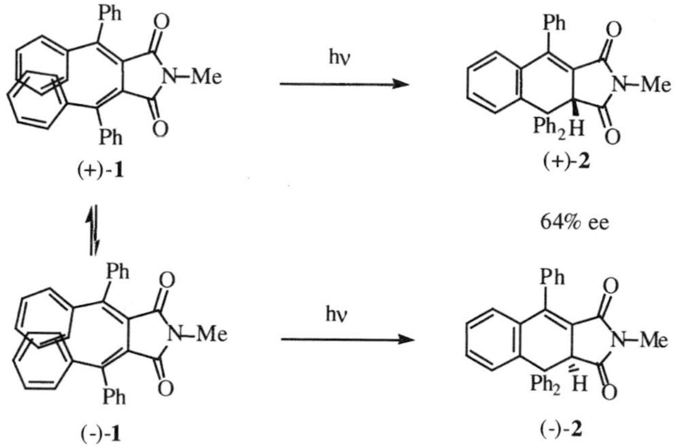

Experimental procedures:

Photoirradiation of powdered (+)-**1** crystal (50 mg) by 100-W high-pressure Hg-lamp for 50 h gave (+)-**2** of 64% ee (50 mg, 100% yield, mp 285–287 °C, $[a]_D$+102 °(*c* 0.05, CHCl$_3$)). The optical purity was determined by HPLC on the chiral solid phase Chiralpak AS.

References: F. Toda, K. Tanaka, *Supramol. Chem.,* **3**, 87 (1994).

Type of reaction: C–C bond formation

Reaction condition: solid-state

Keywords: styrene, charge-transfer complex, chiral crystal, [2+2]photocycloaddition, cyclobutane, spiro compound

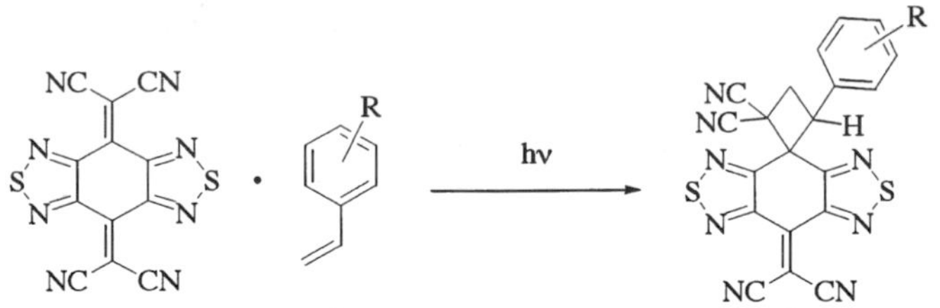

a: R=H
b: R=2-vinyl
c: R=3-vinyl
d: R=4-vinyl

95% ee

Experimental procedures:

Recrystallization of **1** from MeCN containing **2d** in excess afforded the 1:1 CT crystal as purple plates. The 1:1 CT crystals of **2b** and **2c** were obtained as reddish powders by admixing **1** with **2b** and **2c**, respectively. The 1:1 CT crystal of **1** and **2a** was also isolated as an orange powder by the similar method. Upon irradiation of a fine powder of the 1:1 CT complex with wavelength $\lambda > 505$ nm for 1 h, **3a** was obtained in 71% yield. The adduct **3b** was isolated in 91% yield on irradiation for 1 h with light ($\lambda > 540$ nm).

References: T. Suzuki, T. Fukushima, Y. Yamashita, T. Miyashi, *J. Am. Chem. Soc.*, **116**, 2793 (1994).

Type of reaction: C–C bond formation
Reaction condition: solid-state
Keywords: acridine, diphenylacetic acid, photodecarboxylation, two-component molecular crystal, chiral crystal, 7,8-dihydro-7-diphenylmethyl-acridine

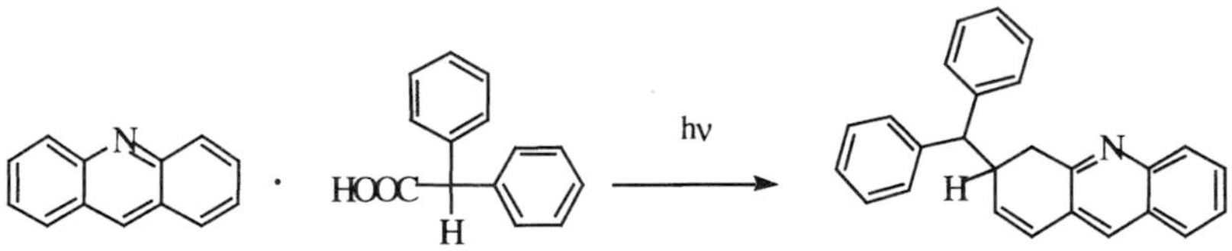

35% ee

Experimental procedures:

Chiral two-component crystal (**1·2**, 800 mg) in preparative scale prepared by seeding was pulverized, placed between two Pyrex plates, and irradiated with a

400-W high-pressure Hg lamp for 3 h under argon at room temperature. The irradiated mixture was submitted to preparative TLC (benzene-ethyl acetate, 7:1) to give 260 mg of a chiral condensation product **3** of 35% ee in 37% yield.

References: H. Koshima, K. Ding, Y. Chisaka, T. Matsuura, *J. Am. Chem. Soc.,* **118,** 12059 (1996).

Type of reaction: C–C bond formation
Reaction condition: solid-state
Keywords: *N,N*-dibenzyl-1-cyclohexenecarbothioamide, chiral crystal, photoirradiation, *β*-thiolactam, spiro compound

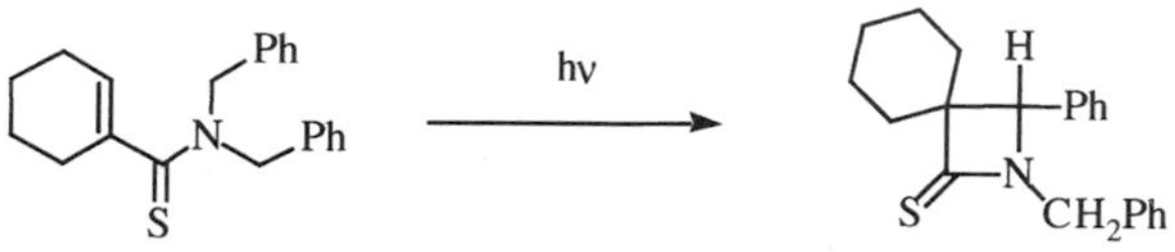

1   **2**

94% ee

Experimental procedures:
Powdered crystals of **1**, well ground and sandwiched by Pyrex glass plates, were irradiated with 500-W Hg lamp at 0 °C for 2 h, which led to the exclusive production of optically active *β*-thiolactam, 1-benzyl-4-phenyl-azetidine-2-thione-3-spiro-1'-cyclohexane **2**, in 96% yield at 58% conversion. The material was purified by column chromatography, and the structure was determined by spectroscopy. As expected, the thiolactam **2** showed optical activity ($[α]_D$+109° *c* 1.0 CHCl$_3$, 94% ee).

References: M. Sakamoto, M. Takahashi, K. Kamiya, K. Yamaguchi, T. Fujita, S. Watanabe, *J. Am. Chem. Soc.,* **118,** 10664 (1996).

Type of reaction: C–C bond formation
Reaction condition: solid-state
Keywords: *cis*-4-*tert*-butylcyclohexyl ketone, Norrish type II photoreaction

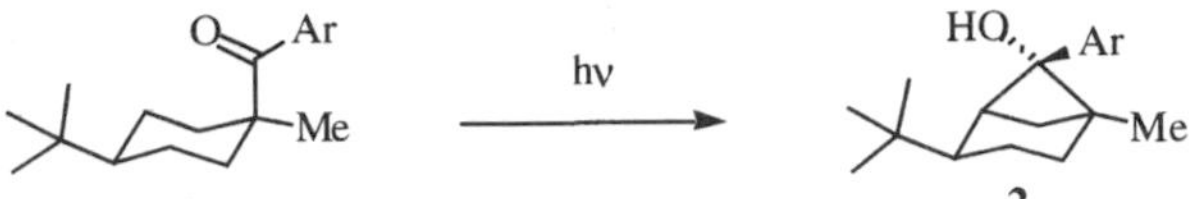

Ar =p-C$_6$H$_4$-COO$^-$ H$_3$N$^+$-R*

H$_3$N-R*	time (min)	ee (%)
(+)-pseudoephedrine	73	79
(+)-α-methylbenzylamine	45	92
(-)-norephedrine	15	96

Experimental procedures:

cis-4-*tert*-butylcyclohexyl carboxylic acid **1** was reacted with various optically active amines, and the resulting crystalline salts were then photolyzed in the solid state. The photolysis mixtures were dissolved in a mixture of ethyl acetate and water and treated with excess ethereal diazomethane to form keto ester. This material was freed of the chiral auxiliary by short path silica gel column chromatography and then analyzed by chial HPLC.

References: M. Leibovitch, G. Olovsson, G. Sundarababu, V. Ramamurthy, J.R. Scheffer, J. Trotter, *J. Am. Chem. Soc.*, **118**, 1219 (1996).

Type of reaction: C–C bond formation
Reaction condition: solid-state
Keywords: *N*-(2,2-dimethylbut-3-enoyl)-*N*-phenylthiocarbamate, chiral crystal, photoreaction, thiolactone

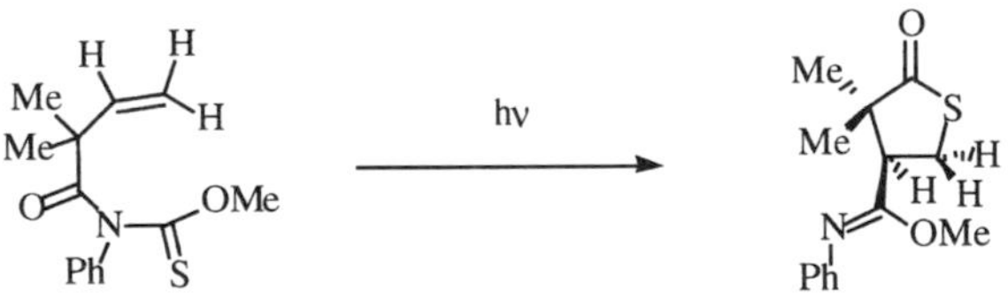

10% ee

Experimental procedures:

The solid state photolysis of **1** gave optically active thiolactone **2** ([a]$_D$+8° (*c* 1.0, CHCl$_3$, 10% ee) in 85% yield when the reaction conversion was 78%.

The enantiomeric purity of **2** was determined by HPLC employing a chiral cell OJ (Daicel).

References: M. Sakamoto, M. Takahashi, T. Arai, M. Shimizu, K, Yanaguchi, T. Mino, S. Watanabe, T. Fujita, *Chem. Commun.*, 2315 (1998).

Type of reaction: C–C bond formation
Reaction condition: solid-state
Keywords: *N*-benzyl-*N*-methylmethacrylthioamide, photoreaction, *β*-thiolactam

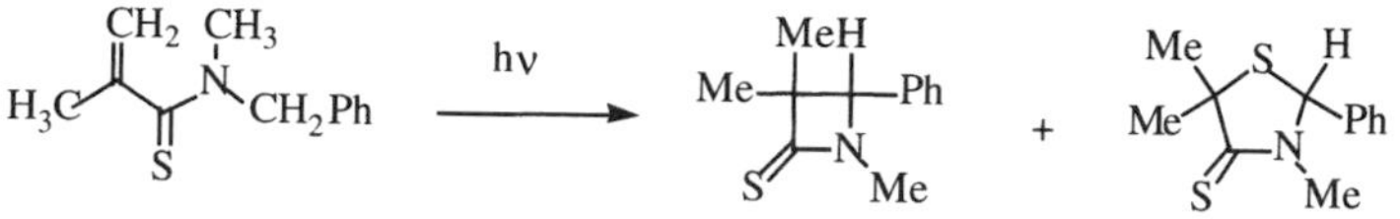

31% ee

Experimental procedures:
The solid-state photoreaction of thioamide **1** was done under an atmosphere purged with dry argon. The solid sample was irradiated for 4 h as a powder prepared by grinding, and placed inside a Pyrex slide. When powdered thioamide **1** was irradiated in the solid state at 0 °C, at up to 19% conversion, optically active *β*-thiolactam **2** of 31% ee was isolated. The optical purity was determined by HPLC using a chiral cell OD column (Daicel Chemical Ind.).

References: M. Sakamoto, M. Takahashi, W. Arai, K. Kamiya, T. Mino, S. Watanabe, T. Fujita, *J. Chem. Soc., Perkin Trans. 1*, 3633 (1999).

Type of reaction: C–C bond formation
Reaction condition: solid-state
Keywords: 1-benzoyl-8-benzylnaphthalene, chiral crystal, photoirradiation, topochemistry, *cis*-1,2-diphenylacenaphthen-1-ol

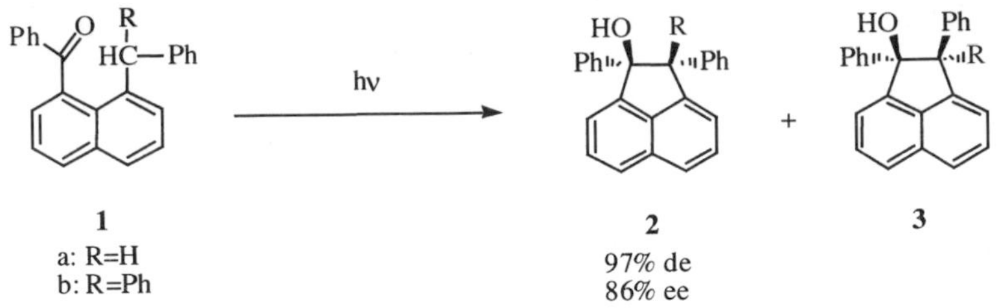
Ph
O
R
HC—Ph
hν
HO
R
Ph
Ph
+
HO
Ph
Ph
R
1
a: R=H
b: R=Ph
2
97% de
86% ee
3

Experimental procedures:
Irradiation of single crystals of **1** in the solid state yields *cis*-1,2-diphenylacenaphthen-1-ol **2** as main product, with up to 97% de and 86% ee. Only a small amount of *trans*-acenaphthenol **3** is also generated in the solid state, presumably due to an increased thermal motion of the molecules in the crystal during the irradiation (40–55 °C).

References: H. Irngartiger, P. W. Fettel, V. Siemund, *Eur. J. Org. Chem.*, 3381 (2000).

4 Carbon–Nitrogen Bond Formation

4.1 Solvent-Free C–N Bond Formation

Type of reaction: C–N bond formation
Reaction condition: solid-state
Keywords: anhydrous hydrazine, hydroquinone, dimethyl terephthalate, inclusion crystal, terephthalic acid dihydrazide

1:1 complex

a: *para*
b: *meta*

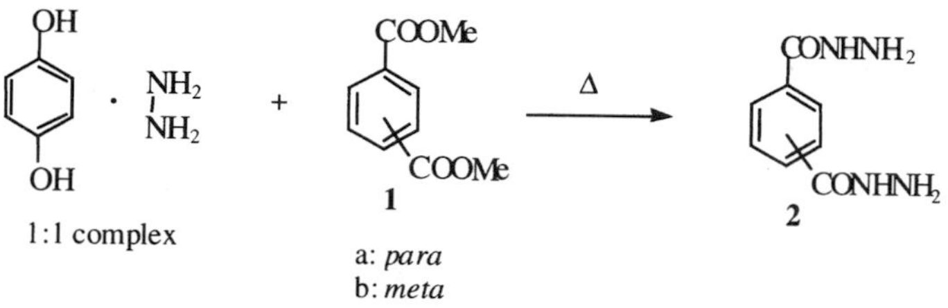

Experimental procedures:

A mixture of powdered dimethyl terephthalate **1a** (19.4 g, 0.1 mol) and 1:1 in-clusion complex of hydroquinone and hydrazine (58.5 g, 0.4 mol) was kept under a nitrogen atmosphere at 100–125 °C for 25 h. To the reaction mixture was added MeOH and MeOH insoluble terephthalic acid dihydrazide **2a** was obtained by filtration (17.1 g, 88.1% yield).

References: F. Toda, S. Hyoda, K. Okada, K. Hirotsu, *Chem. Commun.*, 1531 (1995).

Type of reaction: C–N bond formation
Reaction condition: solid state
Keywords: ω-amino carboxylic acid, cyclodextrins, host-guest chemistry, con-densation, polyamides

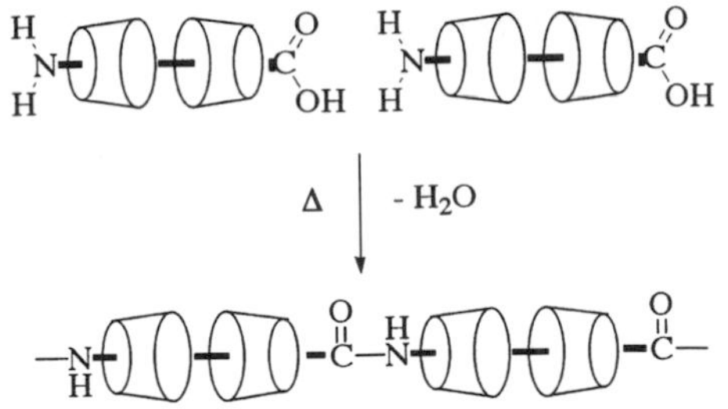

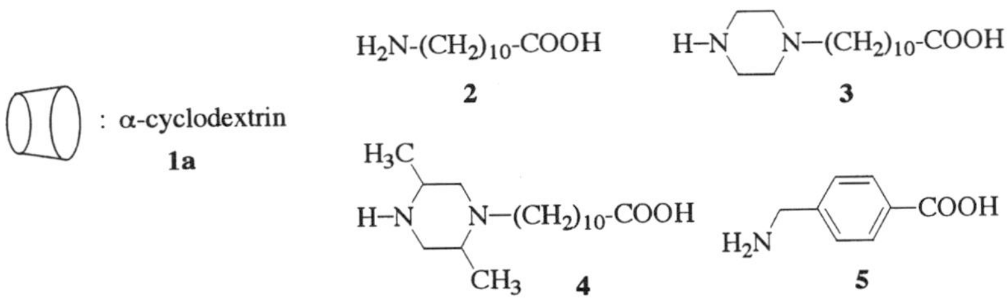

H$_2$N-(CH$_2$)$_{10}$-COOH   H–N◯N–(CH$_2$)$_{10}$-COOH

2   **3**

◯ : α-cyclodextrin

1a

H$_3$C

H–N◯N–(CH$_2$)$_{10}$-COOH   H$_2$N◯-COOH

CH$_3$ **4**   **5**

Experimental procedures:

[2·(1a)$_2$] (203 mg) was tempered at 230 °C and 0.1 mbar for 5 h. Yield: 196 mg (97.4%). [2·(1a)$_2$]n (1.172 g) was stirred in 1% HCl (60 mL) at 80 °C for 25 h. The resulting precipitate was filtered and washed several times with water and methanol. Yield: 103.1 mg (93.9%). For the molecular mass determination, (2)n (90 mg, 0.49 mmol) was added to absolute CH$_2$Cl$_2$ (15 mL) and stirred with tri-fluoroacetic anhydride (0.21 g, 1.0 mmol) at 25 °C for 24 h. After removal of the solvent by distillation, the residue was dried at 0.1 mbar. Yield: 139 mg (96%) N-trifluoroacrylated nylon-11. The soluble product was subjected to size exclu-sion chromatography on Styralgel column (Waters, calibration with polystyrene standards) in absolute THF.

References: M. B. Steinbrunn, G. Wenz, *Angew. Chem. Int. Ed. Engl.,* **35**, 2139 (1996).

Type of reaction: C–N bond formation
Reaction condition: solvent-free
Keywords: aldehyde, ketone, hydrazin derivative, semicarbazide, NaOH, silica gel, hydrazone, semicarbazide

$$\underset{R}{\overset{R'}{>}}\!\!=\!\!O \quad + \quad NH_2G \quad \xrightarrow[\text{NaOH}]{\text{silica gel}} \quad \underset{R}{\overset{R'}{>}}\!\!=\!\!NG$$

1   **2**   **3**

a: R=R'=Ph, G=PhNH
b: R=Me, R'=Ph, G=PhNH
c: R=Me, R'=3,4-(MeO)$_2$C$_6$H$_3$, G=PhNH
d: R=Me, R'=4-PhC$_6$H$_4$, G=PhNH
e: R=Ph, R'=2-pyridyl, G=PhNH
f: R=Me, R'=2-pyridyl, G=PhNH
g: R=Me, R'=4-MeOC$_6$H$_4$, G=PhNH
h: R=Me, R'=2-MeOC$_6$H$_4$, G=PhNH
i: R=R'=Ph, G=4-NO$_2$C$_6$H$_4$NH
j: R=H, R'=4-MeOC$_6$H$_4$, G=4-NO$_2$C$_6$H$_4$NH
k: R=Me, R'=4-ClC$_6$H$_4$, G=4-NO$_2$C$_6$H$_4$NH
l: R=Me, R'=4-PhC$_6$H$_4$, G=4-NO$_2$C$_6$H$_4$NH
m: R=R'=Ph, G=NMe$_2$
n: R=Me, R'=Ph, G=NMe$_2$
o: R=Me, R'=3,4-(MeO)$_2$C$_6$H$_3$, G=NMe$_2$
p: R=Me, R'=4-MeOC$_6$H$_4$, G=NMe$_2$
q: R=H, R'=Ph, G=NH$_2$CONH
r: R=Me, R'=Ph, G=NH$_2$CONH
s: R=Me, R'=4-MeOC$_6$H$_4$, G=NH$_2$CONH
t: R=Me, R'=3,4-(MeO)$_2$C$_6$H$_3$, G=NH$_2$CONH

Experimental procedures:

A mortar was charged with the aldehyde or ketone (1 mmol), hydrazine derivative or semicarbazide (1 mmol), sodium hydroxide (0.04 g, 1 mmol) and silica
gel (0.1 g). The reaction mixture was ground with a pestle in the mortar. When
TLC showed no remaining aldehyde or ketone, the reaction mixture was poured
into a mixture of dichloromethane (20 mL) and 5% HCl (10 mL). The ethereal
layer was washed with saturated NaHCO$_3$, dried (MgSO$_4$), and evaporated by rotary evaporation to give the pure product.

References: A. R. Hajipour, I. Mohammadpoor-Baltork, M. Bigdeli, *J. Chem. Res. (S)*,
570 (1999).

Type of reaction: C–N bond formation
Reaction condition: solid-state
Keywords: 3,5-dihalo-4-aminobenzoylchloride, polymerization, polyamide

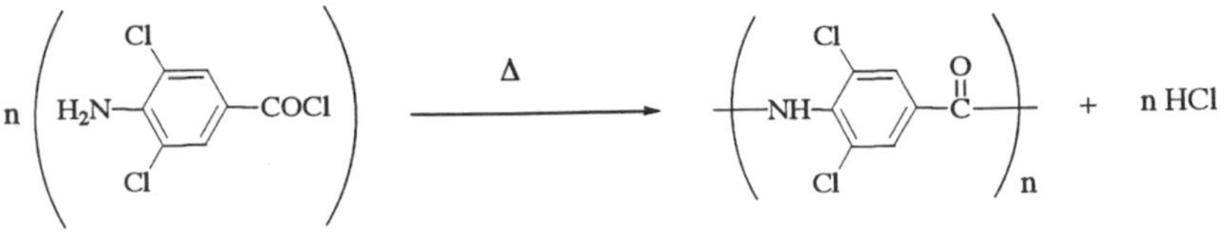

$$n \left(H_2N-\underset{\underset{Cl}{|}}{\overset{\overset{Cl}{|}}{\bigcirc}}-COCl \right) \xrightarrow{\Delta} \left(-NH-\underset{\underset{Cl}{|}}{\overset{\overset{Cl}{|}}{\bigcirc}}-\overset{\overset{O}{\parallel}}{C}- \right)_n + n\ HCl$$

1   2

Experimental procedures:

The crystalline solid was placed in a 1-cm test tube equipped with a side arm. The test tube was heated in a bath of silicone oil and maintained at a constant temperature under dry nitrogen. Depending on the particular experiment the temperature was kept between 95 and 300 °C. The contents of the tube became opaque but did not melt. The solid polymer retained the habit of the monomer crystals. The reaction could be followed by changes in infrared spectra.

References: R.B. Sandor, B.M. Foxman, *Tetrahedron,* **56**, 6805 (2000).

Type of reaction: C–N bond formation
Reaction condition: solid-state
Keywords: aryl isothiocyanate, aromatic primary amine, diaryl thiourea

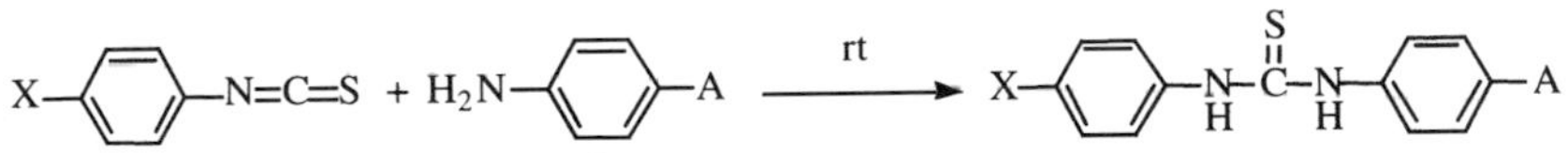

$$X-\bigcirc-N=C=S \ + \ H_2N-\bigcirc-A \xrightarrow{rt} X-\bigcirc-\underset{H}{N}-\overset{\overset{S}{\parallel}}{C}-\underset{H}{N}-\bigcirc-A$$

1   2   3

a: X=Cl; A=CH₃   h: X=Br; A=OCH₃
b: X=Cl; A=OCH₃   i: X=Br; A=α-C₁₀H₇
c: X=Cl; A=α-C₁₀H₇   j: X=Br; A=Br
d: X=Cl; A=Cl   k: X=Br; A=I
e: X=Cl; A=Br   l: X=OEt; A=CH₃
f: X=Cl; A=I   m: X=OEt; A=OCH₃
g: X=Br; A=CH₃   n: X=OEt; A=α-C₁₀H₇

$$X-\bigcirc-NCS \ + \ H_2NHN-\overset{\overset{O}{\parallel}}{C}-Ar \xrightarrow{rt} X-\bigcirc-\underset{H}{N}-\overset{\overset{S}{\parallel}}{C}-NHNH-\overset{\overset{O}{\parallel}}{C}-Ar$$

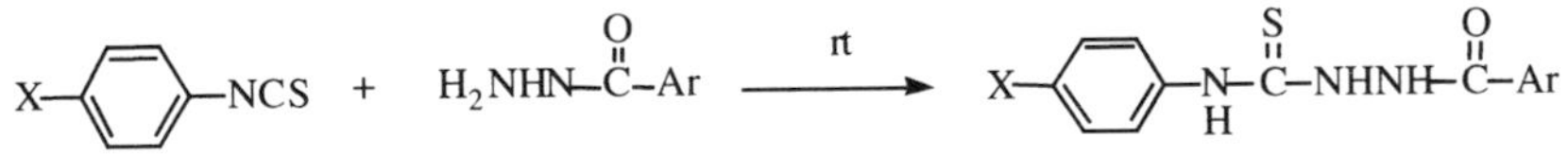

1   4   5

a: X=Cl; Ar=Ph   f: X=EtO; Ar=3-ClC₆H₄
b: X=Br; Ar=Ph   g: X=Cl; Ar=1-C₁₀H₇CH₂
c: X=EtO; Ar=Ph   h: X=Br; Ar=1-C₁₀H₇CH₂
d: X=Cl; Ar=3-ClC₆H₄   i: X=EtO; Ar=1-C₁₀H₇CH₂
e: X=Br; Ar=3-ClC₆H₄   j: X=Cl; Ar=C₆H₅OCH₂
                             k: X=Br; Ar=C₆H₅OCH₂

Experimental procedures:
A mixture of aryl isothiocyanate **1** (1 mmol) and aromatic primary amine **2** (1 mmol) was ground thoroughly in an agate mortar. The reaction was traced with thin-layer chromatography. After the reaction was complete (5–40 min), the crude products were recrystallized with ethanol or acetone and dried under vacuum to yield the pure products.

References: J.-P. Li, Y.-L. Wang, H. Wang, Q.-F. Luo, X.-Y. Wang, *Synth. Commun.*, **31**, 781 (2001); J.-P. Li, Q.-F. Luo, Y.-L. Wang, H. Wang, *Synth. Commun.*, **31**, 1793 (2001).

Type of reaction: C–N bond formation
Reaction condition: solid-state
Keywords: arenes, nitrogen dioxide, nitration, gas-solid reaction

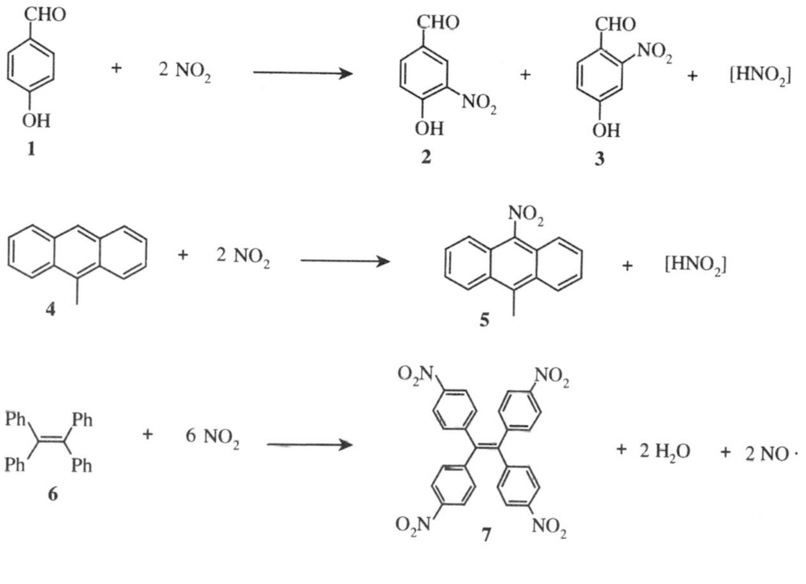

Experimental procedures:
2/3: 4-Hydroxybenzaldehyde **1** (500 mg, 4.1 mmol) was reacted with NO_2 at an initial pressure of 0.3 bar for 12 h. After evaporation of the gases, the product mixture (685 mg, 100%) contained 82% **2** and 18% **3** according to ^1H NMR analysis. The mixture was separated by preparative layer chromatography (SiO_2, EtOAc) to isolate 528 mg (77%) of **2**, mp 143 °C and 116 mg (17%) of **3**, mp 104 °C.

7: Tetraphenylethene **6** (800 mg, 2.41 mmol) was mixed with 800 mg of $MgSO_4 \cdot 2H_2O$ (for the absorbance of the water of reaction). After 12 h of reaction with NO_2 (710 mg, 15.4 mmol) at an initial pressure of 0.3 bar at room temperature, the excess gas and NO were condensed in a cold trap at $-196\,°C$ and analyzed by FT-IR ($NO/NO_2 = 10/2$). The quantitatively formed product **7** was extracted from the drying agent with CH_2Cl_2, yielding pure **7** (1.17 g, 95%), mp 300–302 °C. For larger runs the use of a flow apparatus is advisable, which allows for circulating of the gas and admixing of the calculated amount of oxygen to oxidize the NO formed for use in the running reaction.

Reference: G. Kaupp, J. Schmeyers, *J. Org. Chem.,* **60**, 5494 (1995.)

Type of reaction: C–N bond formation
Reaction condition: solvent-free
Keywords: aromatic compound nitric acid, acetic anhydride, zeolite, nitration, aromatic nitro compound

1

HNO_3-Ac_2O

zeolite H$^+$beta

2

a: X=F
b: X=Cl
c: X=Br
d: X=H
e: X=Me
f: X=Et
g: X=i-Pr
h: X=t-Bu
i: X=Ph
j: X=NO_2

Experimental procedures:
Nitric acid (2.45 g, 90%, 35 mmol) and dried H$^+$beta (1.0 g) were stirred together at 0 °C for 5 min. Addition of acetic anhydride (5 mL, 53 mmol) resulted in an exothermic reaction causing a temporary rise in temperature to ca. 10–15 °C. After a further 5 min the aromatic substrate (35 mmol) was added dropwise and the mixture was then allowed to warm to room temperature and stirred for a further 30 min. The products were obtained by direct vacuum distillation of the mixture, first at 30 mmHg to give the acetic acid by product and then at 0.2 mmHg, to give the nitro compound.

References: K. Smith, A. Musson, G. A. DeBoos, *Chem. Commun.,* 469 (1996).

Type of reaction: C–N bond formation
Reaction condition: solid-state
Keywords: triazene-1-oxide, naphthol, mixed crystals, photoreaction, azo dye

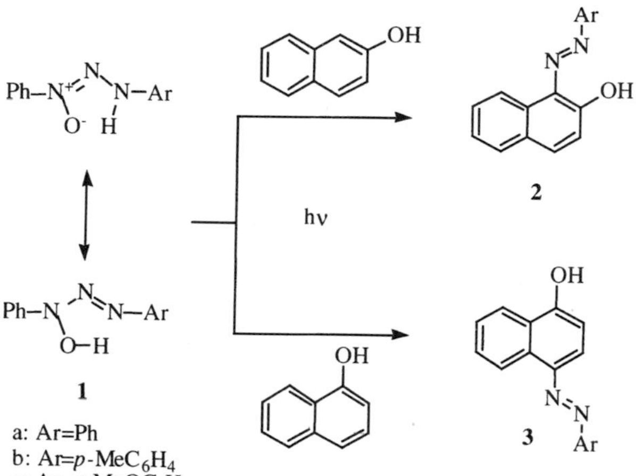

Ph–N⁺=N–N–Ar
       |
   O⁻ H

Ph–N–N=N–Ar
     |
   O–H

1

a: Ar=Ph
b: Ar=p-MeC$_6$H$_4$
c: Ar=p-MeOC$_6$H$_4$
d: Ar=p-ClC$_6$H$_4$

hv

2

3

Experimental procedures:
Mixing of 1 mmol of triazene 1-oxide **1** with 1 mmol of α- or β-naphthol in acetone followed by evaporation gave the mixed crystals. These were finely ground in a mortar, spread on petri dishes and directly exposed to sunlight. TLC examination showed that the reactions were finished within 7–10 h. The temperature was measured as 18 to 25 °C. Chromatographic separation on silica gel with ethyl acetate-toluene (1:10) as eluent gave two zones. The first zone with the higher R$_f$ value (at the start) afforded a multicomponent mixture, which could neither be separated nor could individual components be identified.

References: S. K. Mohamed, A. M. N. El-Din, *J. Chem. Res. (S)*, 508 (1999).

Type of reaction: C–N bond formation
Reaction conditions: solvent-free, solid-state
Keywords: anilines, benzaldehydes, condensation, waste-free, azomethines

Ar—NH$_2$ + Ar'—CHO $\longrightarrow$ Ar—N$\underset{H}{\overset{}{\diagup}}$Ar' + H$_2$O

1   **2**   **3**

a: Ar=4-MeC$_6$H$_4$ a: Ar'=4-MeC$_6$H$_4$
b: Ar=4-MeOC$_6$H$_4$ b: Ar'=4-MeOC$_6$H$_4$
c: Ar=4-NO$_2$C$_6$H$_4$ c: Ar'=4-NO$_2$C$_6$H$_4$
d: Ar=4-ClC$_6$H$_4$ d: Ar'=4-HOC$_6$H$_4$
e: Ar=4-BrC$_6$H$_4$ e: Ar'=4-HO,3-MeOC$_6$H$_3$
f: Ar=4-HOC$_6$H$_4$
g: Ar=4-(4-H$_2$NC$_6$H$_4$)C$_6$H$_4$
h: Ar=1-naphthyl

Experimental procedures:

All reactions were performed by grinding together 10 mmol of the pure aniline **1** with 10 mmol of the pure aldehyde **2** in a mortar and keeping the mixture at room temperature. Some mixtures liquefied intermediately at room temperature, but most of these could be run without melting at lower temperatures. The completion of the reactions was checked by IR spectroscopy in KBr. The water produced in the reaction was removed at 80 °C under vacuum. The yield was 100% at 100% conversion in the twenty studied combinations of **1** and **2**. Chemical analysis was carried out by IR and NMR spectroscopy which gave the expected peaks and signals. Thin layer chromatography and comparison of melting points with literature data confirmed the purity of the products **3**.

References: J. Schmeyers, F. Toda, J. Boy, G. Kaupp, *J. Chem. Soc., Perkin Trans. 2*, 989 (1998); *J. Chem. Soc., Perkin Trans. 2*, 132 (2001).

Type of reaction: C–N bond formation
Reaction condition: solid-state
Keywords: primary amine, carbonyl compound, gas-solid reaction, *N*-alkyl-azomethine

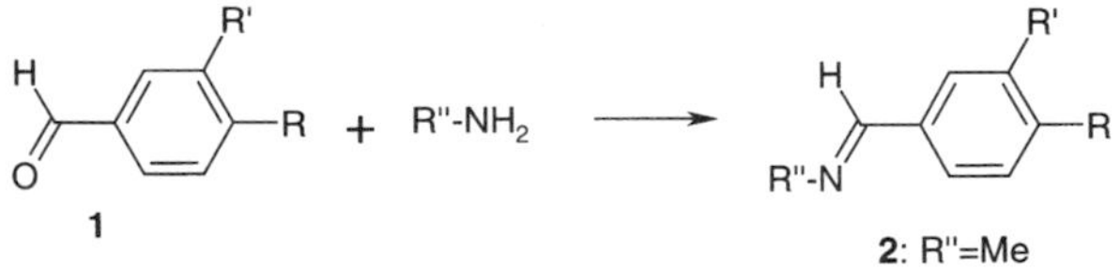

$\underset{O}{\overset{H}{\diagup}}$—C$_6H_3$(R')—R + R''-NH$_2$ $\longrightarrow$ R''-N=CH—C$_6$H$_3$(R')—R

1

a: R=NO$_2$, R'=H   **2**: R''=Me
b: R=OH, R'=H   **3**: R''=Et
c: R=NMe$_2$, R'=H
d: R=OMe, R'=OCH$_2$Ph
e: 9-formylanthracene

Experimental procedures:

1 mmol of the crystalline aromatic aldehyde **1** in an evacuated 50-mL round-bottomed flask were treated with 1 bar of methyl- or ethylamine over night at room temperature. For **1b** 100 mL, 0.25 bar and 0 °C were applied. Excess gas was condensed to a receiver at –196 °C. The water of reaction was removed from the crystals at 0.01 bar and 80 °C. The purity of the weighed products (100% yield) was checked by mp and ^{1}H and ^{13}C NMR.

Reference: G. Kaupp, J. Schmeyers, *J. Boy, Tetrahedron,* **56**, 6899 (2000).

Type of reaction: C–N bond formation
Reaction condition: solid-state
Keywords: dicyclohexylcarbodiimide, [2+2]-cycloaddition, catalysis, waste-free, gas-solid reaction

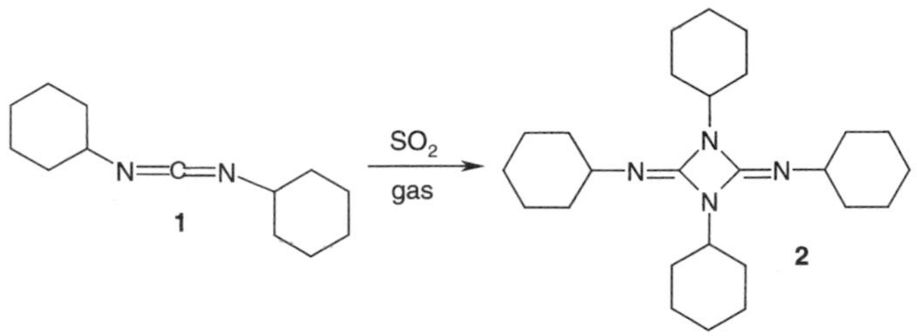

Experimental procedure:

Crystals of dicyclohexylcarbodiimide **1** (1.00 g, 4.8 mmol) were exposed to SO_2 (1 bar) in a 500-mL flask at 0 °C for 21 h. The catalyst gas was pumped off and a quantitative yield of the dimer **2** was obtained (mp 115–118 °C). The purity was verified by various spectroscopic techniques.

Reference: G. Kaupp, D. Lübben, O. Sauerland, *Phosphorus, Sulfur, and Silicon,* **53**, 109 (1990).

Type of reaction: C–N bond formation
Reaction condition: solvent-free
Keywords: aniline, benzaldehyde, large scale, condensation, waste-free, azomethine

Experimental procedures:

A flat steel pan (31×44 cm²) was charged with benzaldehyde **2** (99.5%, 848 g, 7.95 mol) and aniline **1** (99.5%, 744 g, 7.95 mol). The liquids were mixed at 18 °C. The temperature rose to a maximum of 32 °C and fell back to 24 °C when crystallization started with another increase in temperature to a maximum of 35 °C within 12 min when crystallization was virtually complete and water of reaction separated. Next day, the wet crystal cake was crunched with an ordinary house-hold grain-mill and dried in a vacuum at room temperature to give 1.438 kg (100%) of pure benzylidene-aniline **3**.

Reference: G. Kaupp, *Angew. Chem.,* **113**, 4640 (2001); *Angew. Chem. Int. Ed. Engl.,* **40**, 4508 (2001).

Type of reaction: C–N bond formation
Reaction condition: solid-state
Keywords: *p*-hydroxybenzaldehyde, *p*-amino benzoic acid, condensation, large scale, ball-mill, waste-free, solid-solid reaction, azomethine

Experimental procedure:

200 g quantities of a stoichiometric 1:1-mixture of the loosely premixed commercial crystals of **1** and **2**, both at >99% purity, were fed to a stainless-steel 2-L horizontal ball-mill (Simoloyer®) equipped with a hard-metal rotor, steel balls (2 kg; 100Cr6; 5 mm diameter) and water cooling. The temperature was 15 °C at the walls with a maximum of 19 °C in the center of the mill. The rotor was run at 900 min⁻¹ (the power was 610 W) for 15 min for quantitative reaction. 100% conversion and 100% yield was indicated by mp, IR spectrum, chemical analyses and DSC experiments. The product **3**·H₂O was milled out for 10 min leaving

some holdup, but a quantitative recovery was obtained from the second batch and so on. For quantitative recovery of the powdered material in the last batch, an internal air cycle for deposition through a cyclone should be used. The hydrate water was removed from $3 \cdot H_2O$ by heating to 80 °C in a vacuum.

Reference: G. Kaupp, J. Schmeyers, M. R. Naimi-Jamal, H. Zoz, H. Ren, *Chem. Engin. Sci.,* **57**, 763 (2002).

Type of reaction: C–N bond formation
Reaction condition: solvent-free
Keywords: elimination, trapping, radicals, isomerization, laser photochemistry

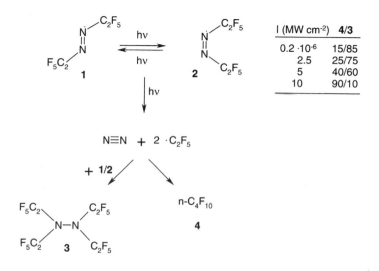

I (MW cm⁻²)	4/3
$0.2 \cdot 10^{-6}$	15/85
2.5	25/75
5	40/60
10	90/10

Experimental procedure:
Samples of **1** (200 mg) were sealed in evacuated Pyrex ampoules (inner diameter 4 mm) and immersed in a 500-mL Pyrex beaker filled with ice and water in such a way that no ice blocked the laser beam. The beam of an excimer laser (Lambda Physics, EMC 201; XeCl; 17 ns pulses; 50 Hz repetition rate; 3 h; $\lambda = 308$ nm) was positioned vertically using two dielectric mirrors and focused to the desired intensity by a quartz-lens with a focal length of 20 cm. For low intensity irradiations, the ampoules were placed in front of a mercury arc at a distance of 5 cm. The product ratio depended on the light intensity. The compounds **1, 2, 3** and **4** were separated by gas chromatography or HPLC on RP18 and spectroscopically characterized after 93–97% conversion to **3** and **4**.

Reference: G. Kaupp, O. Sauerland, *J. Photochem. Photobiol. A: Chem.,* **56**, 375 (1991).

Type of reaction: C–N bond formation
Reaction condition: solid-state
Keywords: thiohydantoin, amine, ring opening, waste-free, gas-solid reaction, thioureido-acetamide

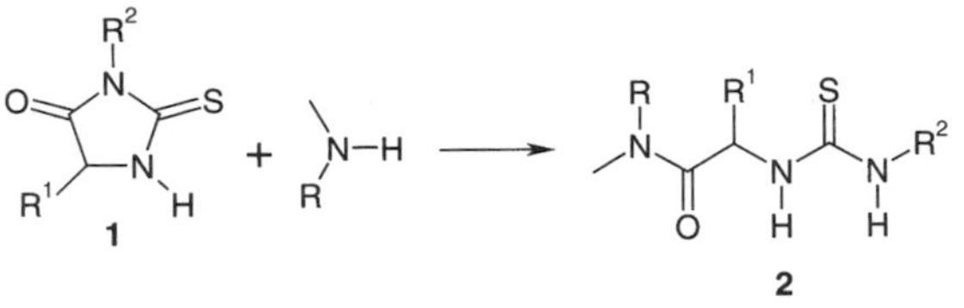

a: $R^1=R^2=R=H$
b: $R^1=R^2=H$, $R=CH_3$
c: $R^1=R=H$, $R^2=Ph$
d: $R^1=H$, $R^2=Ph$, $R=CH_3$
e: $R^1=CH_2Ph$, $R^2=R=H$
f: $R^1=CH_2Ph$, $R^2=H$, $R=CH_3$

Experimental procedures:
The thiohydantoin crystals **1a–f** (2.0 mmol) were treated with the gaseous amine (1 bar) at room temperature in an evacuated 500-mL flask through a vacuum line. After standing for about 12 h the excess gas was pumped off. The yield of solid **2a–f** was 100% in all cases.

Reference: G. Kaupp, J. Schmeyers, *Angew. Chem.*, **105**, 1656 (1993); *Angew. Chem. Int. Ed. Engl.*, **32**, 1587 (1993).

Type of reaction: C–N bond formation
Reaction condition: solid-state
Keywords: solid diazonium salt, barbituric acid, pyrazolinone, azocoupling, solid-solid reaction, hydrazono structure

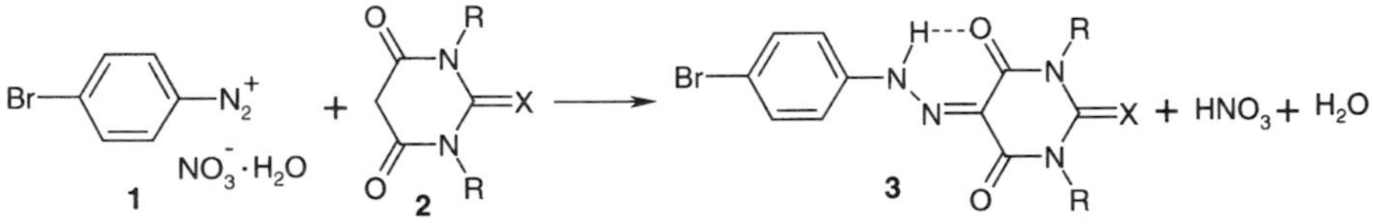

a: R=H, X=O
b: R=Me, X=O
c: R=Et, X=O
d: R=Ph, X=O
e: R=Et, X=S

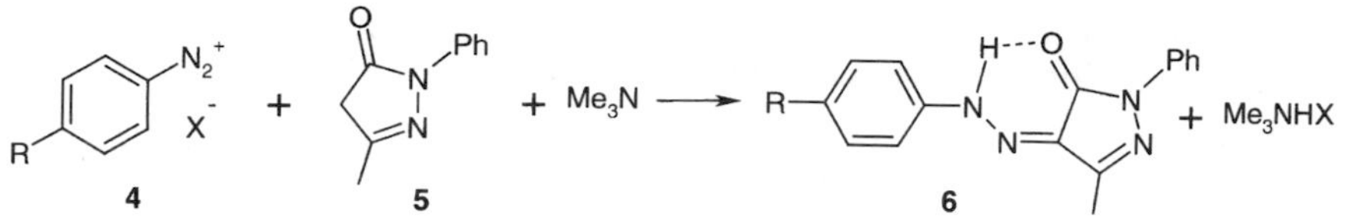

a: R=NO$_2$, X=BF$_4^-$
b: R=OMe, X=NO$_3^-$
c: R=Br, X=NO$_3^-$·H$_2$O
d: R=NO$_2$, X=Cl$^-$

Experimental procedures:

Caution: solid diazonium salts are heat- and shock-sensitive; do not ball-mill!

3: The barbituric acid derivative **2** (0.50 mmol) was ground in an agate mortar. Solid diazonium salt **1** (0.50 mmol) was added and co-ground in 5 portions for 5 min, each. Most of the diazonium band at 2280 cm^{-1} had disappeared, but completion of the reaction was achieved by 24 h ultrasound application in a test tube. After neutralization (0.5 n NaOH, 20 mL), washings (H$_2$O) and drying, the quantitatively obtained products **3a-e** assume the hydrazono structure.

6: The pyrazolone **5** (1.00 mmol) and the solid diazonium salt **4** (1.00 mmol) were cautiously co-ground in an agate mortar for 5 min. The mixture was transferred to a 100 mL flask which was then evacuated. Me$_3$N (0.5 bar) was let in. After 12 h at room temperature, excess gas was recovered in a remote trap at −196 °C. The salt was washed away with water (20 mL) and the residual solid dried. The yield was 98–99% of pure **6a–d** with the hydrazono structure.

Reference: G. Kaupp, A. Herrmann, J. Schmeyers, *Chem. Eur. J.*, **8**, 1395 (2002).

Type of reaction: C–N bond formation
Reaction condition: solid-state
Keywords: solid diazonium salt, *β*-naphthol, azocoupling, solid-solid reaction, hydrazono structure

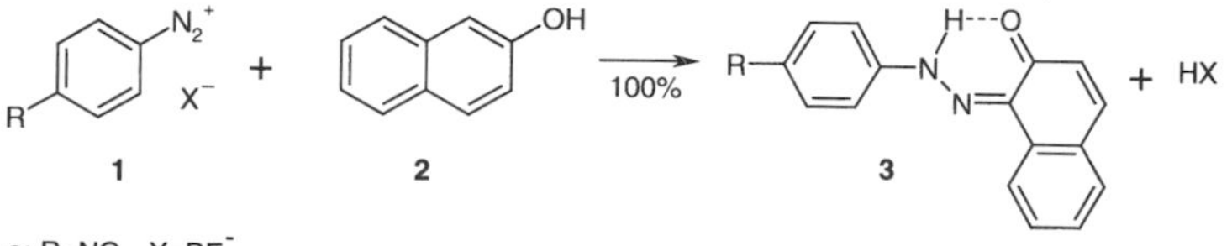

a: R=NO$_2$, X=BF$_4^-$
b: R=H, X=NO$_3^-$
c: R=Cl, X=NO$_3^-$·H$_2$O
d: R=Br, X=NO$_3^-$·H$_2$O

Experimental procedures:
Caution: these reactions might occur violently; use smooth agate mortar and do not ball-mill!
Solid diazonium salt **1** (0.50 mmol) and β-naphthol **2** (0.60 mmol) were separately ground in agate mortars and cautiously mixed. In the case of **1d** $MgSO_4 \cdot 2H_2O$ (0.50 mmol) was added to the mixture. The mixtures rested for 24 h in test tubes and were then exposed to ultrasound for 24 h in a cleaning bath. The quantitatively obtained "azo-dye" salts $3 \cdot HX$ were neutralized and freed from excess **2** by washings with 0.5 n NaOH (20 mL) and water (20 mL). The yields of the neutral dyes with the hydrazono structure **3a–d** were 100, 98, 99 and 99%.

Reference: G. Kaupp, A. Herrmann, J. Schmeyers, *Chem. Eur. J.,* **8**, 1395 (2002).

Type of reaction: C–N bond formation
Reaction condition: solvent-free
Keywords: chloromethylbenzimidazole, urotropin, multimethylation, melt reaction, hexaazapolycycle

Experimental procedure:
2-Chloromethylbenzimidazole **1** (1.66 g, 10.0 mmol), hexamethylenetetramine **2** (1.40 g, 10.0 mmol) and $NaHCO_3$ (1.00 g, 11.9 mmol) were mixed under argon and the melt heated within 5 min from 120 °C to 160 °C. The raw material was extracted with CH_2Cl_2 (to remove excess **2** and other impurities) and water (to remove the salts). The dried residue was pure **3** (0.92 g, 56%) with mp 226 °C.

Reference: G. Kaupp, K. Sailer, *J. Prakt. Chem.,* **338**, 47 (1996).

Type of reaction: C–N bond formation
Reaction condition: solvent-free
Keywords: aldehyde, hydrazine, hydroquinone, ketone, isothiocyanate, solid-solid reaction, azine, thiosemicarbazide, diazepinone

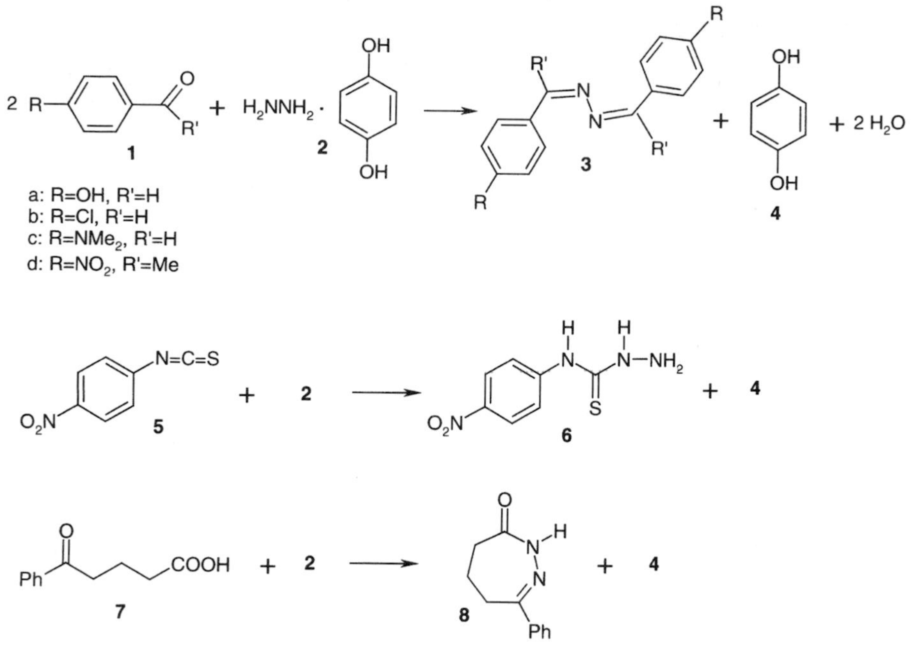

a: R=OH, R'=H
b: R=Cl, R'=H
c: R=NMe$_2$, R'=H
d: R=NO$_2$, R'=Me

Experimental procedures:

The aldehyde/ketone **1** (4.00 mmol), or the ketone **7** (2.00 mmol), or the isothio-cyanate **5** (2.00 mmol) was ball-milled with the solid hydrazine-hydroquinone complex **2** (2.00 mmol) at 25–30 °C for 1 h (3 h in the case of **1d**). The yield was quantitative in all cases, as spectroscopically pure mixtures of **3, 6, 8** with **4** were obtained. Hydroquinone **4** was removed by 5 min trituration with 20 mL of water, filtration and three washings with 2 mL of water, each. The residue was dried in a vacuum to obtain the pure products. **4** was recycled from the aqueous washings by evaporation, addition of 1.0 g of 80% hydrazine hydrate in water per 4 mmol of initially reacted **2** and recrystallization to give 520 mg (91%) of pure **2** after filtration, washing with water and drying.

Reference: G. Kaupp, J. Schmeyers, *J. Phys. Org. Chem.*, **13**, 388 (2000).

Type of reaction: C–N bond formation
Reaction condition: solid-state
Keywords: imines, carbocation, oxonium salt, quaternary salts, waste-free, solid-solid reaction, iminium salts

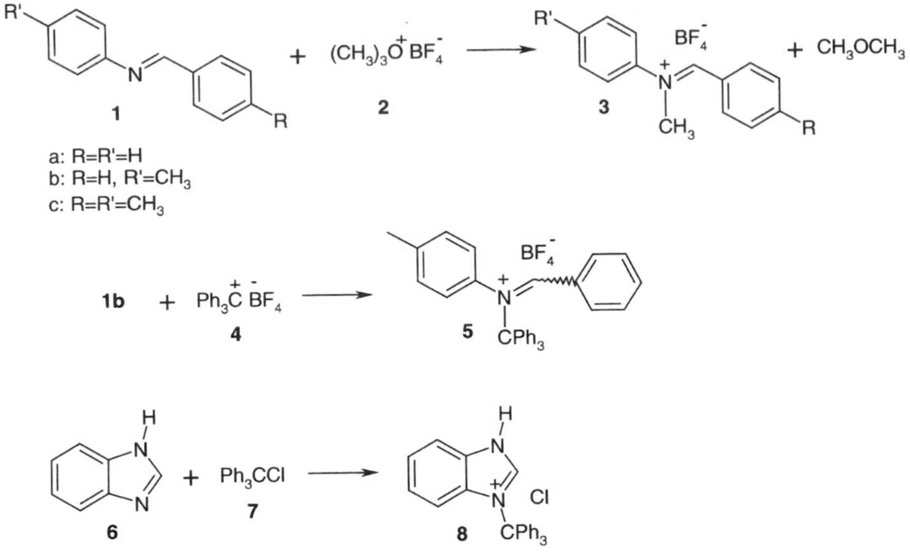

a: R=R'=H
b: R=H, R'=CH$_3$
c: R=R'=CH$_3$

Experimental procedures:

Precisely weighed samples of **2** (ca. 2 mmol), or **4** (ca. 1 mmol), or **7** (1.00 mmol) were placed in a ball mill under argon together with the precise equivalent of **1a–c**, or **1b** , or **6**, respectively. The Teflon gasket was closed with a torque of 50 Nm and ball-milling started for 1 h. The deliquescent salts **3, 5, 8** were quantitatively obtained and collected and stored under dry argon.

Reference: G. Kaupp, J. Schmeyers, J. Boy, *J. Prakt. Chem.,* **342**, 269 (2000).

Type of reaction: C–N bond formation
Reaction condition: solid-state
Keywords: isothiocyanates, amines, waste-free, gas-solid reaction, solid-solid reaction, thiourea

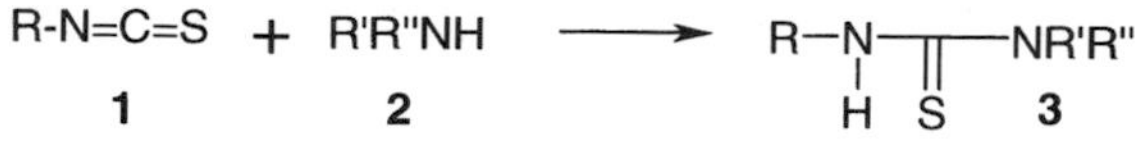

$$R-N=C=S \ + \ R'R''NH \ \longrightarrow \ R-\underset{\underset{H}{|}}{N}-\underset{\underset{S}{\parallel}}{} -NR'R''$$

$$\mathbf{1} \qquad\qquad \mathbf{2} \qquad\qquad\qquad \mathbf{3}$$

	R	R'/R"	T(°C)	
a:	Ph	H/H	-30	gs
b:	4-BrC$_6$H$_4$	Me/H	rt	gs
c:	4-Br C$_6$H$_4$	Me/Me	rt	gs
d:	1-Naph	Me/H	rt	gs
e:	1-Naph	Me/Me	rt	gs
f:	4-NO$_2$C$_6$H$_4$	Me/H	rt	gs
g:	4-NO$_2$C$_6$H$_4$	Me/Me	rt	gs
h:	Me	H/H	0	gs
i:	Me	Me/H	0	gs
j:	Me	Me/Me	0	gs
k:	Me	4-MeOC$_6$H$_4$/H	rt	mo
l:	Me	4-Br C$_6$H$_4$/H	rt	mo
m:	Me	4-Cl C$_6$H$_4$/H	rt	mo
n:	4-S=C=N-C$_6$H$_4$	Me/Me	rt	gs
o:	2-S=C=N- C$_6$H$_4$	2-NH$_2$ C$_6$H$_4$/H	rt	bm

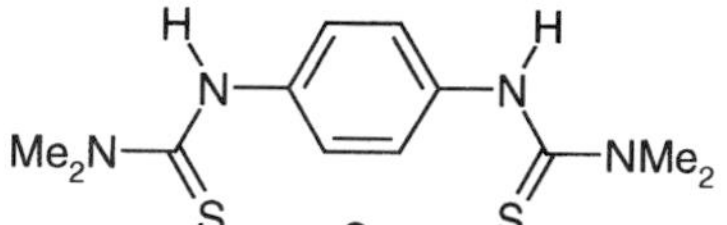

3n

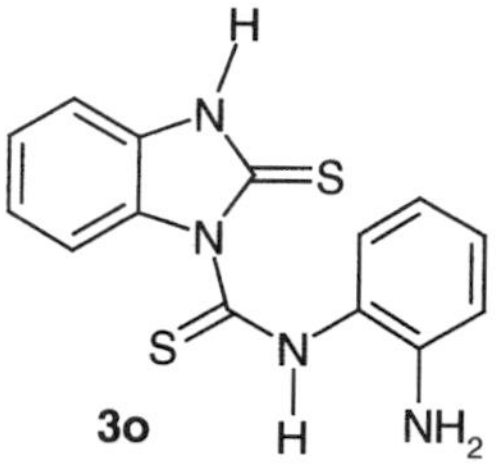

3o

Experimental procedures:

Crystalline isothiocyanates **1** (gs: 2.00 mmol, 1 bar of **2**, 250-mL flask, overnight; mo: 2.00 mmol, stoichiometric mixture, grinding in a mortar, standing for 1 day, intermediate softening; bm: 1 h ball-milling of stoichiometric mixture of **1o** and *o*-phenylendiamine) were, depending on the melting points and eutectica, reacted with the amine gases **2** (1 bar) or with the solid anilines **2** (2.00 mmol). Excess gas was recovered by condensation at –196 °C. The yields were 100% in all cases.

Reference: G. Kaupp, J. Schmeyers, J. Boy, *Tetrahedron*, **56**, 6899 (2000).

Type of reaction: C–N bond formation
Reaction condition: solid-state
Keywords: lactone, carbonic ester, amine, ring opening, waste-free, gas-solid reaction, amide, carbamate

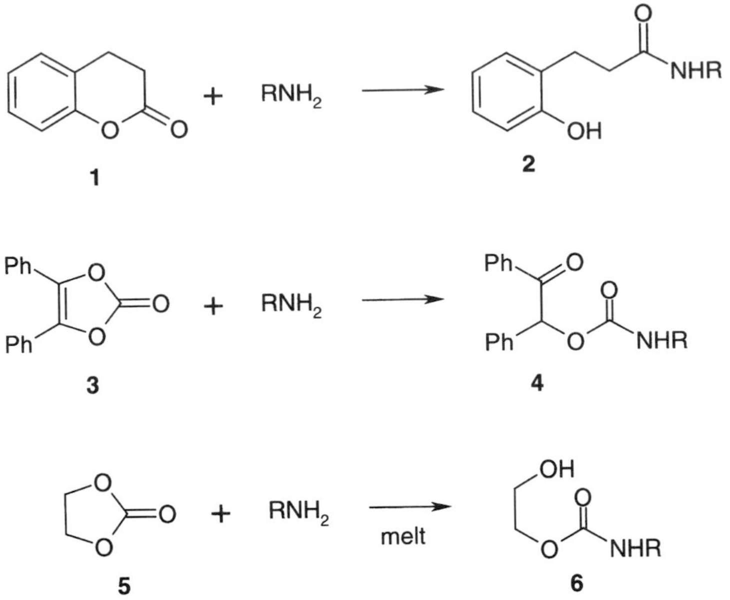

1   +   RNH$_2$ $\longrightarrow$ 2

a: R=H
b: R=Me
c: R=Et

3   +   RNH$_2$ $\longrightarrow$ 4

5   +   RNH$_2$ $\xrightarrow{\text{melt}}$ 6

Experimental procedures:
Compound **1** (2.00 mmol, 0 °C, overnight), **3** (2.00 mmol, rt, 1 day), or **5** (7.5 mmol, rt, 1 h, liquefies) in evacuated 250-mL flasks were treated with 0.8 bar ammonia, or methylamine, or ethylamine (**1** and EtNH$_2$ at 0.2 bar). Excess gas was recovered by condensation at −196 °C. The products **2** and **4** were quantitatively obtained. The liquid products **6** were isolated by distillation in 92, 93 and 92% yield.

Reference: G. Kaupp, J. Schmeyers, J. Boy, *Tetrahedron*, **56**, 6899 (2000).

Type of reaction: C–N bond formation
Reaction condition: solid-state
Keywords: imides, amines, ring opening, cyclization, gas-solid reaction, diamides

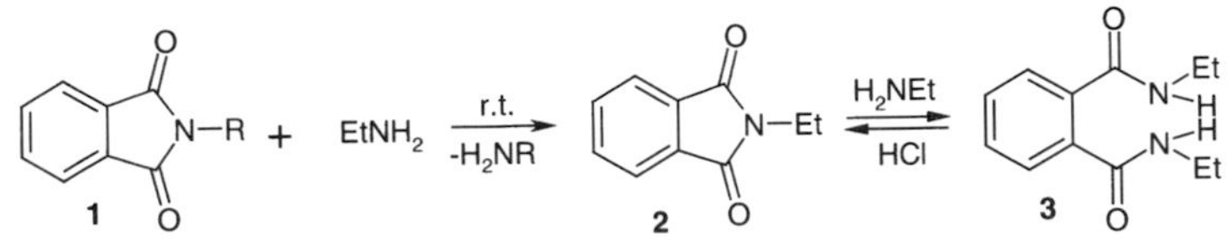

1   +   EtNH$_2$ $\xrightarrow[\text{-H}_2\text{NR}]{\text{r.t.}}$ 2 $\underset{\text{HCl}}{\overset{\text{H}_2\text{NEt}}{\rightleftarrows}}$ 3

a: R=H
b: R=OH
c: R=CO$_2$H

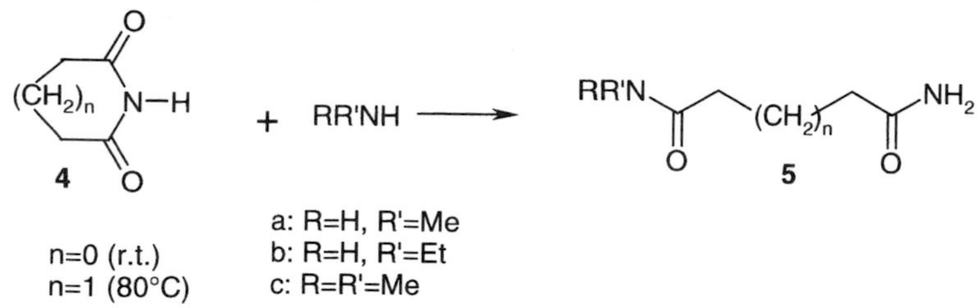

$$\begin{array}{c}\text{(CH}_2)_n \quad N-H \\ 4 \end{array} \quad + \quad RR'NH \quad \longrightarrow \quad \begin{array}{c} RR'N \\ \end{array} (CH_2)_n \begin{array}{c} NH_2 \\ 5 \end{array}$$

n=0 (r.t.)
n=1 (80°C)

a: R=H, R'=Me
b: R=H, R'=Et
c: R=R'=Me

Experimental procedures:

3: The phthalimides **1** (2.00 mmol) were reacted in 500-mL flasks with $EtNH_2$ (0.8 bar, 17.8 mmol) overnight. Excess gas and H_2NR were recovered in a cold trap at −196 °C. Glycine, in the case of **1c**, was washed away with water. Compound **3** was quantitatively obtained in all cases.

2: The diamide **3** (1.00 mmol) was exposed to HCl (1 bar) at room temperature overnight. Ethylammonium chloride was washed away with water and the imide **2** was quantitatively obtained after drying.

5: Crystals of succinimide or glutarimide (200 mg, 2.02 or 1.77 mmol) in an evacuated 500-mL flask were treated at room temperature or 80 °C with 1 bar or 0.8 bar, respectively, of methyl-, or ethyl-, or dimethylamine and left overnight or for 8 h, respectively. Excess gas was recovered in a −196 °C trap. The yield was quantitative in all cases.

Reference: G. Kaupp, J. Schmeyers, J. Boy, *Tetrahedron,* **56**, 6899 (2000).

Type of reaction: C–N bond formation
Reaction condition: solid-state
Keywords: anhydride, methylamine, waste-free, gas-solid reaction, amides, cyclization, imide

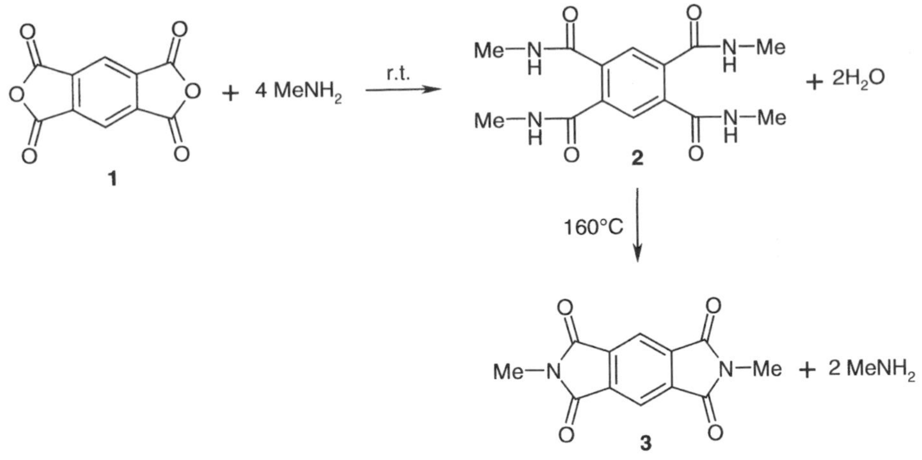

Experimental procedures:

2: Pyromellitic dianhydride **1** (436 mg, 2.00 mmol) was placed in an evacuated 100-mL flask and treated with methylamine from a lecture bottle when the temperature rose to a maximum of 95 °C for 2 min, but melting was avoided. For completion of the reaction, the flask (methylamine, 1 bar) was left overnight. Excess gas and water were evaporated to give a quantitative yield of pure **2**.

3: Pyromellitic tetramethylamide **2** (306 mg, 1.00 mmol) was heated to 160 °C in an evacuated flask for 12 h with occasional evaporation. The pure diimide **3** was quantitatively obtained.

References: G. Kaupp, J. Schmeyers, J. Boy, *Tetrahedron,* **56**, 6899 (2000).

Type of reaction: C–N bond formation
Reaction condition: solid-state
Keywords: large scale, anhydride, ethylamine, waste-free, gas-solid reaction, amides

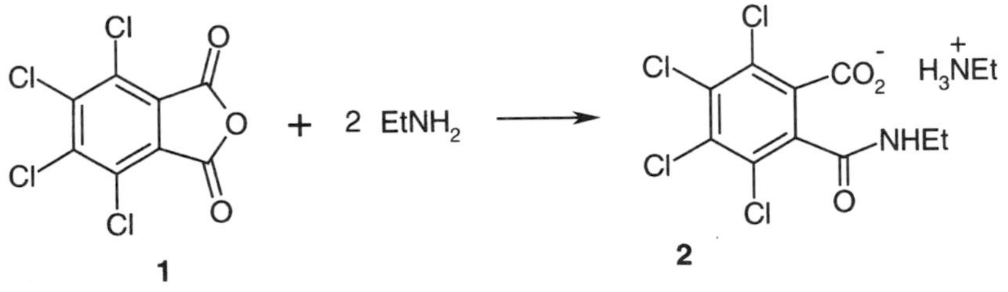

Experimental procedures:
A 2-L flask with tetrachlorophthalic anhydride **1** (500 g) was evacuated at a tight rotatory evaporator. The rotating flask was immersed in a cooling bath at about 14 °C and ethylamine from a steel bottle was continuously let in by adjusting a pressure of 0.5–0.8 bar with a needle valve (while using a security valve under a hood for safety reasons). When the heat production had ceased after 8 h the conversion was ca. 75% and became very slow, due to the aggregation of the 40 μm grains to give 0.5–3 mm particles. After intermediate removal of the gas, the material was ground in a large mortar and the reaction with ethylamine continued at 50 °C for a quantitative yield.

References: G. Kaupp, J. Schmeyers, J. Boy, *Tetrahedron*, **56**, 6899 (2000); *Chemosphere*, **43**, 55 (2001).

Type of reaction: C–N bond formation
Reaction condition: solid-state
Keywords: cyanogen bromide, 2-aminophenol hydrazide, cyclization, waste-free, gas-solid reaction, aminobenzoxazole, aminooxadiazole

a: R=H
b: R=CH$_3$
c: R=OH

Experimental procedures:
Solid **1** (10.0 mmol) was treated with BrCN (11.0 mmol) from a connected flask at a vacuum line. After standing overnight and recovering of the excess gas a quantitative yield of the aminobenzoxazole hydrobromide **2** was obtained. Similarly, the hydrazides **3a–c** (10 mmol) and BrCN (11.0 mmol) gave quantitative yields of **4a-c**.

Reference: G. Kaupp, J. Schmeyers, J. Boy, *Chem. Eur. J.*, **4**, 2467 (1998).

Type of reaction: C–N bond formation
Reaction condition: solid-state
Keywords: cyanogen chloride, cyanogen bromide, amines, gas-solid reaction, cyanamides

R—⟨benzene⟩—NH$_2$ + ClCN + NMe$_3$ ⟶ R—⟨benzene⟩—N(H)(CN) + Me$_3$N$^+$H Cl$^-$
                        (BrCN)   (Br$^-$)

1   **2**

a: R=Me
b: R=OMe
c: R=Cl

R^1—⟨benzimidazole⟩—R^3 + BrCN + NMe$_3$ ⟶ R^1—⟨benzimidazole⟩—R^3
R^2 R^2

3 H   **4** CN

a: R^1=R^2=R^3=H
b: R^1=R^2=H, R^3=Me
c: R^1=R^2=Me, R^3=H

⟨phthalimide⟩N—K + BrCN ⟶ ⟨phthalimide⟩N—CN + KBr

5   **6**

Experimental procedures:
2 and **4**: The crystalline substrate **1a–c** or **3a–c** (10 mmol) was placed in a 500-mL flask which was evacuated and filled with a 1:1-mixture of ClCN and tri-methylamine (11.7 mmol, each), or it was placed in an evacuated 250-mL wide-neck flask connected to a 250-mL wide-neck flask with freshly sublimed BrCN (1.17 g, 11.0 mmol) and trimethylamine (0.5 bar, 11.7 mmol) was added through a vacuum line. A magnetic spin bar was rotated in the flask in order to mix the gases and the system was left overnight at room temperature. Excess gas was pumped to a cold trap at −196 °C. The trimethylammonium chloride (bromide) was removed by washings with water. The yield of **2** or **4** was quantitative in all cases.

  6: Potassium phthalimide **5** (370 mg, 2 mmol) was ball-milled to a particle size of <1 µm (30 min) and then treated with gaseous BrCN (225 mg, 2.10 mmol) for 24 h. Excess gas was pumped to a cold trap at −196 °C and KBr removed by washings with water. The yield of **6** was 340 mg (99%), mp 73 °C.

Reference: G. Kaupp, J. Schmeyers, J. Boy, *Chem. Eur. J.,* **4**, 2467 (1998).

Type of reaction: C–N bond formation
Reaction condition: solid-state
Keywords: barbituric acid, nitrogen dioxide, nitration, waste-free, gas-solid reaction

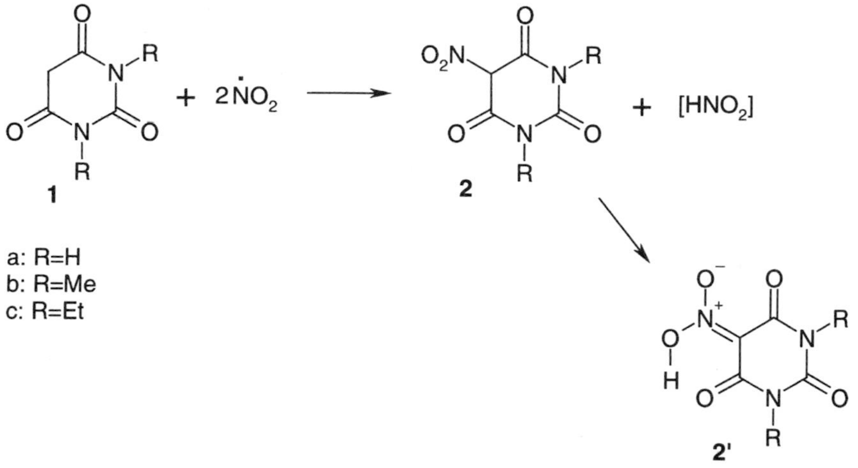

a: R=H
b: R=Me
c: R=Et

Experimental procedures:
Barbituric acid **1a** (500 mg, 3.91 mmol) or **1b** (500 mg, 3.21 mmol), or **1c** (500 mg, 2.72 mmol) in an evacuated flask was treated with NO_2 (6.4 mmol) at an initial pressure of 0.3 bar for 4 h. After condensation of the gases to a cold trap at $-196\,°C$ and drying of the solid products at $80\,°C$ in a vacuum the pure compounds **2a–c**, that assume the aci-nitro form **2′**, were quantitatively obtained.

Reference: G. Kaupp, J. Schmeyers, *J. Org. Chem.*, **60**, 5494 (1995).

Type of reaction: C–N bond formation
Reaction condition: solid-state
Keywords: large scale, ninhydrin, L-proline, cascade reaction, zwitterion, waste-free, solid-solid reaction

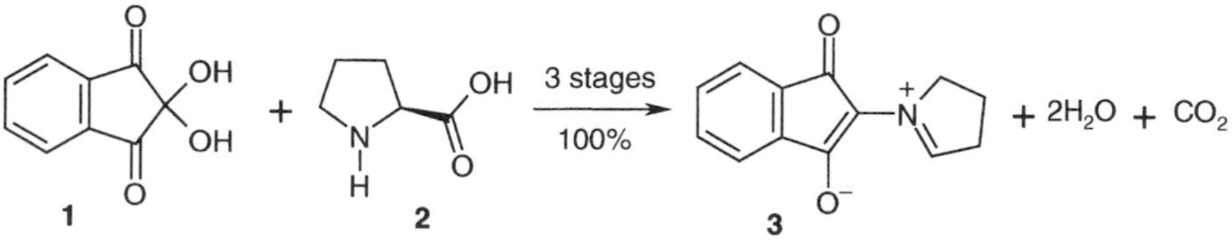

Experimental procedure:
A stoichiometric mixture of ninhydrin and L-proline **2** (200 g) was milled in a 2-L horizontal ball-mill (Simoloyer®) with steel balls (100Cr6, 2 kg, diameter 5 mm) at 1100 min^{-1} for 40 min until the liberation of CO_2 was complete. The temperature varied from 15 °C at the water cooled walls to 21 °C in the center. The power was 800 W. Quantitative reaction to give **3** was secured by weight (146 g, 100%) and by spectroscopic techniques. The product was not separated in a cyclone but the milling-out towards the end was completed with 4 times 250 mL of water, each. This part of the highly disperse (<1 μm) pure azomethine ylide **3** was obtained after centrifugation and drying in a vacuum. The combined water phase contained 0.2 g of **3**.

Reference: G. Kaupp, M.R. Naimi-Jamel, H. Ren, H. Zoz, *Chemie Technik,* **31**, 58 (2002).

Type of reaction: C–N bond formation
Reaction condition: solid-state
Keywords: ninhydrin, *o*-phenylendiamine, *o*-aminothiophenol hydrochloride, (thio)urea, cascade reaction, solid-solid reaction

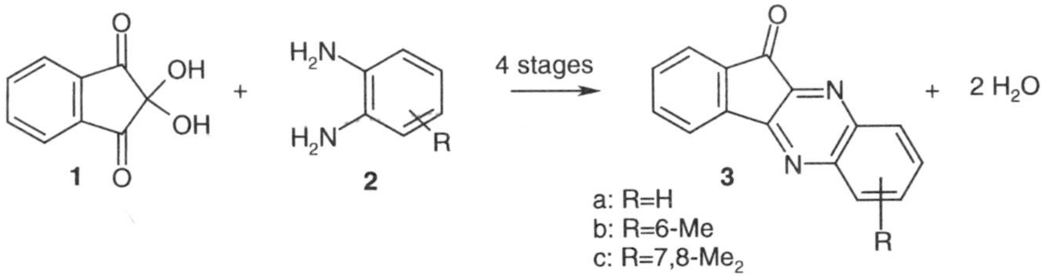

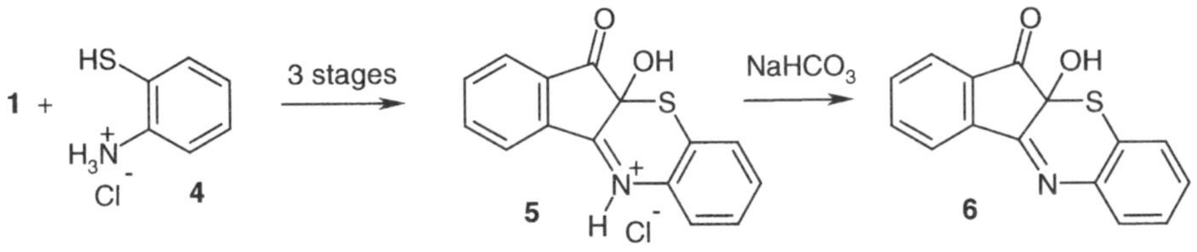

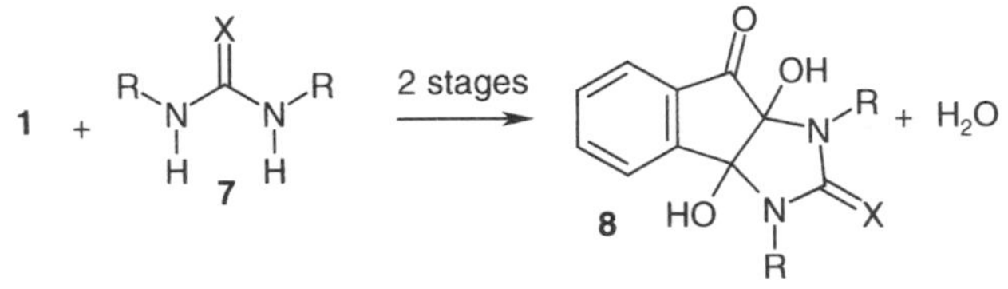

a:	R=H, X=O	(80°C)
b:	R=H, X=S	(25°C)
c:	R=Me, X=O	(25°C)
d:	R=Me, X=S	(20°C)
e:	R=Ph, X=S	(60°C)

Experimental procedures:

Preparation of **3a**: Ninhydrin **1** (356 mg, 2.00 mmol) and *o*-phenylendiamine **2a** (216 mg, 2.00 mmol) were ball-milled for 15 min at –5 °C. Compound **3a** (461 mg, 99%) was obtained after drying in a vacuum at 80 °C. Mp 217–218 °C. Similarly, the compounds **3b–c** were quantitatively obtained.

Preparation of **5, 6**: Ninhydrin **1** (356 mg, 2.00 mmol) and o-aminothiophenol hydrochloride (**4**) (323 mg, 2.00 mmol) were ball-milled for 1 h. The solid salt **5** was triturated with saturated NaHCO₃ solution, washed with water and dried to obtain pure **6** (529 mg, 99%). Mp 227 °C.

Preparation of **8a**: Ninhydrin **1** (534 mg, 3.00 mmol) and urea **7a** (180 mg, 3.00 mmol) were ball-milled at 80 °C for 1 h. After drying at 140 °C in a vacuum for 5 min pure **8a** (696 mg, 100%) was obtained. Mp 218 °C (decomposition). The compounds **8b–e** were similarly obtained in quantitative yield by ball-milling at the temperatures given.

Reference: G. Kaupp, M.R. Naimi-Jamal, J. Schmeyers, *Chem. Eur. J.,* **8**, 594 (2002).

Type of reaction: C–N bond formation
Reaction condition: solid-state
Keywords: *N*-arylmethylenimine, cascade reaction, solid-solid reaction, gas-solid reaction, aminomethylation, cyclization, Troeger base

a: R=H   60% (15%)
b: R=CH₃ 78% (40%)
c: R=OCH₃ 30% (55%)

Experimental procedures:

Gas-solid reaction: The methyleniminium chloride **1a–c** (2.0 mmol) was ball-milled for 5 min and then spread on a watch glass with 15 cm diameter in air of 35% relative humidity at about 22 °C. The characteristic IR-bands of the C=N group of **1a–c** had disappeared after 1 h when the material became deliquescent. The stoichiometric 1:1 mixtures of **2** and **3** and polymerized products were separated by chromatography after washings with $NaHCO_3$ solution to obtain **4a–c** with the yields indicated.

Solid-solid reaction: The methyleniminium chloride **1a–c** (1.0 g) and $MgSO_4 \cdot 7H_2O$ (10 g) as a solid water-supply were co-ground in a mortar for 10 min. The solid mixture was extracted with CH_2Cl_2 (30 mL, 5 times with centrifugation) and the Troeger base **4a, b, c** separated by chromatography.

Reference: G. Kaupp, J. Schmeyers, J. Boy, *J. Prakt. Chem.,* **342**, 269 (2000).

Type of reaction: C–N bond formation
Reaction condition: solid-state
Keywords: solid benzaldehydes, benzhydrazide, isatin, waste-free, solid-solid reaction, acyl hydrazide

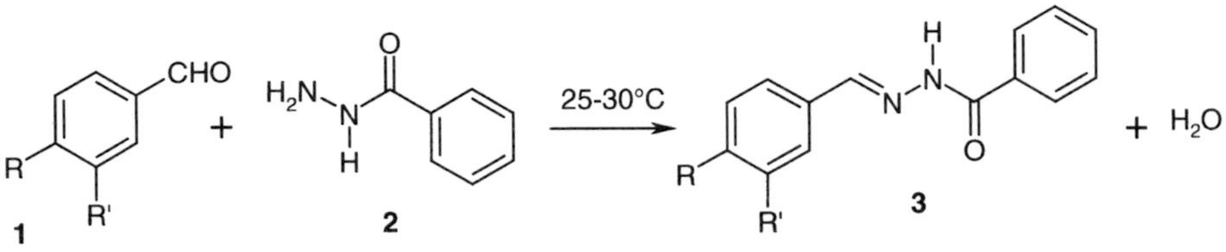

a: R=OH, R'=H
b: R=H, R'=NO$_2$

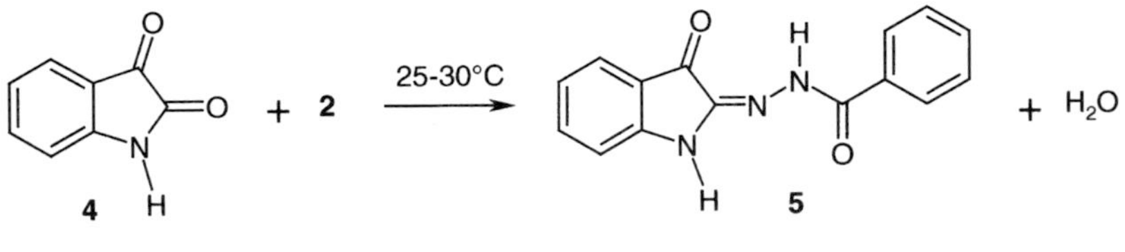

Experimental procedures:
The aldehydes **1** (2.00 mmol) or isatin **4** (2.00 mmol) and benzhydrazide **2** (272 mg, 2.00 mmol) were ball-milled at 25–30 °C for 1 h or 3 h, respectively. After drying at 0.01 bar at 80 °C **3a, b** or **5** were obtained in pure form with quantitative yield.

Reference: G. Kaupp, J. Schmeyers, J. Boy, *J. Prakt. Chem.,* **342**, 269 (2000).

Type of reaction: C–N bond formation

Reaction condition: solid-state

Keywords: enamine, dibenzoylethene, cyclization, cascade reaction, waste-free, solid-solid reaction, pyrrole

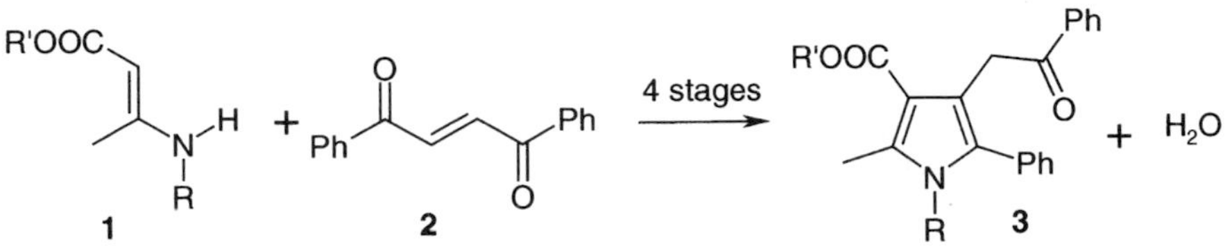

a: R=H, R'=CH₃

b: R=R'=CH₃

c: R=CH₃, R'=C₂H₅

d: R=CH₂Ph, R'=C₂H₅

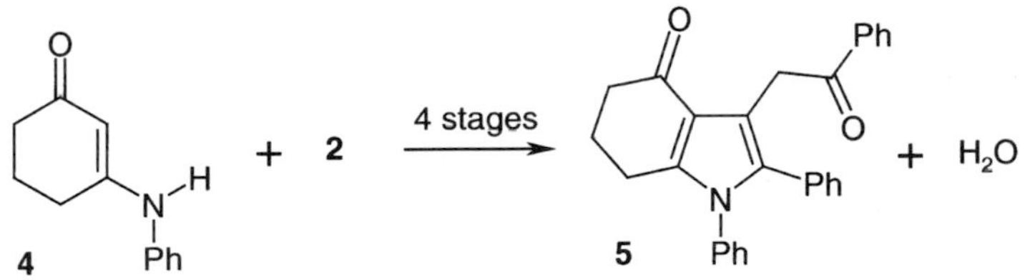

Experimental procedures:

A ball-mill (Retsch MM 2000) with a 10-mL beaker out of stainless steel and two steel balls (6.5 g) was run at 20–30 Hz for 3 h in order to achieve quantitative conversions from stoichiometric mixtures of **1** or **4** and **2**. Milling of **1a–d** or **4** (2.00 mmol) and **2** (2.00 mmol) gave dust-dry powders of **3a, b, d** or **5** in quantitative yield that were heated to 80 °C for removal of the water of reaction. Compound **3c** had to be heated to 150 °C for 5 min in order to complete the elimination of water from its direct precursor. Solution reactions, at 65 °C (3 h) gave inferior yields.

Reference: G. Kaupp, J. Schmeyers, A. Kuse, A. Atfeh, *Angew. Chem.*, **111**, 3073 (1999); *Angew. Chem. Int. Ed. Engl.*, **38**, 2896 (1999).

Type of reaction: C–N bond formation

Reaction condition: solid-state, solvent-free

Keywords: 1,3-cyclohexanedione, anilines, melt-reaction, solid-solid reaction, enamine ketone

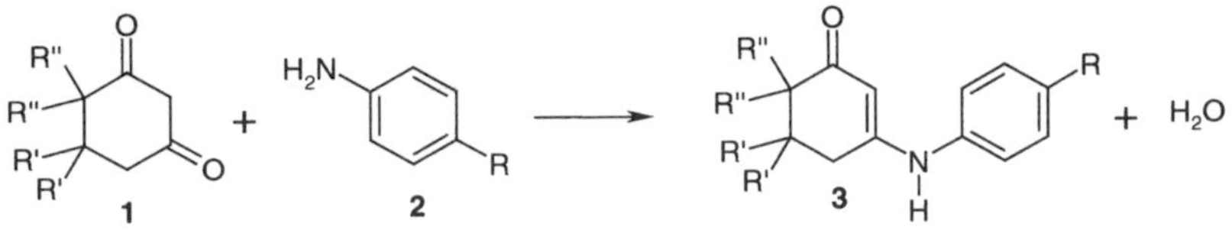

a:	R=R'=R"=H	(melt, 85%)
b:	R=H, R'=Me, R"=H	(melt, 100%)
c:	R=Me, R'=R"=H	(100%)
d:	R=OMe, R'=R"=H	(100%)
e:	R=Cl, R'=R"=H	(100%)
f:	R=R'=Me, R"=H	(100%)
g:	R=OMe,R'=Me, R"=H	(100%)
h:	R=Cl, R'=H, R"=Me	(100%)

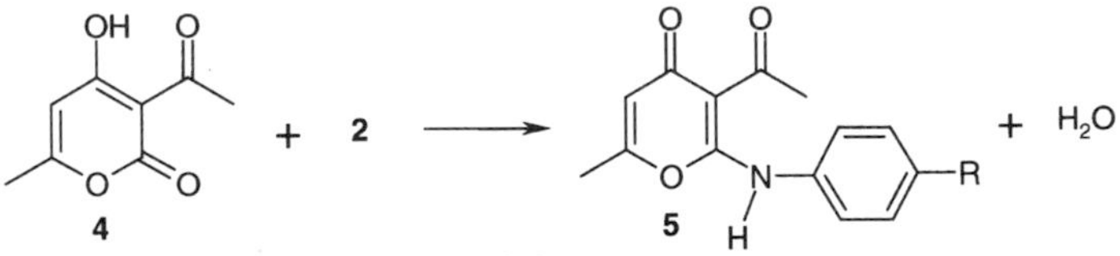

a:	R=H	(melt, 90%)
c:	R=Me	(100%)
d:	R=OMe	(100%)

Experimental procedures:
Melt reactions: 2.00 mmol of the 1,3-dicarbonyl compound **1** or **4** and aniline (186 mg, 2.00 mmol) were heated to 80 °C for 1 h in an evacuated 50-mL flask in a drying oven. The solid product was dried at 0.01 bar at 80 °C. **1a** and **5a** were purified by recrystallization.

Solid-solid reactions: 2.00 mmol of the 1,3-dicarbonyl compound **1** or **4** and 2.00 mmol of the solid aniline derivative **2** were ball-milled at room temperature for 30 min. After drying of the solid hydrates at 0.01 bar at 80 °C the pure enamine ketones **3** or **5** were obtained with quantitative yield.

Reference: G. Kaupp, J. Schmeyers, J. Boy, *J. Prakt. Chem.,* **342**, 269 (2000).

Type of reaction: C–N bond formation
Reaction condition: solid-state
Keywords: Viehe salt, nitroaniline, sulfonamide, hydrazone, solid-solid reaction

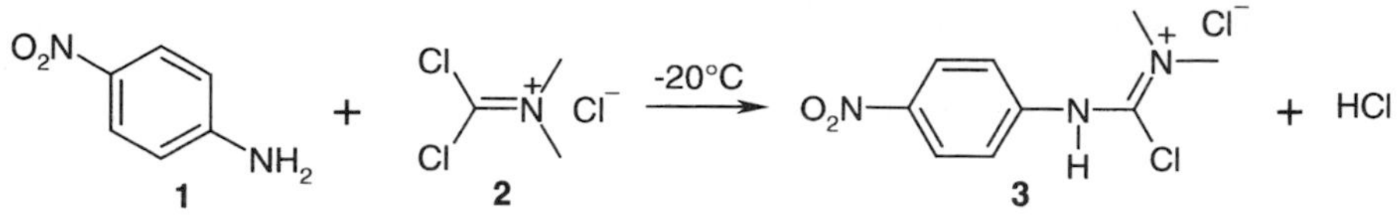

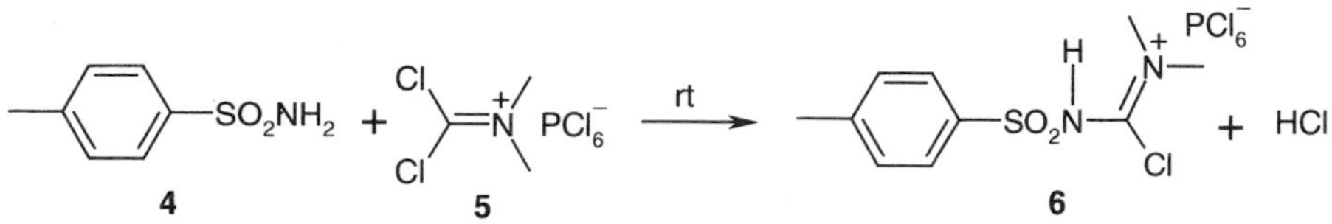

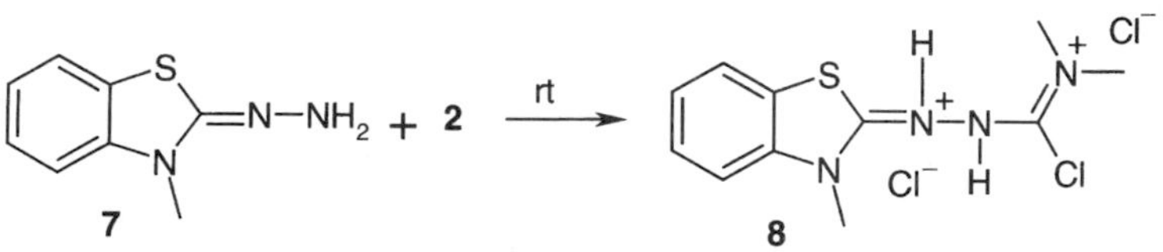

Experimental procedures:

Solid *p*-nitroaniline (690 mg, 5.00 mmol) and Viehe salt **2** (815 mg, 5.00 mmol) were ball-milled at –20 °C for 1 h. Most of the HCl escaped through the Teflon gasket that was tightened against a torque of 50 Nm. Residual HCl was pumped off at room temperature. The yield of pure salt **3** was 1.321 g (100%).

Compound **6** and **8** were quantitatively obtained by similar ball-milling of the appropriate components at room temperature.

Reference: G. Kaupp, J. Boy, J. Schmeyers, *J. Prakt. Chem.,* **340**, 346 (1998).

Type of reaction: C–N bond formation
Reaction condition: solid-state
Keywords: hydroxylamine phosphate, acetone, large scale, exhaust gas purification, flow system, gas-solid reaction, acetone oxime

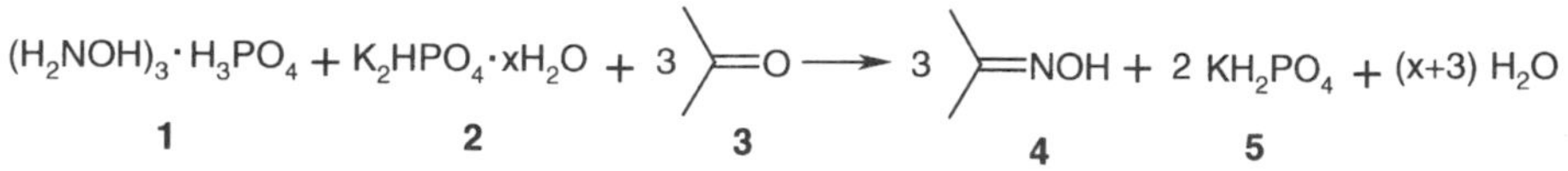

Experimental procedures:

Two heatable glass tubes (l=50 cm, i.d.=2 cm) fitted with glass frits were each loaded with 53.1 g (0.269 mol) of ground (0.14 m^2 g^{-1}) hydroxylaminium phosphate **1** and 51.9 g (0.269 mol) of un-ground (0.10 m^2 g^{-1}) K$_2$HPO$_4$·H$_2$O **2**. Both tubes were externally heated to 80 °C. Air (1 L min^{-1}) was passed through 46.9 g (0.807 mol) of acetone **3** and then over the solids from the top of the tubes. All of the acetone (at a load of 78 g m^{-3}) had reacted within 10 h. Behind the second column was a condenser flask in an ice bath and a filter of activated carbon (5 g, 645 m^2 g^{-1}) to catch the last traces of free acetone oxime **4** which escaped condensation. The product **4** and the water of reaction and crystal water were continuously expelled in gaseous form from the columns and condensed out at 0 °C. Only after >75% conversion, acetone started to escape from the first column and the second column started to react (at <50 g m^{-3} acetone load, such escape started at >90% conversion). After passing all of the acetone, column 1 had reacted to 94% and column 2 to 6%. The condensate consisted of water (18.5 g, 96%) and of crystalline acetone oxime **4** (58.7 g, 99.3%). The missing 400 mg (0.7%) of noncondensed **4** were efficiently and completely absorbed by the activated carbon at the exhaust (400 mg of **4** are absorbed by 3 g of activated carbon). The acetone oxime **4** was separated from water by continuous azeotropic removal with *tert*-butyl methyl ether. The reacted column was contracted by less than 10% and contained crystalline KH$_2$PO$_4$ **5** in analytically pure form as a couple product.

Apart from the quantitative synthesis of **4**, this experiment simulated an exhaust gas purification down to zero emission.

Reference: G. Kaupp, U. Pogodda, J. Schmeyers, *Chem. Ber.*, **127**, 2249 (1994).

Type of reaction: C–N bond formation
Reaction condition: solid-state
Keywords: carbonyl reagent, acetone, hydroxyl amine, semicarbazide, hydrazine, condensation, waste-free, gas-solid reaction, imine

$$\underset{\mathbf{1}}{\diagdown\!\diagup}\!\!=\!\!O \;+\; \underset{\substack{(\cdot HX)\\ \mathbf{2}}}{H_2N\text{-}R} \longrightarrow \underset{\substack{(\cdot HX)\\ \mathbf{3}}}{\diagdown\!\diagup}\!\!=\!\!NR \;+\; H_2O$$

$H_2NOH\cdot HCl$	$(H_2NOH)_2\cdot H_2SO_4$	$(H_2NOH)_3\cdot H_3PO_4$
2a	**2b**	**2c**

$H_2NNH\overset{\overset{\displaystyle O}{\|}}{C}NH_2\cdot HCl$	$H_2NNH\overset{\overset{\displaystyle S}{\|}}{C}NH_2$	$H_2NNH\overset{\overset{\displaystyle O}{\|}}{C}NHNH_2$	$H_2NNH\overset{\overset{\displaystyle S}{\|}}{C}NHNH_2$
2d	**2e**	**2f**	**2g**

$Ph\overset{\overset{\displaystyle O}{\|}}{C}NHNH_2$	$t\text{-}BuO\overset{\overset{\displaystyle O}{\|}}{C}NHNH_2$	$p\text{-}TolNHNH_2\cdot HCl$
2h	**2i**	**2j**

O$_2$N ― NHNH$_2$ ... O$_2$N ... **2k**

benzothiazole =NNH$_2$ ring (N-methyl) **2l**

Experimental procedures:

Semicarbazide hydrochloride **2d** (1.00 g, 9.0 mmol) in an evacuated 1-L flask was connected with a vacuum line to a 100-mL flask containing 1.57 g (27 mmol) of degassed acetone **1**. After 12 h excess gas **1** and the water of the reaction were evaporated or condensed into the smaller flask. The yield of pure **3d** was 1.52 g (100%).

2,4-Dinitrophenylhydrazine **2k** (1.00 g, 5.1 mmol) was air dried from adhering water (which was there for safety reasons) with suction, placed in an evacuated 250-mL flask and connected with a vacuum line to a 100-mL flask which contained 0.58 g (10 mmol) of degassed acetone **1** and left overnight. Excess **1** and the water of the reaction were removed by evaporation. The yield of pure **3k** was 1.20 g (100%).

Similarly, all imines (imine salts) **3a–l** were obtained quantitatively. The reaction of **2a** provided a partially liquefied (deliquescent) **3a**·HCl+H$_2$O that lost the water and solidified upon evaporation. The reaction times with **2g** and **2l** were 2 and 3 days under these modest conditions.

Reference: G. Kaupp, U. Pogodda, J. Schmeyers, *Chem. Ber.*, **127**, 2249 (1994).

Type of reaction: C–N bond formation
Reaction condition: solid-state
Keywords: penicillamine, acetone, cysteine, cyclization, gas-solid reaction, C–S bond formation, thiazolidine

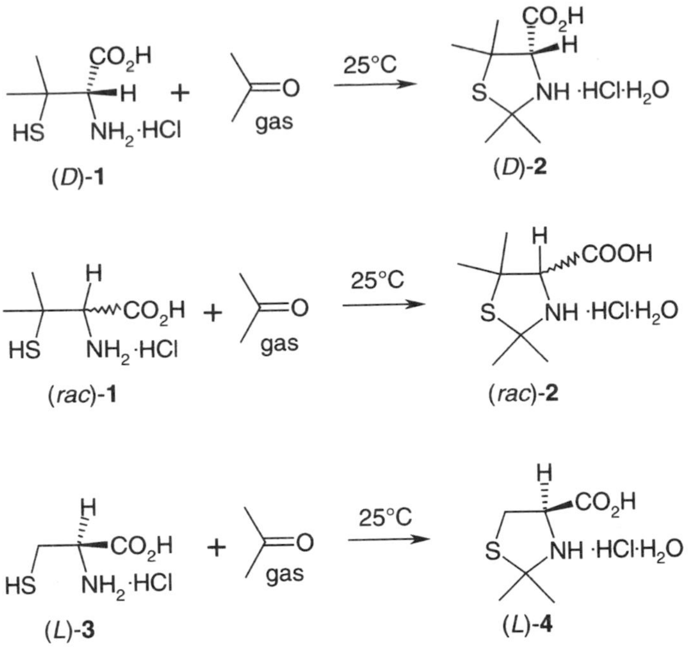

Experimental procedures:
The hydrochlorides D-1, or rac-1, or L-3 (5.00 mmol) in an evacuated 50-mL flask were connected with a vacuum line to a 50-mL flask containing acetone (970 mg, 6.0 mmol). The acetone evaporated as it was consumed. After 12 h, excess gas was pumped off and a quantitative yield of the respective thiazolidine hydrochloride hydrate was obtained.

Reference: G. Kaupp, U. Pogodda, J. Schmeyers, *Chem. Ber.,* **127**, 2249 (1994).

Type of reaction: C–N bond formation
Reaction condition: solid-state
Keywords: acetone, o-phenylendiamine, cyclization, cascade reaction, gas-solid reaction, benzodiazepine, benzothiazolidine

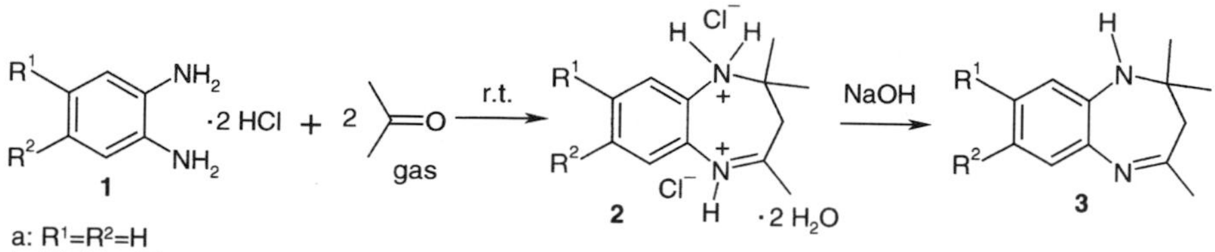

a: R¹=R²=H
b: R¹=Me, R²=H
c: R¹=R²=Me
d: R¹=H, R²=NO₂

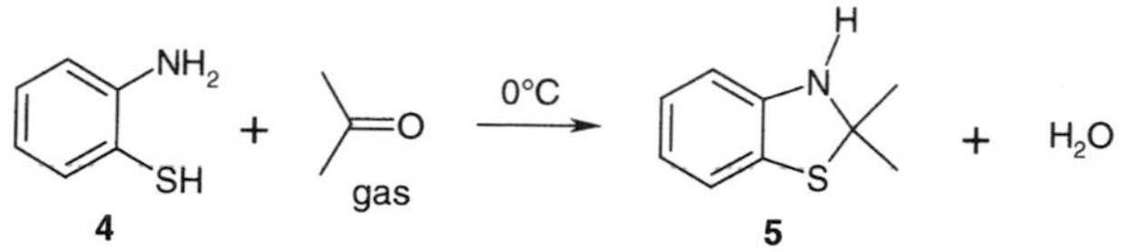

Experimental procedures:

The salt **1**·2HCl was quantitatively obtained from *o*-phenylendiamine and HCl gas at 0.5 bar in a flask that was rotating in an ice bath.

The salt **1**·2HCl (20 g, 80 mmol) was placed in a 10-L desiccator and after evacuation connected to a 100-mL flask containing 9.3–11.6 g (160–200 mmol) of acetone which had previously been degassed in a vacuum and cooled with liquid nitrogen. Upon removal of the cooling bath the acetone evaporated slowly into the desiccator. After 12 h, excess gas was condensed back to the flask (−196 °C) and a 100% yield of **2a**·2H₂O was obtained. The free base **3a** was liberated with NaOH in water.

The reaction of **1d**·2HCl gave **2d**·2HCl·2H₂O (71%) and **3d** therefrom after neutralization. The reactions of **1b, c** and **4** (to give **5**) were performed with 1 g of the starting material. The yield was 100% in all cases.

Reference: G. Kaupp, U. Pogodda, J. Schmeyers, *Chem. Ber.*, **127**, 2249 (1994).

Type of reaction: C–N bond formation
Reaction condition: solvent-free
Keywords: *o*-phenylendiamine, parabanic acid, cyclization, cascade reaction, melt reaction, quinoxaline

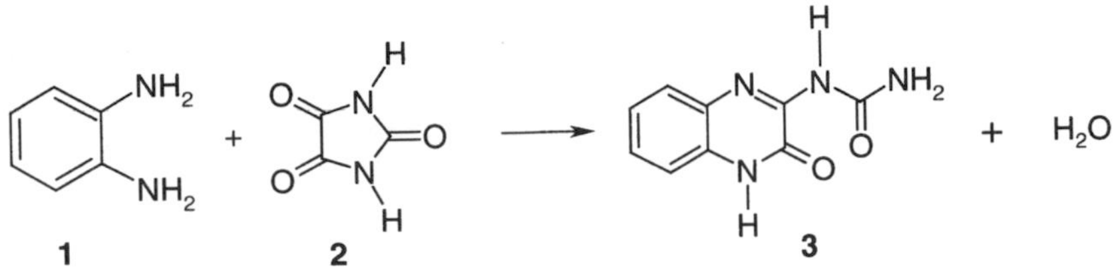

Experimental procedure:
A melt of **1** (216 mg, 2.00 mmol) and parabanic acid **2** (228 mg, 2.00 mmol) was heated to 130–140 °C for 30 min in a vacuum. The raw material contained **3** (78%). It was washed with cold 50% AcOH (10 mL), and the residue was crystallized from AcOH to give 286 mg (70%) **3**.

Reference: G. Kaupp, M.R. Naimi-Jamal, *Eur. J. Org. Chem.*, 1368 (2002).

Type of reaction: C–N bond formation
Reaction condition: solid-state, solvent-free
Keywords: *o*-phenylendiamine, alloxane, cyclization, cascade reaction, solid-solid reaction, melt reaction, 2-quinoxalinone

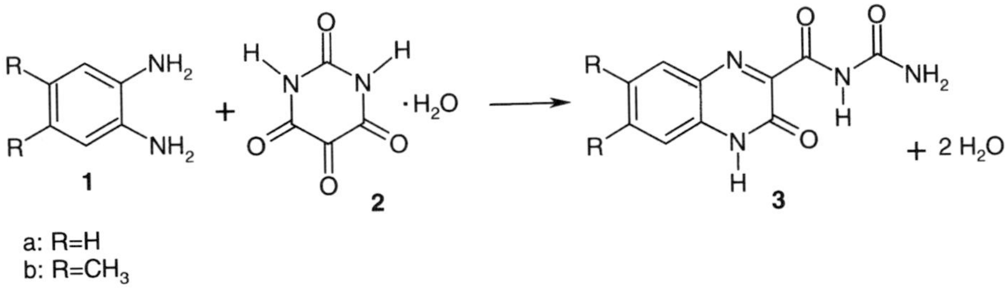

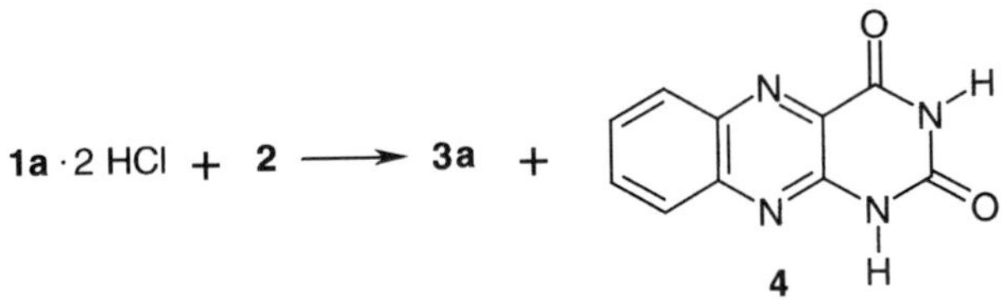

a: R=H
b: R=CH$_3$

Experimental procedures:
Crystalline *o*-phenylendiamine **1a** (108 mg, 1.00 mmol) and alloxane hydrate **2** (160 mg, 1.00 mmol) were ball-milled at room temperature for 1 h to give 232 mg (100%) of pure **3a**.

Similarly, a quantitative yield of **3b** was obtained.

A stoichiometric melt of **1a**·2HCl and **2** was kept at 130–140 °C for 30 min. The mixture was heated with 20 mL of acetic acid when 14% **4** were filtered from the hot solution and 47% **3a** separated from the cold mother liquor.

Reference: G. Kaupp, M.R. Naimi-Jamal, *Eur. J. Org. Chem.,* 1368 (2002).

Type of reaction: C–N bond formation
Reaction condition: solid-state
Keywords: o-phenylendiamine, 2-oxoglutaric acid, cyclization, cascade reaction, waste-free, solid-solid reaction, 2-quinoxalinone

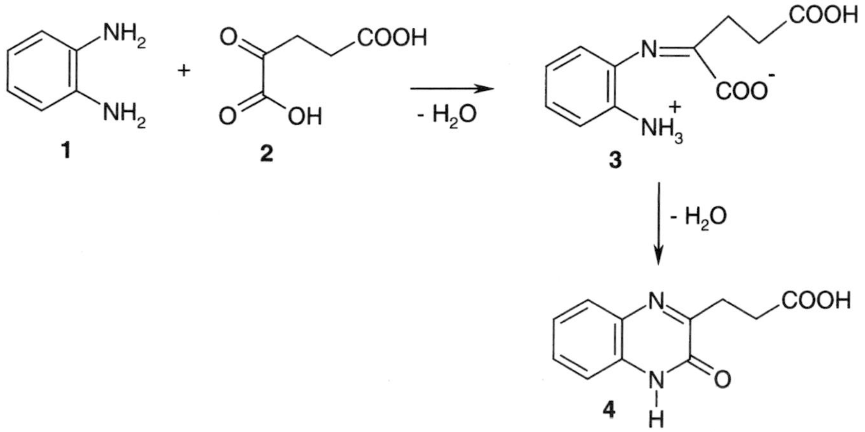

Experimental procedure:
Crystalline o-phenylendiamine **1** (216 mg, 2.00 mmol) was co-ground in a mortar with 2-oxoglutaric acid **2** (242 mg, 2.00 mmol) and heated to 120–125 °C for 30 min in a vacuum. Pure **4** (472 mg, 100%) was obtained directly.

Reference: G. Kaupp, M.R. Naimi-Jamal, *Eur. J. Org. Chem.,* 1368 (2002).

Type of reaction: C–N bond formation
Reaction condition: solid-state
Keywords: o-phenylenediamine, oxalic acid dihydrate, cyclization, quinoxaline-dione, 2,3-dichloroquinoxaline, cascade reaction, solid-solid reaction, melt reaction, bis-benzimidazolyl, fluoflavin

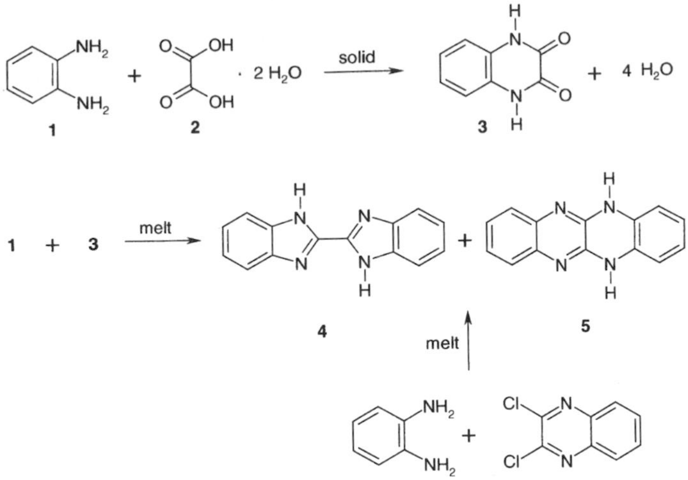

Experimental procedures:
Crystalline *o*-phenylenediamine **1** (108 mg, 1.00 mmol) and oxalic acid dihydrate **2** (126 mg, 1.00 mmol) were co-ground in a mortar. The high melting salt was heated in a vacuum to 150 °C for 8 h, or to 180 °C for 30 min, or to 210–220 °C for 10 min. Pure **3** was obtained in all cases (162 mg, 100%).

The compounds **1** (54 mg, 0.50 mmol) and **3** (81 mg, 0.50 mmol) were heated to 240 °C for 12 h under vacuum. A crystalline 94:6-mixture of **4** and **5** was quantitatively obtained.

The compounds **1** (108 mg, 1.00 mmol) and 2,3-dichloroquinoxaline **6** (199 mg, 1.00 mmol) were heated to 150 °C in a vacuum for 30 min. The crude mixture contained 1·2HCl (39%), **4** (7%) and **5** (53%). After treatment with 1 n NaOH, water and cold ethanol (20 mL) the residue contained **5** and **4** in a 93:7 ratio. Pure fluoflavin **5** was obtained by heating the mixture with 5 mL of ethanol, hot filtration and three washings with 3 mL of hot ethanol.

Reference: G. Kaupp, M.R. Naimi-Jamal, Eur. *J. Org. Chem.*, 1368 (2002).

Type of reaction: C–N bond formation
Reaction condition: solid-state, solvent-free
Keywords: *o*-phenylendiamine, benzil, cyclization, cascade reaction, melt reaction, waste-free, solid-solid reaction, quinoxaline

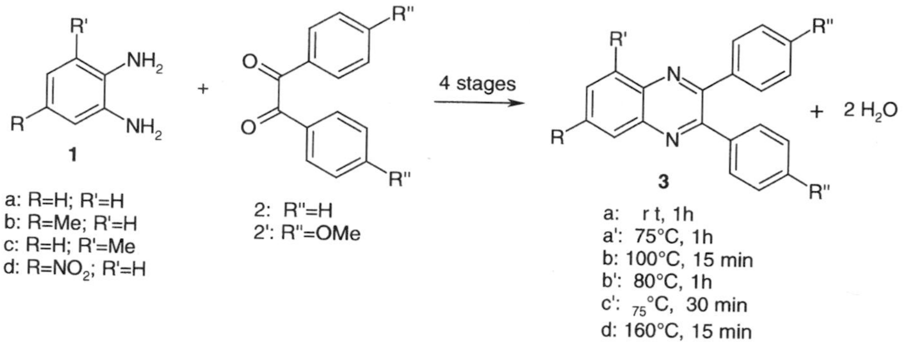

R'
NH2
NH2
R
1
a: R=H; R'=H
b: R=Me; R'=H
c: R=H; R'=Me
d: R=NO2; R'=H
+
O
O
R"
R"
2: R"=H
2': R"=OMe
4 stages
R'
N
N
R
R"
R"
3
+ 2 H2O
a: r t, 1h
a': 75°C, 1h
b: 100°C, 15 min
b': 80°C, 1h
c': 75°C, 30 min
d: 160°C, 15 min

1a +
O
OH
O
4
4 stages
N
N
OH
5
+ 2 H2O

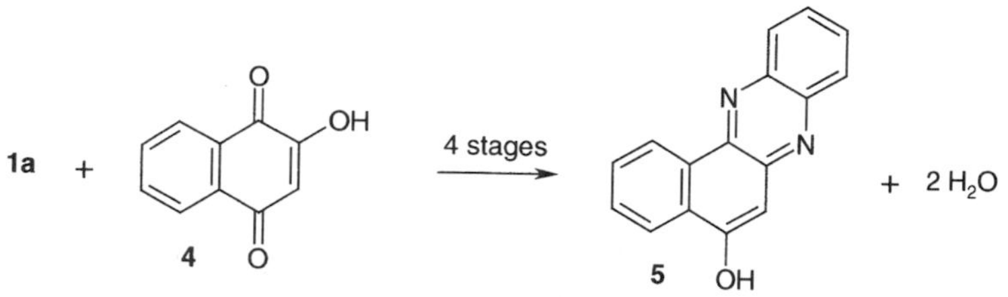

Experimental procedures:
Crystalline *o*-phenylendiamine **1a** (108 mg, 1.00 mmol) and benzil **2** (210 mg, 1.00 mmol) were ball-milled (heatable Retsch MM 2000, 10-mL steel beaker, two 12 mm balls, 20–25 Hz) for 1 h. Pure **3a** (282 mg, 100%) was obtained after removal of the water at 80 °C in a vacuum for 1 h.

Similarly, all the further products **3** were obtained with 100% yield at the temperatures given. **1b** and **1d** were co-ground with **2** at room temperature and heated in an oven for 15 min to 100 °C and 160 °C to obtain pure **3b** and **3d** with 100% yield.

Quantitative yields could also be obtained in stoichiometric melts at considerably higher temperatures in an evacuated 5-mL flask: **3a** at 100 °C, 20 min; **3a'** at 130 °C, 30 min; **3b'** at 130 °C, 30 min.

Compound **5** was similarly obtained with 100% yield in the solid state (70 °C, 15 min) and in the melt (110 °C, 30 min).

Reference: G. Kaupp, M.R. Naimi-Jamal, *Eur. J. Org. Chem.*, 1368 (2002).

Type of reaction: C–N bond formation
Reaction condition: solvent-free
Keywords: aryl ester, *o*-phenylendiamine, cyclization, melt reaction, benzimidazole

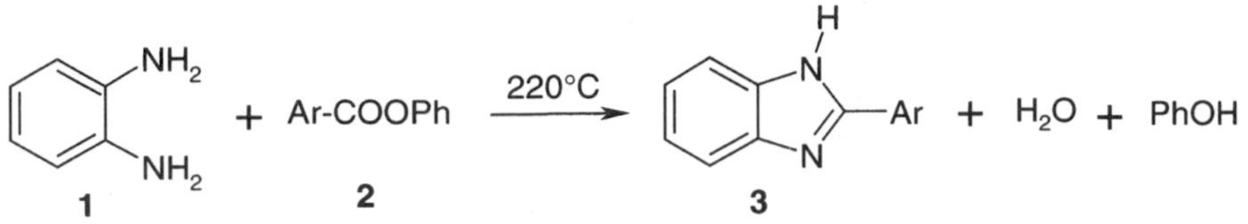

	Ar	yield (%)
a:	H	86
b:	*o*-OH	93
c:	*p*-OH	94
d:	*o*-Me	80
e:	*m*-Me	87
f:	*p*-Me	87
g:	*o*-Cl	87
h:	*p*-Cl	91
i:	*o*-NH$_2$	45-60
j:	*p*-NH$_2$	45-60

Experimental procedures:
Stoichiometric mixtures of *o*-phenylendiamine **1** and (substituted) benzoic ester **2** were heated to 220 °C for 1.5 h. The products **3** were isolated with the yields given.

Reference: T. Ebana, K. Yokota, Y. Takada, *Hokkaido Daigaku Kogakubu Kenkyu Hoko-ku,* **89**, 127–130 (1978), *Chem. Abstr.,* **91**, 157 660 (1979).

Type of reaction: C–N bond formation
Reaction condition: solvent-free
Keywords: 2-aminobenzamide, *o*-phenylenediamine, benzoic acid, ring closure, melt reaction, benzimidazole, quinazolinone

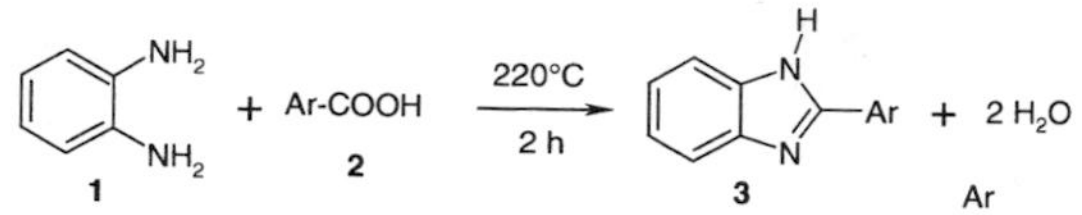

	Ar	yield (%)
a:	Ph	79
b:	p-NO$_2$C$_6$H$_4$	62
c:	m-NO$_2$C$_6$H$_4$	35
d:	o-ClC$_6$H$_4$	65
e:	p-ClC$_6$H$_4$	74
f:	2,4-Cl$_2$C$_6$H$_3$	77
g:	o-MeC$_6$H$_4$	50
h:	m-MeC$_6$H$_4$	60
i:	p-MeC$_6$H$_4$	75
j:	o-OHC$_6$H$_4$	28

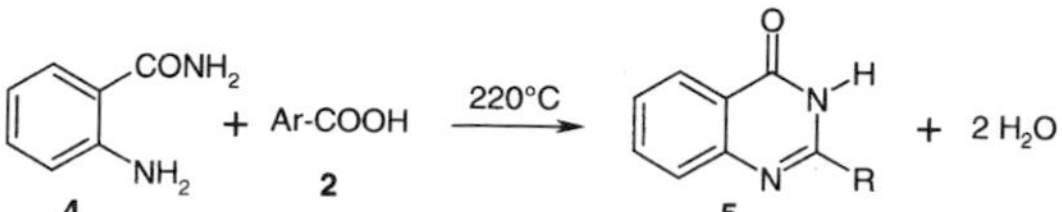

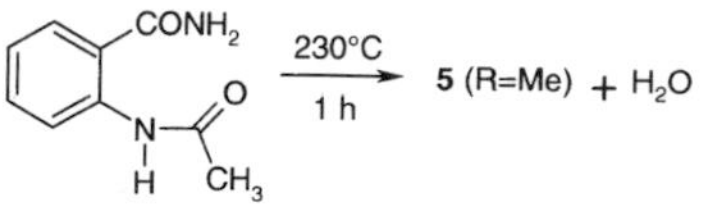

	Ar	yield (%)
a:	Ph	28
b:	p-ClC$_6$H$_4$	29
c:	p-MeOC$_6$H$_4$	25
d:	p-MeC$_6$H$_4$	30
e:	m-NO$_2$C$_6$H$_4$	25
f:	p-NO$_2$C$_6$H$_4$	26

Experimental procedures:

Mixtures of **1** and **2** were fused and kept at 180–220 °C for 1.5-2.5 h. The cooled solids were triturated with 10% aqueous NaHCO$_3$ to yield **3** in the yield indicated. The compounds **5** were similarly obtained at 220 °C for 4 h.

Reference: V.P. Reddy, P.L. Prasunamba, P.S.N. Reddy, C.V. Ratnam, *Indian J. Chem. Sect B.*, **22**, 917 (1983).

Type of reaction: C–N bond formation
Reaction condition: solvent-free
Keywords: pyrazolidinone, benzaldehyde, 1,3-dipol, uncatalyzed, azomethini-mine

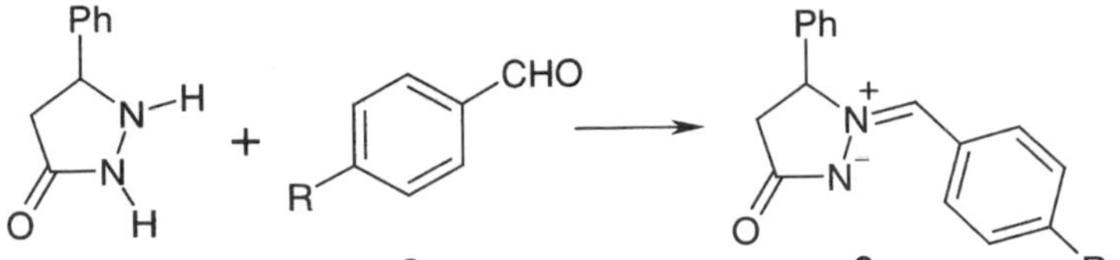

a: R=Cl
b: R=Br
c: R=NO$_2$

Experimental procedures:
A stoichiometric mixture of **1** (5.00 mmol) and **2** (5.00 mmol) was co-ground in a mortar and heated to 80 °C for 1 h. The 1,3-dipol **3** crystallized from an intermediate melt and was obtained quantitatively in pure form: mp **3a**: 192 °C; **3b**: 210–211 °C; **3c**: 187–188 °C.

Reference: G. Kaupp, M. R. Naimi-Jamal, J. Schmeyers, *private communication* (2002).

Type of reaction: C–N bond formation
Reaction condition: solvent-free
Keywords: aldehyde, ketone, calcium oxide, hydroxylamine hydrochloride, oxime

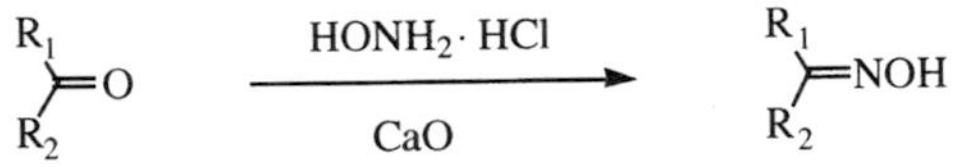

a: R$_1$R$_2$=-(CH$_2$)$_5$-
b: R$_1$R$_2$=-(CH$_2$)$_4$-
c: R$_1$R$_2$=-(CH$_2$)$_6$-
d: R$_1$=R$_2$=Ph
e: R$_1$=Ph; R$_2$=Me
f: R$_1$=4-MeC$_6$H$_4$; R$_2$=H
g: R$_1$=3-MeC$_6$H$_4$; R$_2$=H
h: R$_1$=4-HOC$_6$H$_4$; R$_2$=H
i: R$_1$=2-HOC$_6$H$_4$; R$_2$=H
j: R$_1$=4-ClC$_6$H$_4$; R$_2$=H
k: R$_1$=3-ClC$_6$H$_4$; R$_2$=H
l: R$_1$=2,6-Cl$_2$C$_6$H$_3$; R$_2$=H

Experimental procedures:
A mixture of ketone or aldehyde (1 mmol), fine powder of CaO (0.5 g, 8.9 mmol) were heated in an oil bath for a few minutes. Then hydroxylamine hydrochloride (0.208 g, 3 mmol) was added and the mixture was stirred with a magnetic stirrer in the presence of air for appropriate time. Afterwards, the reaction mixture was mixed with ethylacetate, filtered to remove CaO then mixed with water and extracted. The ethylacetate solution was dried over Na_2SO_4. The solvent was removed in vacuo to give the product.

References: H. Sharghi, M. H. Sarvari, *J. Chem. Res. (S)*, 24 (2000).

Type of reaction: C–N bond formation
Reaction condition: solvent-free
Keywords: aromatic aldehyde, molecular sieve, condensation, α-aryl-*N*-methylnitrone

a: Ar=Ph
b: Ar=4-MeC$_6$H$_4$
c: Ar=2-furyl
d: Ar=2-NO$_2$C$_6$H$_4$
e: Ar=2-HOC$_6$H$_4$

f: Ar=2,4-(MeO)$_2$C$_6$H$_3$
g: Ar=4-NO$_2$C$_6$H$_4$
h: Ar=4-ClC$_6$H$_4$
i: Ar=3-NO$_2$C$_6$H$_4$

Experimental procedures:
A mixture of *N*-methylhydroxylamine hydrochloride (1.2 mmol), finely powdered molecular sieve (3 Å, 0.3 g), and benzaldehyde (1.2 mmol) were grinded thoroughly for 10 min. The reaction mixture was set aside for further 5 min at ambient temperature. To the crude product mixture, 2×10 mL $CHCl_3$ were added, and solid parts were filtered off. The solvent was evaporated to dryness, and the residue was crystallized from petroleum ether (60–80 °C) to afford the α-phenyl-*N*-methylnitrones.

References: M. A. Bigdeli, M. M. A. Nikje, *Monatsh. Chem.*, **132**, 1547 (2001).

Type of reaction: C–N bond formation
Reaction condition: solvent-free
Keywords: *o*-phenylenediamine, ketone, Yb(OTf)$_3$, 2,3-dihydro-1*H*-1,5-benzodiazepine

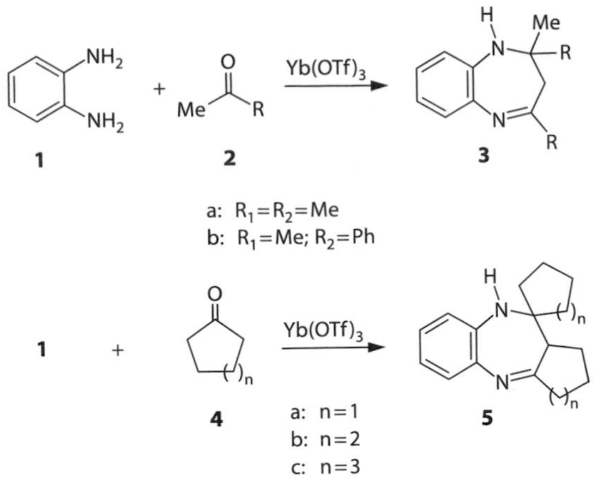

a: $R_1=R_2=Me$
b: $R_1=Me$; $R_2=Ph$

a: n=1
b: n=2
c: n=3

Experimental procedures:

A mixture of *o*-phenylenediamine (1 mmol) and ketone (2.1 mmol) was well stirred with Yb(OTf)$_3$ (0.05 mmol) at room temperature for 4 h. CH$_2$Cl$_2$ (2 mL) was added to crystallized Yb(OTf)$_3$; the catalyst was removed under reduced pressure and the residue was purified by SiO$_2$ gel column chromatography using CH$_2$Cl$_2$-MeOH (95:5) as eluent.

References: M. Curini, F. Epifano, M.C. Marcotullio, O. Rosati, *Tetrahedron Lett.*, **42**, 3193 (2001).

Type of reaction: C–N bond formation
Reaction condition: solid-state
Keywords: glucopyranosyl bromide, silylated uracil, silver trifluoroacetate, *N*-glycosylation, *N*-glucosid

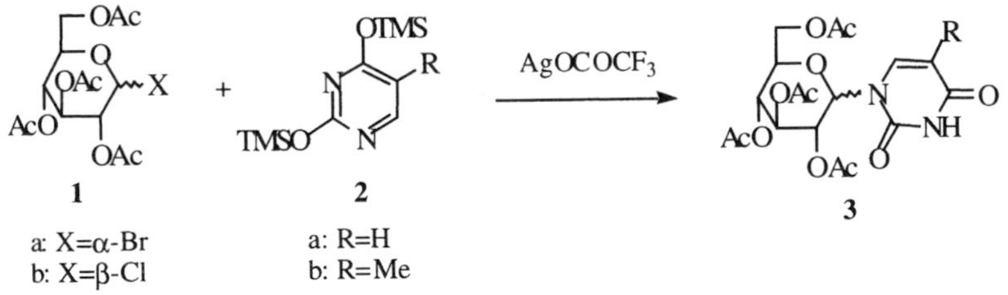

1
a: X=α-Br
b: X=β-Cl

2
a: R=H
b: R=Me

3

Experimental procedures:

A mixture of tetra-*O*-acetyl-*a*-D-glucopyranosyl bromide (0.82 g, 2 mmol) and silver trifluoroacetate (1 g, 4.5 mmol) was ground with mortar and pestle at room temperature for 5 min under a N_2 atmosphere in a glove box. Then the silylated uracil (0.77 g, 3 mmol) was added to the mixture, which was further ground in a ball mill for 2 days. After quenching the reaction by the addition of water, $CHCl_3$ was added and the mixture was filtered. Extraction with $CHCl_3$ and evaporation of the solvent gave a single crude product which was recrystallized from ether-hexane (3:1) to afford 0.43 g (42%) of crystalline product. Mp 147–148 °C.

References: J. Im, J. Kim, S. Kim, B. Hahn, *Tetrahedron Lett.,* **38**, 451 (1997).

Type of reaction: C–N bond formation
Reaction condition: solid-state
Keywords: 11,12-bis(aminomethyl)-9,10-dihydro-9,10-ethenoanthracene, pyrrole

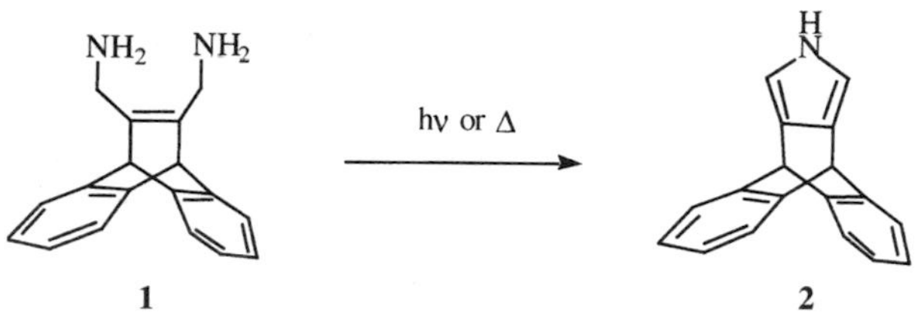

Experimental procedures:
In a quartz flask, 96 mg (0.37 mmol) of **1** were dissolved in tetrahydrofuran and slowly evapolared in vacuo, so that a thin layer of solid remained on the wall. The flask was flushed with argon and the sample was irradiated under argon at room temperature for 16 h. Twofold column chromatography gave 11 mg (0.05 mmol, 50%) of **2** as a pale yellow oil that solidified after several days, mp 182 °C.

References: J.R. Scheffer, H. Ihmels, *Liebigs Ann.,* 1925 (1997).

Type of reaction: C–N bond formation
Reaction condition: solvent-free
Keywords: *a*-nitroalkene, aziridination, 1-(ethoxycarbonyl)-2-nitroaziridine

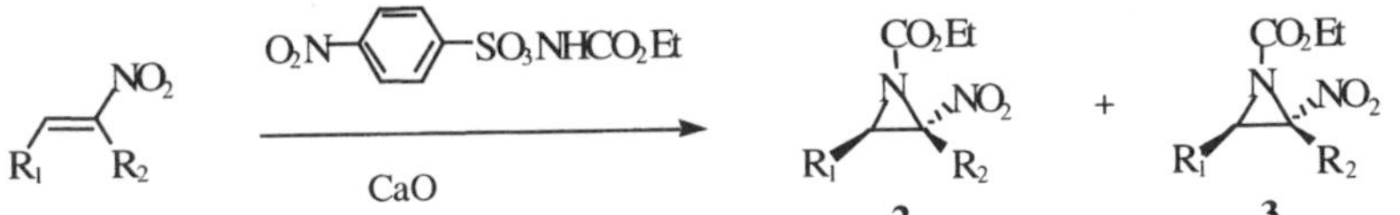

a: $R_1, R_2 = -(CH_2)_4-$
b: $R_1, R_2 = -(CH_2)_6-$
c: $R_1=R_2=Me$
d: $R_1=Et; R_2=Me$
e: $R_1=i\text{-}Pr; R_2=Me$
f: $R_1=$cyclohexyl; $R_2=Me$
g: $R_1=(CH_2)_2Ph; R_2=Me$
h: $R_1=(CH_2)_2Ph; R_2=Et$
i: $R_1=(CH_2)_5Me; R_2=(CH_2)_2CO_2Me$

Experimental procedures:

Equimolar amounts of nitroalkene, CaO and ethyl[(4-nitrobenzenesulfonyl)oxy]-carbamate were ground in a mortar. After 20 min petroleum ether was added to precipitate the salt. After filtration, the crude mixture was concentrated in vacuo and the 1-(ethoxycarbonyl)-2-nitroaziridines were purified by flash chromatography on silica gel (hexane-ethyl acetate, 8:2).

References: S. Fioravanti, L. Pellacani, S. Stabile, P. A. Tardella, *Tetrahedron*, **54**, 6169 (1998).

4.2 Solvent-Free C–N Bond Formation under Microwave Irradiation

Type of reaction: C–N bond formation
Reaction condition: solvent-free
Keywords: 4,6-dimethyl-1,2,3-triazine, enamine, Diels-Alder reaction, N_2 elimination, microwave irradiation, pyridine derivative

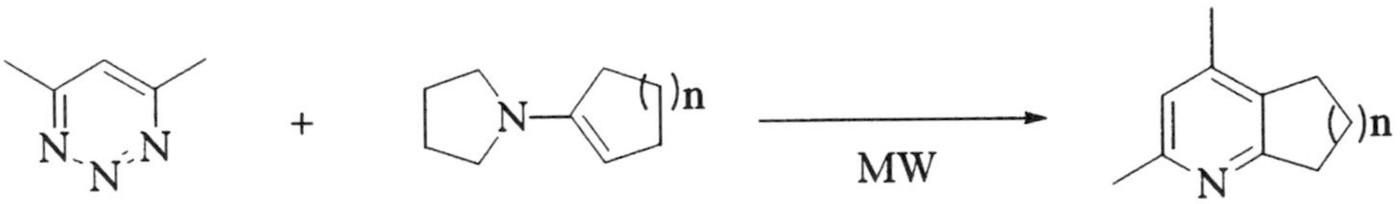

a: n=1
b: n=2
c: n=4
d: n=8

Experimental procedures:

A mixture of triazine and enamine (2:1 molar ratio) was irradiated at atmospheric pressure in a focused microwave reactor (Prolabo MX350) at 270 W for 20 min (max. temperature: 150 °C). The crude reaction mixture was purified by flash chromatography on silica gel (Merck type 60, 230–400 mesh).

References: A. Diaz-Ortiz, A. de la Hoz, P. Pieto, J.R. Carrillo, A. Moreno, H. Neunhoeffer, *Synlett,* 236 (2001).

Type of reaction: C–N bond formation
Reaction condition: solvent-free
Keywords: phthalimide, alkyl halide, microwave irradiation, *N*-alkylation, *N*-alkylphthalimide

R-X= benzyl chloride, 1,2-dibromoethane, 1-bromobutane, 1-chlorodecane
1-bromopentane, 1-iodododecane, 2,3-epoxypropyl chloride

Experimental procedures:

A mixture of phthalimide **1** (0.70 g, 4.8 mmol), alkyl halide (6.0 mmol), tetrabutylammonium bromide – TBAB (0.15 g, 0.45 mmol), and potassium carbonate (2.6 g, 18.8 mmol) was heated in a domestic microwave oven in an open Erlenmeyer flask for an appropriate time. After cooling down, the reaction mixture was extracted with methylene chloride (2×25 mL). Then the extracts were dried with MgSO$_4$, filtered, and the solvent was evaporated to dryness. If necessary before a recrystallization the solid material was purified by means of flash chromatography to afford desired *N*-alkylphthalimide **2**, yield: 49–95%.

References: D. Bogdal, J. Pielichowski, A. Boron, *Synlett,* 873 (1996).

Type of reaction: C–N bond formation
Reaction condition: solvent-free
Keywords: 4-bromophenacyl bromide, azole derivative, *N*-alkylation, microwave irradiation, azaheterocycle

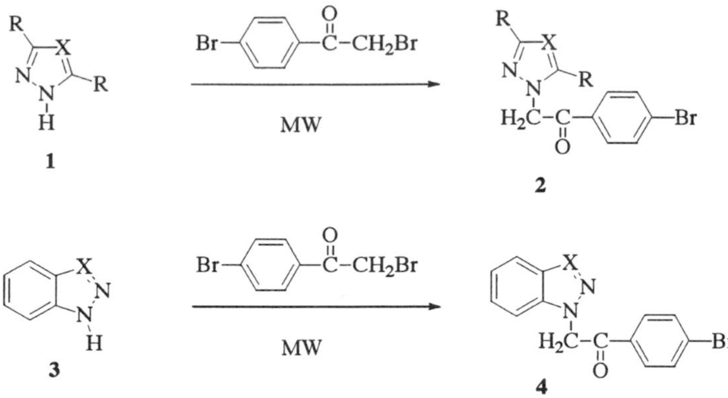

a: X=CH; R=H
b: X=CH; R=CH₃
c: X=N; R=H

Experimental procedures:

The equimolar mixture (3 mmol) of 4-bromophenacyl bromide and the corresponding azole was placed into a Pyrex-glass open vessel and irradiated in a domestic microwave oven. When the irradiation was stopped, the final temperature was measured by introducing a glass thermometer into the reaction mixture and homogenizing it, in order to obtain a temperature value representative of the whole mass. The reaction mixture was washed with cold water and the products were filtered off and conveniently dried, then recrystallized from absolute ethanol.

References: E. Perez, E. Sotelo, A. Loupy, R. Mocelo, M. Suarez, R. Perez, M. Autie, *Heterocycles,* **43**, 539 (1996).

Type of reaction: C–N bond formation
Reaction condition: solvent-free
Keywords: azaheterocycle, alkyl halide, *N*-alkylation, microwave irradiation, *N*-alkyl azaheterocyclic compound

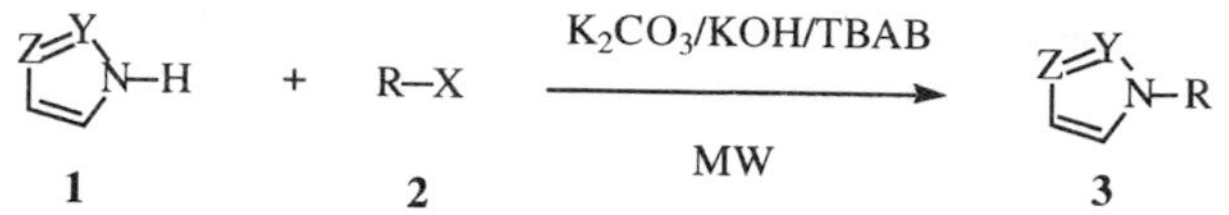

1 **2** **3**

a: Y=Z=CH
b: Y=CH; Z=N
c: Y=N; Z=CH

$R = C_5H_{11}, C_{10}H_{21}, CH_2Ph$

$X = Cl, Br, I$

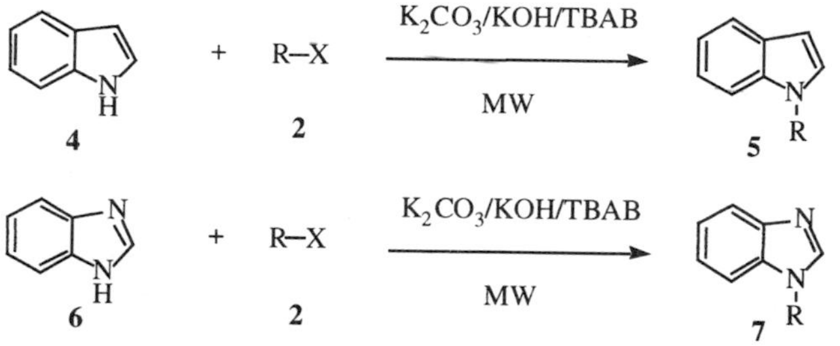

4 **2** **5** R

6 **2** **7** R

Experimental procedures:

A mixture of an azaheterocyclic compound (5.0 mmol), alkyl halide (7.5 mmol), tetrabutylammonium bromide – TBAB (0.17 g, 0.50 mmol), and potassium carbonate (2.8 g, 20 mmol) or mixture of potassium carbonate (2.8 g, 20 mmol) and potassium hydroxide (1.1 g, 20 mmol) was heated in a domestic microwave oven in an open Erlenmeyer flask for an appropriate time. After being cooled down, the reaction mixture was extracted with methylene chloride or THF (2×25 mL). Then the extract was dried with $MgSO_4$, filtered, and the solvent was evaporated to dryness. Liquid compounds were purified on Kugelrohr distillation apparatus, while solid compounds were purified by means of flash chromatography to afford desired *N*-alkyl derivatives of azaheterocycle compound, yield: 58–95%.

References: D. Bogdal, J. Pielichowski, K. Jaskot, *Heterocycles*, **45**, 715 (1997).

Type of reaction: C–N bond formation
Reaction condition: solvent-free
Keywords: carbazole, alkyl halide, *N*-alkylation, microwave irradiation, *N*-alkylcarbazole

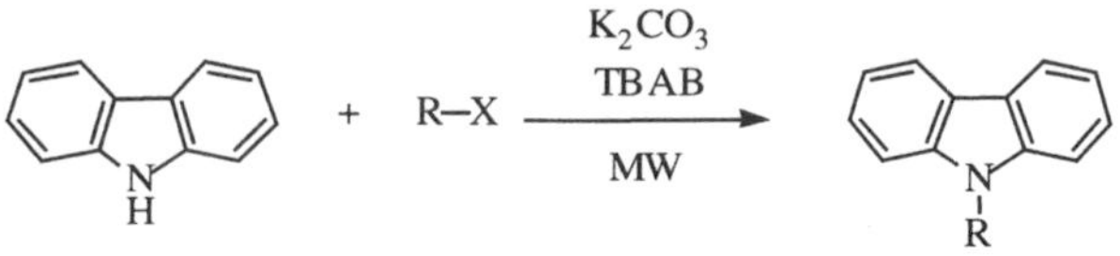

a: RX=PhCH$_2$Cl
b: RX=CH$_2$=CHCH$_2$Cl
c: RX=n-C$_4$H$_9$Br
d: RX=n-C$_5$H$_{11}$Br
e: RX=n-C$_{10}$H$_{21}$Cl
f: RX=n-C$_{10}$H$_{21}$I

Experimental procedures:

A mixture of carbazol (0.84 g, 5.0 mmol), alkyl halide (7.5 mmol), tetrabutylammonium bromide – TBAB (0.17 g, 0.50 mmol), and potassium carbonate (2.8 g, 20 mmol) was heated in a domestic microwave oven in an open Erlenmeyer flask for an appropriate time. After cooling down, the reaction mixture was extracted with CH$_2$Cl$_2$ (2×25 mL). Then the extract was dried with MgSO$_4$, filtered, and the solvent was evaporated to dryness. The solid material was purified by means of flash chromatography to afford the desired *N*-alkylcarbazole, yield: 32–95%.

References: D. Bogdal, J. Pielichowski, K. Jaskot, *Synth. Commun.,* **27**, 1553 (1997).

Type of reaction: C–N bond formation
Reaction condition: solvent-free
Keywords: maleic anhydride, phthalic anhydride, primary amine, TaCl$_5$-silica gel, microwave irradiation, *N*-alkylimide, *N*-arylimide

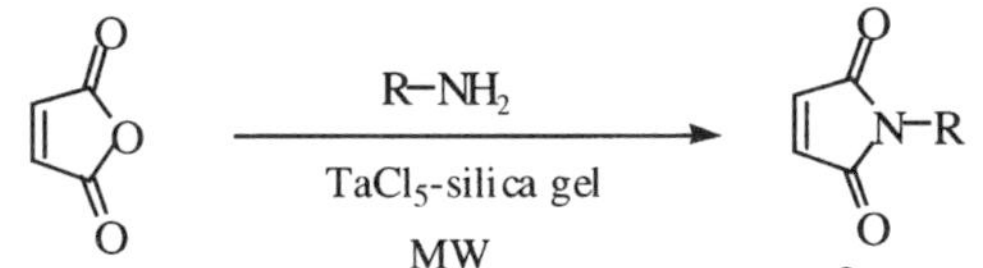

a: R=CH$_2$Ph
b: R=i-Bu
c: R=2-phenylethyl

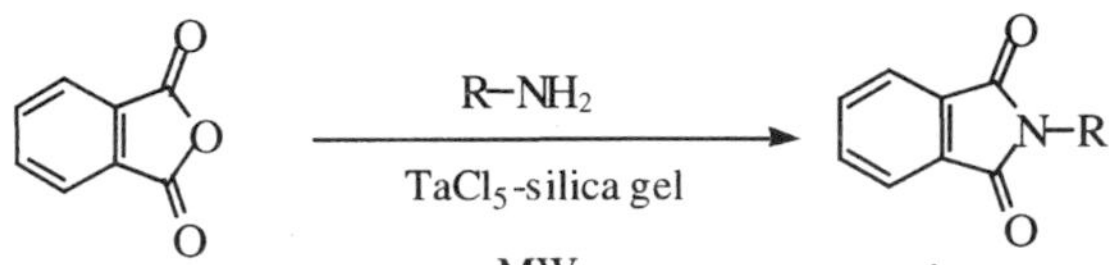

Experimental procedures:

Phthalic anhydride **3** (0.36 g, 2.4 mmol) and benzylamine (0.27 mL, 2.4 mmol) were adsorbed on activated silica gel (100–200 mesh, dried overnight at 100 °C) and stirred at room temperature for 1 h under inert atmosphere. To this was added $TaCl_5$-SiO_2 (0.37 g, 10 mol%) and admixed thoroughly and irradiated in a microwave oven (448 W) for 5 min. The mixture was cooled to room temperature, charged on a small silica pad and eluted with CH_2Cl_2 (30 mL). Removal of volatiles furnished *N*-benzyl phthalimide **4** (0.53 g, 92%).

References: S. Chandrasekhar, M. Takhi, G. Uma, *Tetrahedron Lett.*, **38**, 8089 (1997).

Type of reaction: C–N bond formation
Reaction condition: solvent-free
Keywords: maleic anhydride, phthalic anhydride, primary amine, microwave irradiation, *N*-carboxyalkyl maleimide, phthalimide

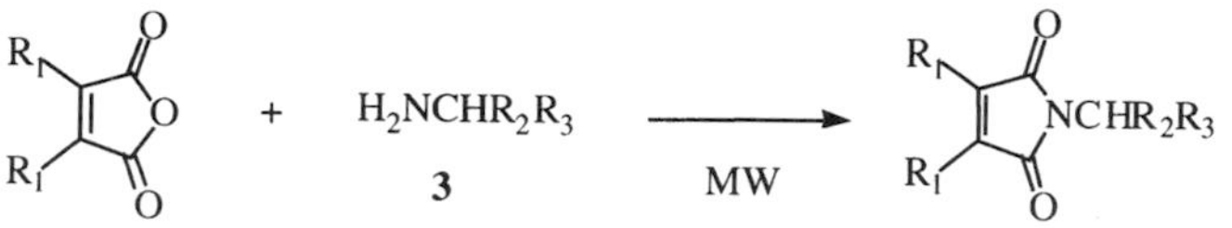

1 R_1=H
2 R_1, R_l=C_4H_4

3

MW

4: R_1=R_2=H; R_3=COOH
5: R_1=H; R_2=Me; R_3=COOH
6: R_1=H; R_2=Ph; R_3=COOMe
7: R_1=R_2=H; R_3=Ph
8: R_1=R_2=H; R_3=C_3H_5
9: R_1, R_l=C_4H_4; R_2=H; R_3=Ph
10: R_1, R_l=C_4H_4; R_2=H; R_3=COOH

Experimental procedures:

A mixture of either maleic anhydride **1** or phthalic anhydride **2** (0.02 mol) and glycine (**3**, R_2=H, R_3=CO_2H, 0.02 mol) was placed in an Erlenmeyer flask fitted with a loose top cap and heated in a commercial microwave oven operating at 2450 MHz by setting the power range to medium high (70% of total power). The reaction mixture turned red. After cooling, the reaction mixture was extracted with chloroform (2×30 mL) and washed with cold water (2×10 mL), dried (Na_2SO_4), filtered and the solvent removed.

References: H. N. Borah, R. C. Boruah, J. S. Sandhu, *J. Chem. Res. (S)*, 272 (1998).

Type of reaction: C–N bond formation
Reaction condition: solvent-free
Keywords: pyrrolidino[60]fullerene, phase transfer reaction, microwave irradiation, *N*-alkylpyrrolidino[60]fullerene

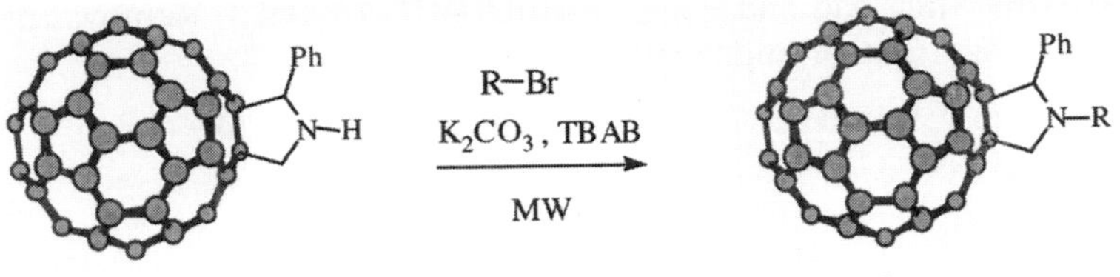

1

2

a: R=C$_6$H$_5$CH$_2$
b: R=4-NO$_2$C$_6$H$_4$CH$_2$
c: R=4-MeO$_2$CC$_6$H$_4$CH$_2$
d: R=*n*-C$_8$H$_{17}$
e: R=H$_2$C=CH-CH$_2$

Experimental procedures:
A mixture of **1** (40 mg, 0.047 mmol), the alkyl or benzyl bromide (10 equiv.), potassium carbonate (22 mg, 0.019 mg) and TBAB (30 mg, 0.094 mmol) was irradiated at 780 W for 10 min. In all cases the final temperature was 80 °C, but working at a smaller scale, adsorption of the radiation is not efficient enough to heat the reaction mixture.

References: P. de la Cruz, A. de la Hoz, L. M. Font, F. Langa, M. C. P. Rodriguez, *Tetrahedron Lett.,* **39**, 6053 (1998).

Type of reaction: C–N bond formation
Reaction condition: solvent-free
Keywords: styrene, , clayan, nitration, microwave irradiation, *β*-nitrostyrene

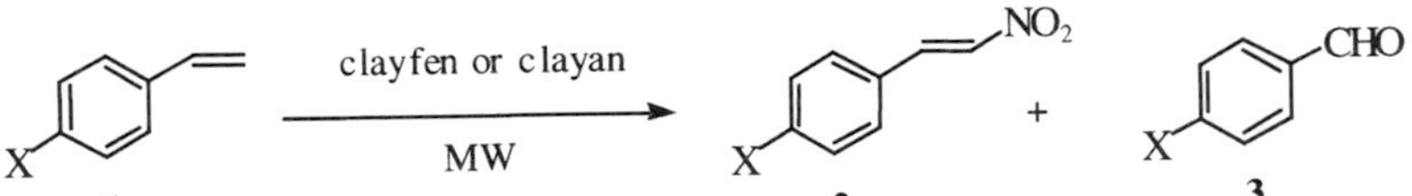

1

2

3

a: X=H
b: X=4-Cl
c: X=4-MeO

Experimental procedures:
In a typical experiment, styrene (180 mg, 1.74 mmol) was admixed with clayfen or clayan (300 mg) in a glass tube. The reaction mixture was placed in an oil bath for 15 min or irradiated for 3 min in an alumina bath inside an unmodified household microwave oven (900 W) at its medium power. On completion of the reaction, followed by TLC examination (hexane-EtOAc, 4:1, v/v), the product was extracted into dichloromethane (45 mL), the combined organic extract dried with anhydrous sodium sulfate and solvent removed under reduced pressure. The relative amounts of product distribution were calculated from GC-MS analysis. Alternatively, the crude material was chromatographed on a silica gel column and eluted with hexane-EtOAc (4:1, v/v) to afford the pure product (147 mg, 57%).

References: R. S. Varma, K. P. Naicker, P. J. Liesen, *Tetrahedron Lett.*, **39**, 3977 (1998).

Type of reaction: C–N bond formation
Reaction condition: solvent-free
Keywords: 4-hydroxy-6-methyl-2(1*H*)-pyridone, primary amine, microwave irradiation, 4-alkylamino-6-methyl-2(1*H*)-pyridone

a: R_1=H; R_2=CH$_2$Ph
b: R_1=CH$_3$; R_2=CH$_2$Ph
c: R_1=H; R_2=CH$_2$CH$_2$Ph
d: R_1=CH$_3$; R_2=CH$_2$CH$_2$Ph
e: R_1=CH$_3$; R_2=CH$_2$(CH$_3$)Ph
f: R_1=CH$_2$Ph; R_2=CH$_2$(CH$_3$)Ph

g: R_1=CH$_2$CH$_2$Ph; R_2=CH$_2$(CH$_3$)Ph
h: R_1=CH$_2$Ph; R_2=CH$_2$Ph
i: R_1=CH$_2$CH$_2$Ph; R_2=CH$_2$CH$_2$Ph
j: R_1=CH$_2$Ph; R_2=CH$_2$CH$_2$Ph
k: R_1=CH$_2$CH$_2$Ph; R_2=CH$_2$Ph

Experimental procedures:
A mixture of 1 mmol of the corresponding 4-hydroxy-2(1*H*)-pyridone **1** and 4.5 mmol of benzyl-, 1-phenylethyl- or 2-phenylethylamine **2** was filled into a 15-mL pressure tube (Aldrich, with threated type A plug, length 10.2 cm and additionally provided with a Teflon ring). Then the reaction tube was placed in the center of an 800-mL beaker which was filled with vermiculite, a polymeric material for covering hazardous compounds in packages. After irradiation in an ordi-

nary domestic microwave oven (Panasonic NN-5206 with rotate plate), the reaction mixture was cooled. Adding ice-cold ethyl acetate the separated solid was collected by filtration and washed twice with ethyl acetate to give TLC pure compounds. The compounds were recrystallized from MeOH, EtOH, or 2-PrOH.

References: D. Heber, E.V. Stoyanov, *Synlett,* 1747 (1999).

Type of reaction: C–N bond formation
Reaction condition: solvent-free
Keywords: alcohol, nitrile, montmorillonite KSF, Ritter reaction, microwave irradiation, amide

a: $R_1=R_2=H$; $R_3=Ph$
b: $R_1=R_2=H$; $R_3=4$-$NO_2C_6H_4$
c: $R_1=H$; $R_2=Me$; $R_3=Ph$
d: $R_1=R_2=H$; $R_3=4$-$MeOC_6H_4$
e: $R_1=R_2=H$; $R_3=4$-$BnOC_6H_4$
f: $R_1=H$; $R_2=R_3=Ph$
g: $R_1=R_2=R_3=Me$
h: $R_1=Me$; $R_2=R_3=Ph$
i: $R_1=R_2=H$; $R_3=PhCH=CH$
j: $R_1=R_2=H$; $R_3=MeCH=CH$
k: $R_1=R_2=H$; $R_3=Me(CH_2)_6CH_2$

a: $R_4=n$-Pr
b: $R_4=Bn$
c: $R_4=Ph$
d: $R_4=EtOCH_2CH_2$
e: $R_4=CH_2=CH$
f: $R_4=t$-Bu
g: $R_4=Cl(CH_2)_2CH_2$
h: $R_4=ClCH_2$

Experimental procedures:
Phenylethanol **1c** (1.2 g, 10 mmol) and 3-ethoxypropionitrile **2d** (1 g, 10 mmol) were admixed with KSF clay (1.2 g, w/w of alcohol) and subjected to microwave irradiation in a Pyrex test tube for 3 min. Then the reaction mass was cooled to room temperature, charged on a short silica gel column (Merck, 200 mesh) and eluted (ethyl acetate-*n*-hexane, 2:8) to afford the pure amide as a pale yellow liquid (2.05 g, 93%).

References: H.M.S. Kumar, B.V.S. Reddy, S. Anjaneyulu, E.J. Reddy, J.S. Yadav, *New J. Chem.,* **23**, 955 (1999).

Type of reaction: C–N bond formation
Reaction condition: solvent-free
Keywords: aniline, aromatic ester, potassium *tert*-butoxide, microwave irradiation, amides

$$R\text{-}CO_2X \quad + \quad R'\text{-}NH_2 \quad \xrightarrow[\text{MW}]{t\text{-BuOK}} \quad R\text{-}CONH\text{-}R'$$

1   **2**   **3**

a: R=Ph   a: R'=Ph
b: R=2-thienyl   b: R'=p-ClC$_6$H$_4$

X=Me, Et

Experimental procedures:

Potassium *tert*-butoxide (1 mmol) was added to a premixed mixture of aniline (1 mmol) and methylbenzoate (1 mmol) in a glass test tube which was placed in an alumina bath (neutral alumina: 125 g, mesh ∼ 150, Aldrich; bath: 5.7 cm diameter) inside an unmodified household microwave oven and irradiated for the specified time at its full power of 900 W. On completion of the reaction, as determined by TLC (hexane-EtOAc, 4 : 1, v/v), the reaction mixture was extracted into ethyl acetate. The combined extracts were dried over anhydrous sodium sulfate and the solvent removed under reduced pressure to afford a residue that upon trituration with hexane gave pure product, benzanilide (83%).

References: R. S. Varma, K. P. Naicker, *Tetrahedron Lett.*, **40**, 6177 (1999).

Type of reaction: C–N bond formation
Reaction condition: solvent-free
Keywords: aryl halide, secondary amine, *N*-alkylation, microwave irradiation, *N*-arylamine

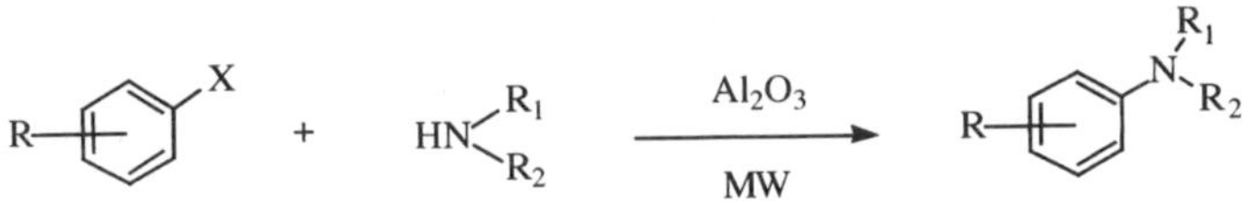

1   2   3

X=Cl, Br

a: R=4-NO$_2$
b: R=3-CN-4-NO$_2$
c: R=2,4-NO$_2$
d: R=2-NO$_2$
e: R=2-Br-5-NO$_2$
f: R=4-CN

a: R$_1$,R$_2$=CH$_2$CH$_2$-O-CH$_2$CH$_2$
b: R$_1$,R$_2$=-(CH$_2$)$_5$-
c: R$_1$,R$_2$=CH$_2$CH(CH$_3$)-O-CH(CH$_3$)CH$_2$
d: R$_1$,R$_2$=CH$_2$CH$_2$-N(Et)-CH$_2$CH$_2$
e: R$_1$,R$_2$=CH$_2$CH$_2$-N(CH$_2$Ph)-CH$_2$CH$_2$
f: R$_1$,R$_2$=CH$_2$CH$_2$-N(Ph)-CH$_2$CH$_2$
g: R$_1$=R$_2$=CH$_2$CH$_2$OH
h: R$_1$,R$_2$=CH$_2$CH$_2$-CH(OH)-CH$_2$CH$_2$

Experimental procedures:
p-Bromonitrobenzene **1a** (10 mmol) and *N*-benzylpiperazine **2e** (12 mmol) were mixed with basic Al$_2$O$_3$ (5 g, Brockmann activity grade 1, 70–290 mesh) in a Pyrex test tube and subjected to microwave irradiation at 450 W (BPL BMO 700T) for a specified time. The reaction progress was monitored by TLC. After complete conversion (TLC), the reaction mass was directly charged on a silica gel column (Aldrich 100–200 mesh) and eluted with ethyl acetate-hexane (3:7) to afford the pure product **3e** as a pale yellow crystalline solid.

References: J. S. Yadav, B. V. S. Reddy, *Green Chem.,* **2**, 115 (2000).

Type of reaction: C–N bond formation
Reaction condition: solvent-free
Keywords: cephalosporin, carboxylic acid, condensation, alumina, microwave irradiation, *N*-acylated cephalosporin

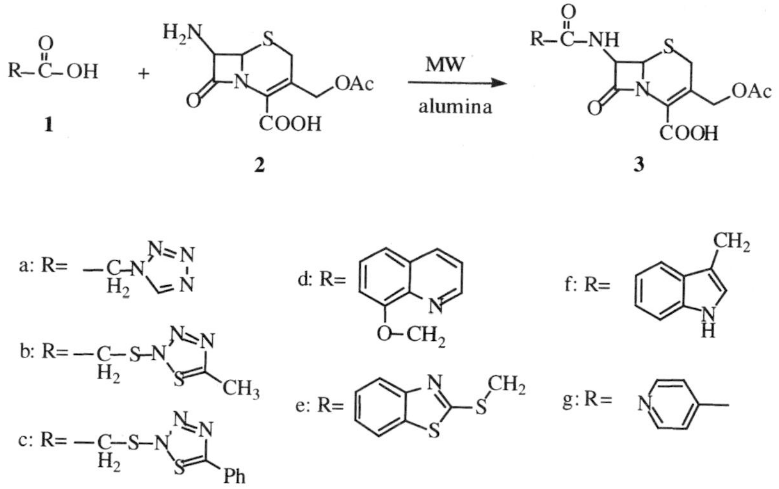

a: R=

b: R=

c: R=

d: R=

e: R=

f: R=

g: R=

Experimental procedures:

Basic alumina (18 g) was added to a 2.72 g **2** (0.01 mol) dissolved in 3 mL aqueous ammonia at room temperature. In another beaker, 10 g basic alumina were added to a solution of **1** (0.01 mol) in acetone. The mixture was dried at room temperature, and the two reactants were thoroughly mixed using a mortar mixer and then placed in an alumina bath inside the microwave oven. Upon completion of the reaction (90–120 s) as monitored by HPLC examination, the mixture was cooled to room temperature, and the product was extracted into a mixture of water and acetic acid (1:4; 4×10 mL). Removal of the solvent under reduced pressure afforded the product which was purified by crystallization from a mixture of acetone-acetic acid.

References: M. Kidwai, P. Misra, K.R. Bhushan, R.K. Saxena, M. Singh, *Monatsh. Chem.*, **131**, 937 (2000).

Type of reaction: C–N bond formation
Reaction condition: solvent-free
Keywords: phthalic anhydride, benzylamine, microwave irradiation, phthalimide

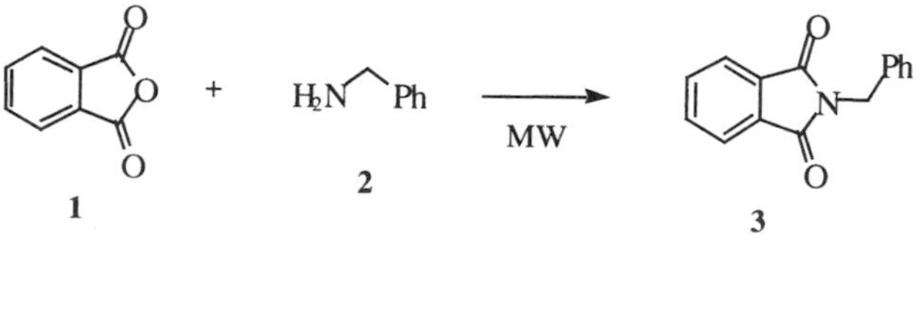

Experimental procedures:

In a typical reaction, 9.15 mmol of benzylamine **2** (1 mL) and 9.15 mmol of phthalic anhydride **1** (1.35 g) were introduced into a Pyrex flask, and then submitted to microwave irradiation under conditions of time and temperature as indicated. At the end of the irradiation, the reaction mixture was cooled to room temperature. The crude solid was recrystallized from ethanol to afford 1.95 g (8.23 mmol, 90%) of pure **3** as white crystals.

References: T. Vidal, A. Loupy, R. N. Gedye, *Tetrahedron*, **56**, 5473 (2000).

Type of reaction: C–N bond formation
Reaction condition: solvent-free
Keywords: amide, aldehyde, secondary amine, microwave irradiation, monoacylaminal

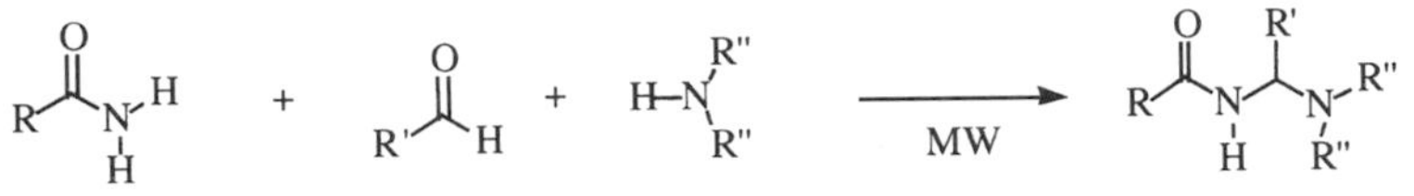

1a: succinimide	**2a:** R'=H	**3a:** piperidine
1b: R=Ph	**2b:** R'=Ph	**3b:** morpholine
1c: R=Me	**2c:** R'=*p*-ClPh	**3c:** pyrrolidine
1d: R=H		**3d:** diethylamine

Experimental procedures:

A mixture of amide (1 mmol), aldehyde (1 mmol and in the case of **2a**, 1.5 mmol as 40% solution in water) and amine (1.5 mmol) were mixed and ground with the solid support in a mortar. The mixture was transferred to a screw-cap Teflon container and irradiated in microwave oven for the required time. The progress of reaction was monitored with TLC. After cooling, the mixture was extracted with ethylacetate and filtered off. Evaporation of the solvent and remaining amine under reduced pressure gives the crude product that if necessary can be purified by recrystallization or chromatography.

References: A. Sharifi, F. Mohsenzadeh, M. R. Naimi-Jamal, *J. Chem. Res. (S)*, 394 (2000).

Type of reaction: C–N bond formation
Reaction condition: solvent-free
Keywords: aromatic/heteroaromatic chloride, piperidine, morpholine, basic alumina, nucleophilic aromatic substitution, microwave irradiation

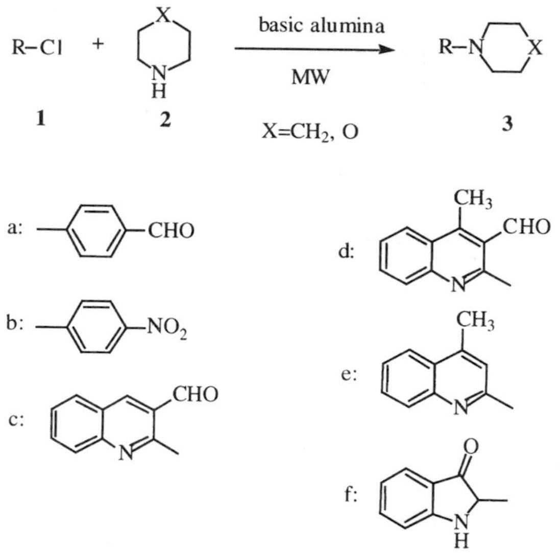

R–Cl + (morpholine/piperidine ring with X top, NH bottom) →[basic alumina / MW] R–N(ring)X

1   2   3

X=CH₂, O

a: (benzene ring)–CHO

b: (benzene ring)–NO₂

c: (quinoline ring)–CHO

d: (quinoline ring with CH₃, CHO, CH₃)

e: (quinoline ring with CH₃)

f: (indolinone ring, N–H, O)

Experimental procedures:

To a solution of aromatic/heteroaromatic chloride **1** (0.01 mol) in ethanol, was added piperidine/morpholine **2** (0.01 mol) followed by basic alumina (20 g) with constant stirring. The reaction mixture was thoroughly mixed and adsorbed material was dried in open vessel and placed in an alumina bath inside the microwave oven for 60–120 s. On completion of the reaction as followed by TLC observation, the mixture was cooled to room temperature and then product was extracted into ethanol (3×15 mL). Recovering the solvent under reduced pressure afforded the product **3**, which was purified through recrystallization from ethanol-chloroform mixture.

References: M. Kidwai, P. Sapra, B. Dave, *Synth. Commun.*, **30**, 4479 (2000).

Type of reaction: C–N bond formation
Reaction condition: solvent-free
Keywords: acyl chloride, *N*-alkyl piperazine, Lawesson's reagent, microwave irradiation, thioamide

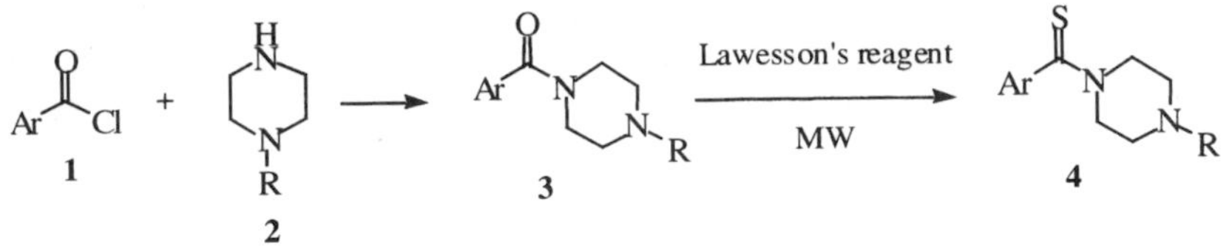

Ar= Ph, 2-MeC$_6$H$_4$, 3-MeC$_6$H$_4$, 2-thienyl, 4-FC$_6$H$_4$CH$_2$

R= Et, PhCH$_2$

Experimental procedures:

The diamine **2** was added to an etherial solution of an acyl chloride **1** in a Teflon-capped vial. Salt precipitaion was instant. The solvent was removed by filtration, where the Teflon cap served as the filter. Lawesson's reagent (1.0–1.5 equiv.) was added to the salt **3** and the resulting mixture of solids was mixed thoroughly and thereafter irradiated for 8 min in a domestic microwave oven (900 W, Whirlpool M401). Solid-phase extraction afforded the thioamides **4** in adequate purities and yields.

References: R. Olsson, H.C. Hansen, C.M. Andersson, *Tetrahedron Lett.,* **41**, 7947 (2000).

Type of reaction: C–N bond formation
Reaction condition: solvent-free
Keywords: 1-benzoyl-3,3-diethylthiourea, amine, KF-alumina, transamidation, microwave irradiation, 1-benzoyl-3-alkylthiourea

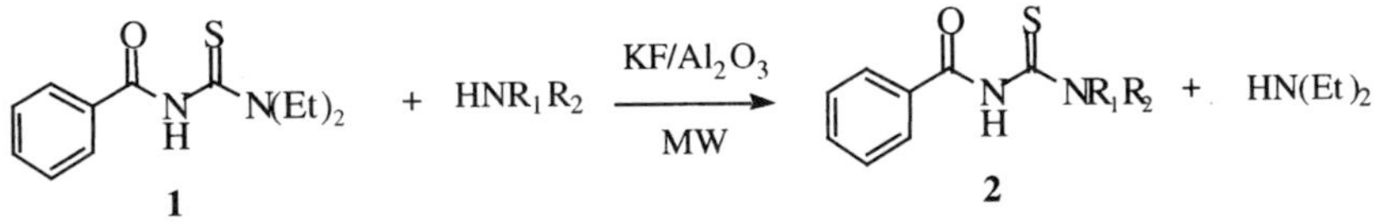

a: R$_1$=H; R$_2$=CH$_2$Ph
b: R$_1$=H; R$_2$=CH$_2$CO$_2$H
c: R$_1$=H; R$_2$=CH$_2$(CH$_2$)$_4$CO$_2$H
d: R$_1$=H; R$_2$=CH(CH$_2$Ph)CO$_2$H
e: R$_1$=R$_2$=CH$_2$Ph
f: R$_1$=H; R$_2$=Ph

Experimental procedures:
Benzoyl-3,3-diethylthiourea **1** (2 mmol) was dissolved in acetone. An equimolar amount of the indicated amines were then added and the mixture then smoothly mixed with 1 g of KF-alumina. The solvent was removed under reduced pressure. The resulting mixture was placed into a Pyrex-glass open vessel and irradiated in a microwave domestic oven at 560 W for the times and final temperature as indicated. The products were extracted from the support with acetone and precipitated with ice water.

References: H. Marquez, E.R. Perez, A.M. Plutin, M. Morales, A. Loupy, *Tetrahedron Lett.*, **41**, 1753 (2000).

Type of reaction: C–N bond formation
Reaction condition: solvent-free
Keywords: dimethoxybenzene, diethyl azodicarboxylate, $InCl_3$-SiO_2, microwave irradiation, aryl hydrazine

a: R=1,2-(MeO)$_2$
b: R=1,2,3-(MeO)$_3$
c: R=1,3-(MeO)$_2$
d: R=1,4-(MeO)$_2$

Experimental procedures:
1,2-Dimethoxybenzene **1a** (5 mmol), diethyl azodicarboxylate (5 mmol) and $InCl_3$-SiO_2 (3 wt equiv. of arene) were admixed in an Erlenmeyer flask and exposed to microwave irradiation at 450 W using BPL, BMO-700 T focused microwave oven for 3 min (pulsed irradiation 1 min with 20 s interval). On completion, the reaction mixture was directly charged on a small silica gel column and eluted with a mixture of ethyl acetate-hexane (2:8) to afford pure hydrazine **2a** in 88% yield as a pale yellow solid.

References: J.S. Yadav, B.V.S. Reddy, G.M. Kumar, C. Madan, *Synlett*, 1781 (2001).

Type of reaction: C–N bond fomation
Reaction condition: solvent-free
Keywords: aldehyde, ketone, primary amine, diethyl phosphite, montmorillonite KSF, microwave irradiation, α-amino phosphonate

$$
\underset{\textbf{1}}{\underset{R}{\overset{O}{\|}}{\overset{\|}{C}}R'} + \underset{\textbf{2}}{R_2-NH_2} + \underset{\textbf{3}}{HO-\underset{OEt}{\overset{OEt}{P}}} \xrightarrow[\text{MW}]{\text{montmorillonite KSF}} \underset{\textbf{4}}{\underset{R'}{\overset{HN-R_2}{R-\underset{|}{\overset{|}{C}}-PO(OEt)_2}}}
$$

R = aryl, alkyl, naphthyl, cinnamyl
R' = H or alkyl
R_2 = alkyl or aryl

Experimental procedures:

Aldehyde/ketone (5 mmol), amine (5 mmol), diethyl phosphite (5 mmol) and montmorillonite (1.5 g, Aldrich, montmorillonite, KSF) were admixed in a Pyrex test tube and exposed to microwave irradiation at 450 W using a (BPL, BMO, 700T, indicates the commercial name of microwave oven) focused microwave oven for an appropriate time (pulsed irradiation 1 min with 20 s interval). After complete conversion of the reaction, as indicated by TLC, the reaction mixture was directly charged on a small silica gel column and eluted with a mixture of ethyl acetate-hexane (3:7) to afford corresponding pure α-amino phosphonate.

References: J. S. Yadav, B. V. S. Reddy, C. Madan, *Synlett,* 1131 (2001).

Type of reaction: C–N bond formation
Reaction condition: solvent-free
Keywords: alkyne, amine, paraformaldehyde, alumina, Mannich condensation, microwave irradiation, aminoalkyne

$$
\underset{\textbf{1}}{R_1-C\equiv C-H} + (CH_2O)_n + HNR_2R_3 \xrightarrow[\text{MW}]{CuI/Al_2O_3} \underset{\textbf{2}}{R_1-C\equiv C-CH_2NR_2R_3}
$$

a: R_1=(CH$_2$)$_7$CH$_3$; R_2=R_3=CH$_2$Ph
b: R_1=(CH$_2$)$_5$CH$_3$; R_2=R_3=CH$_2$Ph
c: R_1=4-CH$_3$C$_6$H$_4$; R_2=R_3=CH$_2$Ph
d: R_1=C$_6$H$_5$; R_2, R_3=-(CH$_2$)$_5$-
e: R_1=2-FC$_6$H$_4$; R_2, R_3=-(CH$_2$)$_2$-O-(CH$_2$)$_2$-
f: R_1=2-ClC$_6$H$_4$; R_2, R_3=-(CH$_2$)$_2$-N(Ph)-(CH$_2$)$_2$-
g: R_1=4-BrC$_6$H$_4$; R_2=R_3n-Bu
h: R_1=(CH$_2$)$_7$CH$_3$; R_2=CH$_2$Ph; R_3=H

Experimental procedures:

Dibenzylamine (0.197 g, 1.00 mmol), and 1-decyne **1a** (0.138 g, 1.00 mmol) were added to a mixture of cuprous iodide (0.572 g, 3.00 mmol), paraformalde-hyde (0.090 g, 3.00 mmol) and alumina (1.00 g) contained in a dry 10-mL round-bottomed flask. The mixture was stirred at room temperature to ensure efficient mixing. The flask was then fitted a septum (punctured by an 18 gauge needle to serve as a pressure release valve), placed in the microwave oven (1000 W, sharp model R-4A38) and irradiated at 30% power for 4 min. After cooling, ether (4 mL) was added and the slurry stirred at room temperature to ensure product removal from the surface. The mixture was vacuum filtered using a sintered glass funnel and the product purified by flash chromatography to yield 0.312 g of dibenzyl(undec-2-ynyl)amine **2a** (90%).

References: G.W. Kabalka, L.Wang, R.M. Pagni, *Synlett,* 676 (2001).

Type of reaction: C–N bond formation
Reaction condition: solvent-free
Keywords: hydrazin derivative, (α-hetero)-β-dimethylaminoacrylate, microwave irradiation, β-hydrazino acrylate

1
a: X=NH
b: X=O
c: X=S

2

3
a: X=NH; R_1=R_2=Me
b: X=NH; R_1=H; R_2=Ph
c: X=NH; R_1=H; R_2=CO_2Me
d: X=O; R_1=R_2=Me
e: X=O; R_1=H; R_2=Me
f: X=S; R_1=R_2=Me

Experimental procedures:
Standard Procedure Using a Focused Microwave Oven for Compounds **3a, b, d–f**:
A mixture of ethyl 3-dimethylamino acrylate **1a–c** (2 mmol) and the appropriate hydrazine **2a–d** was placed in a cylindrical quartz tube. The tube was then introduced into a synthewave 402 Prolabo microwave reactor. Microwave irradiation was carried out at the appropriate reaction time. The mixture was allowed to cool down. After addition of CH_2Cl_2 (10 mL) in the reactor and removal of solvent in vacuo, the crude residue was purified by chromatography on silica gel 60 F 254 (Merck) or by recrystallization with the appropriate solvent. Solvent evaporation gave the desired ethyl α-hetero-β-hydrazino acrylate **3** as a viscous oil which crystallized on standing.

References: N. Meddad, M. Rahmouni, A. Derdour, J.P. Bazureau, J. Hamelin, *Synthesis,* 581 (2001).

Type of reaction: C–N bond formation
Reaction condition: solvent-free
Keywords: 1-methylimidazole, alkyl halide, microwave irradiation, ionic liquids, 1-alkyl-3-methylimidazolium halides

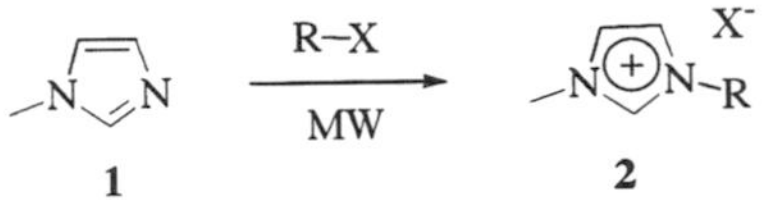

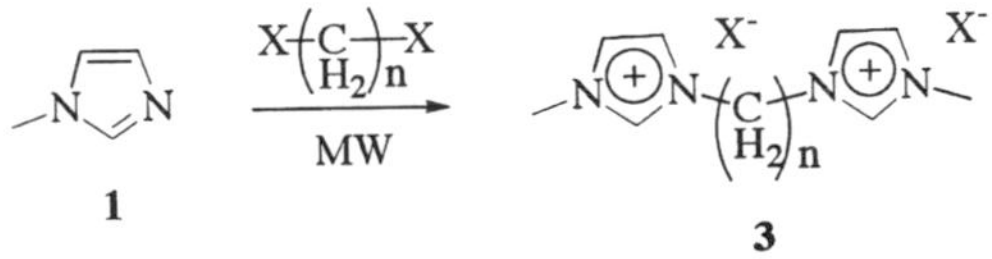

X = Cl, Br, I

Experimental procedures:
Bromobutane (2.2 mmol) and 1-methylimidazole (2 mmol) are placed in a test tube, mixed thoroughly on a vortex mixer (Fisher, model 231) and the mixture is heated intermittently in an unmodified household MW oven (Panasonic NN-S740WA-1200W) at 240 W (30 s irradiation with 10 s mixing) until a clear single phase is obtained. The bulk temperature recorded is in the range 70 to 100 °C. The resulting ionic liquid is then cooled, washed with ether (3×2 mL) to remove unreacted starting materials and product dried under vacuum at 80 °C to afford 86% of 1-butyl-3-methylimidazolium bromide.

References: R.S. Varma, V.V. Namboodiri, *Chem. Commun.*, 643 (2001).

Type of reaction: C–N bond formation
Reaction condition: solvent-free
Keywords: primary amine, *N*-hydroxymethylene pyrazole derivative, microwave irradiation, *N,N*-bis(pyrazole-1-yl-methyl)alkylamine

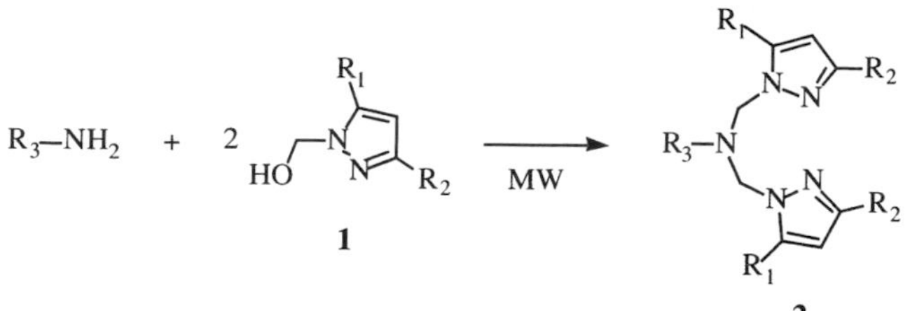

1a: $R_1=R_2=Me$
1b: $R_1=Me$; $R_2=CO_2Me$
1c: $R_1=Me$; $R_2=CO_2Et$

2a: $R_1=R_2=Me$; $R_3=C_6H_{11}$
2b: $R_1=Me$; $R_2=CO_2Me$; $R_3=C_6H_{11}$
2c: $R_1=Me$; $R_2=CO_2Et$; $R_3=C_6H_{11}$
3a: $R_1=R_2=Me$; $R_3=CH_2CH=CH_2$

Experimental procedures:

A mixture of 1 equiv. of the appropriate amine (10 mmol) with 2 equiv. (20 mmol) of compounds **1a–c** is introduced into a Pyrex tube, which was then placed in a microwave reactor and irradiated by microwave (60 W) in the absence of solvent for 20 min. The nitrogen containing derivatives **2a–c, 3a** and **4a** were extracted with dichloromethane and washed with water to eliminate the residual amine. The solvent of the dried organic solution was eliminated under reduced pressure. The white solids **2a–c** were washed with diethylether and dried. The yellow oils **3a–4a** were analyzed as such.

References: R. Touzani, A. Ramdani, T. Ben-Hadda, S.E. Kadiri, O. Maury, H.L. Bozec, P.H. Dixneuf, *Synth. Commun.,* **31,** 1315 (2001).

Type of reaction: C–N bond formation
Reaction condition: solvent-free
Keywords: a,a,a-tris(hydroxymethyl)methylamine, carboxylic acid, microwave irradiation, 2-oxazoline

RCOOH + HOH$_2$C–C(CH$_2$OH)(CH$_2$OH)–NH$_2$ $\xrightarrow{\text{MW}}$

1   **2**   **3**

a: R=Ph
b: R=2-furyl
c: R=heptadecenyl

Experimental procedures:
0.1 mol of *a,a,a*-tris(hydroxymethyl)methylamine was added to 0.1 mol of the finely pulverized carboxylic acid. The mixture, occupying about 10% of the volume of an open Pyrex vessel, was irradiated during 2–3 min at 850 W. The reaction mixture was recrystallized from the appropriate solvent and filtered, and the dried product was compared by IR, elemental analysis and melting point with the compound obtained by traditional methods.

References: A. L. Marrero-Terrero, A. Loupy, *Synlett*, 245 (1996).

Type of reaction: C–N bond formation
Reaction condition: solvent-free
Keywords: *N*-acyl imidate, imidazolidine ketone aminal, microwave irradiation, 2,3-dihydro imidazol[1,2-*c*]pyrimidine

1   **2**   **MW**   **3**

a: X=CN; R$_1$=Me; R$_2$=Me
b: X=CN; R$_1$=Me; R$_2$=Et
c: X=CN; R$_1$=Et; R$_2$=Et
d: X=COOEt; R$_1$=Me; R$_2$=Me
e: X=COOEt; R$_1$=Me; R$_2$=Et

Experimental procedures:
A mixture of 1.5 mmol of *N*-acyl imidate **1** and 3.9 mmol of imidazolidine ketene aminals **2** was placed in a Pyrex glass. Then, the tube was introduced into a Synthewave 420 Prolabo microwave oven. Microwave irradiation was carried out with a suitable power for an appropriate time. The reaction temperature is monitored by a computer coupled with the microwave oven. After, the mixture was

cooled to room temperature, the crude residue was purified by recrystallization or by distillation under reduced pressure with a Buchi GKR-50 Kugelrohr apparatus.

References: M. Rahmouni, A. Derdour, J.P. Bazureau, J. Hamelin, *Synth. Commun.*, **26**, 453 (1996).

Type of reaction: C–N bond formation
Reaction condition: solvent-free
Keywords: orthoester, *o*-substituted aminoaromatics, KSF clay, microwave irradiation, benzimidazole, benzoxazole, benzthiazole

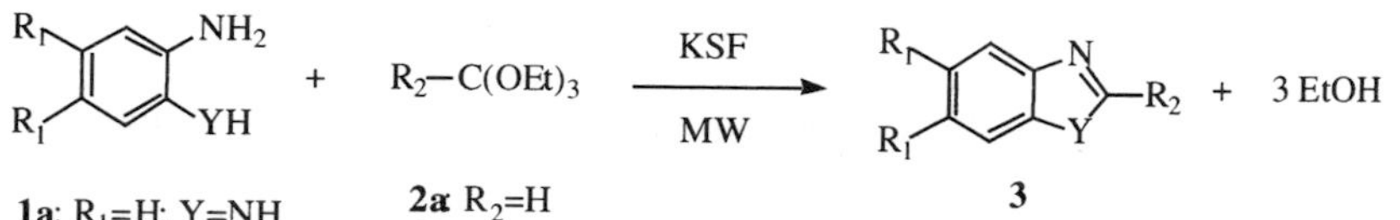

1a: R_1=H; Y=NH
1b: R_1=Me; Y=NH
1c: R_1=H; Y=O
1d: R_1=H; Y=S

2a R_2=H
2b R_2=Me

3

Experimental procedures:
A mixture of 2-aminoaromatic **1** (10 mmol), orthoester **2a** or **2b** (10 mmol) and KSF clay (2 g) in a Pyrex tube under nitrogen was mixed and then irradiated in a focused microwave cavity (cavity E013 of MES) for 5 min with a power of 60 W. The mixture was extracted with CH_2Cl_2 (3×20 mL) and the solvent was evaporated in vacuum. The residue was distilled or recrystallized.

References: D. Villemin, M. Hammadi, B. Martin, *Synth. Commun.*, **26**, 2895 (1996).

Type of reaction: C–N bond formation
Reaction condition: solvent-free
Keywords: thiourea, *α*-chloroketone, microwave irradiation, iminothiazoline

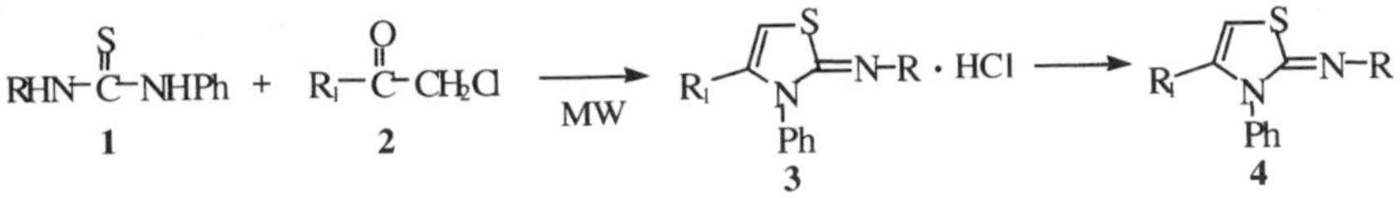

a: R=Me; R_1=Ph
b: R=Me; R_1=PhCH$_2$
c: R=Me; R_1=i-Pr
d: R=Me; R_1=t-Bu
e: R=Me; R_1=naphtyl

f: R=Ph; R_1=Ph
g: R=Ph; R_1=PhCH$_2$
h: R=Ph; R_1=i-Pr
i: R=Ph; R_1=t-Bu
j: R=Ph; R_1=naphtyl

Experimental procedures:
The thiourea (1.2 equiv.) and the α-chloro ketone (1 equiv.) were mixed, placed in quartz tube, and introduced into a Synthewave 402 (Prolabo) single mode apparatus (4 cm, ϕ, reactor) with a temperature monitored at 80 °C. The mixture was irradiated for the time given in Table 1. Thiazolinium salts precipitate in the tube. Products were isolated by crystallization from ether. The free iminothiazoline is obtained by treatment with aqueous ammonia followed by extraction with CH$_2$Cl$_2$. Reactions under classical heating are carried out without solvent in an oil bath previously set at 80 °C during the same time.

References: S. Kasmi, J. Hamelin, H. Benhaoua, *Tetrahedron Lett.,* **39**, 8093 (1998).

Type of reaction: C–N bond formation
Reaction condition: solvent-free
Keywords: urea, phthalic anhydride, microwave irradiation, metallophthalocyanine

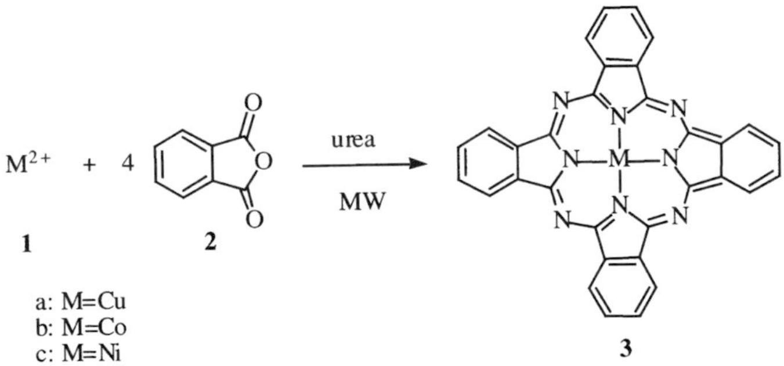

a: M=Cu
b: M=Co
c: M=Ni
d: M=Fe

Experimental procedures:
The preparation of copper phthalocyanines is representative of the general procedure employed. Phthalic anhydride (26.50 g, 0.18 mol), urea (955.40 g, 0.92 mol), copper chloride (5.00 g, 0.05 mol), and ammonium molybdate (0.75 g,

3.80 mmol) were ground together until homogeneous, placed in a beaker and ir-radiated in a microwave oven at high power for 6 min. Upon completion of the reaction the product was ground and washed with 5% caustic soda, water, 2% hydrochloric acid and again with water respectively. The dried phthalocyanine thus obtained weighed 23.3 g, 91% of the theoretical amount based on phthalic anhydride. [Cu(pc)] was subsequently recrystallized two times from concentrated H_2SO_4. Usually during recrystallization the solution of phthalocyanine in concen-trated H_2SO_4 was poured into distilled water. After complete recrystallization the [Cu(pc)] obtained was purified by Soxhlet extraction using in succession metha-nol and methylene chloride, yield 86%.

References: A. Shaabani, *J. Chem. Res. (S)*, 672 (1998).

Type of reaction: C–N bond formation
Reaction condition: solvent-free
Keywords: *N,N*-dimethylurea, paraformaldehyde, primary amine, microwave ir-radiation, triazone, 4-oxo-oxadiazinane

a: R=Me
b: R=Et
c: R=*n*-Pr
d: R=*i*-Pr
e: R=*n*-Bu
f: R=*s*-Bu

Experimental procedures:
General Procedure for Preparation of Triazones **5a–f**. 264 mg (3 mmol) *N,N*-di-methylurea, 1 g paraformaldehyde, 3 mmol primary amine **4a–f** and 2 g montmo-rillonite K-10 were irradiated by microwave in a Teflon vessel. The reaction mix-ture was filtered and washed with water. The organic phase was separated and dried with Na_2SO_4 and concentrated by vacuum distillation. Purification of the

crude material by chromatography on a short column (silica gel, 70–230 mesh) and elution with dichloromethane, or vacuum distillation afforded triazones **5a–f**. The obtained yields were in the range of 67–84%.

References: S. Balalaie, M.S. Hashtroudi, A. Sharifi, *J. Chem. Res. (S)*, 392 (1999).

Type of reaction: C–N bond formation
Reaction condition: solvent-free
Keywords: isatoic anhydride, L-proline, microwave irradiation, dilactam, pyrrolo[2,1-c][1,4]benzodiazepine-5,11-dione

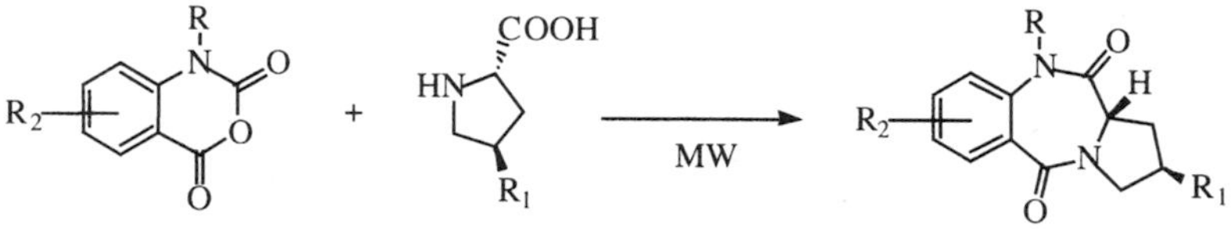

R = H, Me; R_1 = H, OH; R_2 = H, Me, OH, OMe

Experimental procedures:
A mixture of isatoic anhydride **1** (R=R_2=H; 0.163 g, 1.0 mmol) and L-proline **2** (R_1= H; 0.115 g, 1.0 mmol) is placed in a test-tube inside the alumina bath (heat sink) and the contents are irradiated in a microwave oven at full power (600 W) for 3 min (3×1 min) with 1 min intervals. The reaction is monitored by TLC (ethyl acetate-hexane, 6:4), and the reaction mixture is cooled to room temperature and charged on silicagel pad and eluted with ethyl acetate-hexane (7:3) to afford the PBD dilactam **3** in 87% yield.

References: A. Kamal, B.S.N. Reddy, G.S.K. Reddy, *Synlett,* 1251 (1999).

Type of reaction: C–N bond formation
Reaction condition: solvent-free
Keywords: aldehyde, isocyanide, 2-aminopyridine, montmorillonite K-10, microwave irradiation, imidazol[1,2-a]annulated pyridine

R-CHO

1

a: R=Ph
b: R=4-MeC$_6$H$_4$
c: R=4-MeOC$_6$H$_4$
d: R=Pr

R$_1$-NC

2

a: R=PhCH$_2$
b: R=C$_6$H$_{11}$
c: R=(CH$_3$)$_3$CCH$_2$C(CH$_3$)$_2$CH$_2$
d: R=Bu

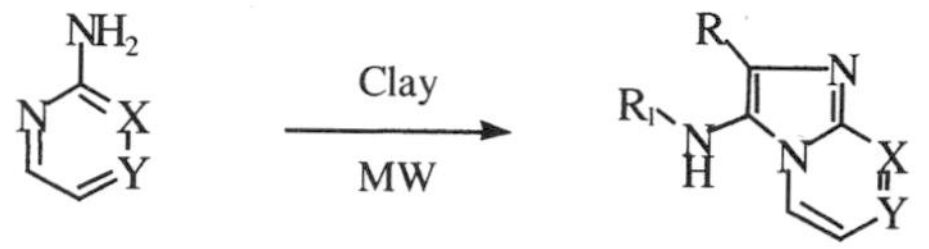

a: X=Y=C
b: X=C, Y=N
c: X=N, Y=C

3

4

Experimental procedures:

A mixture of benzaldehyde **1a** (106 mg, 1 mmol) and 2-aminopyridine **3a** (94 mg, 1 mmol) was irradiated in an unmodified household microwave oven for 1 min (at full power of 900 W) in the presence of montmorillonite K-10 clay (50 mg). After addition of benzyl isocyanide **2a** (117 mg, 1 mmol), the reaction mixture was further irradiated successively (2 min) at 50% power level for a duration of 1 min followed by a cooling period of 1 min. The resulting product was dissolved in dichloromethane (2×5 mL) and the clay was filtered off. The solvent was removed under reduced pressure and the crude product was purified either by crystallization or by passing it through a small bed of silica gel using EtOAc-hexane (1:4, v/v) as eluent to afford **4a**.

References: R.S. Varma, D. Kumar, *Tetrahedron Lett.,* **40**, 7665 (1999).

Type of reaction: C–N bond formation
Reaction condition: solvent-free
Keywords: benzil, aromatic aldehyde, primary amine, NH$_4$OAc, zeolite HY, silica gel, microwave irradiation, imidazole

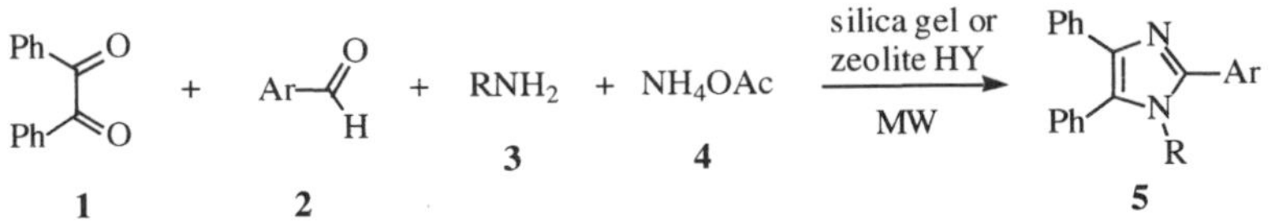

Ph—C(=O)—C(=O)—Ph **1** + Ar—CHO **2** + RNH₂ **3** + NH₄OAc **4** →[silica gel or zeolite HY / MW] Ph,Ph—imidazole—Ar (N-R) **5**

a: Ar=Ph; R=Me
b: Ar=4-MeC₆H₄; R=Me
c: Ar=4-BrC₆H₄; R=Me
d: Ar=Ph; R=Et
e: Ar=4-MeC₆H₄; R=Et
f: Ar=4-BrC₆H₄; R=Et
g: Ar=Ph; R=i-Pr
h: Ar=Ph; R=CH₂=CHCH₂
i: Ar=4-BrC₆H₄; R=CH₂=CHCH₂

j: Ar=Ph; R=i-Bu
k: Ar=4-MeC₆H₄; R=i-Bu
l: Ar=Ph; R=C₆H₁₁
m: Ar=Ph; R=C₆H₁₁
n: Ar=Ph; R=Ph
o: Ar=4-MeC₆H₄; R=Ph
p: Ar=Ph; R=PhCH₂
q: Ar=4-MeC₆H₄; R=PhCH₂

Experimental procedures:

Benzil **1** (421 mg, 2 mmol), aldehyde **2** (2 mmol), ammonium acetate **4** (309 mg, 2 mmol), primary amine **3** (2 mmol) and 2 g silica gel or zeolite HY (prepared from zeolite NH₄Y in an oven at 600 °C for 5 h that afforded zeolite HY) were mixed and then transferred into a beaker (250 mL) and irradiated for 6 min. The progress of the reactions were monitored by TLC using CH₂Cl₂ as eluent. The mixture was extracted into CH₂Cl₂ (3×30 mL), then filtered and washed with H₂O. The organic phase was removed by a rotary evaporator. Further purification by column chromatography (CH₂Cl₂ as eluent) and recrystallization gave the desired products **5**.

References: S. Balalaie, A. Arabanian, *Green Chem.*, **2**, 274 (2000).

Type of reaction: C–N bond formation
Reaction condition: solvent-free
Keywords: benzil, aromatic aldehyde, NH₄OAc, zeolite HY, microwave irradiation, silica gel, triarylimidazole

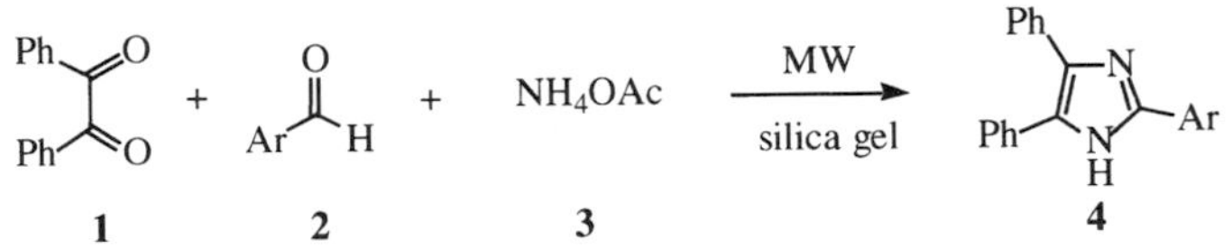

Ph—C(=O)—C(=O)—Ph **1** + Ar—CHO **2** + NH₄OAc **3** →[MW / silica gel] Ph,Ph—imidazole—Ar (N-H) **4**

a: Ar=Ph
b: Ar=4-CH₃C₆H₄
c: Ar=4-CH₃OC₆H₄
d: Ar=4-NO₂C₆H₄

e: Ar=4-ClC₆H₄
f: Ar=2-HOC₆H₄
g: Ar=2,6-Cl₂C₆H₃
h: Ar=4-(CH₃)₂NC₆H₄

Experimental procedures:

841 mg benzil (4 mmol), 4 mmol aldehyde, 617 mg ammonium acetate (8 mmol), and 4 g silica gel or zeolite HY (prepared from zeolite NH_4Y in an oven at 600 °C for 5 h that afforded zeolite HY) were mixed thoroughly in a mortar. Then the reaction mixture was transferred into a beaker (250 mL) and irradiated with microwaves for 6 min. The progress of reaction was monitored by TLC using CH_2Cl_2-EtOAc (90:10) as the eluent. The mixture was extracted with CH_2Cl_2 (3×30 mL), filtered, and washed with H_2O. The organic phase was removed by means of a rotary evaporator. Further purification by column chromatography (eluent CH_2Cl_2-EtOAc, 98:2) on silica gel gave the desired products.

References: S. Balalaie, A. Arabanian, M.S. Hashtroudi, *Monatsh. Chem.*, **131**, 945 (2000).

Type of reaction: C–N bond formation
Reaction condition: solvent-free
Keywords: 3-propynylthio-1,2,4-triazin-5-one, heterocyclization, microwave irradiation, thiazole

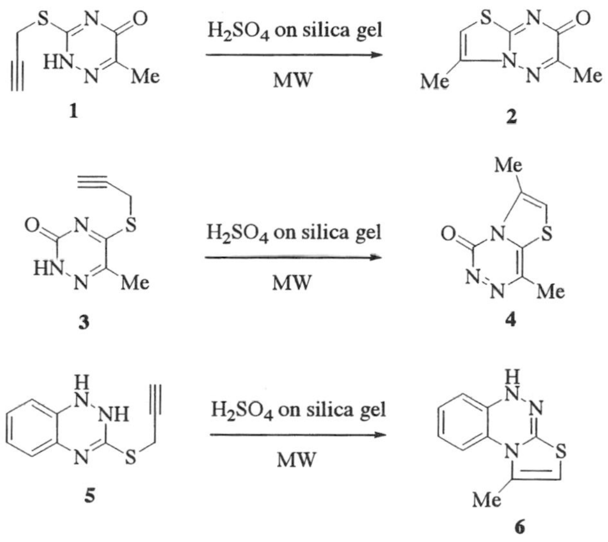

Experimental procedures:

Compound **1** (0.71 g, 0.005 mol) was mixed with 2.1 g of finely ground sulfuric acid adsorbed on silica gel. The reaction mixture was exposed to microwave irradiation for 5 min. After the completion of the reaction (monitored by TLC using CHCl₃-MeOH, 95:5) the crude product was extracted by MeOH, treated with active charcoal and filtered. On evaporation of solvent, fairly pure products can be obtained which were crystallized from EtOH. Yield: 85%, mp 224–225 °C.

References: M. M. Heravi, N. Montazeir, M. Rahimizadeh, M. Bakavolia, M. Ghassemzadeh, *J. Chem. Res. (S)*, 464 (2000).

Type of reaction: C–N bond formation
Reaction condition: solvent-free
Keywords: alkyl isocyanate, microwave irradiation, cyclotrimerization, isocyanurate

a: R=Ph
b: R=4-ClC$_6$H$_4$
c: R=4-MeOC$_6$H$_4$
d: R=1-naphthyl
e: R=Et
f: R=cyclohexyl

Experimental procedures:
Preparation of tri-1-naphtyl isocyanurate **3d**: A mixture of naphthyl isocyanate **1d** (3.38 g, 20 mmol) and sodium piperidinedithiocarbamate **2** (18.3 mg, 0.1 mmol) or sodium nitrite (55.2 mg, 0.8 mmol) was taken in a Teflon tube with screw cap and was then placed inside the microwave oven and irradiated for 5 min at 385 W. After 5 min (during this period it gently cools to about room temperature) it was irradiated again at the same power for another 5 min. The reaction mixture was allowed to cool to room temperature and the resultant residue washed with 50 mL ether. Trituration of the crude residue with water led to the title compound which was recrystallized from ethanol to give 81% of **3d**.

References: M. S. Khajavi, M. G. Dakamin, H. Hazarkhani, *J. Chem. Res. (S)*, 145 (2000).

Type of reaction: C–N bond formation
Reaction condition: solvent-free
Keywords: carboxylic acid, thiosemicarbazide, microwave irradiation, thiadiazole

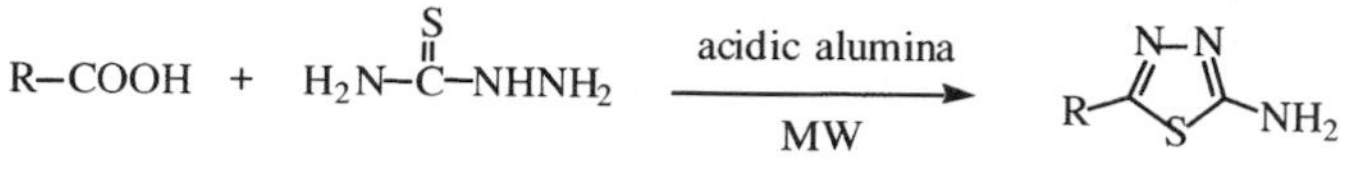

$$R-COOH \quad + \quad H_2N-\overset{\overset{\displaystyle S}{\|}}{C}-NHNH_2 \quad \xrightarrow[\text{MW}]{\text{acidic alumina}} \quad R-\overset{N-N}{\underset{S}{\bigtriangleup}}-NH_2$$

1

a: R=Me
b: R=C_7H_{15}
c: R=C_9H_{19}
d: R=$C_{11}H_{23}$

Experimental procedures:

Acidic alumina (18 g) was added to the solution of carboxylic acids (0.01 mol) and thiosemicarbazide (0.01 mol) in CH_2Cl_2 (5 mL) at room temperature. The reaction mixture was mixed, adsorbed material was dried, kept inside alumina bath and irradiated in microwave oven for 40–80 s. The mixture was cooled and then product was extracted into acetone (4×10 mL). The product was collected by evaporating the solvent under reduced pressure and recrystallized from a mixture of methanol and acetone.

References: M. Kidwai, P. Misra, K. R. Bhushan, B. Dave, *Synth. Commun.*, **30**, 3031 (2000).

Type of reaction: C–N bond formation
Reaction condition: solvent-free
Keywords: aromatic aldehyde, 1,2-dicarbonyl compound, primary amine, acidic alumina, ammonium acetate, microwave irradiation, imidazoles

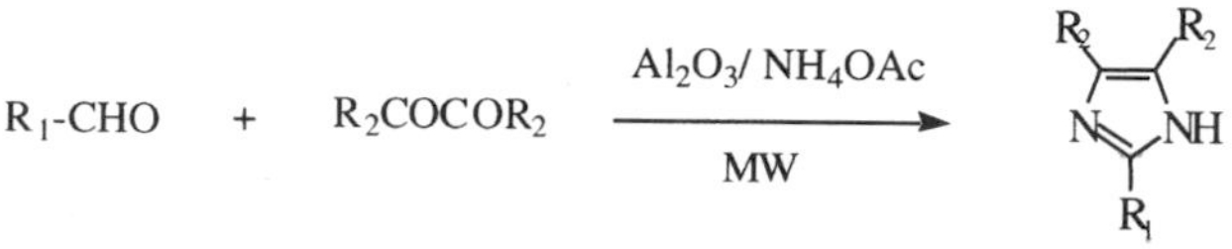

$$R_1-CHO \quad + \quad R_2COCOR_2 \quad \xrightarrow[\text{MW}]{\text{Al}_2\text{O}_3/\text{ NH}_4\text{OAc}} \quad \overset{R_2 \quad R_2}{\underset{\underset{R_1}{N \quad NH}}{\bigvee}}$$

1

a: R_1=R_2=Ph
b: R_1=2-Thienyl; R_2=Ph
c: R_1=4-ClC$_6$H$_4$; R_2=Ph
d: R_1=2-Thienyl; R_2=4-MeC$_6$H$_4$

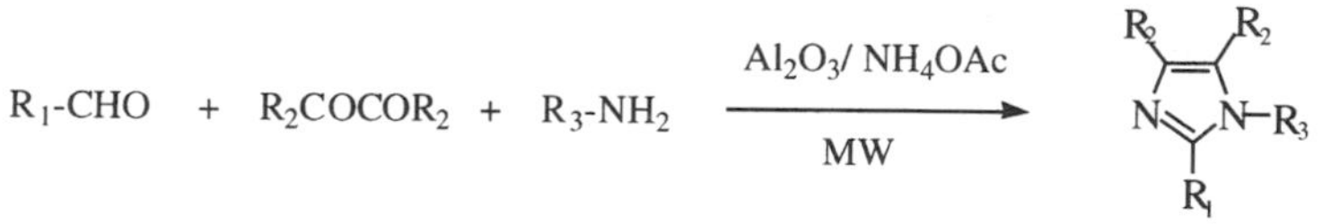

$$R_1-CHO \quad + \quad R_2COCOR_2 \quad + \quad R_3-NH_2 \quad \xrightarrow[\text{MW}]{\text{Al}_2\text{O}_3/\text{ NH}_4\text{OAc}} \quad \overset{R_2 \quad R_2}{\underset{\underset{R_1}{N \quad N-R_3}}{\bigvee}}$$

2

a: R_1=R_2=Ph; R_3=PhCH$_2$CH$_2$
b: R_1=2-Thienyl; R_2=Ph; R_3=PhCH$_2$CH$_2$
c: R_1=Ph; R_2=4-MeC$_6$H$_4$; R_3=PhCH$_2$CH$_2$
d: R_1=4-EtC$_6$H$_4$; R_2=Ph; R_3=3-ClC$_6$H$_4$CH$_2$

Experimental procedures:

A mixture of acidic alumina (9.3 g) and ammonium acetate (4.4 g) was ground in a mortar until a homogeneous powder was formed. A solution of 0.5 mmol of 1,2-dicarbonyl compound and 0.5 mmol of aldehyde in 2 mL of diethyl ether or methylene chloride (synthesis of **1a–c**), or a solution of 0.5 mmol of 1,2-dicarbonyl compound, 0.5 mmol of aldehyde and 0.5 mmol of amine in 2 mL of ethyl ether or methylene chloride (synthesis of **2a–c**) was added to 2.5 g of the alumina/ammonium acetate mixture in a 20-mL glass vial. The solvent was allowed to evaporate and the dry residue was irradiated in a domestic Kenmore microwave oven at 130 W (10% power) for 20 min in the open vial. The mixture was then allowed to cool to room temperature and washed with a mixture of acetone-triethylamine (7:3, v/v), 3×10 mL to extract the product. The combined washes were filtered and the solvent was evaporated under reduced pressure. The resulting solid residue was purified by flash chromatography using FlashElute system from elution solutions on a 90 g silica cartridge. A mixture of hexane-ethyl acetate (4:1, v/v) was used as eluent for the purification of **1a–d, 2a**, and the solvent system hexane-ethyl acetate (7:1, v/v) was used for the isolation of **2b–d**.

References: A.Y. Usyatinsky, Y.L. Khmelnitsky, *Tetrahedron Lett.*, **41**, 5031 (2000).

Type of reaction: C–N bond formation
Reaction condition: solvent-free
Keywords: graphite, microwave irradiation, benzimidazoquinazoline, tetraazabenzo[*a*]indeno[1,2-*c*]anthracen-5-one

a: R=H
b: R=Me

Experimental procedures:
A mixture of compound **1** (0.2 g, 0.75 mmol), anthranilic acid (0.374 g, 4.3 mmol) and graphite (2 g) was placed in the microwave oven in a 70-mL quartz vial. The irradiation was programmed at 105 W for 90 min (after a period of 2–3 min the temperature reached a plateau, 170 °C, and remained constant). After cooling, the graphite powder was filtered and washed with dichloromethane. The organic solution was washed with a saturated solution of sodium bicarbonate, and the crude product recrystallized in ethanol.

References: M. Soukri, G. Guillaumet, T. Besson, D. Aziane, M. Aadil, E.M. Essassi, M. Akssira, *Tetrahedron Lett.*, **41**, 5857 (2000).

Type of reaction: C–N bond formation
Reaction condition: solvent-free
Keywords: 4,4′-bipyridinium diquaternary salts, acetylenedicarboxylate, 1,3-dipolar cycloaddition, microwave irradiation, 7,7′-bis-indolizine

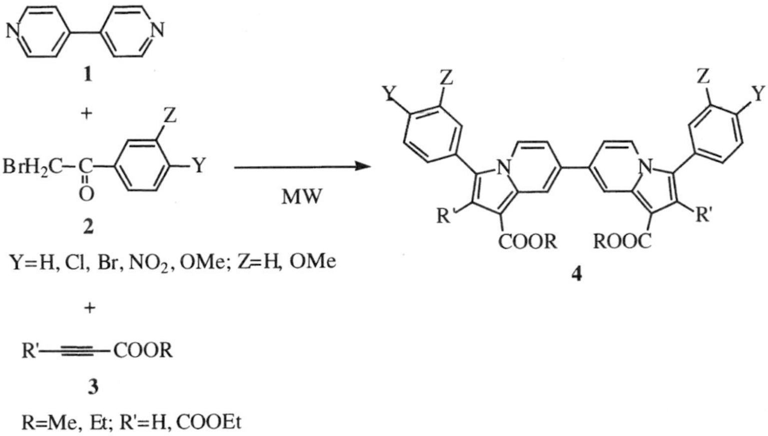

Y=H, Cl, Br, NO$_2$, OMe; Z=H, OMe

R′═══COOR

3

R=Me, Et; R′=H, COOEt

Experimental procedures:
4,4′-Bipyridinium diquaternary salts (1.0 mmol) and acetylenedicarboxylate (1.0 mmol) were dissolved into 1 mL of acetone and adsorbed on KF-Merck Alumina 70–230 mesh. The solvent was then evaporated and the mixture was irradiated at atmospheric pressure in a focused microwave reactor Prolabo MX350 with measurement and control of power and temperature by infrared detection for the time and at power indicated in Table 1. All derivatives were characterized by IR, ^{1}H and ^{13}C NMR spectroscopy, and elemental analyses.

References: R. M. Dinica, I. I. Druta, C. Pettinari, *Synlett,* 1013 (2000).

Type of reaction: C–N bond formation
Reaction condition: solvent-free
Keywords: *N,N*-diarylthiourea, chloroacetylchloride, K$_2$CO$_3$, microwave irradiation, thiohydantoin

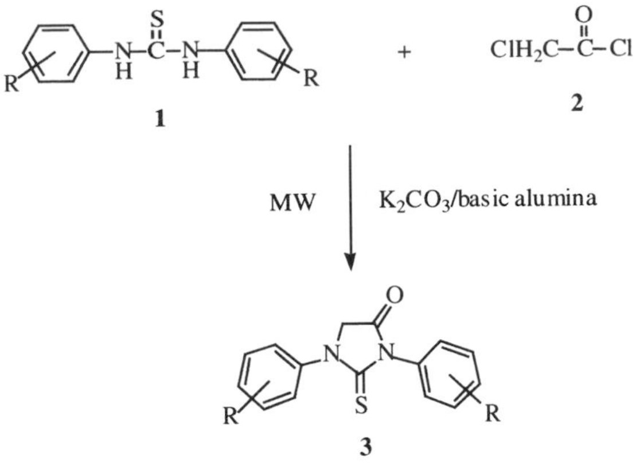

1 + $ClH_2C-\overset{\overset{\displaystyle O}{\|}}{C}-Cl$ **2**

MW | K$_2$CO$_3$/basic alumina

3

R=H, *o*-Me, *o*-MeO, *p*-Cl, *p*-Br

Experimental procedures:

To a methanolic solution of 0.01 mol of *N,N*-diarylthiourea **1** and 0.01 mol of chloroacetylchloride **2** was added 15 g of K$_2$CO$_3$/basic alumina. The reaction mixture was air dried and irradiated in domestic microwave oven (800 W) for 23 min. On completion of the reaction, as monitored by TLC (every 30 s) the reaction mixture was treated with cold water/ethanol. The product obtained from filtration/evaporation was dried and recrystallized from methanol.

References: M. Kidwai, R. Venkataramanan, B. Dave, *Green Chem.*, **3**, 278 (2001).

Type of reaction: C–N bond formation
Reaction condition: solvent-free
Keywords: thalidomide, microwave irradiation, *N*-phtaloyl-*L*-glutamic acid

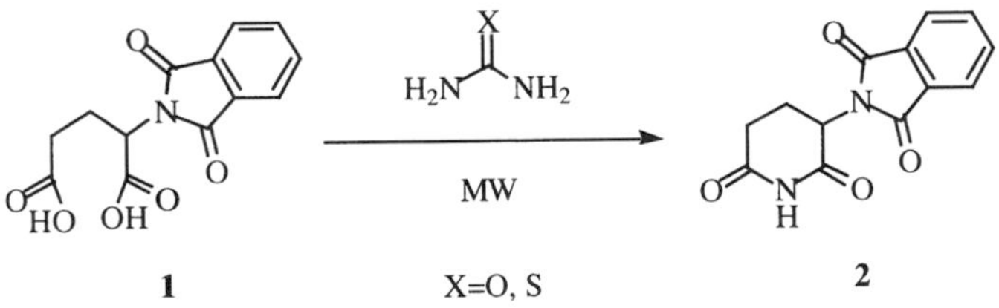

1   X=O, S   **2**

Experimental procedures:

N-Phthaloyl-*L*-glutamic acid **1** (3.0 g, 10.8 mmol) was mixed with thiourea (0.882 g, 11.6 mmol) and then introduced into a Pyrex test tube. The mixture was irradiated in a domestic oven (1000 W output) with irradiation control set at 70% for 15 min. The crude reaction mixture was dissolved in THF and purified by column chromatography on silica gel (THF-hexane, 1:1) to afford a solid product of (±)-thalidomide **2** (2.375 g, 85% yield). Mp 269–271 °C.

References: J. A. Seijas, M. P. Vazquez-Tato, C. Gonzalez-Bande, M. M. Martinez, B. Pacios-Lopez, *Synthesis*, 999 (2001).

Type of reaction: C–N bond formation
Reaction condition: solvent-free
Keywords: 2-aminobenzenethiol, β-ketoester, β-diketone, basic alumina, microwave irradiation, 4*H*-1,4-benzothiazin

a: R=H; R'=OC$_2$H$_5$
b: R=CH$_3$; R'=OCH$_3$
c: R=CH$_3$; R'=CH$_3$
d: R=OCH$_3$; R'=OC$_2$H$_5$
e: R=OC$_2$H$_5$; R'=OC$_2$H$_5$
f: R=Br; R'=CH$_3$
g: R=Cl; R'=CH$_3$

h: R=Cl; R'=OC$_2$H$_5$
i: R=CH$_3$; R'=*p*-CH$_3$OC$_6$H$_4$
j: R=CH$_3$; R'=*p*-CH$_3$C$_6$H$_4$
k: R=Br; R'=*p*-CH$_3$OC$_6$H$_4$
l: R=Br; R'=*p*-CH$_3$C$_6$H$_4$
m: R=Cl; R'=*p*-CH$_3$OC$_6$H$_4$
n: R=Cl; R'=*p*-CH$_3$C$_6$H$_4$

Experimental procedures:
Appropriate 2-aminobenzenethiol **1** (2 mmol) and β-ketoester or β-diketone **2** (2 mmol) were introduced in a beaker (100 mL) and dissolved in chloroform (5 mL). Basic alumina (S.D. Fine Chemicals Pvt. Ltd., 5 g) was then added and swirled for a while followed by removal of solvent under gentle vacuum. The dry powder thus obtained was irradiated in a microwave oven at power output of 520 W for an appropriate time (monitored by TLC). The inorganic support (which can be reused 3–4 times without any loss of activity) was separated by filtration after eluting the product with acetone (5×20 mL). The filtrate was dried over sodium sulfate and the product, obtained after removal of solvent, was recrystallized from methanol as colored crystals in good yields.

References: S. Paul. R. Gupta, A. Loupy, B. Rani, A. Dandia, *Synth. Commun.*, **31**, 711 (2001).

Type of reaction: C–N bond formation
Reaction condition: solvent-free
Keywords: potassium thiocarbamate, ethyl bromoacetate, basic alumina, microwave irradiation, thiadiazolyl thiazolothione

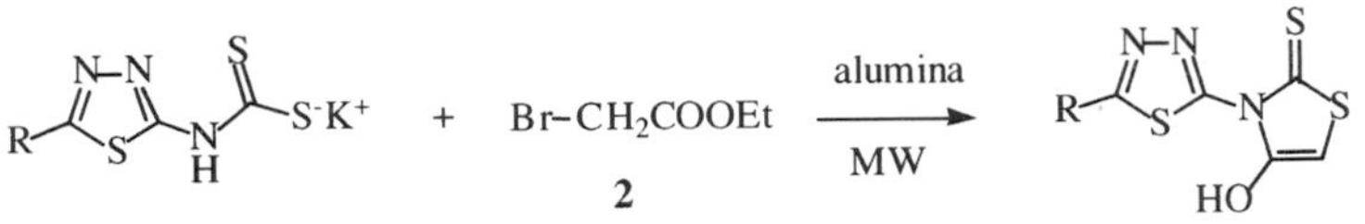

a: R=CH$_3$
b: R=C$_7$H$_{15}$
c: R=C$_9$H$_{19}$
d: R=C$_{11}$H$_{23}$

Experimental procedures:
Basic alumina (18 g) was added to the solution of ethyl bromoacetate **2** (0.01 mol) in dichloromethane (20 mL) and to the solution of potassium thiocarbamate **1** (0.01 mol) in hot acetone (5 mL) at room temperature in two separate beakers. Adsorbed material was dried and the two reactants mixed properly using a mortar and pestle. The reaction mixture was kept inside an alumina bath and irradiated in a microwave oven for 100–110 s. The mixture was cooled and the product was extracted into CH$_2$Cl$_2$ (5×4 mL). The product was collected by evaporating the solvent and recrystallizing from a mixture of methanol and CH$_2$Cl$_2$.

References: M. Kidwai, P. Misra, K.R. Bhushan, *Synth. Commun.,* **31**, 817 (2001).

Type of reaction: C–N bond formation
Reaction condition: solvent-free
Keywords: hydrazide, phenacyl bromide, alumina, montmorillonite, silica gel, microwave irradiation, pyrazine

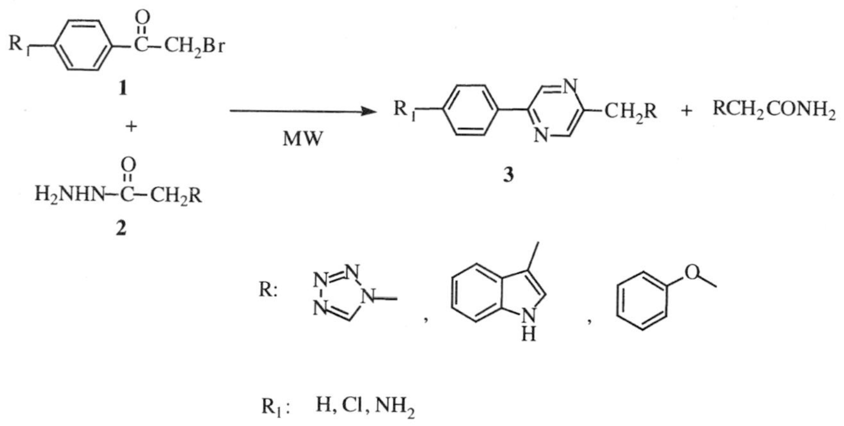

R$_\Gamma$—〈C$_6$H$_4$〉—$\overset{\overset{O}{\|}}{C}$—CH$_2$Br

1

+

MW ⟶ R$_\Gamma$—〈 〉—〈pyrazine〉—CH$_2$R + RCH$_2$CONH$_2$

3

$\overset{\overset{O}{\|}}{H_2NHN-C-CH_2R}$

2

R: 〈triazole N=N〉 , 〈indole〉 , 〈phenoxy〉

R$_1$: H, Cl, NH$_2$

Experimental procedures:

To a solution of substituted hydrazides **2** (0.02 mol) in ethanol added inorganic solid support (10 g) at room temperature. Similarly inorganic support (8 g) was added in the solution of phenacyl bromide/4-chlorophenacyl bromide/4-amino phenacyl bromide **1** (0.01 mol) in dichloromethane. Adsorbed material was mixed properly, kept inside alumina bath and subjected to microwave irradiation for 60–120 s. On completion of the reaction as followed by TLC examination, the mixture was cooled to room temperature and then product was extracted into ethanol (2×15 mL). The filtrate was concentrated and recovering the solvent under reduced pressure afforded the product which was purified through recrystallization from ethanol-DMF mixture.

References: M. Kidwai, P. Sapra, K.R. Bhushan, P. Misra, *Synth. Commun.*, **31**, 1639 (2001).

Type of reaction: C–N bond formation
Reaction condition: solvent-free
Keywords: *β*-chlorovinylaldehyde, hydrazine, microwave irradiation, pyrazolo[3,4-*b*]quinoline, pyrazolo[3,4-*c*]pyrazole, pyrazole

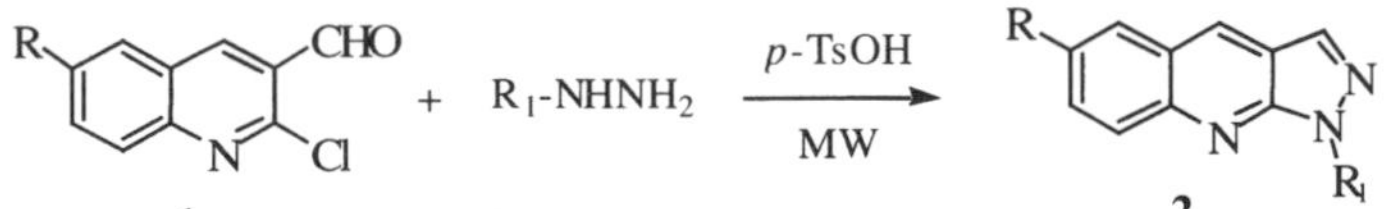

1

2

a: R=H; R$_1$=H
b: R=H; R$_1$=Ph
c: R=OMe; R$_1$=H
d: R=OMe; R$_1$=Ph
e: R=Me; R$_1$=H
f: R=Me; R$_1$=Ph

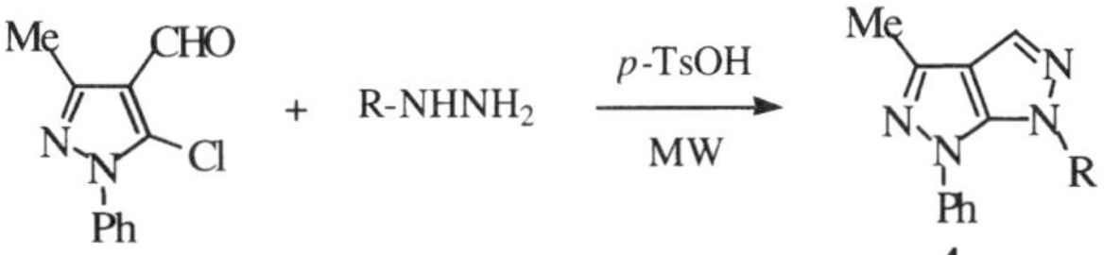

3

4

a: R=H
b: R=Ph

Experimental procedures:

Appropriate β-chlorovinylaldehydes **1** or **3** (1 mmol) and hydrazine hydrate (4 mmol) or phenylhydrazine (1.25 mmol) were introduced in a beaker (50 mL). A small amount of *p*-TsOH (100 mg) was added and properly mixed with the help of glass rod. The so-obtained paste was irradiated in a microwave oven at a power output of 300 W for the appropriate time. After irradiation, cold water (25 mL) was added. The obtained solid was collected, washed with water, dried and recrystallized from adequate solvent to afford the desired products in good yields (78–97%).

References: S. Paul, M. Gupta, R. Gupta, A. Loupy, *Tetrahedron Lett.,* **42**, 3827 (2001).

Type of reaction: C–N bond formation
Reaction condition: solvent-free
Keywords: α,β-unsaturated ester, amine, Michael 1,4-addition, microwave irradiation, β-amino ester

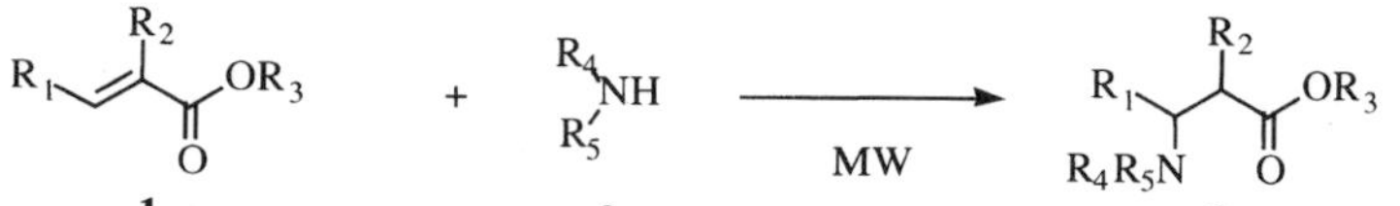

a: R_1=Ph; R_2=H; R_3=Me
b: R_1=Me; R_2=H; R_3=Me
c: R_1=Ph; R_2=H; R_3=s-Bu
d: R_1=H; R_2=Me; R_3=Me
e: R_1=H; R_2=Me; R_3=n-Bu

a: R_4,R_5= -(CH$_2$)$_2$-O-(CH$_2$)$_2$-
b: R_4= PhCH$_2$; R_2=H
c: R_4= PhCHMe; R_2=H
d: R_4,R_5= -(CH$_2$)$_5$-

Experimental procedures:

The reactions of esters of a,β-unsaturated acids with a 1.5 fold excess of amines were carried out without a solvent in a commercial microwave oven (Funai MO785VT) in a reaction vessel for several minutes. The power of microwave irradiation depends on the stability of the starting compound under microwave irradiation conditions. The Michael adduct was separated from the unreacted starting substances by column chromatography on silica gel with a hexane-ethyl acetate mixture (2:1) as the eluent.

References: N.N. Romanova, A.G. Gravis, G.M. Shaidullina, I.F. Leshcheva, Y.G. Bundel, *Mendeleev. Commun.,* 235 (1997).

Type of reaction: C–N bond formation
Reaction condition: solvent-free
Keywords: 2'-aminochalcone, microwave irradiation, 2-aryl-1,2,3,4-tetrahydro-4-quinolone

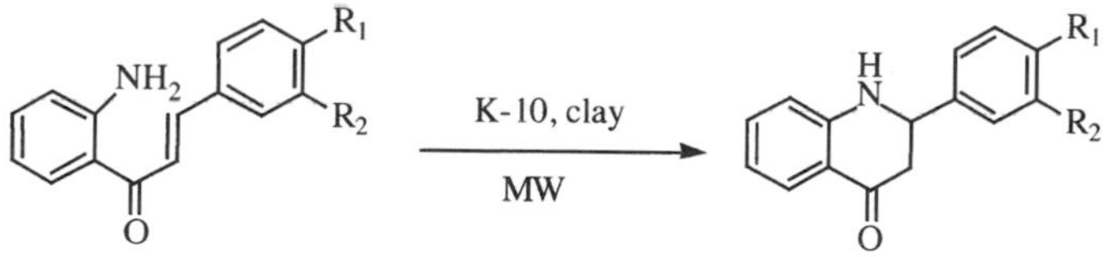

a: R_1=R_2=H
b: R_1=H; R_2=CH$_3$
c: R_1=OMe; R_2=H
d: R_1=Cl; R_2=H
e: R_1=Br; R_2=H
f: R_1=NO$_2$; R_2=H
g: R_1=Me; R_2=2-HOC$_6$H$_4$
h: R_1=R_2=OMe

Experimental procedures:
Montmorillonite K-10 clay (1.0 g) is mixed with **1a** (0.1 g, 0.45 mmol) in solid state using a pestle and mortar or alternatively with a solution of **1a** in dichloromethane (2 mL). The adsorbed material is transferred to a glass tube and is inserted in an alumina bath (alumnia: 100 g, mesh 65–325, Fisher scientific; bath: 5.7 cm diameter) inside the microwave oven. The compound is irradiated for 1.5 min (the temperature of alumina bath reached 110 °C at the end of this period) and the completion of the reaction is monitored by TLC examination. The product is extracted into dichloromethane (2×15 mL) and clay is filtered off. Removal of the solvent under reduced pressure affords 2-phenyl-1,2,3,4-tetrahydro-4-quinolone **2a**, in 80% yield, mp 148–150 °C.

References: R. S. Varma, R. K. Saini, *Synlett,* 857 (1997).

Type of reaction: C–N bond formation
Reaction condition: solvent-free
Keywords: 4,5-dihydroxy-2-pentenoic ester, amine, Michael 1,2-addition, asymmetric induction, microwave irradiation, 3-amino-4,5-dihydroxy-pentanoic ester

1
a: R=Et
b: R=Me

2
a: R₁=H; R₂=CH₂Ph
b: R₁,R₂=-(CH₂)₅-
c: R₁=H; R₂=(S)-PhCHMe

Experimental procedures:
The reactions of esters of *trans-(S)*-**1a** with amines **2a–c** were carried out in a commercial microwave oven (Funai MO 785 VT) in the absence of solvent for 12 min at a power of 510 W and the temperature of the reaction mixture not higher than 60 °C. The reaction run was monitored by thin layer chromatography on silica gel, and reaction products were isolated by column chromatography on silica gel using a heptane-ethyl acetate (2:1) mixture as the eluent.

References: N. N. Romanova, A. G. Gravis, I. F. Leshcheva, Y. G. Bundel, *Mendeleev. Commun.,* 147 (1998).

Type of reaction: C–N bond formation
Reaction condition: solvent-free
Keywords: styrene, aniline, potassium *tert*-butoxide, hydroamination, microwave irradiation, β-phenylethylamine

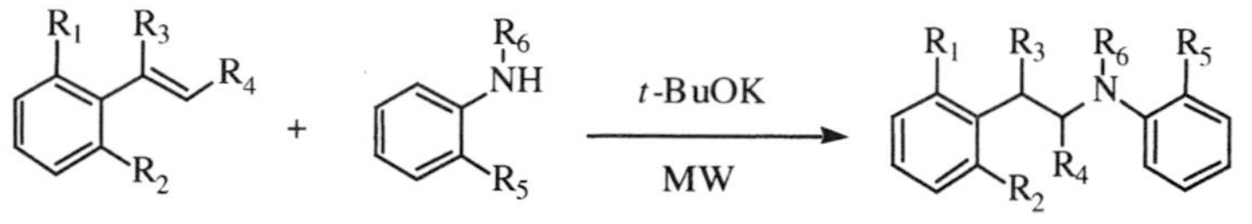

a: $R_1=R_2=R_3=R_4=R_5=R_6=H$
b: $R_1=R_2=R_3=R_4=R_6=H$; $R_5=Me$
c: $R_1=R_2=R_3=R_4=R_5=H$; $R_6=NH_2$
d: $R_1=R_2=R_3=R_5=R_6=H$; $R_4=Me$
e: $R_1=R_2=R_4=R_5=R_6=H$; $R_3=Me$
f: $R_1=Cl$; $R_2=R_3=R_4=R_5=R_6=H$
g: $R_1=OMe$; $R_2=Ox$; $R_3=R_4=R_5=R_6=H$

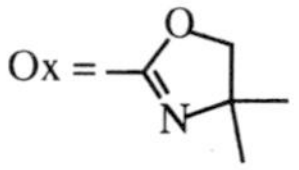

Experimental procedures:
A mixture of styrene (100 mg, 0.962 mmol), aniline (895 mg, 9.62 mmol) and potassium *tert*-butoxide (107 mg, 0.962 mmol) in a open test tube was heated in a domestic microwave (1000 W, 70% of total power) until no starting styrene was observed by TLC (10 min). The crude reaction mixture was purified by chromatography column on silica gel to afford *N*-phenyl-2-phenylethylamine (154 mg, 81%).

References: J. A. Seijas, M. P. Vazquez-Tato, M. M. Martinez, *Synlett,* 875 (2001).

Type of reaction: C–N bond formation
Reaction condition: solvent-free
Keywords: benzaldehyde, aniline, ketone, secondary amine, Envirocat EPZG, microwave irradiation, imine, enamine

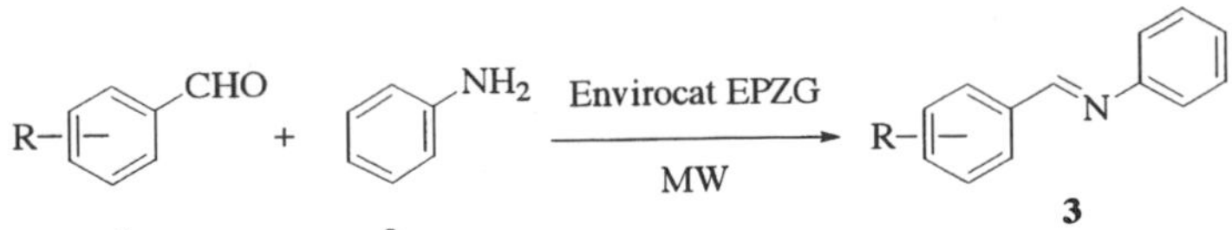

1

a: R=H   d: R=p-Me
b: R=o-OH   e: R=p-NMe$_2$
c: R=p-OH   f: R=p-OMe

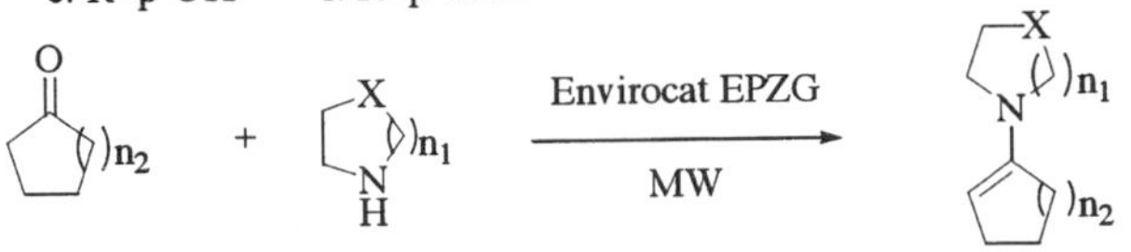

4   **5**

a: n_1=1; X=CH$_2$
b: n_1=2; X=O, CH$_2$
c: n_2=1,2

Experimental procedures:

A mixture of benzaldehyde **1** (0.53 g, 5 mmol), aniline **2** (0.47 g, 5 mmol) and Envirocat EPZGR (100 mg) is placed in a small beaker inside the alumina bath (heat sink) and the contents are irradiated in a MW oven at full power for 1.0 min; the reaction is monitored by TLC (hexane-AcOEt, 10:1). The product is extracted into dichloromethane (10 mL) and Envirocat EPZGR is removed by filtration and washed with dichloromethane (3×10 mL). Removal of the solvent on a rotary evaporator affords the benzylidene aniline **3** in 97% yield.

References: R.S. Varma, R. Dahiya, *Synlett,* 1245 (1997).

Type of reaction: C–N bond formation
Reaction condition: solvent-free
Keywords: aldehyde, ketone, amine, montmorillonite K-10 clay, microwave irradiation, imine, enamine

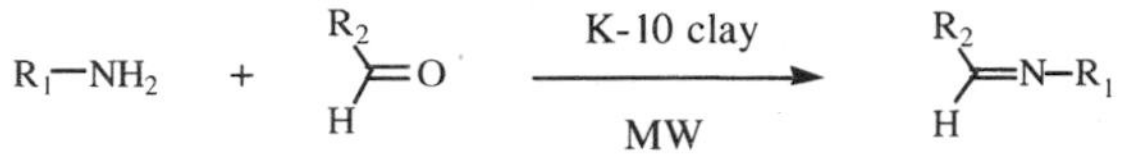

a: R_1=Ph; R_2=Ph
b: R_1=Ph; R_2=p-HOC$_6$H$_4$
c: R_1=Ph; R_2=o-HOC$_6$H$_4$
d: R_1=Ph; R_2=p-Me$_2$C$_6$H$_3$
e: R_1=Ph; R_2=p-MeOC$_6$H$_4$

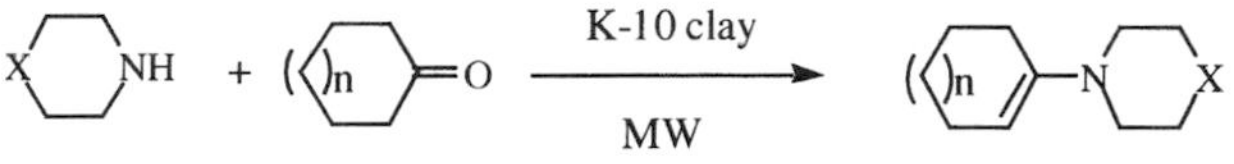

a: X=NH; n=1
b: X=CH$_2$; n=1
c: X= – ; n=1
d: X=CH$_2$; n=0

Experimental procedures:

To an equimolar (1 mmol) mixture of benzaldehyde **2a** (106 mg) and aniline **1a** (93 mg) placed in an open glass container, montmorillonite K-10 clay (20 mg) is added and the reaction mixture is irradiated in a microwave oven at full power for 3 min. Upon completion of the reaction, as followed by TLC examination, the product is extracted into dichloromethane (3×10 mL). Removal of the solvent under reduced pressure affords the benzylidene aniline **3a** in 98% yield.

References: R. S. Varma, R. Dahiya, S. Kumar, *Tetrahedron Lett.,* **38**, 2039 (1997).

Type of reaction: C–N bond formation
Reaction condition: solvent-free
Keywords: hydroxylamine hydrochloride, aldehyde, ketone, microwave irradiation, oxime

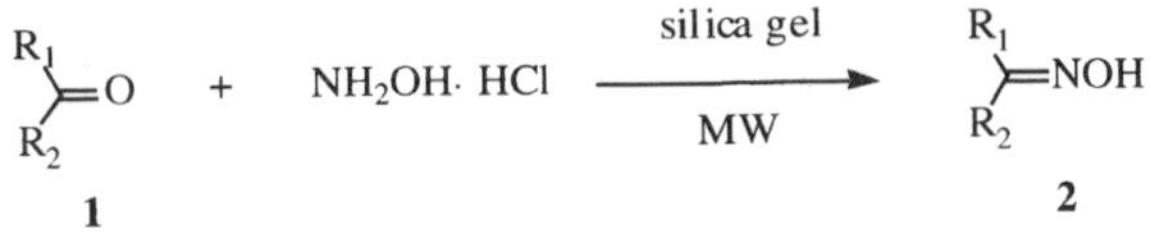

$$\underset{1}{\overset{R_1}{\underset{R_2}{\diagdown}}}{=}O \quad + \quad NH_2OH \cdot HCl \quad \xrightarrow[\quad MW \quad]{silica \ gel} \quad \underset{2}{\overset{R_1}{\underset{R_2}{\diagdown}}}{=}NOH$$

a: R_1=Ph; R_2=H

b: R_1=2-MeOC$_6$H$_4$; R_2=H

c: R_1=4-BrC$_6$H$_4$; R_2=H

d: R_1=4-ClC$_6$H$_4$; R_2=H

e: R_1=4-NO$_2$C$_6$H$_4$; R_2=H

f: R_1=4-MeOC$_6$H$_4$; R_2=H

g: R_1=4-Me$_2$NC$_6$H$_4$; R_2=H

h: R_1=2-Cl-6-FC$_6$H$_3$; R_2=H

i: R_1=PhCH=CH; R_2=H

j: R_1=3,4-(MeO)$_2$C$_6$H$_3$; R_2=H

k: R_1=Ph; R_2=Ph

l: R_1=Ph; R_2=H

m: R_1=PhCO; R_2=Ph

n: R_1=Ph; R_2=Me

Experimental procedures:

In a typical experiment, a mixture of 0.106 g (1 mmol) of benzaldehyde **1a** and (2 mmol, 0.140 g) of hydroxylamine hydrochloride was ground thoroughly in a mortar and supported on silica gel (silica gel 60, 230–400 mesh, Merck) (1 g). The mortar was covered with a watch glass and put inside a Samsung microwave (2450 MHz, 900 W). The compound was irradiated for 2 min and the completion of the reaction is monitored by TLC examination. After completion of the reaction, the mortar was removed from the oven and the mixture was cooled to room temperature, 10 mL of 5% aqueous HCl were added and the solution was extracted with CH$_2$Cl$_2$ (2×5 mL). The extracts were combined, dried (CaCl$_2$) and evaporation of solvent under vacuum gave benzaldehyde oxime **2a**, which was >98% pure, (TLC, ^{1}H NMR, IR and, melting point). The product could be further purified by crystallization from *n*-hexane.

References: A.R. Hajipour, S.E. Mallakpour, G. Imanzadeh, *J. Chem. Res. (S)*, 228 (1999).

Type of reaction: C–N bond formation

Reaction condition: solvent-free

Keywords: aromatic aldehyde, sulfonamide, montmorillonite K-10, microwave irradiation, *N*-sulfonylimine

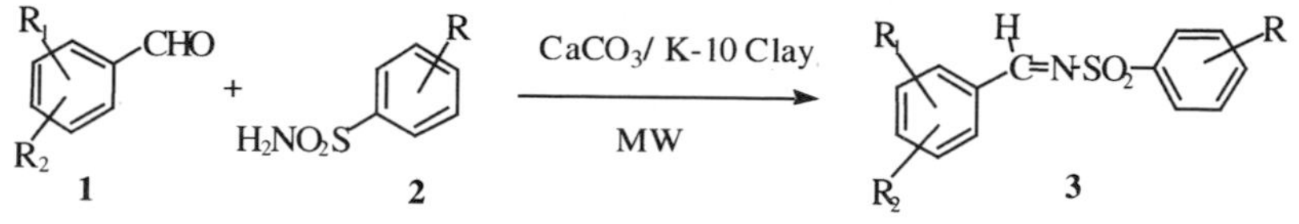

a: $R_1=R_2=R=H$
b: $R_1=R_2=H$; $R=4\text{-Me}$
c: $R_1=R_2=H$; $R=4\text{-Cl}$
d: $R_1=R_2=H$; $R=2\text{-COOMe}$
e: $R_1=4\text{-Br}$; $R_2=H$; $R=4\text{-Me}$
f: $R_1=4\text{-OAc}$; $R_2=H$; $R=4\text{-Me}$
g: $R_1=3\text{-OMe}$; $R_2=4\text{-OMe}$; $R=4\text{-Me}$
h: $R_1=3\text{-OMe}$; $R_2=4\text{-OCHMe}_2$; $R=4\text{-Me}$
i: $R_1=3\text{-OAc}$; $R_2=4\text{-OAc}$; $R=H$

Experimental procedures:

Aromatic aldehydes (10 mmol) and trimethylorthoformate (20 mmol) was added to a mixture of sulfonamide (10 mmol), finely powdered calcium carbonate (9 g) and K-10 clay (2 g). The solid homogenized mixture was placed in a modified reaction tube which was connected to a removable cold finger and sample collector to trap the ensuing methanol and methyl formate. The reaction tube is inserted into Maxidigest MX 350 (Prolabo) microwave reactor equipped with a rotational mixing system. After irradiation for a specified period, the contents were cooled to room temperature and mixed thoroughly with ethyl acetate (2×20 mL). The solid inorganic material was filtered off and solvent was evaporated to afford the residue which was crystallized from the mixture of hexane and ethyl acetate.

References: A. Vass, J. Dudas, R. S. Varma, *Tetrahedron Lett.*, **40**, 4951 (1999).

Type of reaction: C–N bond formation
Reaction condition: solvent-free
Keywords: aldehyde, ketone, primary amine, sodium borohydride, reductive amination, K-10 clay, microwave irradiation, secondary amine

$$R_2 \bigvee_{R_1} = O \ + \ H_2N-R_3 \xrightarrow[\text{MW}]{\text{K-10 clay}} \ R_1 \bigvee^{R_2} \diagdown_{N} \diagup^{R_3} \xrightarrow[\text{MW}]{\text{NaBH}_4\text{-K-10 clay}} \ R_1 \bigvee^{R_2} \diagdown_{\underset{H}{N}} \diagup^{R_3}$$

$$\textbf{1} \qquad\qquad \textbf{2} \qquad\qquad\qquad\qquad \textbf{3} \qquad\qquad\qquad\qquad \textbf{4}$$

a: R_1=Ph; R_2=H; R_3=Ph
b: R_1=Ph; R_2=H; R_3=C_7H_{15}
c: R_1=Ph; R_2=H; R_3=4-$HOC_6H_4CH_2CH_2$
d: R_1=PhCH=CH; R_2=H; R_3=Ph
e: R_1=PhCH=CH; R_2=H; R_3=4-$NO_2C_6H_4$
f: R_1=4-ClC_6H_4; R_2=H; R_3=Ph
g: R_1=4-ClC_6H_4; R_2=H; R_3=4-HOC_6H_4
h: R_1=4-$MeOC_6H_4$; R_2=H; R_3=Ph
i: R_1=4-$MeOC_6H_4$; R_2=H; R_3=4-HOC_6H_4
j: R_1=4-$NO_2C_6H_4$; R_2=H; R_3=Ph
k: R_1=Ph; R_2=Me; R_3=Ph
l: R_1=Ph; R_2=Me; R_3=PhCH$_2$
m: R_1, R_2=-$(CH_2)_5$-; R_3=Ph

Experimental procedures:

A mixture of *p*-chlorobenzaldehyde **1f** (0.7 g, 5 mmol), aniline **2f** (0.455 g, 5 mmol) and montmorillonite K-10 clay (0.1 g) contained in a 25-mL beaker was placed in an alumina bath inside the microwave oven and irradiated for 2 min. The in situ generated Schiff's base was mixed thoroughly with freshly prepared NaBH$_4$-clay (5.0 mmol of NaBH$_4$ on 1.72 g of reagent) and water (1 mL). The reaction mixture was again irradiated for 30 s (65 °C). Upon completion of the reaction, monitored on TLC, the product was extracted into methylene chloride (3×15 mL). The removal of solvent under reduced pressure provided pure *N*-phenyl-*p*-chlorobenzylamine **4f** in 90% yield. The identity of the product was confirmed by formation of the hydrochloride salt, mp 209–211 °C.

References: R. S. Varma, R. Dahiya, *Tetrahedron*, **54**, 6293 (1998).

Type of reaction: C–N bond formation
Reaction condition: solvent-free
Keywords: aromatic ketone, hydrazine, Wolff-Kishner reduction, microwave irradiation, hydrazone, hydrocarbon

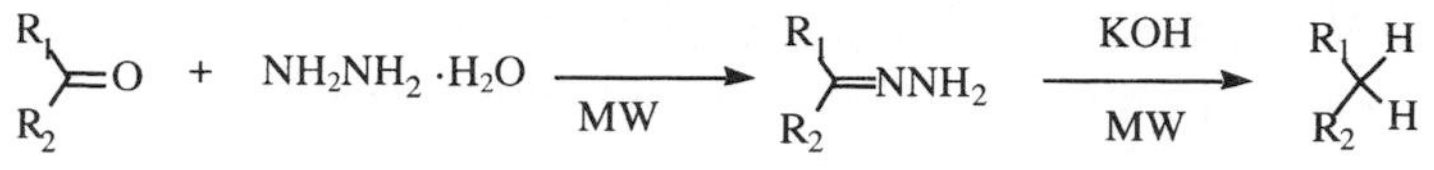

$$R_1\!\!\diagdown\!\!=\!O + NH_2NH_2\cdot H_2O \xrightarrow[MW]{} R_1\!\!\diagdown\!\!=\!NNH_2 \xrightarrow[MW]{KOH} R_1\!\!\diagdown\!\!\begin{smallmatrix}H\\H\end{smallmatrix}$$

$$\begin{array}{cccc} \mathbf{1} & \mathbf{2} & \mathbf{3} & \mathbf{4} \end{array}$$

a: $R_1 = R_2 = Ph$
b: $R_1 = R_2 = 4\text{-}MeOC_6H_4$
c: $R_1 = 4\text{-}MeOC_6H_4$; $R_2 = Me$
d: $R_1 = Ph$; $R_2 = Me$

Experimental procedures:

A mixture of benzophenone (1.84 g, 10 mmol) and 80% hydrazine hydrate (1 g, 20 mmol) in toluene (15 mL) was taken in an Erlenmeyer flask and placed in a commercial microwave oven operating at 2450 MHz frequency. After irradiation of the mixture for 20 min, (monitored by TLC) it was cooled to room temperature, extracted with chloroform and dried over anhydrous Na_2SO_4. Removal of solvent gave the benzophenone hydrazone in 95% yield. For the Wolff-Kishner reductions, a mixture of hydrazone **3a** (5 mmol) and KOH (2 g) were taken in an Erlenmeyer flask and placed in a microwave oven. Irradiation for 30 min and usual workup gave the corresponding diphenylmethane in 95% yield.

References: S. Gadhwal, M. Baruah, J. S. Sandhu, *Synlett*, 1573 (1999).

Type of reaction: C–N bond formation
Reaction condition: solvent-free
Keywords: hydrazine sulfate, aldehyde, ketone, $CaCl_2$, microwave irradiation, azine

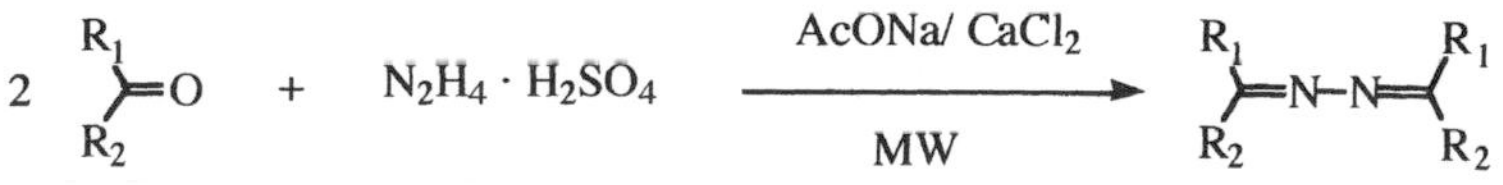

$$2\; R_1\!\!\diagdown\!\!=\!O + N_2H_4\cdot H_2SO_4 \xrightarrow[MW]{AcONa/\,CaCl_2} R_1\!\!\diagdown\!\!=\!N\!-\!N\!=\!\!\diagdown\!R_1$$

$$\begin{array}{cc} \mathbf{1} & \mathbf{2} \end{array}$$

a: $R_1 = R_2 = Ph$
b: $R_1 = Ph$; $R_2 = H$
c: $R_1 = Ph$; $R_2 = Me$
d: $R_1 = 2\text{-}HOC_6H_4$; $R_2 = H$

Experimental procedures:

A mixture of 0.212 g (2 mmol) of benzaldehyde **1b** and 0.13 g (1 mmol) of hydrazine sulfate was placed in an open glass container: sodium acetate anhydrous/calcium chloride (0.3 g + 0.3 g) was added and the mixture was irradiated in a microwave oven for 30 s. When the reaction was complete, the mixture was cooled to room temperature: water (15 mL) added and the solution was extracted with chloroform (3×5 mL). The combined extracts were dried on $MgSO_4$ and evaporation of solvent under vacuum gave benzalazine **2b** which was >98% pure.

References: H. Loghmani-Khouzani, M.M.M. Sadeghi, J. Safari, M.S. Abdorrezaie, M. Jafarpisheh, *J. Chem. Res. (S)*, 80 (2001).

Type of reaction: C–N bond formation
Reaction condition: solvent-free
Keywords: hydrazone, benzopyrandione derivative, microwave irradiation, heterocyclic hydrazone

1 a: R=H
b: R=Me

R_1=Ph
R_1=2,5-$F_2C_6H_3$
R1=3-$CF_3C_6H_4$

2

3 4

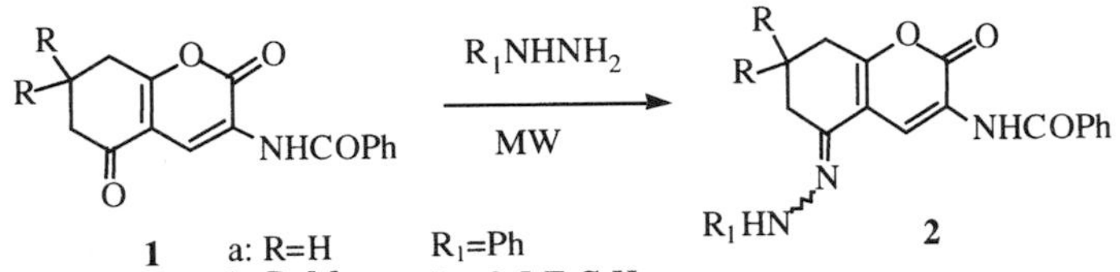

Experimental procedures:
A household microwave oven operating at 2450 MHz was used at its full power, 650 W. A neat mixture of benzopyran derivative **1** or **3** (1 mmol) and hydrazine (1.2–3 mmol) in a 10-mL glass beaker was thoroughly mixed for about 5 min, then it was placed in an alumina bath inside the household microwave oven and irradiated. Maximum temperature reached in the alumina after 10 min was about 150 °C. After cooling, methanol (ca. 4 mL) was added to the mixture and the separated solid was filtered off and washed with a small amount of methanol to give the products **2** and **4**.

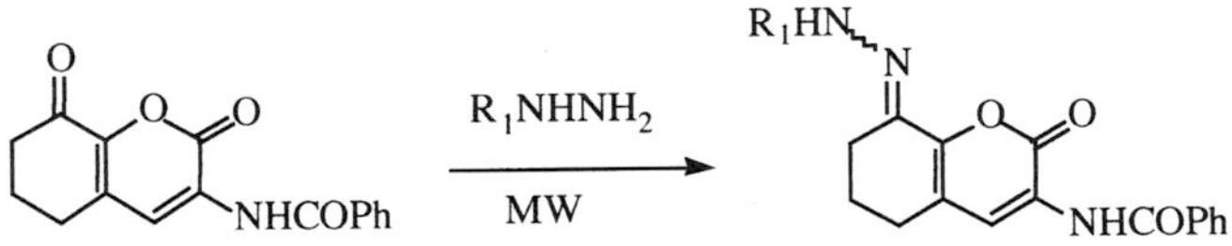

References: M. Jeselnik, R.S. Varma, S. Polanc, M. Kocevar, *Chem. Commun.*, 1716 (2001).

Type of reaction: C–N bond formation
Reaction condition: solvent-free
Keywords: aromatic aldehyde, hydroxylamine, semicarbazide, tosylhydrazine, microwave irradiation, oxime, semicarbazone, tosylhydrazone

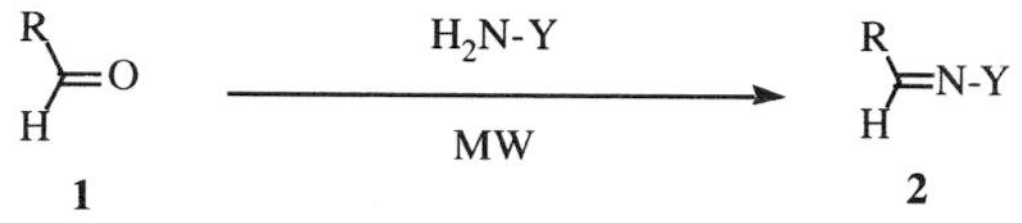

a: R=4-FC$_6$H$_4$; Y=OH
b: R=4-ClC$_6$H$_4$; Y=OH
c: R=4-BrC$_6$H$_4$; Y=OH
d: R=2,4-Cl$_2$C$_6$H$_3$; Y=OH
e: R=2-HOC$_6$H$_4$; Y=OH
f: R=4-HOC$_6$H$_4$; Y=OH
g: R=4-MeOC$_6$H$_4$; Y=OH
h: R=3-MeO,4-HOC$_6$H$_3$; Y=OH

i: R=4-NO$_2$C$_6$H$_4$; Y=OH
j: R=4-NO$_2$C$_6$H$_4$; Y=NHCONH$_2$
k: R=4-ClC$_6$H$_4$; Y=NHCONH$_2$
l: R=4-HOC$_6$H$_4$; Y=NHCONH$_2$
m: R=PhCH=CH; Y=NHCONH$_2$
n: R=4-NO$_2$C$_6$H$_4$; Y=NHTs
o: R=4-HOC$_6$H$_4$; Y=NHTs
p: R=4-MeOC$_6$H$_4$; Y=NHTs

Experimental procedures:
A paste of a mixture of aromatic aldehyde (5 mmol) and hydroxylamine (5 mmol), semicarbazide (5 mmol), or tosylhydrazine (5 mmol) in methanol (few drops) was exposed to microwaves (heating and cooling at intervals of 15 s and 1 min for the specific time). After completion of the reaction (TLC) the product was extracted with 3×5 mL CH$_2$Cl$_2$. Removal of the solvent under reduced pressure gave pure product in excellent yields.

References: B.P. Bandger, V.S. Sadavarte, L.S. Uppalla, R. Govande, *Monatsh. Chem.,*
132, 403 (2001).

Type of reaction: C–N bond formation
Reaction condition: solvent-free
Keywords: aromatic aldehyde, hydroxylamine hydrochloride, basic Al$_2$O$_3$, microwave irradiation, oxime

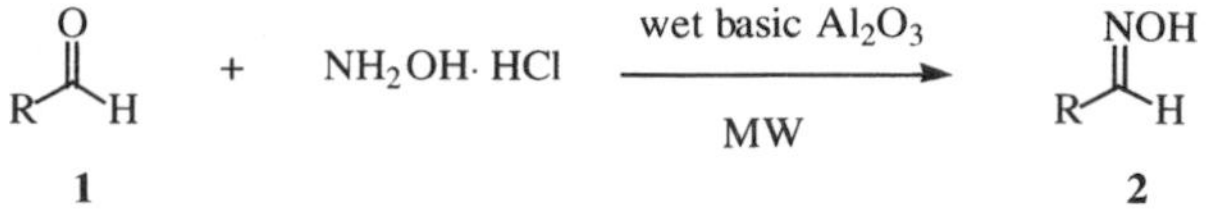

a: R=3-NO$_2$C$_6$H$_4$
b: R=4-MeOC$_6$H$_4$
c: R=2-furyl
d: R=4-HOC$_6$H$_4$
e: R=3-MeO, 4-HOC$_6$H$_3$
f: R=4-NO$_2$C$_6$H$_4$
g: R=Ph
h: R=4-ClC$_6$H$_4$
i: R=2-BrC$_6$H$_4$
j: R=3,4-(MeO)$_2$C$_6$H$_3$
k: R=2-thienyl

Experimental procedures:

A mixture of aldehyde (1 mmol) and hydroxylamine hydrochloride (2 mmol) was impregnated on wet basic Al$_2$O$_3$ (1.5 equiv.) placed in a 25-mL Erlenmeyer flask and subjected to microwave irradiation at the required power level for 1–10 min. On cooling, the reaction mixture was extracted with diethyl ether (2×10 mL) and the combined organic layers were washed with water (2×10 mL), brine and dried over anhydrous Na$_2$SO$_4$. Evaporation of the solvent in vacuo afforded pure oximes in 47–99% yield.

References: G.L. Kad, M. Bhandari, J. Kaur, R. Rathee, J. Singh, *Green Chem.,* **3**, 275 (2001).

Type of reaction: C–N bond formation
Reaction condition: solvent-free
Keywords: aromatic aldehyde, primary amine, malonic acid monoethyl ester, Rodionov reaction, microwave irradiation, β-aryl-β-amino acid ester, ethyl cinnamate

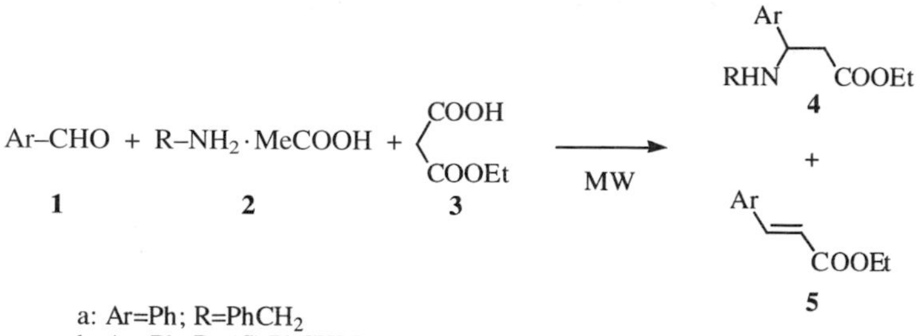

Ar–CHO + R–NH$_2$·MeCOOH +

$$\begin{array}{c} \text{COOH} \\ \diagdown \\ \text{COOEt} \end{array}$$

1   **2**   **3**   MW

$$\begin{array}{c} \text{Ar} \\ | \\ \text{RHN} \qquad \text{COOEt} \\ \mathbf{4} \end{array}$$

+

$$\begin{array}{c} \text{Ar} \\ \diagup\!\!= \\ \text{COOEt} \\ \mathbf{5} \end{array}$$

a: Ar=Ph; R=PhCH$_2$
b: Ar=Ph; R=(S)-PhCHMe
c: Ar=o-O$_2$NC$_6$H$_4$; R=(S)-PhCHMe
d: Ar=p-O$_2$NC$_6$H$_4$; R=PhCH$_2$
e: Ar=p-O$_2$NC$_6$H$_4$; R=(S)-PhCHMe
f: Ar=p-Me$_2$NC$_6$H$_4$; R=(S)-PhCHMe

Experimental procedures:

An equimolar mixture of an aldehyde **1**, an amine acetate, and a malonic acid derivative **3** (without solvent) was irradiated in an open glass vessel in a microwave oven (Funai MO785VT, 170 W) for several minutes (until the release of CO$_2$ was complete). The reaction was monitored by TLC. The reaction products were separated by column chromatography on silica gel with a benzene-ethyl acetate eluent (2:1). Note that an increase in the microwave power up to 350 W resulted in a decrease in the yield of the target amino ester.

References: N.N. Romanova, A.G. Gravis, P.V. Kudan, Y.G. Bundel, *Mendeleev. Commun.*, 26 (2001).

Type of reaction: C–N bond formation
Reaction condition: solvent-free
Keywords: active methylene compound, LiBr, nitroso arene, Ehrlich-Sachs reaction, microwave irradiation, imine

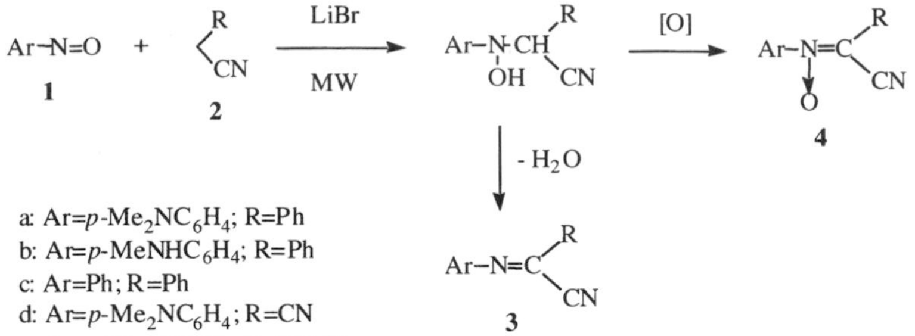

a: Ar=p-Me$_2$NC$_6$H$_4$; R=Ph
b: Ar=p-MeNHC$_6$H$_4$; R=Ph
c: Ar=Ph; R=Ph
d: Ar=p-Me$_2$NC$_6$H$_4$; R=CN
e: Ar=p-Me$_2$NC$_6$H$_4$; R=CO$_2$Et
f: Ar=p-Me$_2$NC$_6$H$_4$; R=COPh
g: Ar=p-Me$_2$NC$_6$H$_4$; R=CONH$_2$
h: Ar=p-Me$_2$NC$_6$H$_4$; R=CONHPh
i: Ar=p-Me$_2$NC$_6$H$_4$; R=p-MeC$_6$H$_4$
j: Ar=p-Me$_2$NC$_6$H$_4$; R= naphthyl
k: Ar=Ph; R=COOMe

Experimental procedures:

N,N-Dimethyl-4-nitrosoaniline (**1a**, 1.50 g, 0.01 mol), benzyl cyanide **2a** (1.17 g, 0.01 mol), and LiBr (0.43 g, 5 mmol) were mixed in an Erlenmeyer flask and placed in a commercial microwave oven (operating at 2450 MHz frequency and 70 W). The reaction mixture was irradiated for 2 min, allowed to cool to room temperature, washed with water, and extracted with CH$_2$Cl$_2$ (2×34 mL). Removal of solvent and purification by column chromatography of the residue using petroleum ether-CHCl$_3$ (4:1) as eluent, afforded exclusively the imine **3a** in 90% yield without formation of any N-oxide **4**. Crystallization from EtOH gave the pure imine, red crystals, mp 90 °C.

References: D. D. Laskar, D. Prajapati, J. S. Sandhu, *Synth. Commun.,* **31**, 1427 (2001).

Type of reaction: C–N bond formation
Reaction condition: solvent-free
Keywords: aldoxime, H$_2$SO$_4$/SiO$_2$, microwave irradiation, nitrile

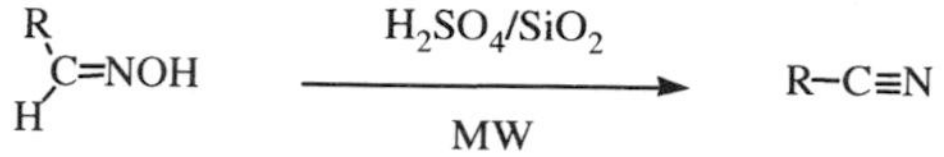

R
 \C=NOH $\xrightarrow[\text{MW}]{\text{H}_2\text{SO}_4/\text{SiO}_2}$ R–C≡N
H/

1　　　　　　　　　　　　　　　　　**2**

a: R=Ph
b: R=4-MeC$_6$H$_4$
c: R=4-MeOC$_6$H$_4$
d: R=4-HOC$_6$H$_4$
e: R=2,4,6-(Me)$_3$C$_6$H$_2$
f: R=2,4,6-(MeO)$_3$C$_6$H$_2$
g: R=4-NO$_2$C$_6$H$_4$

h: R=2,6-Cl$_2$C$_6$H$_3$
i: R=n-Pr
j: R=n-C$_6$H$_{13}$
k: R=9-anthryl
l: R=2-thienyl

Experimental procedures:

The aldoxime (10 mmol) was thoroughly mixed with H$_2$SO$_4$/SiO$_2$ catalyst (5 g) in an agate mortar. The resulting fine powder was taken in a Pyrex Erlenmeyer flask (25 mL) and irradiated in a microwave oven. The inorganic support was separated by filtration after eluting the product with CH$_2$Cl$_2$ (20 mL). The filtrate was dried (Na$_2$SO$_4$) and the crude product obtained after evaporation of solvent was purified by crystallization, distillation or column chromatography.

References: H. M. S. Kumar, P. K. Mohanty, M. S. Kumar, J. S. Yadav, *Synth. Commun.*, **27**, 1327 (1997).

Type of reaction: C–N bond formation
Reaction condition: solvent-free
Keywords: aldehyde, hydroxylamine hydrochloride, peroxymonosulfate, alumina, microwave irradiation, nitrile

R–CHO + NH$_2$OH·HCl $\xrightarrow[\text{MW}]{\text{peroxymonosulfate-alumina}}$ R–C≡N

1　　　　　　　　　　　　　　　　　**2**

a: R=Ph
b: R=4-ClC$_6$H$_4$
c: R=3,4,5-(MeO)$_3$C$_6$H$_2$
d: R=2-NO$_2$C$_6$H$_4$
e: R=2,4-Cl$_2$C$_6$H$_3$
f: R=3,4-(CH$_2$O)C$_6$H$_3$
g: R=cyclohexyl
h: R=PhCH$_2$
i: R=anthryl

Experimental procedures:
Trimethoxybenzaldehyde **1c** (0.196 g, 1.0 mmol), hydroxylamine hydrochloride (0.077 g, 1.1 mmol) and peroxymonosulfate (0.61 g, 1.0 mmol) doped on a neutral alumina (1.0 g) were mixed thoroughly on a vortex mixer. The reaction mixture was placed in an alumina bath inside a commercial microwave oven (operating at 2450 MHz frequency) and irradiated for a period of 7 min. After completion of the reaction (monitored by TLC) the inorganic support was separated by filtration, after eluting the product with dichloromethane (2×15 mL). The solvent was removed and the residue on purification by column chromatography on silica gel gave the corresponding trimethoxybenzonitrile **2a** in 95% yield and there was no evidence for the formation of any side products.

References: D.S. Bose, A.V. Narsaiah, *Tetrahedron Lett.,* **39**, 6533 (1998).

Type of reaction: C–N bond formation
Reaction condition: solvent-free
Keywords: aldehyde, hydroxylamine hydrochloride, silica gel, microwave irradiation, nitrile

$$R-CHO \xrightarrow[\text{MW}]{\substack{NH_2OH \cdot HCl \\ NaHSO_4 \cdot SiO_2}} R-CN$$

$$\mathbf{1} \qquad\qquad\qquad \mathbf{2}$$

a: $R=C_6H_5$
b: $R=4\text{-}(HO)C_6H_4$
c: $R=4\text{-}(MeO)C_6H_4$
d: $R=3\text{-}(MeO)\text{-}4\text{-}(HO)C_6H_3$
e: $R=3\text{-}(HO)C_6H_4$
f: $R=3,4\text{-}(MeO)_2C_6H_3$

g: $R=3,4\text{-}(CH_2O_2)C_6H_3$
h: $R=4\text{-}(NO_2)C_6H_4$
i: $R=C_6H_5CH=CH$
j: $R=C_7H_{15}$
k: $R=C_8H_{17}$

Experimental procedures:
3,4-Dimethoxybenzaldehyde **1f** (166 mg, 1 mmol) and $NH_2OH \cdot HCl$ (91 mg, 1.3 mmol) were mixed thoroughly with $NaHSO_4$-SiO_2 catalyst (100 mg, prepared by the reported method). The mixture was kept inside a microwave oven (BPL, BMO, 700T, 466 watt) and irradiated for 1 min. Then it was removed from the oven, cooled and shaken with $CHCl_3$ (10 mL). The mixture was filtered and concentrated. The residue was purified by column chromatography over silica gel using EtOAc as eluent to produce 3,4-dimethoxybenzonitrile **2f** (158 mg, 97%).

References: B. Das, P. Madhusudhan, B. Venkataiah, *Synlett*, 1569 (1999).

Type of reaction: C–N bond formation
Reaction condition: solvent-free
Keywords: aldehyde, hydroxylamine hydrochloride, ammonium acetate, microwave irradiation, nitrile

$$NH_2OH \cdot HCl$$
$$NH_4OAc$$

R–CHO $\xrightarrow{\quad MW \quad}$ R–CN

1   **2**

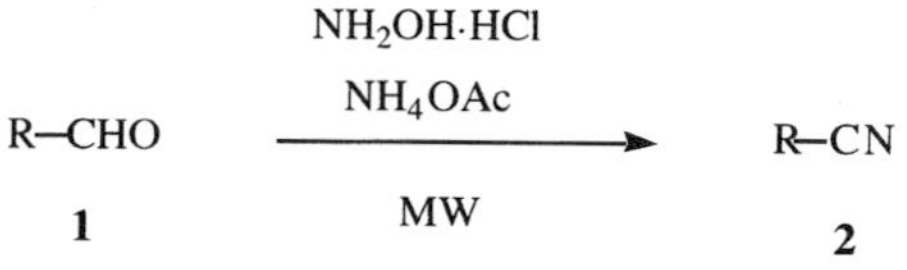

a: $R=4\text{-}HOC_6H_4$
b: $R=4\text{-}MeOC_6H_4$
c: $R=4\text{-}NO_2C_6H_4$
d: $R=3\text{-}HOC_6H_4$
e: $R=3\text{-}MeOC_6H_4$
f: $R=3\text{-}MeO,4\text{-}HOC_6H_3$
g: $R=3,4\text{-}(MeO)_2C_6H_3$
h: $R=3,4\text{-}(OCH_2O)C_6H_3$
i: $R=C_6H_5CH=CH$
j: $R=C_6H_5CH_2CH_2$
k: $R=C_{15}H_{31}$

Experimental procedures:
Methoxybenzaldehyde **1b** (136 mg, 1 mmole) and $NH_2OH \cdot HCl$ (84 mg, 1.2 mmole) were mixed thoroughly with NH_4OAc (108 mg, 1.4 mmole). The resulting powder was taken in a test tube, kept in an alumina bath inside a microwave oven and irradiated for 1 min. The mixture was removed from the oven, cooled and shaken with $CHCl_3$ (15 mL). After filtration, the filtrate was concentrated and the residue was subjected to column chromatography over silica gel using hexane-EtOAc (7:3) as eluent to afford 4-methoxybenzonitrile **2b** (119 mg, 90%).

References: B. Das, C. Ramesh, P. Madhusudhan, *Synlett,* 1599 (2000).

Type of reaction: C–N bond formation
Reaction condition: solvent-free
Keywords: aldoxime, *p*-toluenesulfonyl chloride, pyridine, alumina, Beckmann rearrangement, microwave irradiation, nitrile

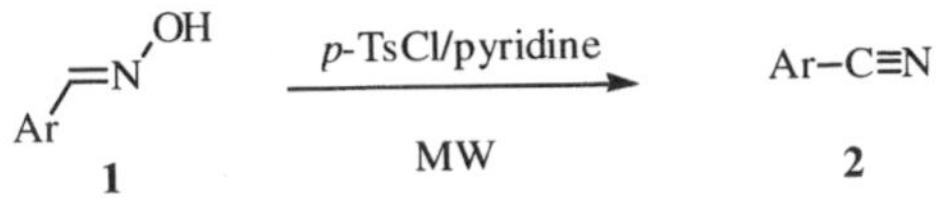

$$\underset{\substack{Ar \\ \textbf{1}}}{\overset{\displaystyle =N'^{OH}}{\diagup}} \quad \xrightarrow[\text{MW}]{p\text{-TsCl/pyridine}} \quad \underset{\textbf{2}}{Ar-C\equiv N}$$

a: Ar=4-ClC$_6$H$_4$
b: Ar=4-BrC$_6$H$_4$
c: Ar=2,6-Cl$_2$C$_6$H$_3$
d: Ar=3,4,5-(MeO)$_3$C$_6$H$_2$
e: Ar=2-OHC$_6$H$_4$

Experimental procedures:

A solution of *p*-chlorobenzaldehyde oxime **1a** (1.16 g, 7.5 mmol), *p*-toluenesulfonyl chloride (2.86 g, 15 mmol) and pyridine (1.2 mL, 15 mmol) in CHCl$_3$ (20 mL) was mixed with 3 g alumina and concentrated under vacuum to give a yellowish solid. The mixture was introduced in a Pyrex flask and submitted to microwave irradiation. After completion of the reaction (TLC, 3–15 min) 50 mL CH$_3$OH was added. The mixture was filtered and concentrated. To the crude mixture, chloroform (25 mL) was added. The filtrate was washed with brine (2×25 mL), and then with 2×25 mL of 5% sodium bicarbonate. The collected organic phase was dried with MgSO$_4$, filtered and evaporated to give the crude product with good chemical purity. Flash chromatography (cyclohexane-ethyl acetate) gave analytically pure sample in 64% (mp 89–90 °C) isolated yield.

References: M. Ghiaci, K. Bakhtiari, *Synth. Commun.*, **31**, 1803 (2001).

Type of reaction: C–N bond formation
Reaction condition: solvent-free
Keywords: aromatic aldehyde, hydroxylamine hydrochloride, dibutyltin oxide, microwave irradiation, nitrile

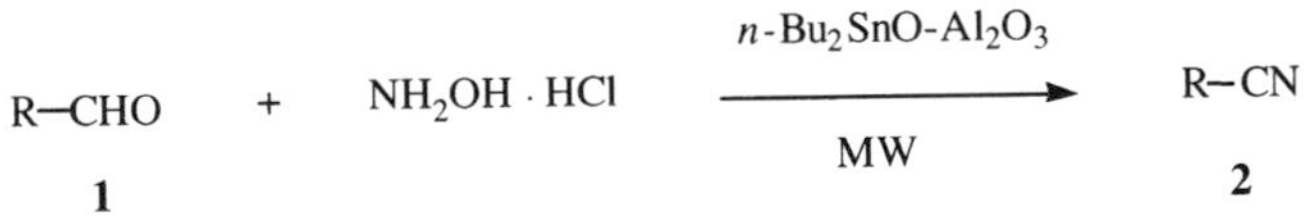

$$\underset{\textbf{1}}{R-CHO} \quad + \quad NH_2OH\cdot HCl \quad \xrightarrow[\text{MW}]{n\text{-}Bu_2SnO\text{-}Al_2O_3} \quad \underset{\textbf{2}}{R-CN}$$

a: R=Ph
b: R=3,4-Cl$_2$C$_6$H$_3$
c: R=3,4-(OCH$_2$O)C$_6$H$_3$
d: R=3,4-(MeO)$_2$C$_6$H$_3$
e: R=3,4,5-(MeO)$_3$C$_6$H$_2$
f: R=4-NO$_2$C$_6$H$_4$
g: R=4-MeC$_6$H$_4$
h: R=4-BrC$_6$H$_4$
i: R=2-naphthyl
j: R=9-anthryl

Experimental procedures:
Piperonal **1c** (5 mmol), hydroxylamine hydrochloride (6 mmol), and *n*-dibutyltin oxide (0.5 mmol) were admixed with alumina (1 g) and exposed to microwave irradiation at 450 W (BPL, BMO-700T) for an appropriate time. After complete conversion, as indicated by TLC, the reaction mixture was filtered and the residue washed with dichloromethane (2×15 mL) which was then concentrated in vacuo. The resulting solid was recrystallized in ethanol to give 3,4-methylenedioxybenzonitrile **2c** in 92% yield as a white solid, mp 68–70 °C. The liquid products were purified by column chromatography on silica gel (Merck, 100–200 mesh, ethyl acetate-hexane, 1:9) to afford the corresponding nitriles in pure form.

References: J. S. Yadav, B. V. Reddy, C. Madan, *J. Chem. Res. (S)*, 190 (2001).

Type of reaction: C–N bond formation
Reaction condition: solvent-free
Keywords: epoxide, ammonium acetate, aminolysis, microwave irradiation, *β*-amino alcohol

a: R=CH$_2$OPh
b: R=CH$_2$O-2-MeC$_6$H$_4$
c: R=CH$_2$O-3,4-Me$_2$C$_6$H$_3$
d: R=CH$_2$O-4-MeOC$_6$H$_4$
e: R=CH$_2$O-4-*i*-PrC$_6$H$_4$
f: R=CH$_2$O-4-*t*-BuC$_6$H$_4$
g: R=CH$_2$O-4-ClC$_6$H$_4$
h: R=CH$_2$O-2-naphthyl

Experimental procedures:
The epoxide **1a** (1.5 g, 10 mmol) and ammonium acetate (1.15 g, 15 mmol) were admixed in a Pyrex test tube and subjected to microwave irradiation for 40 s. After complete conversion, as indicated by TLC, the reaction mass was diluted with water (20 mL) and extracted with ethyl acetate (2×25 mL), washed with brine (20 mL) and dried over Na$_2$SO$_4$. The organic extract was purified by column chromatography on silica gel (Aldrich 100–200 mesh, hexane-ethyl acetate, 7:3) affording **2a** (1.25 g, 75% yield) as an oily liquid.

References: G. Sabitha, B. V. S. Reddy, S. Abraham, J. S. Yadav, *Green Chem.*, **1**, 251 (1999).

Type of reaction: C–N bond formation

Reaction condition: solvent-free

Keywords: epoxide, primary amine, montmorillonite K-10, aminolysis, microwave irradiation, β-amino alcohol

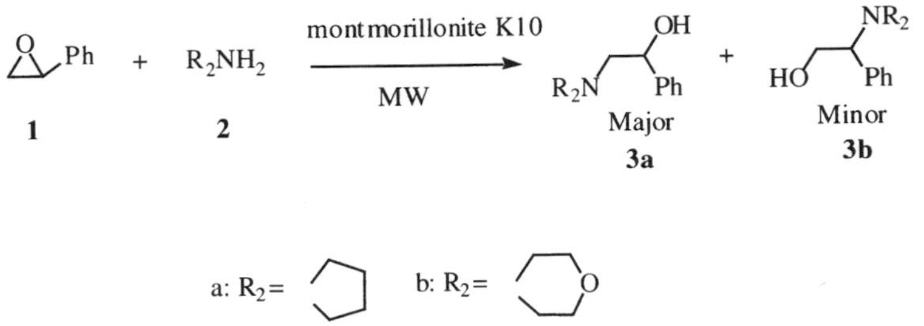

Experimental procedures:

A mixture of epoxide **1** (1 mmol), amine **2** (1 mmol) and montmorillonite K-10 clay (0.2 g) was placed in a Teflon container with a screw cap. Then the mixture was irradiated with high power (900 W) in a conventional microwave oven for 1 min. After the mixture was cooled to room temperature, it was washed with dichloromethane (2×10 mL). The solvent was evaporated and the products were identified.

References: M.M. Mojtahedi, M.R. Saidi, M. Bolourtchian, *J. Chem. Res. (S)*, 128 (1999).

5   Carbon–Oxygen Bond Formation

5.1   Solvent-Free C–O Bond Formation

Type of reaction: C–O bond formation
Reaction condition: solid-state
Keywords: phenol, carboxylic acid, P_2O_5/SiO_2, ester

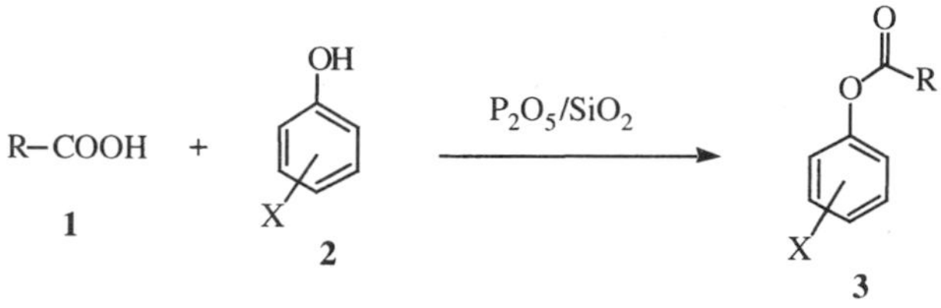

a: R=Me; X=H
b: R=Me; X=p-Me
c: R=Me; X=m-Me
d: R=Me; X=p-NO$_2$
e: R=Me; X=1- naphthyl
f: R=Me; X=2- naphthyl
g: R=Ph; X=H
h: R=Ph; X=p-Me
i: R=Ph; X=m-Me
j: R=Ph; X=o-Me
k: R=Ph; X=p-NO$_2$
l: R=Ph; X=1- naphthyl

Experimental procedures:
A mixture of 1.42 g (0.01 mol) phosphorus pentoxide and 2.5 g of chromatography grade silica gel was placed in a flask and stirred for 30 min. A mixture of equimolar amounts (5 mmol) of the carboxylic acid and phenol was added. Usually an immediate color change was observed. After stirring for 6 h at the temperature indicated in the Table, methylene chloride (50 mL) was added. The mixture was stirred for 1 min and then filtered. The spent reagent was washed twice with methylene chloride (10 mL). The combined organics were washed with aqueous NaOH solution, water and dried over sodium sulfate, and the solvent was removed under reduced pressure.

References: H. Eshghi, M. Rafei, M.H. Karimi, *Synth. Commun.,* **31**, 771 (2001).

Type of reaction: C–O bond formation
Reaction condition: solid-state
Keywords: alcohol, TsOH, etherification, ether

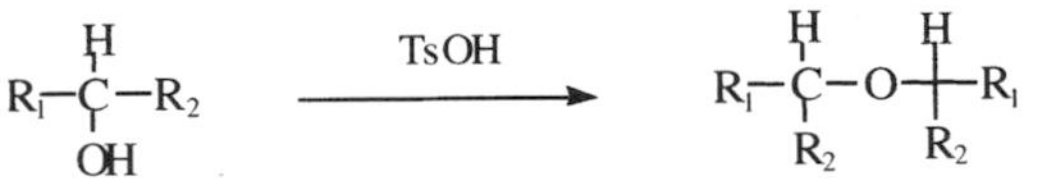

a: $R_1=R_2=Ph$
b: $R_1=Ph$; $R_2=2\text{-}ClC_6H_4$
c: $R_1=Ph$; $R_2=4\text{-}BrC_6H_4$
d: $R_1=Ph$; $R_2=2\text{-}NO_2C_6H_4$
e: $R_1=Ph$; $R_2=2\text{-}MeC_6H_4$
f: $R_1=Ph$; $R_2=3\text{-}MeC_6H_4$
g: $R_1=Ph$; $R_2=4\text{-}MeC_6H_4$
h: $R_1=R_2=4\text{-}MeC_6H_4$

Experimental procedures:
A mixture of powdered 4-methylbenzhydrol **1g** and an equimolar amount of TsOH was kept at room temperature for 10 min, and the reaction mixture was extracted with ether. The ether solution was worked up by the usual method and distillation of the crude product gave the corresponding ether **2g** in 96% yield.

References: F. Toda, H. Takumi, M. Akehi, *Chem. Commun.*, 1270 (1990).

Type of reaction: C–O bond formation
Reaction condition: solid-state
Keywords: 2′-hydroxy-4′,6′-dimethylchalcone, Michael-type addition, flavanone

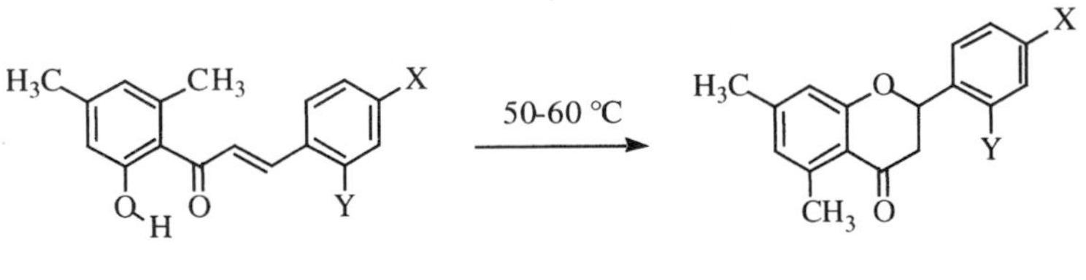

a: X=H; Y=H
b: X=Cl; Y=H
c: X=Br; Y=H
d: X=H; Y=Cl

Experimental procedures:
Finely powdered chalcones (100–500 mg) were heated at 50–60 °C with occasional stirring for 7–20 days. Chalcones **1a–c** were found to gradually change their color during heating but chalcone **1d** remained stable. The flavanones **2a–c**

were separated from the reaction mixtures by column chromatography (silica gel, hexane-5% EtOAc) in nearly 70% isolated yield. All products were chracterized by NMR and IR spectroscopy and also compared with the authentic samples.

References: B. S. Goud, K. Panneerselvam, D. E. Zacharias, G. R. Desiraju, *J. Chem. Soc., Perkin Trans. 2*, 325 (1995).

Type of reaction: C–O bond formation
Reaction condition: solid-state
Keywords: thiocarbonylimidazole, alcohol, thiocarbonylimidazolide

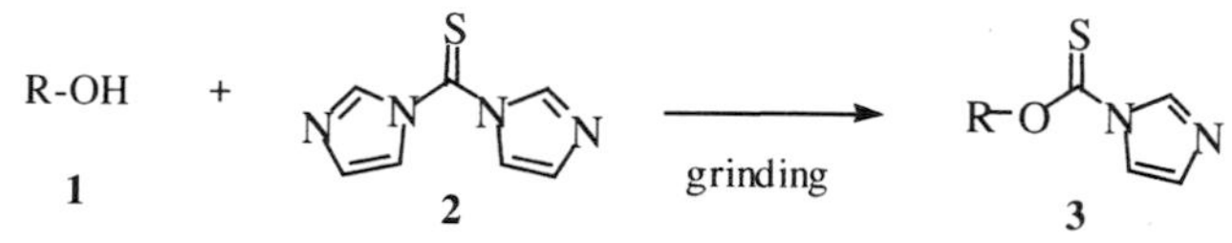

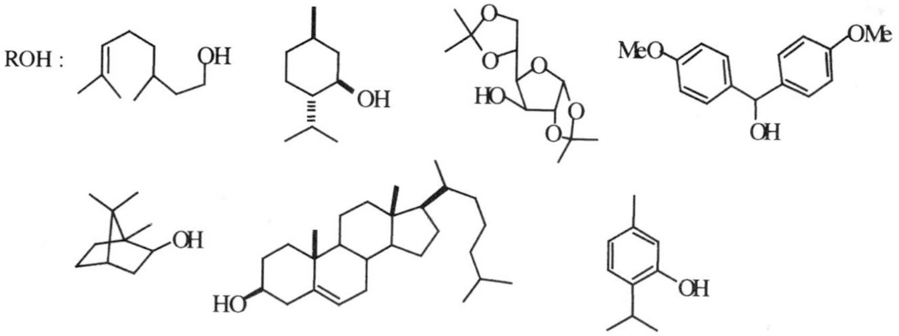

Experimental procedures:
A mixture of alcohol **1** and 1.2 equiv. of thiocarbonylimidazole **2** in a mortar was ground well with a pestle at room temperature in ambient atmosphere. Progress of the reaction was monitored by TLC using a glass capillary tube which kept ethanol inside beforehand. In some cases, termination of the reaction was indicated by liquefaction of solid. After grinding occasionally for 2–3 h, the resulting mixture was dissolved in ethyl acetate. The solution was passed through a short column of silica gel. Evaporation of the solvent followed by purification of the residue by medium pressure liquid chromatography provided thiocarbonylimidazolide **3**.

References: H. Hagiwara, S. Ohtsubo, M. Kato, *Tetrahedron*, **53**, 2415 (1997).

Type of reaction: C–O bond formation
Reaction condition: solid-state
Keywords: 9-thienothienylfluoren-9-ol, DDQ, gas-solid reaction, etherification

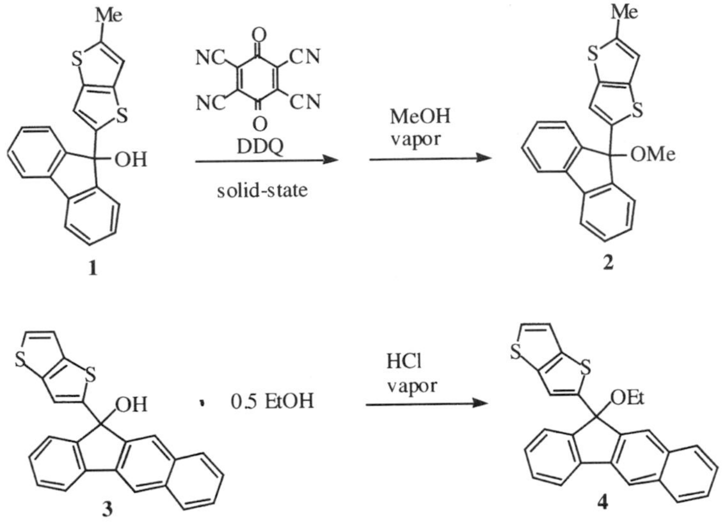

Experimental procedures:
The substrate (10–20 mg, ca. 0.04 mmol) was coground with ca. 9.1 mg (0.04 mmol) of DDQ in a mortar with a pestle for about 10 min. The resulting colored solids were transferred to a small vial without a cap, and it was placed in a closed vessel, in which a shallow pool of methanol (1 mL) was maintained at the bottom. The vessel was kept in a refrigerator for 5–10 h. Product analysis was carried out by means of the ^{1}H NMR spectra for the sample after working up with a saturated NaHCO$_3$ solution.

References: M. Tanaka, K. Kobayashi, *Chem. Commun.*, 1965 (1998); M. Tanaka, N. Tanifuji, K. Kobayashi, *J. Org. Chem.*, **66**, 803 (2001).

Type of reaction: C–O bond formation
Reaction condition: solvent-free
Keywords: 3-phenylpropionaldehyde, acetonyltriphenylphosphonium bromide, cyclotrimerization

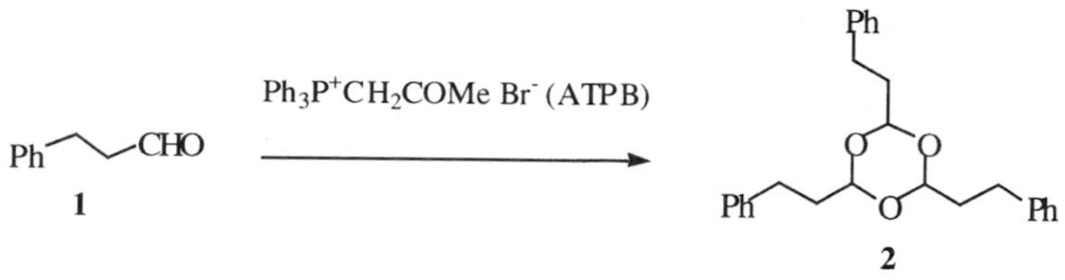

Ph₃P⁺CH₂COMe Br⁻ (ATPB)

Ph⌒⌒CHO

1

2

Experimental procedures:

A mixture of the 3-phenylpropionaldehyde **1** (385 mg, 2.87 mmol) and ATPB (115 mg, 0.287 mmol) was stirred at room temperature for 24 h. The crude mixture were subjected to silica gel column chromatography to isolate the desired product **2** (300 mg) in 78% yield.

References: Y. Hon, C. Lee, *Tetrahedron*, **57**, 6181 (2001).

Type of reaction: C–O bond formation
Reaction condition: solid-state
Keywords: alcoholate, CO_2, large scale, carboxylation, gas-solid reaction, carbonic acid half ester salts

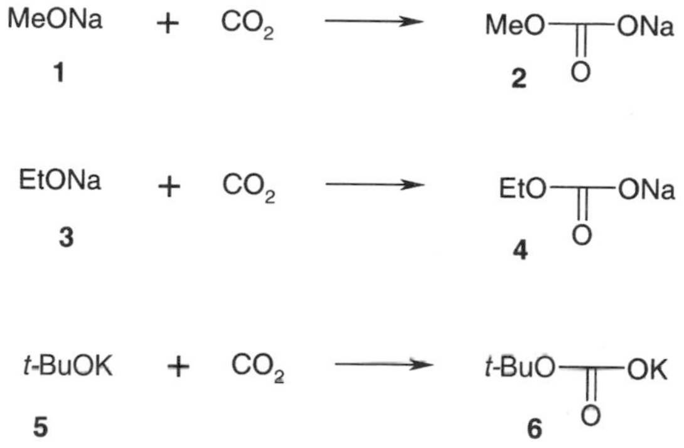

MeONa + CO_2 ⟶ MeO⎓ONa

1   **2** O

EtONa + CO_2 ⟶ EtO⎓ONa

3   **4** O

t-BuOK + CO_2 ⟶ *t*-BuO⎓OK

5   **6** O

Experimental procedures:

A chromatography column (diameter 4 cm, height 60 cm) with a D3-frit and gas outlet through a drying tube was charged upon glass wool with commercial 97.5% **1** (250 g, 4.63 mol) or 95% **3** (250 g, 3.68 mol) or 95% **5** (250 g, 2.23 mol) and covered with glass wool. Initially a mixture of CO_2 (250 mL min⁻¹) and N_2 (2.25 L min⁻¹) was applied from the bottom. It created a warm zone of about 50 °C that passed the column in about 1 h. After that, the N_2-stream was halved and the reaction continued until the heat production ceased and such halving was repeated twice again. Finally the N_2-flow was removed and the now cold column left closed with pure CO_2 overnight. The weight increase (without correction for losses due to initially present ROH contents) was 191 g (97%), 154 g (95%) and 90 g (92%), and the fill volume increase 29, 45 and 56%, respectively. The pH values of aqueous solutions of the products varied from 8.5 to 9. Titration after decomposition with aqueous H_2SO_4 gave content values between 97 and 102% of the versatile reagents **2**, **4**, or **6** for alkylations, carboxylations and acylations.

References: G. Kaupp, D. Matthies, C. de Vrese, *Chem. Ztg.*, **113**, 219 (1989); G. Kaupp, *Merck Spectrum*, **3**, 42–45 (1991).

Type of reaction: C–O bond formation
Reaction condition: solid-state
Keywords: phenolate, cyanogen chloride, cyanogen bromide, cyanate, gas-solid reaction

a: $R^1=R^2=H$, $R^3=OMe$
b: $R^1=H$, $R^2=R^3=OMe$
c: $R^1=R^2=H$, $R^3=CHO$
d: $R^1=NO_2$, $R^2=R^3=H$

Experimental procedures:
Solid potassium 4-methoxyphenolate **1a** (10 mmol) was treated with ClCN (1 bar, 11.2 mmol) or with BrCN (1.17 g, 11.0 mmol from a remote flask at a vacuum line) and left overnight. Excess gas was recovered in a cold trap. KCl (KBr) was washed away with water and the quantitatively obtained product **2a** dried in a vacuum.

Similarly, the aryl cyanates **2b–d** were quantitatively obtained.

Reference: G. Kaupp, J. Schmeyers, J. Boy, *Chem. Eur. J.*, **4**, 2467 (1998).

Type of reaction: C–O bond formation
Reaction condition: solid-state
Keywords: Viehe salt, pyrocatechol, solid-solid reaction, keteneiminium salt

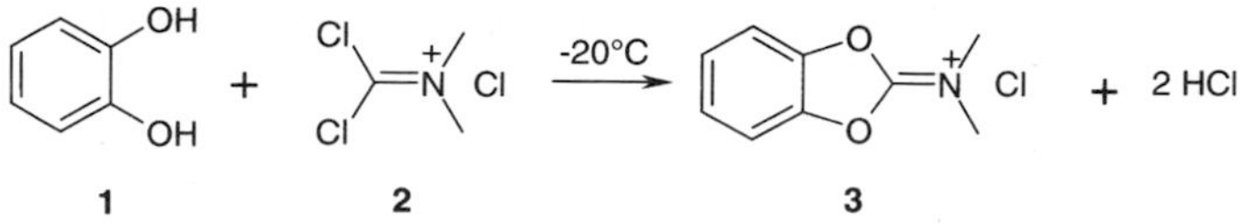

Experimental procedure:

Solid pyrocatechol **1** (550 mg, 5.00 mmol) and Viehe salt **2** (815 mg, 5.00 mmol) were ball-milled at –20 °C for 1 h. Most of the liberated HCl escaped through the Teflon gasket that was tightened against a torque of 50 Nm. Residual HCl was pumped off at room temperature. The yield of **3** was 998 mg (100%).

Refrence: G. Kaupp, J. Boy, J. Schmeyers, *J. Prakt. Chem.*, **340**, 346 (1998).

Type of reaction: C–O bond formation
Reaction condition: solid-state
Keywords: aldehyde, (–)-ephedrine, (+)-pseudoephedrine, optically activ oxazolidine

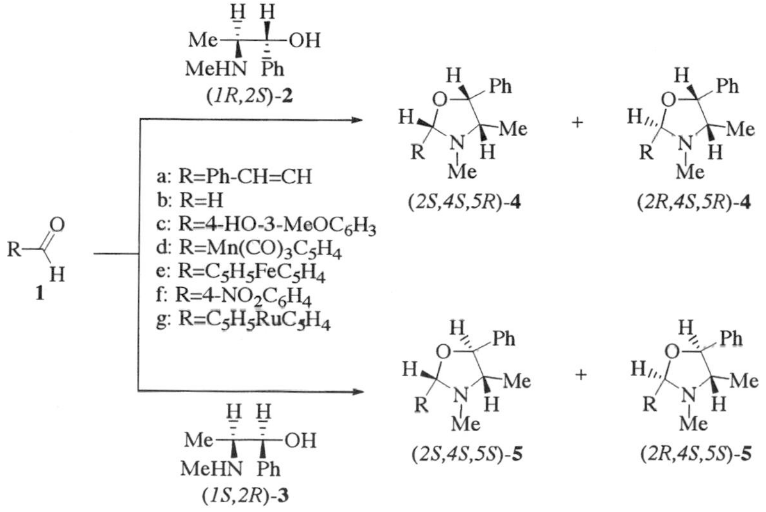

Experimental procedures:

After an enantiomerically pure β-amino alcohol (1 mmol) had been mixed with the appropriate aldehyde (1 mmol) the mixture was kept in the dark at room temperature and periodically stirred. The course of the reaction was monitored regularly by recording the NMR spectra of samples of the reaction mixture. In all cases the yield of the products was 100%.

References: N. S. Khruscheva, N. M. Loim, V. I. Solkolov, V. D. Makhaev, *J. Chem. Soc., Perkin Trans. 1*, 2425 (1997).

Type of reaction: C–O bond formation
Reaction condition: solid-state
Keywords: halogenoacetate, polymerization, polyglycolide

$$X-\overset{\overset{H}{|}}{\underset{\underset{H}{|}}{C}}\overset{O^-}{\underset{O}{\diagdown}} \quad M^+ \quad \overset{\Delta}{\longrightarrow} \quad \left[\overset{\overset{H}{|}}{\underset{\underset{H}{|}}{C}}\overset{O}{\underset{}{\overset{\|}{C}}}-O \right]_n \quad + \quad MX$$

$$\mathbf{1} \qquad\qquad\qquad\qquad \mathbf{2}$$

a: M=Na; X=Cl   f: M=Na; X=Br
b: M=K; X=Cl   g: M=Ag; X=Br
c: M=NH$_4$; X=Cl   h: M=Na; X=I
d: M=Ag; X=Cl   i: M=K; X=I
e: M=0.5 Ca^{2+}; X=Cl

Experimental procedures:
Ca. 5–20 g of mortared halogenoacetate **1** was annealed in a round-bottom flask in an oil bath. The sample should be heated carefully to avoid self-heating that may lead to combustion. Reaction conditions: Sodium chloroacetate: 60 min at 180 °C; potassium chloroacetate: 60 min at 150 °C; ammonium chloroacetate: 30 min at 110 °C; silver chloroacetate: 60 min at 180 °C; sodium bromoacetate: 60 min at 160 °C; silver bromoacetate: 10 min at 100 °C; potassium iodoacetate: 60 min at 180 °C. The sample changed from a white to a yellow or grayish powder but retained its morphology, except ammonium chloroacetate. The polyglycolide **2** was separated from the also formed metal halide by multiple washing with water. Yield: quantitative. Polyglycolide: White powder, insoluble in common solvents, partially soluble in 1,1,1,3,3,3-hexafluoro-2-propanol.

References: M. Epple, H. Krischnick, *Chem. Ber.*, **129**, 1123 (1996).

Type of reaction: C–O bond formation
Reaction condition: solid-state
Keywords: chlorocarboxylic acid, oligomerization, polymerization, polyglycolide

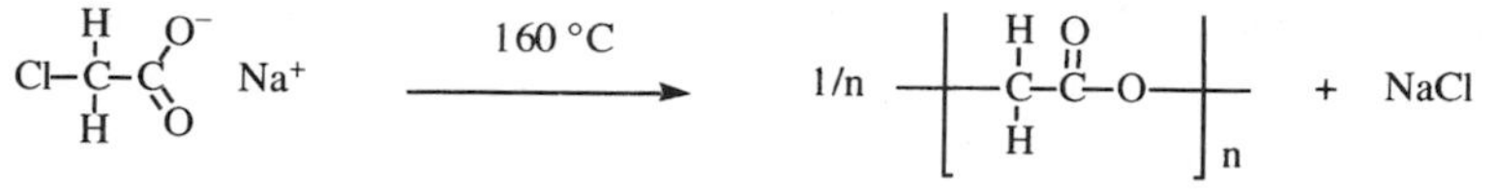

$$1 \quad \xrightarrow{\;160\,°C\;} \quad 2 \quad + \quad NaCl$$

Experimental procedures:

Ca. 5 g of ground halogenocarboxylate **1** is heated in an open round-bottom flask in an oil bath. Heating to the reaction temperatures should occur carefully; otherwise, decomposition and combustion may occur. Reaction conditions: sodium 2-chloropropionate **1**: 30 min at 160 °C; sodium 3-chloropropionate **2**: 45 min at 120 °C; sodium 2-chlorobutyrate **3**: 30 min at 160 °C. The reaction mixture becomes liquid during the reaction. After reaction is complete, a viscous brown-yellow oil remains. The organic part of the reaction mixture is extracted with CHCl$_3$, and the undissolved NaCl is filtered off. Removal of CHCl$_3$ in vacuo leaves a viscous yellow oil.

References: M. Epple, H. Kirshnick, *Liebigs Ann.,* 81 (1997).

5.2   Solvent-Free C–O Bond Formation under Photoirradiation

Type of reaction: C–O bond formation
Reaction condition: solid-state
Keywords: nitrone, photocyclization, inclusion complex, oxaziridine

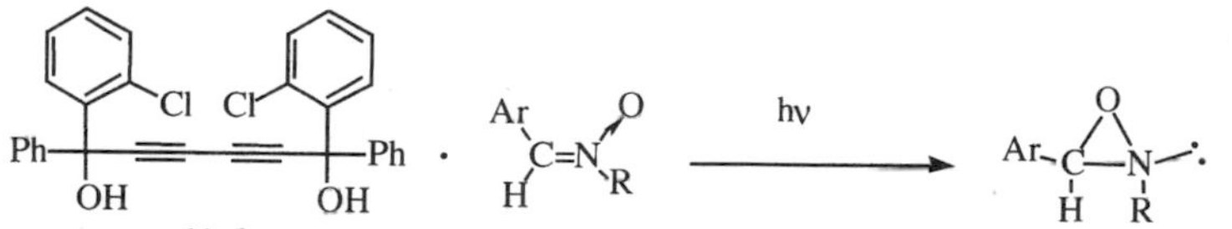

a: Ar=Ph; R=*i*-Pr
b: Ar=Ph; R=*t*-Bu
c: Ar=4-ClC$_6$H$_4$; R=*t*-Bu
d: Ar=2-ClC$_6$H$_4$; R=*t*-Bu
e: Ar=3,4-CH$_2$O$_2$C$_6$H$_3$; R=*i*-Pr
f: Ar=3,4-CH$_2$O$_2$C$_6$H$_3$; R=*t*-Bu
g: Ar=Ph; R=*i*-PrMeCH

Experimental procedures:
The 1:1 inclusion complex of (–)-**1** and **2a** was powdered and irradiated by a high pressure Hg-lamp for 24 h at room temperature, and the reaction mixture was chromatographed on silica gel (benzene) to give (+)-**3** in 56% yield.

References: F. Toda, K. Tanaka, *Chem. Lett.*, 2283 (1987).

Type of reaction: C–O bond formation
Reaction condition: solid-state
Keywords: bis(9-hydroxyfluoren-9-yl)thieno[3,2-*b*]thiophene, clathrate, photosolvolysis, etherification

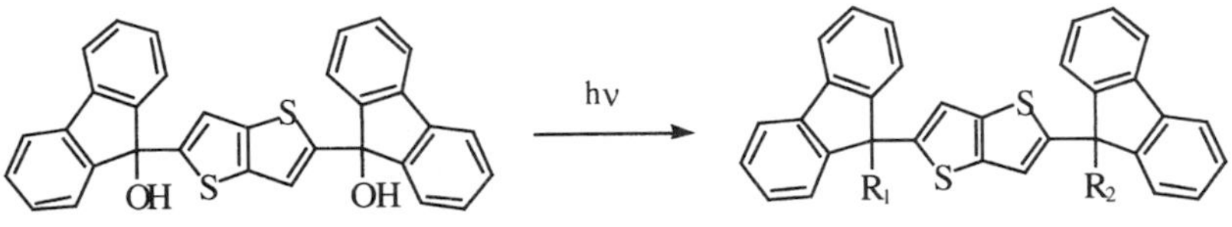

1 · 2 EtOH

2

a: R_1=OH; R_2=OEt
b: R_1=R_2=OEt

Experimental procedures:
Upon recrystallization from ethanol **1** afforded the clathrate crystal with a host-guest molar ratio of 1:2. Grinded **1**·2EtOH crystals were irradiated by means of a high-pressure mercury lamp at ambient temperature for 6 h. The photoproducts were chromatographed on a silica gel column to give monoether **2a** (43%) and diether **2b** (21%) along with unreacted **1** (36%).

References: N. Hayashi, Y. Mazaki, K. Kobayashi, *Tetrahedron Lett.*, **35**, 5883 (1994).

Type of reaction: C–O bond formation
Reaction condition: solid-state
Keywords: 2-benzoylbenzamide, absolute asymmetric synthesis, single-crystal-to-single-crystal reaction, photoirradiation, 3-(*N*-methylanilino)-3-phenylphthalide

a: $R_1=R_2=Me$, $R_3=H$
b: $R_1=Me$, $R_2=Ph$, $R_3=H$
c: $R_1=Me$, $R_2=Ph$, $R_3=Me$

80% ee

Experimental procedures:

Solid samples placed in the bottom of the test tube were cooled in a cooling apparatus and were irradiated by Pyrex-filtered light transmitted by using a flexible light guide from a 250-W ultra-high-pressure mercury lamp. When the powdered **1b** was irradiated at 15 °C for 2 h, a quantitative amount of 3-(*N*-methylanilino)-3-phenylphthalide **2b** ($[\alpha]_D 21°$ (*c* 1.0, CHCl$_3$)) was obtained.

References: M. Sakamoto, N. Sekine, H. Miyoshi, T. Mino, T. Fujita, *J. Am. Chem. Soc.*, **122**, 10210 (2000).

5.3 Solvent-Free C–O Bond Formation under Microwave Irradiation

Type of reaction: C–O bond formation
Reaction condition: solvent-free
Keywords: *n*-octyl bromide, benzoic acid, Aliquat 336, microwave irradiation, *n*-octyl benzoate

$$PhCO_2^- K^+ \quad + \quad n\text{-}C_8H_{17}Br \quad \xrightarrow[\text{MW}]{\text{Aliquat 336}} \quad PhCO_2 n\text{-}C_8H_{17} \quad + \quad KBr$$

Experimental procedures:

To 10 mmoles of potassium carboxylate were added in a Pyrex flask 10 mmol of *n*-octyl bromide and 1 mmol of tetraalkylammonium salt (Aliquat 336 or Bu$_4$NBr). After shaking, the flask was introduced in the microwave oven (or in an oil bath for control experiments) for the indicated time. The temperature was measured by introducing a Quick digital thermometer in the sample just at the end of each irradiation. Organic products were recovered by a simple elution with 50 mL diethyl ether or methylene chloride and subsequent filtration over

Florisil to remove mineral salts and catalyst. Products were analyzed by GC, characterized by ^{1}H and ^{13}C NMR, IR and MS after purification.

References: A. Loupy, P. Pigeon, M. Ramdani, *Tetrahedron,* **52**, 6705 (1996).

Type of reaction: C–O bond formation
Reaction condition: solvent-free
Keywords: 2,5-furandimethanol, alkyl halide, phase transfer catalyst, microwave irradiation, *O*-alkylation, etherification

$$HOH_2C-\overset{}{\underset{O}{\text{furan}}}-CH_2OH \ + \ 2\,RX \ \xrightarrow[\text{MW}]{2\,KOH} \ ROH_2C-\overset{}{\underset{O}{\text{furan}}}-CH_2OR$$

1   **2**

a: RX=CH$_3$(CH$_2$)$_{16}$CH$_2$Br
b: RX=CH$_3$(CH$_2$)$_{14}$CH$_2$Br
c: RX=CH$_3$(CH$_2$)$_{10}$CH$_2$Br
d: RX=CH$_3$(CH$_2$)$_6$CH$_2$Br
e: RX=PhCH$_2$Br

Experimental procedures:
In a Pyrex cylindrical reactor adapted to the Synthewave system, 10 mmol of FDM (1.28 g) were mixed with 25 mmol of alkyl halide, 2 mmol of Aliquat 336 (0.8080 g) and 25 mmol of powdered KOH (1.6 g, containing about 15% of water). The mixture was then homogenized and submitted to monomode micro-waves with mechanical stirring for the adequate time. At the end of the reaction, the mixture was cooled down to room temperature and diluted with 20 mL of methylene chloride or diethyl ether. The solution was filtered (KOH in excess, generated salts). The filtrate was then concentrated and poured dropwise into 300 mL of methanol under intense stirring. The diethers **2** precipitate, therefore free from excess of reactants, catalyst and monoethers which are all soluble in methanol. After filtration and drying under vacuum, the product was recrystal-lized from adequate solvent.

References: M. Majdoub, A. Loupy, A. Petit, S. Roudesli, *Tetrahedron,* **52**, 617 (1996).

Type of reaction: C–O bond formation
Reaction condition: solvent-free
Keywords: styrene, TaCl$_5$-SiO$_2$, paraformaldehyde, microwave irradiation, Prins reaction, 1,3-dioxane derivative

$$R\text{–}\underset{R}{\overset{R_1}{\diagdown}}=\underset{H}{\overset{R_2}{\diagup}} + \text{HCHO} \xrightarrow[\text{MW}]{\text{TaCl}_5\text{-SiO}_2} \underset{\substack{R_1 \ R_2}}{R\text{–}} \text{(dioxane)}$$

$$\mathbf{1} \qquad\qquad\qquad\qquad\qquad\qquad\qquad \mathbf{2}$$

R_1=H or Me
R_2=H or Me or Ph

Experimental procedures:
Styrene (1.04 g, 0.01 mol) and paraformaldehyde (0.9 g, 0.01 mol) were adsorbed on activated silica gel (2 g, 100–200 mesh, dried overnight at 110 °C) and stirred at room temperature for 1 h under N_2 atmosphere. To this $TaCl_5$-SiO_2 (0.37 g, 10 mol%) was added and admixed thoroughly and irradiated in a microwave oven (600 W) for 3 min. This mixture was cooled to room temperature, charged on a small silica pad and eluted with *n*-hexane-ethyl acetate mixture (80:20). Removal of volatiles furnished 1,3-dioxane derivative **2** (1.48 g, 90%).

References: S. Chandrasekhar, B.V.S. Reddy, *Synlett*, 851 (1998).

Type of reaction: C–O bond formation
Reaction condition: solvent-free
Keywords: *o*-hydroxydibenzoylmethane, montmorillonite K-10, microwave irradiation, flavone

a: R_1=R_2=H
b: R_1=H; R_2=CH_3
c: R_1=H; R_2=OCH_3
d: R_1=H; R_2=NO_2
e: R_1=OCH_3; R_2=H
f: R_1=OCH_3; R_2=CH_3
g: R_1 = R_2=OCH_3

Experimental procedures:
1-(2-Hydroxy-5-methoxyphenyl)-3-(4-methylphenyl)propane-1,3-dione **1f** (0.2 g, 0.70 mmol) was dissolved in a small amount of dichloromethane (1 mL) and adsorbed on montmorillonite K-10 clay (1.0 g). The contents, in a test-tube, were

placed in an alumina bath inside the microwave oven and irradiated for 1.5 min. The crude product was extracted in dichloromethane (2×15 mL) and then crystallized from methanol to afford **2f** in 80% yield, mp 161–162 °C.

References: D. Karmakar, D. Prajapati, J.S. Sandhu, *J. Chem. Res. (S)*, 382 (1998).

Type of reaction: C–O bond formation
Reaction condition: solvent-free
Keywords: *p*-toluenesulfinic acid, alcohol, silica gel, microwave irradiation, alkyl *p*-toluenesulfinate

a: R=Me
b: R=Et
c: R=*n*-Pr
d: R=*n*-Bu
e: R=*i*-Bu
f: R=CH$_3$(CH$_2$)$_2$CH(CH$_3$)
g: R=cyclohexyl
h: R=PhCH$_2$
i: R=PhCH$_2$CH$_2$
j: R=L-menthyl

Experimental procedures:
A mortar was charged with the alcohol (1 mmol), *p*-toluenesulfinic acid (1 mmol, 0.16 g), and silica gel (0.3 g). The reaction mixture was ground with a pestle in the mortar for 1 min, then the mortar was covered with a watch glass, placed in a microwave oven and irradiated for the time specified in Table 1. When TLC showed no remaining *p*-toluenesulfinic acid, the reaction mixture was poured into a mixture of ether (20 mL) and H$_2$O (5 mL). The ethereal layer was washed with saturated NaHCO$_3$, dried (MgSO$_4$), and evaporated to dryness using a rotary evaporator to give the pure product.

References: A.R. Hajipour, S.E. Mallakpour, A. Afrousheh, *Tetrahedron*, **55**, 2311 (1999).

Type of reaction: C–O bond formation
Reaction condition: solvent-free
Keywords: 2-alkanone, copper(II) organosulfonate, microwave irradiation, 3-organosulfonyloxy-2-alkanone

1

a: R=Me
b: R=Et
c: R=Pr
d: R=*i*-Bu
e: R=*i*-bu

R_1= Me, 4-MeC$_6$H$_4$, 1-naphthyl

Experimental procedures:
A mixture of 2-alkanone (1.0 mmol) and a copper(II) organosulfonate (3.0 mmol) was irradiated in an open glass scintillation vial (10 mL) in a domestic microwave oven for 3 min. After reaction mixture was allowed to cool, the residue was extracted with dichloromethane (2×50 mL). The extract was washed with water (2×25 mL), dried over MgSO$_4$, and the solvent was evaporated to dryness. The residue was purified by silica gel flash column chromatography using ethyl acetate-hexane (1:3) as eluent to give an analytically pure product.

References: J.C. Lee, S.H. Oh, I.G. Song, *Tetrahedron Lett.,* **40**, 8877 (1999).

Type of reaction: C–O bond formation
Reaction condition: solvent-free
Keywords: sodium carboxylate, halide, layered double hydroxide, carboxylate alkylation, microwave irradiation

1   2   3

a: R=Ph   a: R' =C$_7$H$_{15}$
b: R=Me   b: R' =Ph

Experimental procedures:
The intercalated LDH were obtained by anionic exchange: 1 g of ZnCr-NO$_3$ precursor was dispersed into a 1 M solution of the corresponding sodium carboxylate (acetate or benzoate). The reaction mixture was continuously stirred for 24 h at room temperature under a nitrogen flow. Then the precipitate was recovered by centrifugation and the resulting solid washed three times with deionized water and, finally, dried at room temperature. These intercalation compounds were impregnated with a stoichiometric amount of halide, either 1-bromooctane or benzyl bromide, and placed in a Teflon reactor. 100 mg of ZnCr-Ac·3H$_2$O

(ZnCr-Bn · 3H$_2$O) were mixed with 29.9 mL (25.9 mL) of benzyl bromide or with 43.4 mL (37.6 mL) of 1-bromooctane. The reaction took place by heating either in a domestic microwave oven (600 W) or in a conventional oven at 100 °C. After the reaction was completed, the resulting products were extracted with methanol and analyzed by GC-MS.

References: A. L. Garcia-Ponce, V. Prevot, B. Casal, E. Ruiz-Hitzky, *New J. Chem.*, **24**, 119 (2000).

Type of reaction: C–O bond formation
Reaction condition: solvent-free
Keywords: α-phenoxy, acetophenon, KSF clay, microwave irradiation, cyclodehydration, 3-phenylbenzofuran

R$_1$=H, Me, *i*-Pr, *t*-amyl, Cl, Br
R$_2$=Me, Ph

Experimental procedures:
The α-phenoxy acetophenone 2 (212 mg, 1 mmol) was dissolved in minimum amount of dichloromethane, adsorbed over montmorillonite KSF clay (substrate:clay=1:2 w/w). It was transferred into a test tube and subjected to microwave irradiation (BPL make, BMO 700T, 650 W, high power). The reaction was monitored by TLC. After completion of the reaction (6 min) it was leached with dichloromethane (3×10 mL). The solvent was evaporated under reduced pressure and purified through column chromatography using ethyl acetate-hexane (9:1) to give 3-phenylbenzofuran (165 mg, 85%).

References: H. M. Meshram, K. C. Sekhar, Y. S. S. Ganesh, J. S. Yadav, *Synlett,* 1273 (2000).

Type of reaction: C–O bond formation
Reaction condition: solvent-free
Keywords: carbonyl compound, iodobenzene diacetate, microwave irradiation, α-organosulfonyloxylation

$$R'' = p\text{-}CH_3C_6H_4 \ (\text{-OTs}), \ p\text{-}NO_2C_6H_4 \ (\text{-ONs})$$

Experimental procedures:
A mixture of carbonyl compound **1**, iodobenzene diacetate (1.5 equiv.), and orga-nosulfonic acid (4.5 equiv.) was placed in a test tube placed inside the alumina bath and irradiated for 10–40 s at ten-second intervals in a domestic microwave oven at full power (600 W). After cooling at room temperature the mixture was extracted with dichloromethane (2×20 mL), washed with H_2O, and dried over $MgSO_4$. The mixture was evaporated and the residue was column chromato-graphed with ethyl acetate-hexane (1:3) to give the desired α-organosulfonyloxy carbonyl compound **2**.

References: J.C. Lee, J.-H. Choi, *Synlett,* 234 (2001).

Type of reaction: C–O bond formation
Reaction condition: solvent-free
Keywords: 3,4-dihydroxybenzaldehyde, alkyl halide, phase transfer catalyst, mi-crowave irradiation, 1,8-cineole

a: $R=C_{12}H_{25}$
b: $R=C_{14}H_{29}$
c: $R=C_{16}H_{33}$
d: $R=C_{18}H_{37}$

Experimental procedures:
In a 40-mL Pyrex tube, 1.38 g (10 mmol) of 3,4-dihydroxybenzaldehyde **1** were mixed together with 1.12 g (20 mmol) of powdered KOH, 2.8 g (20 mmol) of an-hydrous K_2CO_3, 0.32 g (1 mmol) of TBAB and 22 mmol of the appropriate alkyl halide. The reaction mixture was introduced into the microwave reactor and irra-diated for 10 min at 130 °C under stirring (maintaining the temperature by power modulation from 15 to 300 W). After cooling to room temperature, the organic

product was recovered by elution with dichloromethane (50 mL) and subsequent filtration on Florisil with a Gooch filter, to remove the mineral salts and catalyst. The solution was evaporated under reduced pressure and the crude product **2** was purified by recrystallization from 95% ethanol.

References: M. Rohr, C. Geyer, R. Wandeler, M.S. Schneider, E.F. Murphy, A. Baiker, *Green Chem.*, **3**, 123 (2001).

Type of reaction: C–O bond formation
Reaction condition: solvent-free
Keywords: alcohol, amine, thiol, acetic anhydride, microwave irradiation, acetylation

$$R-XH \quad + \quad Ac_2O \quad \xrightarrow[MW]{} \quad R-XAc \quad + \quad AcOH$$

1   **2**

a: R=CH$_3$(CH$_2$)$_{17}$; X= O
b: R=PhCH=CHCH$_2$; X= O
c: R=PhCH$_2$; X= O
d: R=2,4-Cl$_2$C$_6$H$_3$; X= O
e: R=4-MeOC$_6$H$_4$; X= O
f: R=2-NO$_2$C$_6$H$_4$; X= O
g: R=*t*-Bu; X= O
h: R=Ph$_3$C; X= O
i: R=4-ClC$_6$H$_4$; X= S
j: R=4-BrC$_6$H$_4$; X= S
k: R=4-OHC$_6$H$_4$; X=NH

Experimental procedures:
A mixture of 4-methoxybenzyl alcohol **1e** (5 mmol) and acetic anhydride (5 mmol) in a beaker covered with watch glass was irradiated by microwaves for 10 min (heating and cooling at the interval of 1 min). After completion of the reaction (TLC), the product was extracted with ether (3×15 mL). The ether layer was washed with 10% NaOH and then dried with anhydrous sodium sulfate. Removal of the solvent under reduced pressure gave 4-methoxybenzyl acetate **2e** in excellent yield (95%).

References: B.P. Bandgar, S.P. Kasture, V.T. Kamble, *Synth. Commun.*, **31**, 2255 (2001).

Type of reaction: C–O bond formation
Reaction condition: solvent-free
Keywords: a-D-glucopyranoside, microwave irradiation, esterification

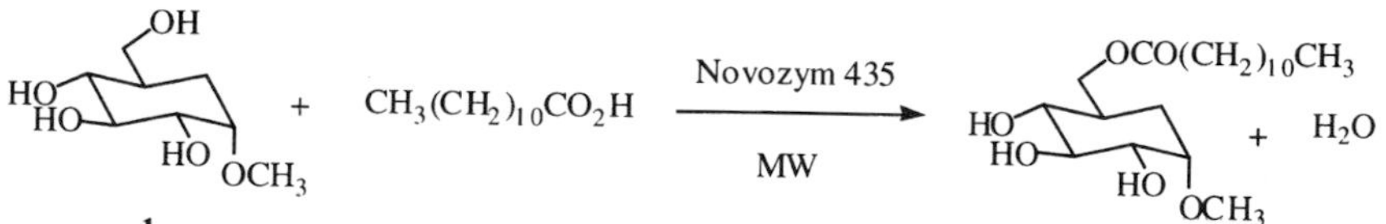

Experimental procedures:

a-D-Glucopyranoside **1** (1–1.5 mmol) was dissolved in the minimal amount of water (0.5–1 mL), impregnated on Novozym 435 (1 g) and dried by different modes (A, B, C as described in the text). Dodecanoic acid (1–5 mmol) in diethyl ether (1–1.5 mL) was added; the mixture was placed in the microwave reactor and irradiated as indicated in Tables 1 and 2, or introduced into a thermostated oil bath. After cooling to room temperature, the mixture was washed first with pentane in order to remove the excess of dodecanoic acid and then the products were dissolved in methanol and analyzed by GC after silylation or by HPLC. To the crude products $CHCl_3$ was added and the starting glucopyranosides were washed out with water. The organic phase was dried over Na_2SO_4 and the pure esters were precipitated with pentane. The positions of acylation were established from ^{13}C NMR spectra according to the Yoshimoto method.

References: M. Gelo-Pujic, E. Guibe-Jampel, A. Loupy, S. A. Galema, D. Mathe, *J. Chem. Soc., Perkin Trans. 1*, 2777 (1996).

6 Carbon–Sulfur Bond Formation

6.1 Solvent-Free C–S Bond Formation

Type of reaction: C–S bond formation
Reaction condition: solvent-free
Keywords: sodium dithionite, sodium formaldehyde sulfoxylate, benzyl halide, dibenzyl sulfone

$$2\ PhCH_2\!-\!X\ +\ Na_2S_2O_4\ \xrightarrow{\ \Delta\ }\ (PhCH_2)_2SO_2\ +\ SO_2\ +\ 2\ NaX$$

$$2\ PhCH_2\!-\!X\ +\ NaSO_2CH_2OH\ \xrightarrow{\ \Delta\ }\ (PhCH_2)_2SO_2\ +\ HCHO\ +\ 2\ NaX$$

$$X = Cl, Br$$

Experimental procedures:
Benzyl bromide (17.1 g, 0.1 mol) was added to a mixture of solid sodium dithionite (10.44 g, 0.06 mol) and Aliquat 336 (1.2 g, 0.03 mol). The mixture was vigorously shaken for 5 min and then heated in an oil bath for 20 h at 120 °C. Dibenzyl sulfone was removed by filtration through Florisil with 50 mL of methylene chloride. The solvent was evaporated and the crude solid crystallized from a 1:1 mixture of ethanol and toluene to give the pure sulfone as a white solid (7.54 g, 61%), mp 148–149 °C.

An analogous procedure was carried out with benzyl bromide (17.1 g, 0.1 mol), sodium formaldehyde sulfoxylate dihydrate (10.35 g, 0.06 mol) and sodium carbonate (10.35 g, 0.075 mol) to obtain the pure dibenzyl sulfone (9.35 g, 76% after recrystallization).

References: A. Loupy, J. Sansoulet, A.R. Harris, *Synth. Commun.,* **19**, 2939 (1989).

Type of reaction: C–S bond formation
Reaction condition: solid-state
Keywords: benzyltriethylammonium tetrathiomolybdate, alkyl halide, disulfide

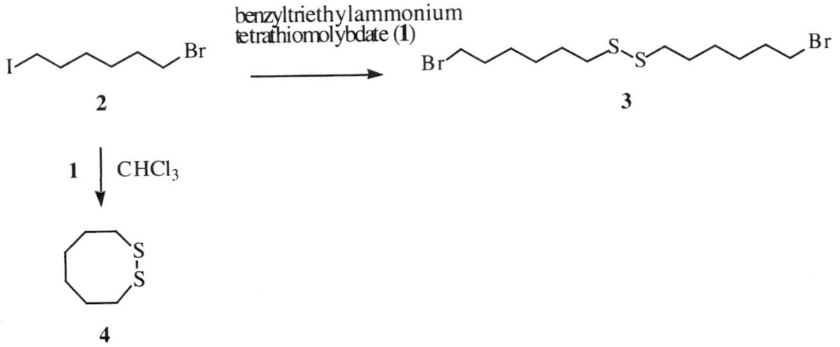

benzyltriethylammonium
tetrathiomolybdate (**1**)

I~~~~~Br   ⟶   Br~~~~~S-S~~~~~~Br

2   **3**

1 | CHCl₃

4

Experimental procedures:

1-Bromo-6-iodohexane **2** (0.29 g, 1 mmol) was added in one portion to benzyl-triethylammonium tetrathiomolybdate **1** (1.4 g, 2.2 mmol) in an agate mortar and the mixture was ground continuously for 5 min. The color of the mixture changed immediately from dark red to black and the initial viscid product became powderly within minutes. The mixture was ground occasionally for 20 min and then extracted with dichloromethane or diethyl ether; flash column chromatography of the extract on silica gel gave the product **3** as a viscous oil (0.1 g, 52%).

References: A.R. Ramesha, S. Chandresekaran, *J. Chem. Soc., Perkin Trans. 1*, 767 (1994).

Type of reaction: C–S bond formation
Reaction condition: solvent-free
Keywords: aldehyde, ketone, thiol, lithium trifluoromethanesulfonate, dithioacetalization, dithiane

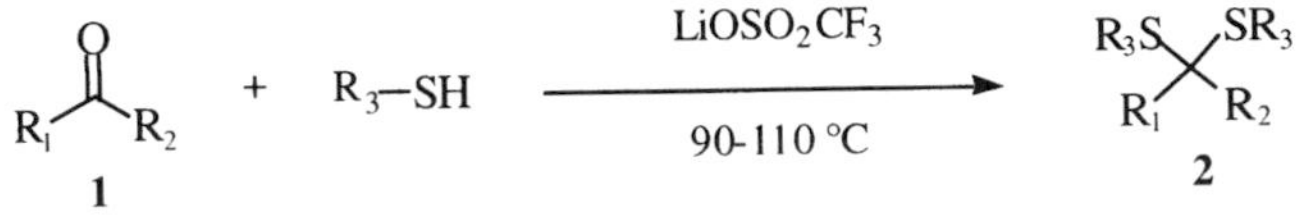

$$\underset{\textbf{1}}{\underset{R_1\quad R_2}{\overset{O}{\|}}} + R_3\text{–SH} \xrightarrow[90-110\,°C]{\text{LiOSO}_2\text{CF}_3} \underset{\textbf{2}}{\underset{R_1\quad R_2}{\overset{R_3S\quad SR_3}{}}}$$

a: R_1=C_6H_5; R_2=H; R_3=-(CH$_2$)$_3$-
b: R_1=C_6H_5; R_2=H; R_3=-C_6H_5
c: R_1=C_6H_5; R_2=H; R_3=$C_6H_4CH_2$
d: R_1=C_6H_5; R_2=H; R_3=C_6H_{11}
e: R_1=4-MeC_6H_4; R_2=H; R_3=HS(CH$_2$)$_3$SH
f: R_1=4-ClC_6H_4; R_2=H; R_3=-(CH$_2$)$_3$-
g: R_1=4-MeOC_6H_4; R_2=H; R_3=-(CH$_2$)$_3$-
h: R_1=C_6H_5CH=CH; R_2=H; R_3=-(CH$_2$)$_3$-
i: R_1; R_2=-(CH$_2$)$_5$-; R_3=-(CH$_2$)$_3$-

Experimental procedures:
To a stirred mixture of the carbonyl compound (10 mmol) and dithiol (11–17 mmol) or monothiol (21 mmol), anhydrous LiOTf (0.5–3 mmol) was added. The mixture was heated at 90 °C (or 110 °C for ketones) while stirring was continued, and the progress of the reaction was followed by TLC. After completion of the reaction, CHCl$_3$ (100 mL) was added and the mixture was washed successively with 10% NaOH solution (2×25 mL), brine (15 mL), and water (15 mL). The organic layer was separated and dried over anhydrous Na$_2$SO$_4$. Evaporation of the solvent under reduced pressure gave almost pure product. Further purification was achieved by column chromatography on silica gel or recrystallization from an appropriate solvent to give the desired product in good to excellent yields. Most of the products are known and gave satisfactory physical data compared with those of authentic samples.

References: H. Firouzabadi, B. Karimi, S. Eslami, *Tetrahedron Lett.,* **40**, 4055 (1999).

Type of reaction: C–S bond formation
Reaction condition: solvent-free
Keywords: aldehyde, thiol, lithium bromide, dithioacetalization, 1,3-dithiane, 1,3-dithiolane

R_1=Ph, 4-MeC$_6$H$_4$, 3-MeC$_6$H$_4$, 3-ClC$_6$H$_4$, 4-ClC$_6$H$_4$, 4-MeOC$_6$H$_4$, 1-naphthyl, PhCH=CH
R_2= Ph, PhCH$_2$, c-C$_6$H$_{11}$, HS-(CH$_2$)$_2$-SH, HS-(CH$_2$)$_3$-SH

Experimental procedures:
To a stirred mixture of the carbonyl compound **1** (10 mmol) and dithiol (11 mmol) or monothiol (20–21 mmol) was added anhydrous LiBr (2.5–4.0 mmol). The mixture was heated to 75–80 °C and the progress of the reaction was followed by TLC. After completion of the reaction (15–50 min), CH$_2$Cl$_2$ (100 mL) was added and the mixture was washed successively with 10% NaOH solution (2×25 mL), brine (15 mL), and H$_2$O (15 mL). The organic layer was separated and dried (Na$_2$SO$_4$). Evaporation of the solvent under reduced pressure gave almost pure product. Further purification was achieved by column chromatography on silica gel or recrystallization from appropriate solvent to give the desired products in good to excellent yields.

References: H. Firouzabadi, N. Iranpoor, B. Karimi, *Synthesis*, 58 (1999).

Type of reaction: C–S bond formation
Reaction condition: solvent-free
Keywords: ketone, aldehyde, ethane-1,2-dithiol, neutral alumina surface, dithio-acetal, 1,3-dithiolane

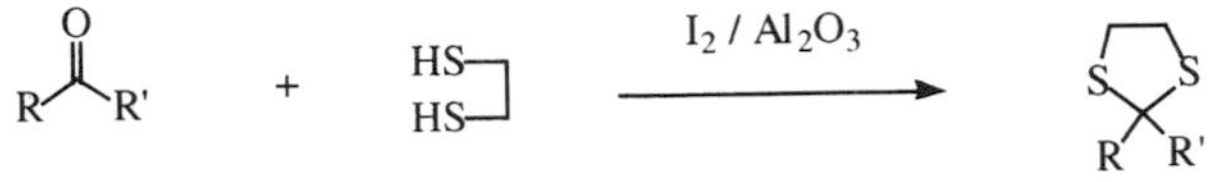

a: R=Ph; R'=H
b: R=4-MeOC$_6$H$_4$; R'=H
c: R=4-ClC$_6$H$_4$; R'=H
d: R=4-NO$_2$C$_6$H$_4$; R'=H
e: R=PhCH=CH; R'=H
f: R=MeCH=CH; R'=H
g: R=2-furyl; R'=H
h: R=2-HOC$_6$H$_4$; R'=H
i: R=C$_{16}$H$_{33}$; R'=H
j: R=Me; R'=CH$_2$COOMe
k: R=Ph; R'=Me
l: R, R'=-(CH$_2$)$_5$-
m: R= R =Ph

Experimental procedures:

To a freshly prepared catalyst (1 g, 0.2 mmol of iodine) under stirring, a mixture of benzaldehyde **1a** (2 mmol) and ethane-1,2-dithiol (2.2 mmol) was added and stirring continued for 10 min or till the reaction was complete. For TLC monitoring, a small amount of the solid reaction mixture was taken out with a spatula and washed with a little amount of ethyl acetate to get a solution. On completion the reaction mixture was loaded on a short column of silica gel (60 to 120 mesh) and eluted with ethyl acetate. The organic layer was washed with a dilute solution of sodium thiosulfate followed by water and dried over anhydrous sodium sulfate. Evaporation of the solvent under reduced pressure and purification of the residue by column chromatography yielded the pure product.

References: N. Deka, J.C. Sarma, *Chem. Lett.*, 794 (2001).

Type of reaction: C–S bond formation
Reaction condition: solvent-free
Keywords: aminal, sulfur, dehydrogenation, C-H activation, thiourea, carbenium thiocyanate

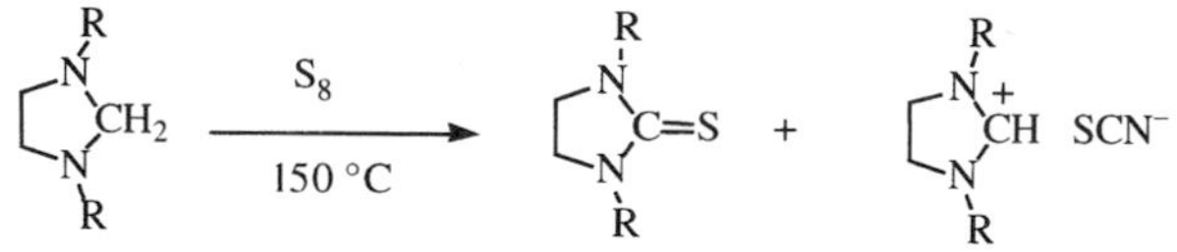

1   2   3

a: R=Me
b: R=Et
c: R=*i*-Pr
d: R=*t*-Bu
e: R=Ph

Experimental procedures:

Sulfur (1.22 g, 4.75 mmol) was added to the aminal (19.0 mmol) in a Swagelock 50-mL stainless steel cylinder. The vessel was sealed and placed in an oil bath (150 °C) for 12 h. The cold reaction mixture was extracted with methanol (3×10 mL). The combined methanol extracts were filtered and the methanol was evaporated in vacuo. The thiourea were obtained by extraction of the solid residue with toluene (2×10 mL). The toluene extract was filtered through a short column of neutral Al_2O_3. The carbenium salts were isolated by dissolving the toluene-insoluble fraction in water (5 mL). The remaining orange solution was decanted or filtered from the brown, insoluble impurities, and extracted with toluene (5 mL portions) until the organic phase remained colorless. The carbenium thiocyanates were obtained by slow evaporation of the aqueous phase in the form of large colorless crystals.

References: M.K. Denk, S. Gupta, J. Brownie, S. Tajammul, A.J. Lough, *Chem. Eur. J.*, **7**, 4477 (2001).

Type of reaction: C–S bond formation
Reaction condition: solid-state
Keywords: thioureideo-acetamide, phenacyl bromide, cascade reaction, cyclization, solid-solid reaction, heterocycle

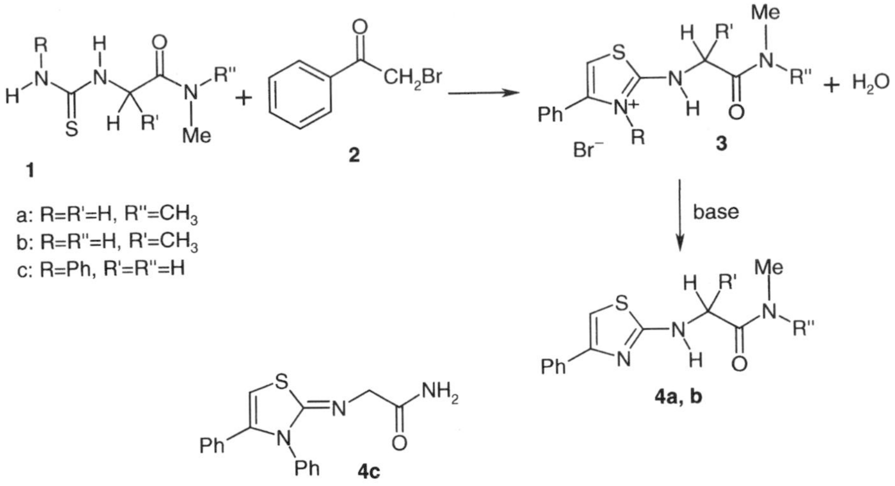
1
2
3
4a, b
4c
base
+ H2O
a: R=R'=H, R''=CH3
b: R=R''=H, R'=CH3
c: R=Ph, R'=R''=H

Experimental procedures:
The solid thioureido-acetamide **1a–c** (2.00 mmol) and solid phenacyl bromide **2** (398 mg, 2.00 mmol) were ball-milled in a Retsch MM 2000 mill with a 10-mL beaker (two balls out of stainless steel) at room temperature for 30 min. After drying at 0.01 bar at 80 °C quantitative yields of the pure (hydro)bromides **3a–c** were obtained in all cases. The salts were washed with 5% Na_2CO_3 solution to afford the free bases **4a–c** in pure form in 98, 99 and 98% yield, respectively.

References: J. Schmeyers, G. Kaupp, *Tetrahedron,* **58**, 7241 (2002).

Type of reaction: C–S bond formation
Reaction condition: solid-state
Keywords: thioureideo-acetamide, acetylene dicarboxylate, cascade reaction, ring closure, heterocycle

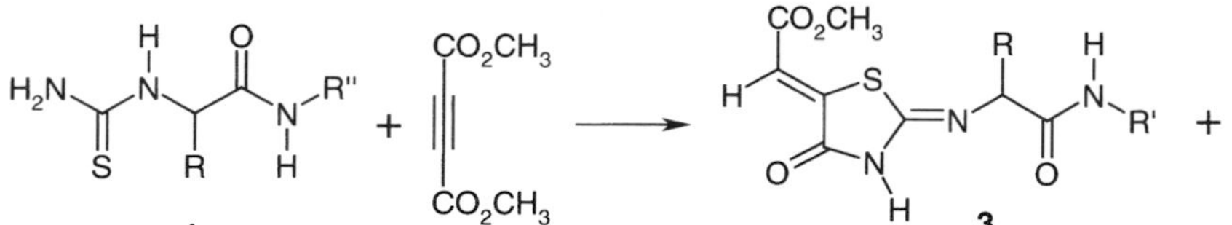

	3 / 4	isol. yield
a: R=R'=Me	1.0 / 0.0	76%
b: R=H, R'=Me	2.0 / 1.0	54% / 26%
c: R=H, R'=Et	1.5 / 1.0	50% / 33%

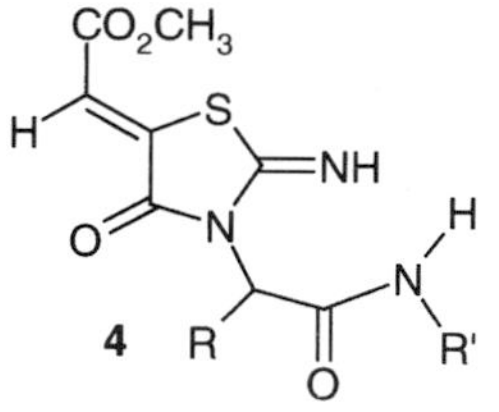

Experimental procedures:

The thioureido-acetamide **1b** (2.00 mmol) and **2** (2.00 mmol) were weighed into a 10-mL ball-mill vessel. The mixture became immediately solid and was ball-milled at 25 Hz for 1 h. **3b** remained insoluble in hot ethyl acetate whereas **4b** dissolved and was separated by filtration. **3b** was recrystallized from DMSO (**3b**, yield 54%). The ethyl acetate extract was evaporated to dryness and the crude solid recrystallized from ethanol (**4b**, yield 26%).

In ethanol solution the **3b:4b**-ratio was found to be 0.43:1.0.

The compounds **3a** and **3c**, **4c** were similarly obtained.

References: J. Schmeyers, G. Kaupp, *Tetrahedron*, **58**, 7241 (2002).

Type of reaction: C–S bond formation
Reaction condition: solvent-free
Keywords: 2-aminothiophenol, cinnamic acid, phenylpropynoic acid, condensation, cyclization, benzothiazepinone, benzothiazole

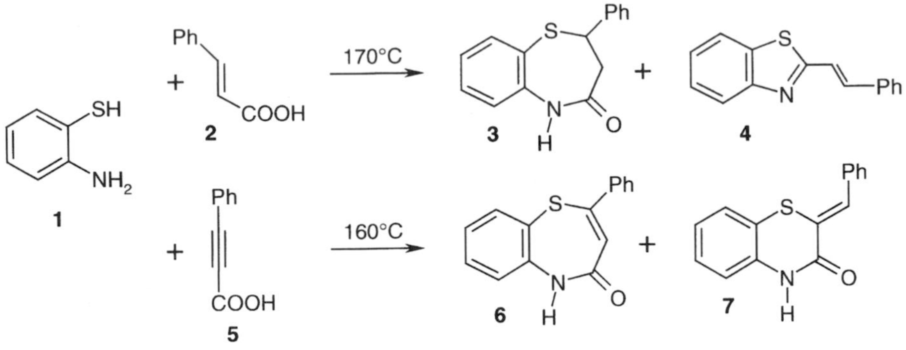

Experimental procedures:

2-Aminothiophenol **1** (8.0 g, 64 mmol) and trans-cinnamic acid **2** (10.0 g, 67.5 mmol) were heated to 170 °C under argon for 2 h. 7.9 g (48%) of **3** crystallized from ethanol. The mother liquor contained **1, 2, 3** and **4**. The latter (900 mg, 6%) was isolated by column chromatography on SiO$_2$ with dichloromethane.

The analogous reaction of **1** (21.7 g, 0.17 mol) and **5** (25 g, 0.17 mol) at 160 °C for 1 h gave a mixture that was cooled to 100 °C and added to 80 mL of boiling acetonitrile. The first crop of yellow crystals (3.4 g) contained **6** and **7** in a 1:1-ratio. The cooled mother liquor separated 7.0 g (16%) **6**. The chromatographic separation of **6** and **7** was inefficient.

References: G. Kaupp, E. Gründken, D. Matthies, *Chem. Ber.,* **119**, 3109 (1986).

Type of reaction: C–S bond formation
Reaction condition: solid-state
Keywords: mercaptobenzimidazol, imidazoline-2-thiol, phenacylbromide, thiuronium salt, waste-free, solid-solid reaction

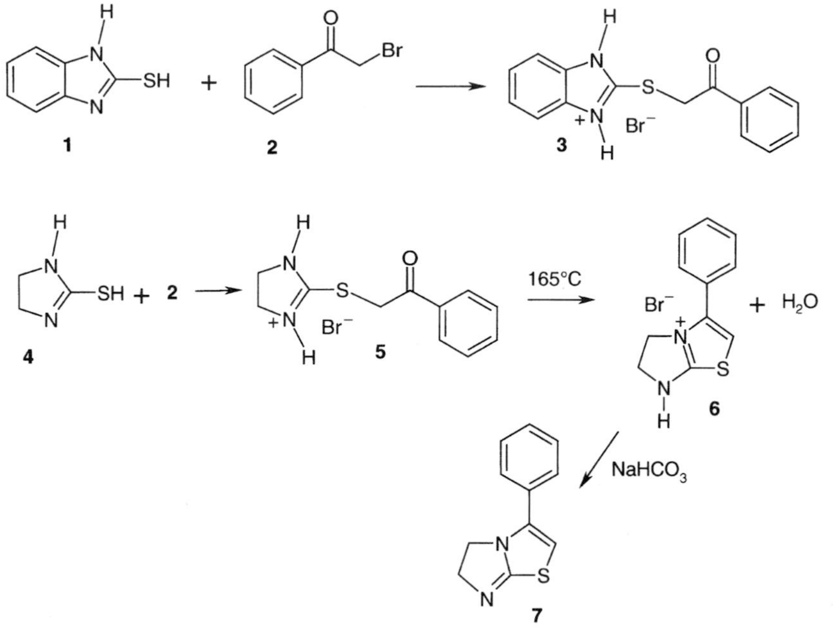

Experimental procedures:

Solid 2-mercaptobenzimidazole **1** (300 mg, 2.00 mmol) or imidazoline-2-thiol **4** (204 mg, 2.00 mmol) and phenacylbromide **2** (398 mg, 2.00 mmol) were ball-milled at room temperature for 1 h. The yield of pure product **3** or **5** was 100%.

The salt **5** (301 mg, 1.00 mmol) was heated to 165 °C for 60 min. After evacuation, the pure bicyclic thiazoliumbromide **6** (280 mg, 100%) was obtained. The free base **7** may be obtained from **6** by extraction from NaHCO$_3$ solution with CH$_2$Cl$_2$. Mp 112–113 °C.

References: G. Kaupp, J. Schmeyers, J. Boy, *J. Prakt. Chem., ***342**, 269 (2000).

Type of reaction: C–S bond formation
Reaction condition: solid-state
Keywords: thiolate, cyanogen chloride, cyanogen bromide, gas-solid reaction, thiocyanate

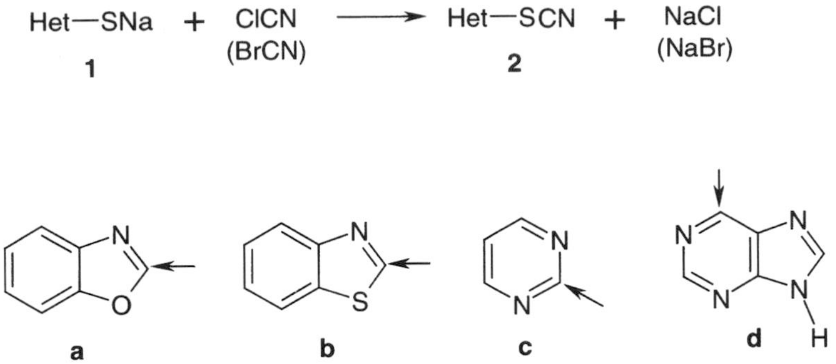

Het—SNa + ClCN $\longrightarrow$ Het—SCN + NaCl
1   (BrCN)   **2**   (NaBr)

a   b   c   d   H

Experimental procedures:

Solid sodium thiolate **1a–d** (10 mmol) was treated with ClCN (1 bar, 11.2 mmol) or with BrCN (1.17 g, 11.0 mmol from a remote flask at a vacuum line) and left overnight. Excess gas was recovered in a cold trap at −196 °C. NaCl (NaBr) was washed away with water and the quantitatively obtained thiocyanates **2a–d** were dried in a vacuum.

References: G. Kaupp, J. Schmeyers, J. Boy, *Chem. Eur. J.*, **4**, 2467 (1998).

Type of reaction: C–S bond formation
Reaction condition: solid-state
Keywords: cysteine, paraformaldehyde, cyclization, waste-free, solid-solid reaction, thiazolidine

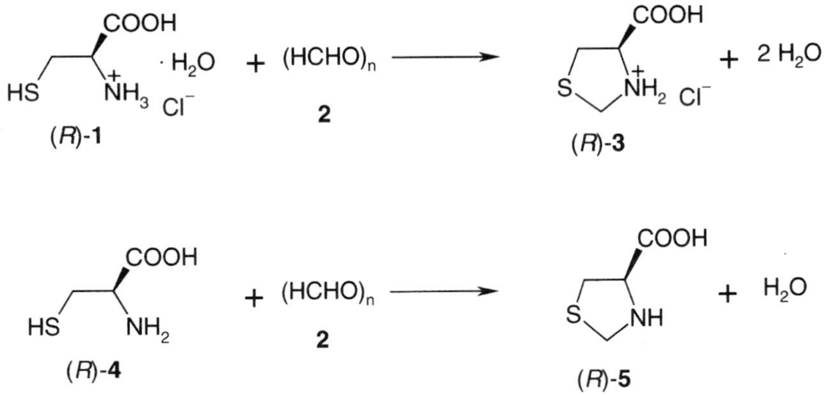

Experimental procedures:

(*R*)-**3**: Solid L-cysteine hydrochloride monohydrate (*R*)-**1** (351 mg, 2.00 mmol) and paraformaldehyde (**2**) (60 mg, 2.00 mmol) were ball-milled at room temperature for 1 h. After drying at 0.01 bar at 80 °C, pure (*R*)-**3** (338 mg, 100%) was obtained. (Ref. 1)

(*R*)-**5**: Solid L-cysteine **4** (242 mg, 2.00 mmol) and paraformaldehyde (62 mg of 97% purity, 2.00 mmol) were ball-milled at room temperature for 1 h. After drying at 0.01 bar at 80 °C, pure (*R*)-**5** (263 mg, 100%) was obtained. Mp 196 °C. (Ref. 2)

References:
(1) G. Kaupp, J. Schmeyers, J. Boy, *J. Prakt. Chem.*, **342**, 269 (2000).
(2) G. Kaupp, J. Schmeyers, J. Boy, *Tetrahedron*, **56**, 6899 (2000).

Type of reaction: C–S bond formation
Reaction condition: solid-state
Keywords: mercaptobenzothiazole, chloroacetone, gas-solid reaction, thiuronium salt

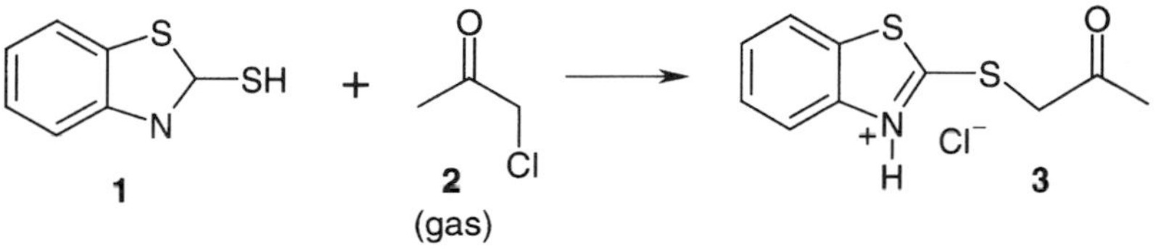

Experimental procedure:
Solid 2-mercaptobenzothiazole **1** (334 mg, 2.00 mmol) in an evacuated 100-mL flask was connected at a vacuum line to a 100-mL flask that contained chloroacetone **2** (500 mg, 5.4 mmol). The whole setup was heated to 60 °C and left for 12 h. Excess gas was removed by evaporation. 520 mg (100%) of pure **3** were obtained.

References: G. Kaupp, J. Schmeyers, J. Boy, *J. Prakt. Chem.*, **342**, 269 (2000).

Type of reaction: C–S bond formation
Reaction condition: solid-state
Keywords: thiourea, bromoacetophenone, cyclization, waste-free, solid-solid reaction, 2-aminothiazole

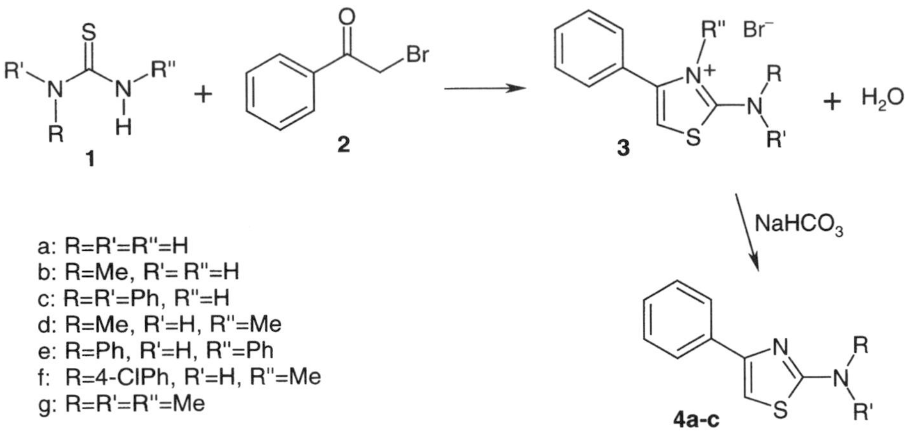

1
2
3
4a-c
NaHCO3
+ H2O
a: R=R'=R"=H
b: R=Me, R'= R"=H
c: R=R'=Ph, R"=H
d: R=Me, R'=H, R"=Me
e: R=Ph, R'=H, R"=Ph
f: R=4-ClPh, R'=H, R"=Me
g: R=R'=R"=Me

Experimental procedures:
The solid thiourea **1** (2.00 mmol) and phenacylbromide **2** (398 mg, 2.00 mmol) were ball-milled at room temperature for 30 min. After drying at 0.01 bar at 80 °C quantitative yields of the pure products **3** were obtained in all cases. The free bases **4a–c** can be obtained by trituration with $NaHCO_3$ solution.

References: G. Kaupp, J. Schmeyers, J. Boy, *J. Prakt. Chem.*, **342**, 269 (2000).

Type of reaction: C–S bond formation
Reaction condition: solid-state
Keywords: Viehe salt, benzimidazole thiol, waste-free, solid-solid reaction

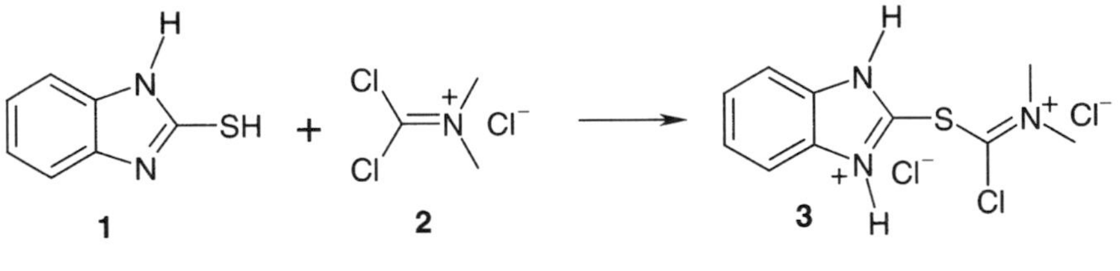

1
2
3

Experimental procedure:
Solid 2-mercaptobenzimidazole **1** (750 mg, 5.00 mmol) and Viehe salt **2** (815 mg, 5.00 mmol) were ball-milled for 1 h. The yield of pure salt **3** was 1.515 g (100%).

References: G. Kaupp, J. Boy, J. Schmeyers, *J. Prakt. Chem.*, **340**, 346 (1998).

Type of reaction: C–S bond formation
Reaction condition: solid-state
Keywords: methanethiol, *N,S*-acetal, substitution, gas-solid reaction

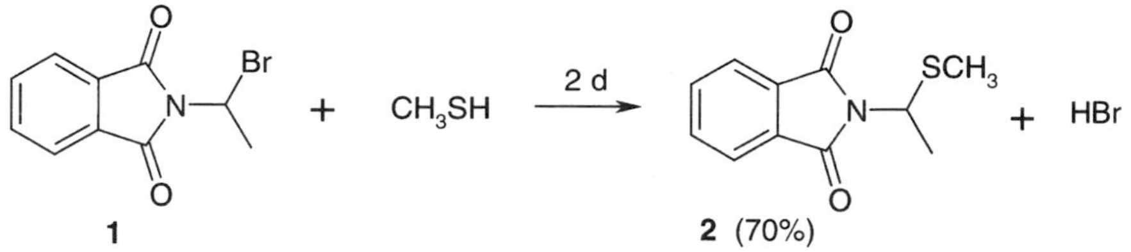

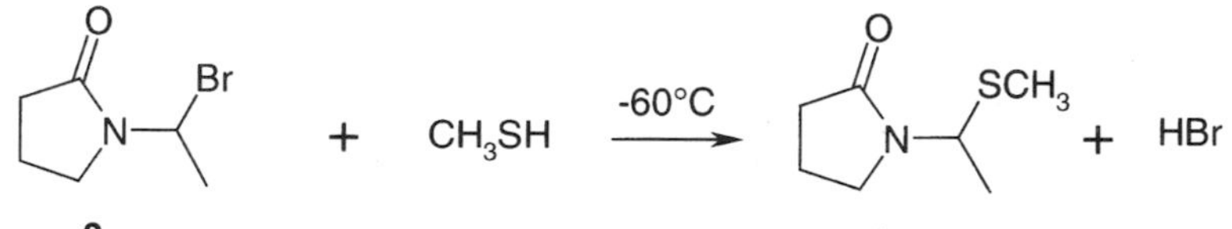

Experimental procedures:
2: Freshly prepared **1** (from *N*-vinylphthalimid and HBr gas) (1.0 g, 3.9 mmol) in an evacuated 500-mL flask was exposed to CH₃SH (0.5 bar, 11 mmol) at room temperature for 2 days. The gases were pumped off. The conversion was 70% and the product **2** was isolated by preparative TLC (basic SiO₂, benzene-ethylacetate, 4:1).

4: N-Vinylpyrrolidone (1.00 g, 9.0 mmol) was crystallized on Raschig coils (3 g) at –60 °C in an evacuated 500-mL flask. HBr gas (1 bar, 22 mmol) was applied for 2 h. Excess gas was evacuated and the still cooled flask now containing solid **3** was connected at a vacuum line to a 1-L flask that was previously filled with CH₃SH (0.3 bar, 13.4 mmol). After 19 h at –60 °C, all gases were evacuated and the crystals which soften above –30 °C were chromatographed at basic SiO₂ with ethyl acetate to give oily **4** (600 mg, 42%).

References: G. Kaupp, D. Lübben, O. Sauerland, *Phosphorus, Sulfur, and Silicon,* **53**, 109 (1990).

Type of reaction: C–S bond formation
Reaction condition: solid-state
Keywords: *S*-vinyl compound, methanethiol, *anti*-Markovnikov addition, gas-solid reaction, thermal, photochemical

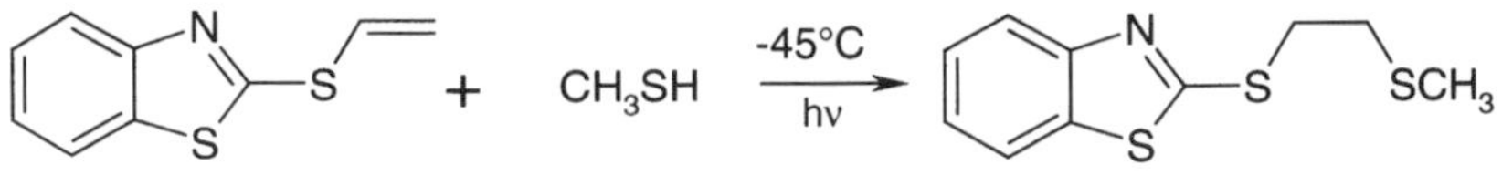

1   **2 (25%)**

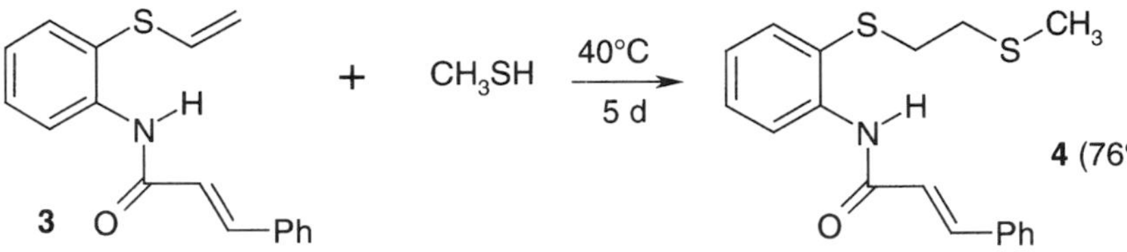

3   **4 (76%)**

Experimental procedures:

2: Liquid **1** (0.50 g, 2.6 mmol) was crystallized in an evacuated 500-mL flask at −45 °C. Methanethiol at a pressure of 0.3 bar was let in and illumination was performed through the cooling bath out of methanol in a Pyrex vessel using a high-pressure Hg-lamp (Hanau, 150 W) for 5 h. Product **2** was isolated as an oil by preparative TLC. There was no addition in the absence of light.

4: Crystalline **3** (200 mg, 0.71 mmol) was exposed to methanethiol (1 bar) in a 250-mL flask at 40 °C for 5 days. The solid product **4** was separated from unreacted **3** by preparative TLC (SiO$_2$, CH$_2$Cl$_2$); mp 106–107 °C.

References: G. Kaupp, D. Lübben, O. Sauerland, *Phosphorus, Sulfur, and Silicon*, **53**, 109 (1990).

Type of reaction: C–S bond formation
Reaction condition: solid-state
Keywords: *N*-vinyl compounds, methanethiol, addition, Markovnikov, *anti*-Markovnikov, gas-solid reaction, thermal, photochemical

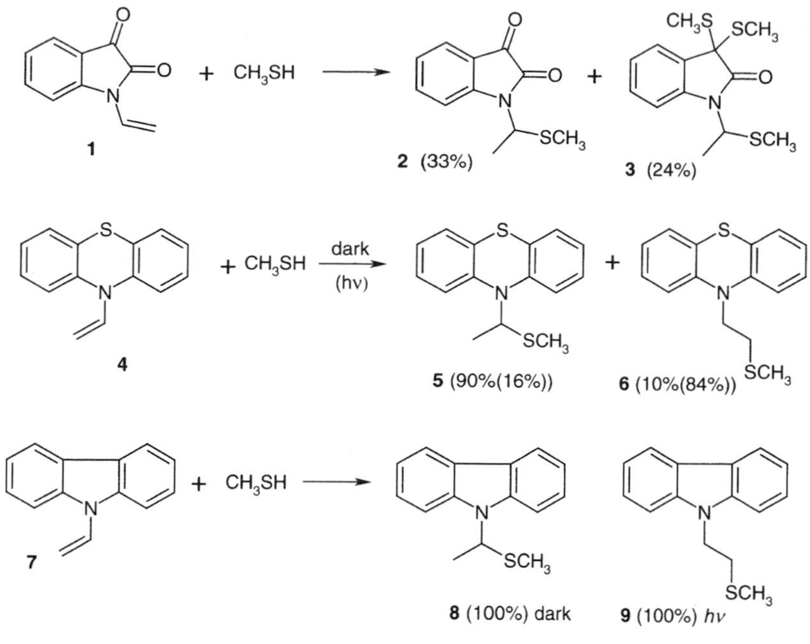

Experimental procedures:

Crystalline *N*-vinylization **1** (200 mg, 1.16 mmol) was exposed to 10 mmol methanethiol gas (0.9 bar) at a vacuum line for 21 h. The products **2** and **3** were isolated by preparative TLC on SiO$_2$ with methylene chloride.

Crystalline *N*-vinylphenothiazine **4** or *N*-vinylcarbazole **7** reacted correspondingly at room temperature for 2 d or 18 h in the dark. Irradiations were performed with a 500-W tungsten lamp in a 250-mL round-bottomed flask under 0.9 bar CH$_3$SH (**4**: 1.8 mmol, 11 h, rt; **7**: 5.2 mmol, 2 h, 0 °C). The purity of the products was determined by ^{1}H NMR spectroscopy.

References: G. Kaupp, D. Matthies, *Chem. Ber.,* **120**, 1897 (1987); G. Kaupp, D. Lübben, O. Sauerland, *Phosphorus, Sulfur, and Silicon,* **53**, 109 (1990).

6.2 Solvent-Free C–S Bond Formation under Microwave Irradiation

Type of reaction: C–S bond formation
Reaction condition: solvent-free
Keywords: alkyl halide, sodium phenyl sulfinate, alkylation, alumina, microwave irradiation, sulfone

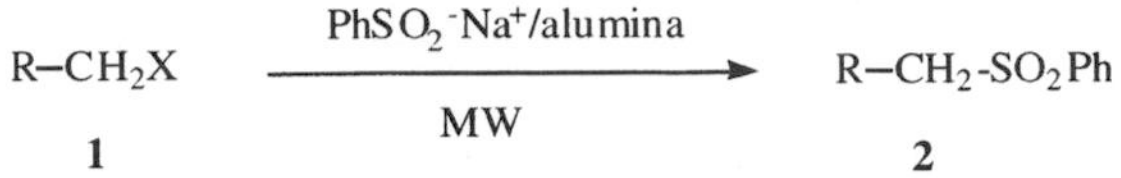

$$\text{R-CH}_2\text{X} \xrightarrow[\text{MW}]{\text{PhSO}_2^-\text{Na}^+/\text{alumina}} \text{R-CH}_2\text{-SO}_2\text{Ph}$$

    1   **2**

a: R=H
b: R=Ph
c: R=CN
d: R=COPh
e: R=COOEt
f: R=CH=CH$_2$

Experimental procedures:

The mixture of sodium phenyl sulfinate (6.566 g, 40 mmol) and neutral chroma-tographic alumina (20 g) was evaporated with a rotary evaporator under vacuum. The solid was dried 2 h at 110 °C under vacuum. The alkyl halide (5 mmol) was absorbed onto sodium phenyl sulifinate on alumina (6.6 g, 1.52 mmol of sodium phenyl sulfinate). The mixture was irradiated by microwave (160 W, 5 min). After cooling at room temperature the mixture was extracted with acetonitrile. The product was recrystallized in ethanol after evaporation of solvent.

References: D. Villemin, A.B. Alloum, *Synth. Commun.,* **20**, 925 (1990).

Type of reaction: C–S bond formation
Reaction condition: solvent-free
Keywords: active methylene compound, *S*-methyl methanesulfonothioate, alumi-na-KF, microwave irradiation, dithioacetal

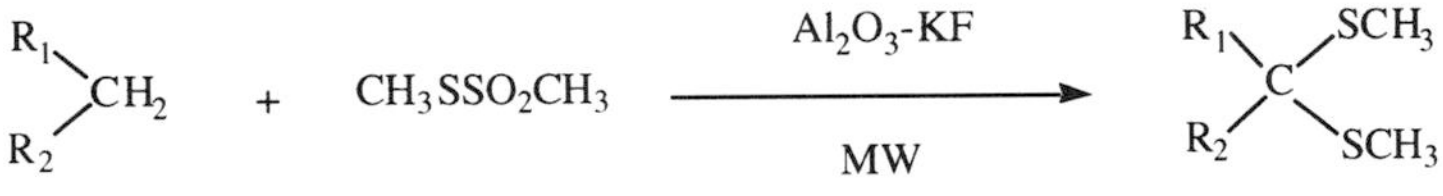

$$\begin{array}{c} \text{R}_1 \\ \diagdown \\ \text{R}_2 \end{array}\!\!\text{CH}_2 \ + \ \text{CH}_3\text{SSO}_2\text{CH}_3 \xrightarrow[\text{MW}]{\text{Al}_2\text{O}_3\text{-KF}} \begin{array}{c} \text{R}_1 \diagdown \ \ \diagup \text{SCH}_3 \\ \text{C} \\ \text{R}_2 \diagup \ \ \diagdown \text{SCH}_3 \end{array}$$

    1   **2**

a: R$_1$=R$_2$=COOEt
b: R$_1$=COOEt; R$_2$=CN
c: R$_1$=COOEt; R$_2$=Ph
d: R$_1$=COOEt; R$_2$=CH$_3$CO
e: R$_1$=COOEt; R$_2$=PO(OEt)$_2$
f: R$_1$=PO(OEt)$_2$; R$_2$=PO(OEt)$_2$

Experimental procedures:

The acidic methylene compound (5 mmol) and *S*-methyl methanesulfonothioate (10 mmol) were adsorbed on Al$_2$O$_3$-KF (5 g or 8 g for the phosphonates) and were either irradiated under microwave in an open Erlenmeyer flask or left at room temperature. The reactions were monitored by TLC on silica gel (eluent

CH$_2$Cl$_2$-methanol, 99:1). The products were extracted with CH$_2$Cl$_2$ (3×20 mL). The solvent was distilled and the residue purified by chromatography (eluent CH$_2$Cl$_2$-methanol, 99:1) or by Kugelrohr distillation.

References: D. Villemin, A.B. Alloum, F. Thibault-Starzyk, *Synth. Commun.*, **22**, 1359 (1992).

Type of reaction: C–S bond formation
Reaction condition: solvent-free
Keywords: carbonyl compound, Lawesson's reagent, microwave irradiation, thioketone, thiolactone, thioamide, thionoester, thioflavonoid

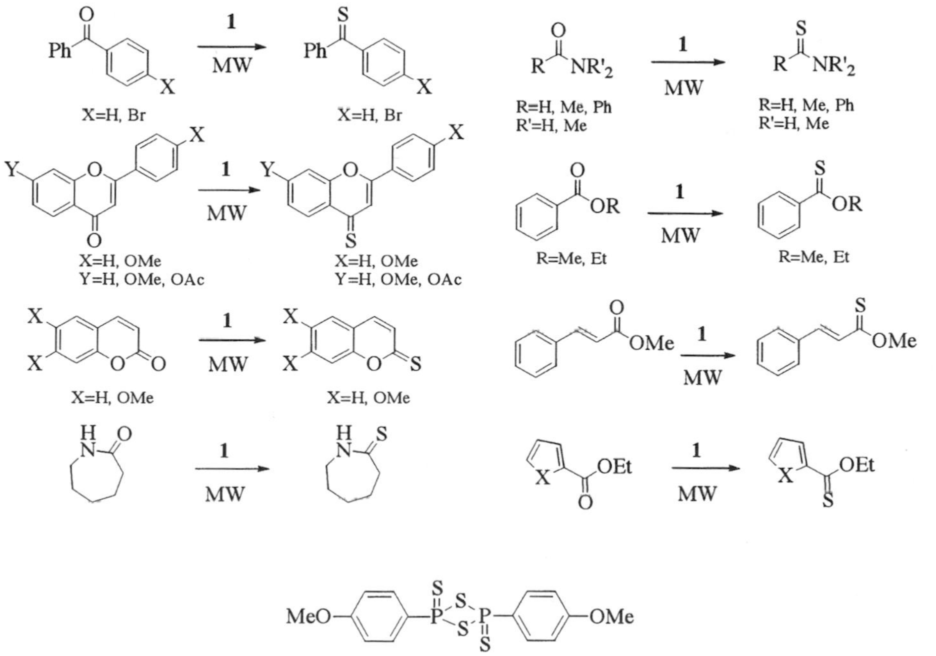

Experimental procedures:
A mixture of 6,7-dimethoxycoumarin (1 mmol) and Lawesson's reagent **1** (0.5 mmol, in the case of esters, 0.8 mmol), was taken in a glass tube and mixed thoroughly with a spatula. The glass tube was then placed in an alumina bath inside the microwave oven (900 W) and irradiated for 3 min. On completion of the reaction, followed by TLC examination, the colored material was dissolved in dichloromethane and adsorbed on silica gel and purified by silica gel column chro-

matography using hexane as initial eluent followed by ethyl acetate-hexane (1:9 v/v) which afforded the pure 6,7-dimethoxythiocoumarin in 94% yield.

References: R. S. Varma, D. Kumar, *Org. Lett.,* **1**, 697 (1999).

Type of reaction: C–S bond formation
Reaction condition: solvent-free
Keywords: 1,3,4-oxadiazole-2-thiol, 1,3,4-thiadiazole-2-thiol, alumina, microwave irradiation, quinolone, sulfide

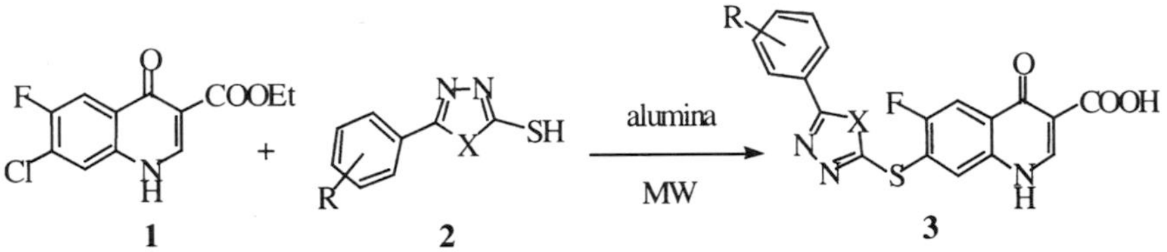

a: X=S; R=H	f: X=O; R=H
b: X=S; R=4-Cl	g: X=O; R=4-Cl
c: X=S; R=2-Cl	h: X=O; R=2-Cl
d: X=S; R=4-NO₂	i: X=O; R=4-NO₂
e: X=S; R=4-CN	j: X=O; R=4-CN

Experimental procedures:
5-Aryl-1,3,4-thiadiazole/oxadiazole-2-thiols **2** (0.01 mol) and 0.01 mol ethyl carboxylate **1** were adsorbed separately on basic alumina. Both adsorbed materials were mixed intimately and irradiated in the microwave oven for 60–90 s. After completion of the reaction the mixture was cooled, and the product was extracted with 4×10 mL CH₂Cl₂. Recovering the solvent under reduced pressure afforded the products **3** which were recrystallized from EtOH-CH₂Cl₂.

References: M. Kidwai, P. Misra, B. Dave, K. R. Bhushan, R. K. Saxena, M. Singh, *Monatsh. Chem.,* **131**, 1207 (2000).

Type of reaction: C–S bond formation
Reaction condition: solvent-free
Keywords: *p*-toluenesulfonyl chloride, aromatic compound, Zn dust, sulfonylation, microwave irradiation, sulfone

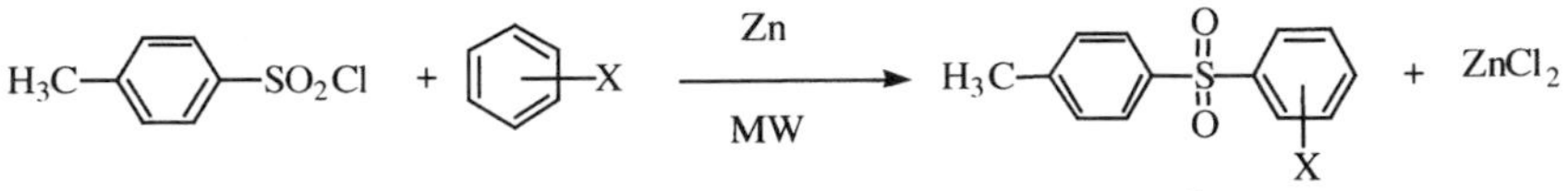

1   2   3

a: X=H
b: X=Ph
c: X=Me
d: X=i-Bu
e: X=OMe
f: X=Cl
g: X=Br
h: X=NO$_2$
i: X=2,4,6-Me$_3$
j: X=2,4-Me$_2$

Experimental procedures:

A mixture of toluene **2c** (5 mmol), *p*-toluenesulfonyl chloride **1** (5 mmol), and Zn dust (5 mmol) was exposed to microwaves for 45 s or heated at 110 °C for 60 min. After completion of reaction (TLC), the product **3** was extracted with ether (3×10 mL). Removal of the solvent under reduced pressure gave product in good yield and in pure form. In most of the cases, the reaction works very well on 25 mmol scale.

References: B.P. Bandger, S.P. Kasture, *Synth. Commun.*, **31**, 1065 (2001).

Type of reaction: C–S bond formation
Reaction condition: solvent-free
Keywords: heterocyclic ketone, P$_2$S$_5$-SiO$_2$, microwave irradiation, heterocyclic thione

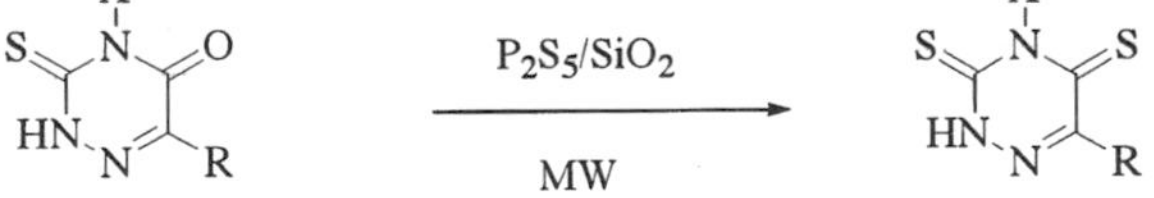

1   2

a: R=Me; X=NH$_2$
b: R=Me; X=H
c: R=H; X=NH$_2$
d: R=H; X=H

3

a: R=Me
b: R=H

Experimental procedures:
Phosphorus pentasulfide (P$_2$S$_5$) and a weight equivalent of silica gel (silica gel 60, particle size 0.2–0.6 mm for column chromatography) were crushed together so as to form an intimate mixture. In a typical reaction 2–3 equiv. of supported P$_2$S$_5$ was mixed with the appropriate heterocycles with oxo group in a beaker. The beaker was exposed to microwave irradiation (900 W) for the indicated time (Table). The residue was extracted with boiling ethanol, the solvent was evaporated to dryness and the crude product was crystallized from an appropriate solvent to afford the corresponding thio heterocycles.

References: M. M. Heravi, G. Rajabzadeh, M. Rahimizadeh, M. Bakavoli, M. Ghassemzadeh, *Synth. Commun.*, **31**, 2231 (2001).

Type of reaction: C–S bond formation
Reaction condition: solvent-free
Keywords: monothiol, dithiol, aldehyde, CdI$_2$, microwave irradiation, dithioacetalization, dithioacetal, 1,3-dithiolane

1   2

a: R$_1$=Ph; R$_2$=H; R$_3$=-(CH$_2$)$_3$-
b: R$_1$=Ph; R$_2$=H; R$_3$=-(CH$_2$)$_2$-
c: R$_1$=Ph; R$_2$=H; R$_3$=Ph
d: R$_1$=4-MeC$_6$H$_4$; R$_2$=H; R$_3$=-(CH$_2$)$_3$-
e: R$_1$=4-ClC$_6$H$_4$; R$_2$=H; R$_3$=-(CH$_2$)$_3$-
f: R$_1$=PhCH=CH; R$_2$=H; R$_3$=PhCH$_2$
g: R$_1$=PhCH=CH; R$_2$=H; R$_3$=-(CH$_2$)$_2$-
h: R$_1$=PhCH=CH; R$_2$=H; R$_3$=Ph
i: R$_1$=CH$_3$CO(CH$_2$)$_4$; R$_2$=H; R$_3$=-(CH$_2$)$_3$-
j: R$_1$=CH$_3$CO(CH$_2$)$_4$; R$_2$=H; R$_3$=-(CH$_2$)$_2$-

Experimental procedures:

A mixture of benzaldehyde (1.06 g, 10 mmol), ethanedithiol (0.94 g, 10 mmol) and commercial grade cadmium iodide (1.85 g, 5 mmol) was mixed thoroughly in an Erlenmeyer flask and placed in a commercial microwave oven operating at 2450 MHz frequency. The whole operation was carried out in a fume-cupboad. After irradiation of the mixture for 75 s (monitored by TLC) it was cooled to room temperature and extracted with dichloromethane. Evaporation of the solvent gave almost pure products. Further purification was achieved by column chromatography on silica gel using chloroform-petroleum ether (1:5) as eluent.

References: D. D. Laskar, D. Prajapati, J. S. Sandhu, *J. Chem. Res. (S),* 313 (2001).

Type of reaction: C–S bond formation
Reaction condition: solvent-free
Keywords: 1,4-dicarbonyl compound, Lawesson's reagent, microwave irradiation, 2-alkoxythiophene, 1,3-thiazole, 1,3,4-thiadiazole

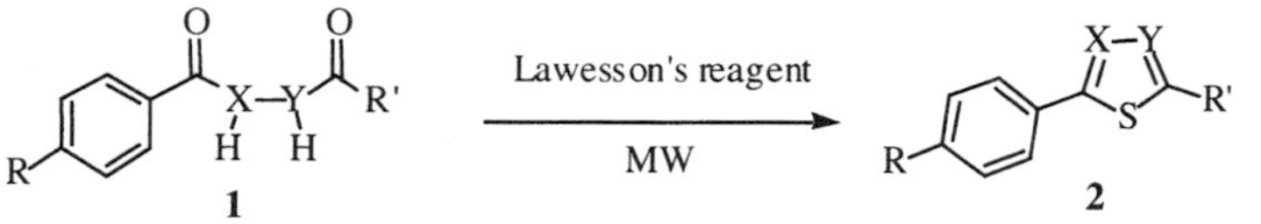

a: R=Br, R'=C_2H_5O; X=Y=CH
b: R=Br; R'=n-C_4H_9O; X=Y=CH
c: R=Br; R'=n-$C_6H_{13}O$; X=Y=CH
d: R=Br; R'=n-$C_8H_{17}O$; X=Y=CH
e: R=Br; R'=$C_6H_{13}CH(CH_3)O$; X=Y=CH
f: R=Br; R'=n-$C_{10}H_{21}O$; X=Y=CH

g: R=Br; R'=n-$C_{12}H_{25}O$; X=Y=CH
h: R=CH_3O; R'=4-BrC_6H_4O; X=Y=CH
i: R=H; R'=Ph; X=Y=CH
j: R=Br; R'=n-$C_{12}H_{25}O$; X=N; Y=CH
k: R=Br; R'=n-$C_{13}H_{27}$; X=Y=N
l: R=Br; R'=n-$C_{13}H_{27}$; X=Y=N

Experimental procedures:

The 1,4-dicarbonyl compound (1.0 mmol) and Lawesson's reagent (0.486 g, 1.20 mmol) were mixed thoroughly in a glass reaction tube. The tube was placed in a beaker with a cover glass and irradiated using a commercial conventional microwave oven with a rotating tray (Samsung, MW6940W, 1000 W) for 3–6 min until the evolution of gas ceased and the reaction mixture became an orange or yellow transparent liquid. It was then allowed to cool, and the resulting viscous mixture was dissolved in CH_2Cl_2 and evaporated on silica gel. Flash column chromatography on activated basic alumina (50–200 mm) or silica (200–425 mesh) provided the corresponding five-membered ring *S*-heterocycle, which was dried under vacuum (1.1 mm Hg, P_2O_5, 24 h). Caution: these reactions must be performed in an efficient fume hood due to the generation of toxic H_2S gas.

References: A. A. Kiryanov, P. Sampson, A. J. Seed, *J. Org. Chem.,* **66**, 7925 (2001).

7 Carbon–Phosphorus Bond Formation

7.1 Solvent-Free C–P Bond Formation

Type of reaction: C–P bond formation
Reaction condition: solvent-free
Keywords: diethyl hydrogen phosphite, aromatic aldehyde, magnesia, surface-mediated reaction, diethyl 1-hydroxyarylmethylphosphonate

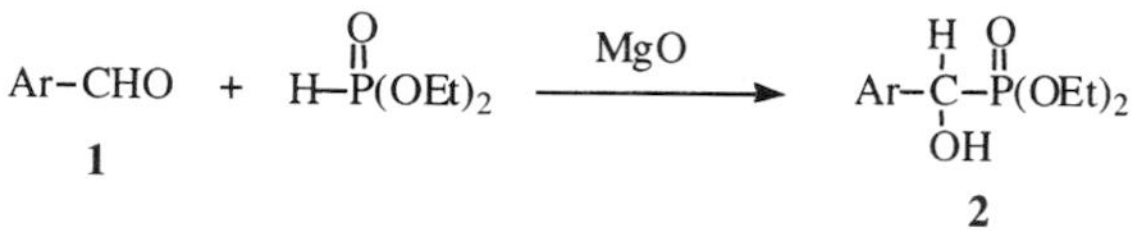

$$Ar\text{-CHO} + H\text{-}\overset{O}{\overset{\|}{P}}(OEt)_2 \xrightarrow{\quad MgO \quad} Ar\text{-}\overset{H}{\underset{OH}{\overset{|}{\underset{|}{C}}}}\text{-}\overset{O}{\overset{\|}{P}}(OEt)_2$$

1   **2**

```
a: Ar=Ph
b: Ar=4-MeC₆H₄
c: R=4-MeOC₆H₄
d: R=3-MeOC₆H₄
e: R=3-HOC₆H₄
f: R=4-ClC₆H₄
g: R=2-ClC₆H₄
h: R=4-O₂NC₆H₄
```

a: Ar=Ph
b: Ar=4-MeC_6H_4
c: R=4-$MeOC_6H_4$
d: R=3-$MeOC_6H_4$
e: R=3-HOC_6H_4
f: R=4-ClC_6H_4
g: R=2-ClC_6H_4
h: R=4-$O_2NC_6H_4$

Experimental procedures:

Magnesia (0.8 g, 0.02 mol) was added to a mixture of diethyl hydrogen phosphite (2.76 g, 0.02 mol) and the aldehyde (0.02 mol). This mixture was stirred at room temperature for 2 min–4 h. The solid mixture was washed with CH_2Cl_2 (4×25 mL) and the crude product was isolated in a pure state by simple filtration chromatography through a short plug of silica gel and then was crystallized from CH_2Cl_2-hexane. For solid aldehydes, prior to addition of magnesia, the mixture of diethyl hydrogen phosphite and the aldehyde must be stirred and heated at 60 °C.

References: A. R. Sardarian, B. Kaboudin, *Synth. Commun.*, **27**, 543 (1997).

7.2   Solvent-Free C–P Bond Formation under Microwave Irradiation

Type of reaction: C–P bond formation
Reaction condition: solvent-free
Keywords: primary amine, diethyl phosphite, aldehyde, alumina, microwave irradiation, 1-aminoalkyl phosphonate

$$\underset{\textbf{1}}{R-\overset{O}{\overset{\|}{C}}-H} \quad + \quad \underset{\textbf{2}}{R'-NH_2} \quad + \quad H-\overset{O}{\overset{\|}{P}}(OEt)_2 \quad \xrightarrow[\text{MW}]{Al_2O_3} \quad \underset{\underset{NHR'}{\overset{\textstyle \quad}{}}}{\underset{\textbf{3}}{R-\overset{H}{\overset{\|}{C}}-\overset{O}{\overset{\|}{P}}(OEt)_2}}$$

a: R=R'=Ph
b: R=p-MeC$_6$H$_4$; R'=Ph
c: R=p-Me$_2$CHC$_6$H$_4$; R'=Ph
d: R=p-NO$_2$C$_6$H$_4$; R'=Ph
e: R=n-C$_4$H$_9$; R'=Ph
f: R=C$_6$H$_5$CH=CH; R'=Ph
g: R=p-MeC$_6$H$_4$; R'=m-NO$_2$C$_6$H$_4$
h: R=p-Me$_2$CHC$_6$H$_4$; R'=m-NO$_2$C$_6$H$_4$
i: R=Ph; R'=m-NO$_2$C$_6$H$_4$
j: R=Ph; R'=cyclohexyl
k: R=PhCH=CH; R'=cyclohexyl
l: R=PhCH=CH; R'=HOCH$_2$CH$_2$
m: R=Ph; R=HOCH$_2$CH$_2$

Experimental procedures:

30 mmol of amine was added to a mixture of an aldehyde (30 mmol) and alumina (Al$_2$O$_3$, acidic, 5.75 g). Diethyl phosphite was added, then the mixture was irradiated by microwave for 3–6 min using 720 W (a kitchen-type microwave was used in all experiments). The reaction mixture was extracted with CH$_2$Cl$_2$ (200 mL), the extracts were dried (Na$_2$SO$_4$) and the solvent was evaporated to give the crude products. Pure products were obtained by crystallization from CH$_2$Cl$_2$-n-hexane or by distillation under reduced pressure in 70–95% yield.

References: B. Kaboudin, R. Nazari, *Tetrahedron Lett.*, **42**, 8211 (2001).

Type of reaction: C–P bond formation
Reaction condition: solvent-free
Keywords: aldehyde, diethyl phosphite, alumina/ammonium formate, microwave irradiation, 1-aminophosphonate

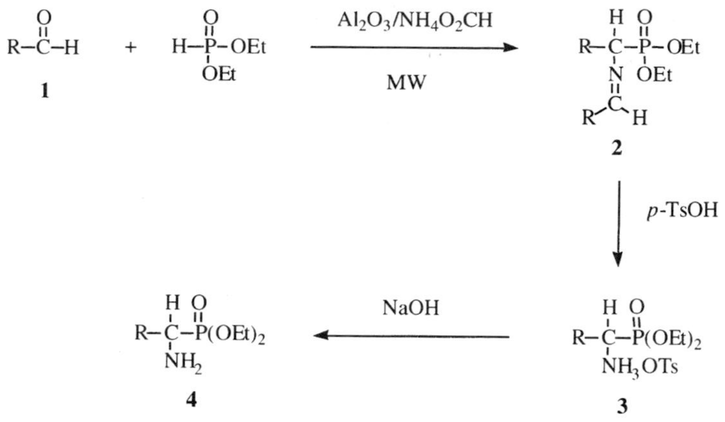

$$R-\overset{\overset{O}{\|}}{C}-H \quad + \quad H-\overset{\overset{O}{\|}}{\underset{OEt}{P}}-OEt \quad \xrightarrow[\text{MW}]{Al_2O_3/NH_4O_2CH} \quad R-\overset{\overset{H}{|}}{\underset{\overset{|}{N}}{C}}-\overset{\overset{O}{\|}}{\underset{OEt}{P}}-OEt$$

1 **2**

$$R-\overset{\overset{H}{|}}{\underset{NH_3OTs}{C}}-P(OEt)_2 \quad \xleftarrow{\;} \text{(from 2 via } p\text{-TsOH)}$$

down arrow labeled *p*-TsOH

$$R-\overset{\overset{H}{|}}{\underset{NH_2}{C}}-\overset{\overset{O}{\|}}{P}(OEt)_2 \quad \xleftarrow{\text{NaOH}} \quad R-\overset{\overset{H}{|}}{\underset{NH_3OTs}{C}}-\overset{\overset{O}{\|}}{P}(OEt)_2$$

4 **3**

$R = CH_3, n\text{-}C_4H_9, n\text{-}C_5H_{11}, C_6H_5, p\text{-}CH_3OC_6H_4, o\text{-}ClC_6H_4, m\text{-}ClC_6H_4, p\text{-}ClC_6H_4,$
$p\text{-}FC_6H_4, m\text{-}CH_3C_6H_4, Ph\text{-}CH=CH\text{-}, furfuryl, \alpha\text{-naphthyl}, \beta\text{-naphthyl}$

Experimental procedures:

The reagent (30 mmol) was prepared by the combination of ammonium formate (30 mmol, finely ground) and alumina (Al_2O_3, acidic, 5.75 g) in a mortar by grinding them together until a fine, homogeneous powder is obtained (5–10 min). The aldehyde (60 mmol) is added to this mixture and the whole was irradiated by microwave for 3–6 min using 720 W (a kitchen-type microwave was used in all experiments). The reaction mixture is washed with diethyl ether (200 mL). *p*-TsOH-H_2O (30 mmol) was added to the etherial solution with stirring. After completion of the reaction (1 h), the solid was filtered and neutralized with NaOH (10%). Chromatography through a plug of silica gel with EtOAc-*n*-hexane (1:9) and evaporation of the solvent under reduced pressure gave the pure product as oil in 63–78% yields.

References: B. Kaboudin, *Chem. Lett.*, 880 (2001).

Type of reaction: C–P bond formation
Reaction condition: solvent-free
Keywords: alkyl chloride, trialkyl phosphite, alumina, microwave irradiation, Arbuzov rearrangement, dialkyl alkylphosphonate

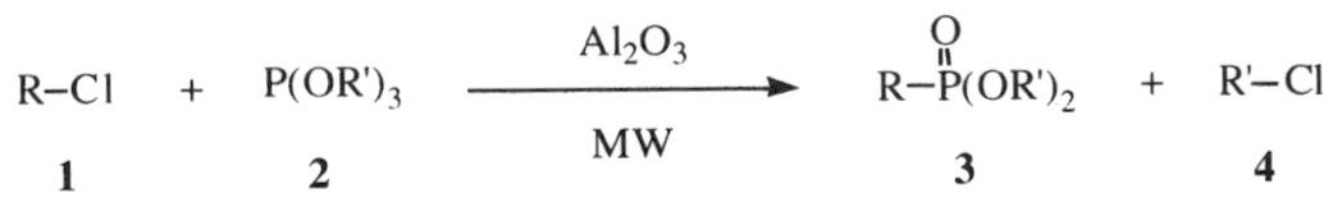

$$\text{R–Cl} \quad + \quad \text{P(OR')}_3 \quad \xrightarrow[\text{MW}]{\text{Al}_2\text{O}_3} \quad \text{R–}\overset{\overset{\text{O}}{\|}}{\text{P}}\text{(OR')}_2 \quad + \quad \text{R'–Cl}$$

  1   **2**   **3**   **4**

a: R=PhCH$_2$; R'= Et
b: R=CCl$_3$; R'= Et
c: R=CH$_2$=CHCH$_2$; R'= Et
d: R=PhCOCH$_2$CH$_2$; R'= Et
e: R=t-Bu; R'= Et
f: R=PhCH$_2$; R'= Me
g: R=CCl$_3$; R'= Me
h: R=CH$_2$=CHCH$_2$; R'= Me
i: R=PhCOCH$_2$CH$_2$; R'= Me
j: R=t-Bu; R'= Me

Experimental procedures:

Alumina (Al$_2$O$_3$, neutral, 2 g) was added to a mixture of trialkyl phosphite (2.5 mmol) and the alkyl chloride (2 mmol). This mixture was irradiated by microwave for 5–10 min. The reaction mixture was washed with CH$_2$Cl$_2$ (200 mL), dried (CaCl$_2$), and the solvent evaporated to give the crude products. Pure product was obtained by distillation under reduced pressure in 73–90% yields.

References: B. Kaboudin, M.S. Balakrishna, *Synth. Commun.,* **31**, 2773 (2001).

8 Carbon–Halogen Bond Formation

8.1 Solvent-Free C–X Bond Formation

Type of reaction: C-halogen bond formation
Reaction condition: solid-state
Keywords: cinnamic acid, stilbene, halogenation, addition, gas-solid reaction

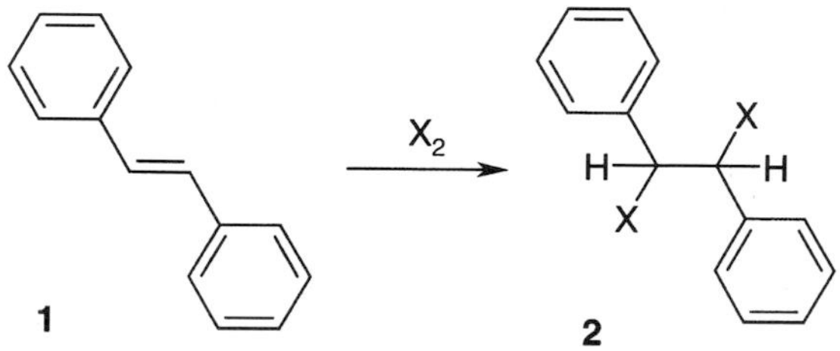

	meso/dl	yield
a: X=Br pulverized crystals:	62:38	20%
milled crystals:	25:75	100%
b: X=Cl pulverized crystals:	39:61	39%

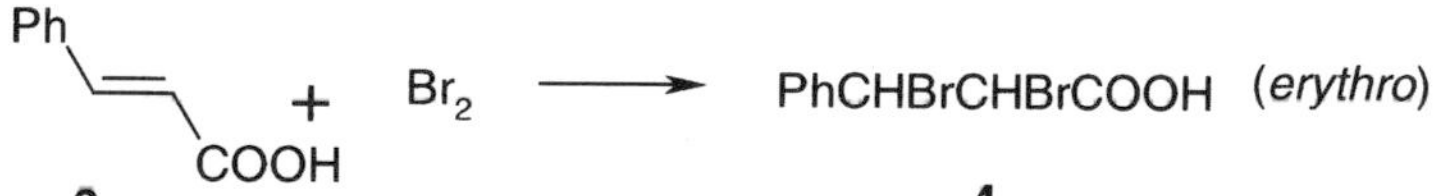

Experimental procedures:

2: 1.00 g (5.6 mmol) of powdered *trans*-stilbene **1** was treated with Cl_2 gas (500 mL, 22 mmol) at room temperature for 6 h. After removing excess Cl_2 gas, the reaction mixture was chromatographed on silica gel (benzene-*n*-hexane, 1:10) to give a 61:39 mixture of *dl*- and *meso*-**2b** (0.55 g, 39%) along with recovered **1** (0.60 g, 60%). The yield of **2b** was increased to 100% and the cis-addition strongly enhanced, if the crystals of **1** were ball-milled to µm size prior to a very slow addition of the Cl_2 at 0°C (0.05 bar, 3 h) and then 0.5 bar Cl_2 (24 h).

Similar results were obtained if powdered or ball-milled *trans*-stilbene **1** was treated with the stoichiometric amount of Br_2 gas to give *meso*- and *dl*-**2a**.

4: Pulverized *trans*-cinnamic acid **3** (2.0 mmol, α or β-modification) and Br$_2$ gas (5 mmol) gave a 100% yield of the erythro-adduct 4.

References: G. Kaupp, D. Matthies, *Chem. Ber.*, **120**, 1897 (1987); *Mol. Cryst. Liq. Cryst.*, **161**, 119 (1988); G. Kaupp, *Mol. Cryst. Liq. Cryst.*, **242**, 153 (1994); G. Kaupp, A. Kuse, *Mol. Cryst. Liq. Cryst.*, **313,** 361 (1998).

Type of reaction: C–halogen bond formation
Reaction condition: solid-state
Keywords: phenol, *N*-bromosuccinimide, bromination

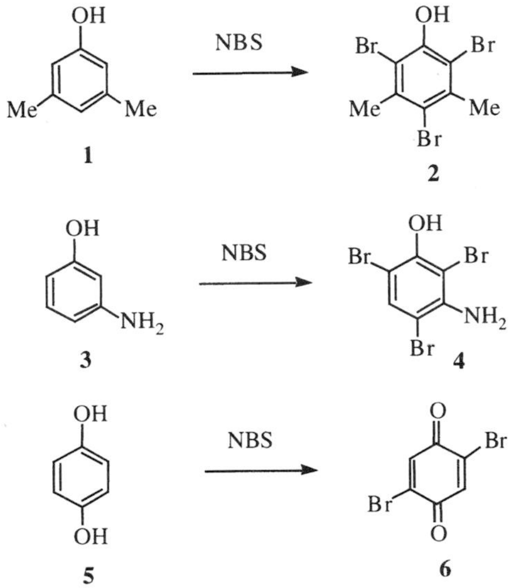

Experimental procedures:
In a typical experiment, the phenol (1 mol) and freshly crystallized NBS (3 mol) were ground in a mortar. The color of the mixture changed immediately from white to yellow-brown. The solid mixture was treated with CCl$_4$ after an appropriate time (ranging from 1 min to 2 h), the succinimide removed by filtration and the product chromatographed on silica gel (hexane-EtOAc).

References: B. S. Goud, G. R. Desiraju, *J. Chem. Res. (S)*, 244 (1995).

Type of reaction: C–halogen bond formation
Reaction condition: solid-state
Keywords: N-bromosuccinimide, aniline, phenol, bromination

$$\text{Y–C}_6\text{H}_3(\text{X})(\text{NH}_2) \quad \xrightarrow{\text{NBS}} \quad \text{Y–C}_6\text{H}_2(\text{X})(\text{NH}_2)(\text{Br})$$

1 → **2**

a: X=Me; Y=Br
b: X=NO$_2$; Y=OMe
c: X=Y=Cl

$$\text{X–C}_6\text{H}_3(\text{Y})(\text{OH}) \quad \xrightarrow{\text{NBS}} \quad \text{X–C}_6\text{H}(\text{Y})(\text{OH})(\text{Br})_2$$

1 → **2**

a: X=COMe; Y=H
b: X=NO$_2$; Y=H
c: X=Cl; Y=Me

Experimental procedures:
In general the substrate (ca. 1.0 g) and freshly powdered NBS (1:1 molar equivalents) were mixed very gently for a few seconds. After the specified reaction
time the mixture was dissolved in petroleum ether and ethyl acetate (10:1) and
the mixture was separated using column chromatography. Either grinding of the
mixture or monitoring the reaction for a longer time usually resulted in a paste,
especially if very low melting reactants or products were involved. The reaction
being highly exothermic, reaction temperature and time were varied not only to
ensure crystallinity but also to optimize the selectivity and yield. All products
were characterized with NMR and mass spectra.

References: J.A.R.P. Sarma, A. Nagaraju, *J. Chem. Soc., Perkin Trans. 2*, 1113 (2000);
J.A.R.P. Sarma, A. Nagaraju, K.K. Majumdar, P.M. Samuel, I. Das, S. Roy, A.J.
McGhie, *J. Chem. Soc., Perkin Trans. 2*, 1119 (2000).

Type of reaction: C–halogen bond formation
Reaction condition: solid-state
Keywords: N-vinyl compounds, S-vinyl compounds, halogenation, addition, gas-
solid reaction

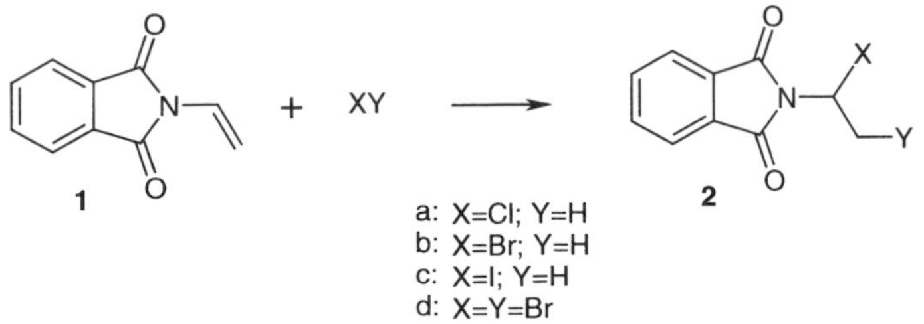

1   +   XY $\longrightarrow$ 2

a: X=Cl; Y=H
b: X=Br; Y=H
c: X=I; Y=H
d: X=Y=Br

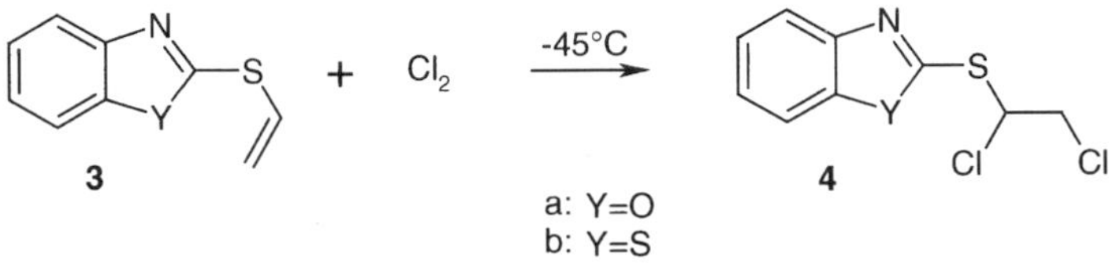

3   +   Cl$_2$ $\xrightarrow{-45°C}$ 4

a: Y=O
b: Y=S

Experimental procedures:

2: Ground crystals of *N*-vinylphthalimide **1** (100 mg, 0.58 mmol) were exposed to HBr gas (100 mL, 1 bar). The excess gas was evacuated after 30 min and a quantitative yield of **2b** was obtained. HCl and HI reacted similarly to give **2a** and **2c**, while the quantitative addition of Br$_2$ took 16 h to produce quantitatively **2d**.

The analogous reactions (**a–d**) were obtained with *N*-vinylsaccharin instead of **1** on a gram scale.

4: The liquid **3a** or **3b** (0.50 g, 2.8 or 2.6 mmol) was spread over Raschig coils (2 g) in a 500-mL flask and it crystallized after evacuation at –45 °C. Cl$_2$ gas (0.5 bar, 11 mmol) was let in. After 5 h, excess gas was evaporated and the yellow solid **4a** or **4b** obtained directly pure with quantitative yield.

References: G. Kaupp, D. Matthies, *Chem. Ber.*, **119**, 2387 (1986); *Mol. Cryst. Liq. Cryst.*, **161**, 119 (1988); G. Kaupp, D. Lübben, O. Sauerland, *Phosphorus, Sulfur, and Silicon*, **53**, 109 (1990).

Type of reaction: C–halogen bond formation
Reaction condition: solid-state
Keywords: 1,1-bisarylethylene, triphenylethylene, halogenation, stilbene, elimination, gas-solid reaction, 1-halogeno-2,2-diarylethene, 1-halogeno-1,2,2-triphenylethene

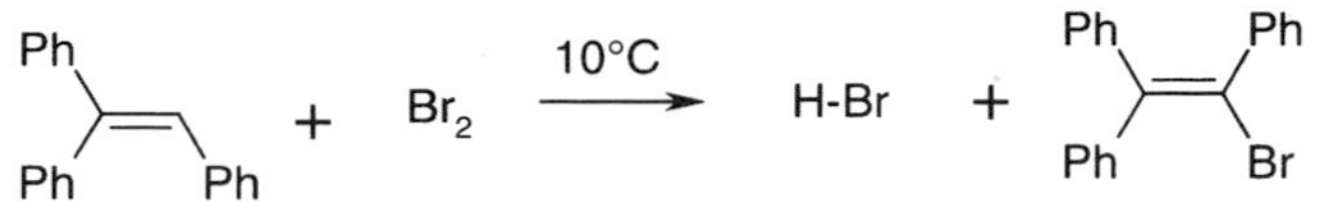

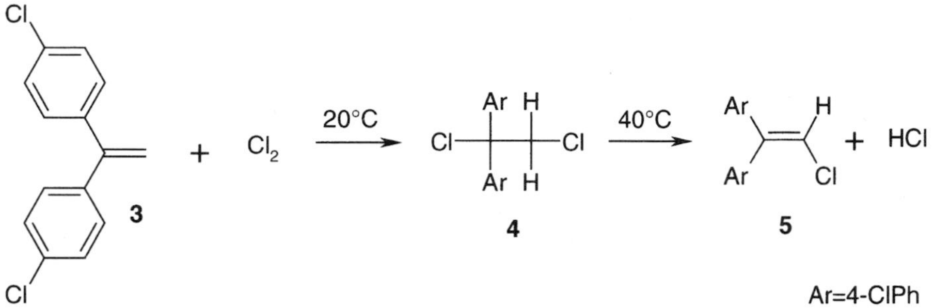

Ar=4-ClPh

Experimental procedures:

Powdered triphenylethylene **1** (1.00 g, 3.9 mmol) or 1,1-bis-(4-chlorophenyl)-eth-ylene **3** (1.37 g, 5.5 mmol) was treated with Br2 gas (4.24g, 26.5 mmol from a connected flask) or 500 mL Cl$_2$ gas (1 bar, 22 mmol) at 10 °C for 64 h (in the dark) or at 20 °C for 3 h. The intermediate 1,2-dihalogeno-1,1-diarylethanes eliminated quantitatively HBr or HCl (**4** at 40 °C, 3h) to give a 94% or a 100% yield of the products **2** or **5**. Crystalline **5** was obtained pure, **2** was separated from residual **1** by crystallization from *n*-heptane.

Stirring would certainly have enhanced both the rate of reaction and the yield in the case of **1**.

References: G. Kaupp, D. Matthies, *Chem. Ber.,* **120**, 1897 (1987); *Mol. Cryst. Liq. Cryst.,* **161**, 119 (1988); G. Kaupp, A. Kuse, *Mol. Cryst. Liq. Cryst.,* **313**, 361 (1998).

Type of reaction: C–halogen bond formation
Reaction condition: solid-state
Keywords: tetraphenylethylene, Br$_2$, substitution, autocatalysis, waste free, gas-solid reaction

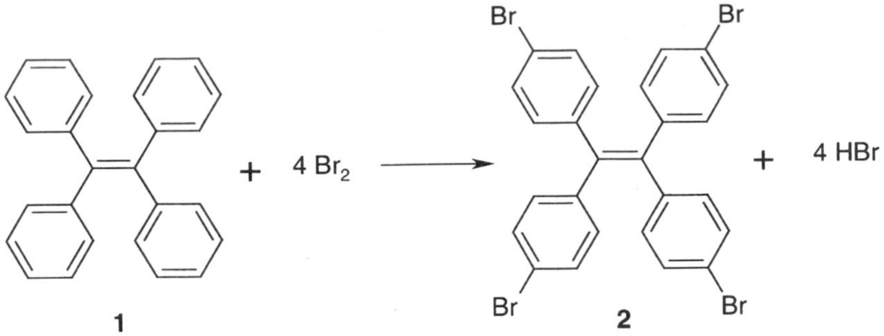

Experimental procedure:
Tetraphenylethylene **1** (1.00 g, 3.0 mmol) was treated with Br$_2$ (3.2 g, 20 mmol) in a previously evacuated 2-L flask that was rotated around a horizontal axis at room temperature for 12 h. The HBr/Br$_2$-mixture was condensed at a vacuum line to a recipient at −196 °C from where pure HBr was distilled off during thawing to −78 °C under vacuum for further use of the gases. The solid product **2** (1.95 g, 100%) was obtained in pure form.

References: G. Kaupp, A. Kuse, *Mol. Cryst. Liq. Cryst.*, **313**, 361 (1998).

Type of reaction: C–halogen bond formation
Reaction condition: solid-state
Keywords: cholesterol ester, stereospecific halogenation, waste-free, gas-solid reaction

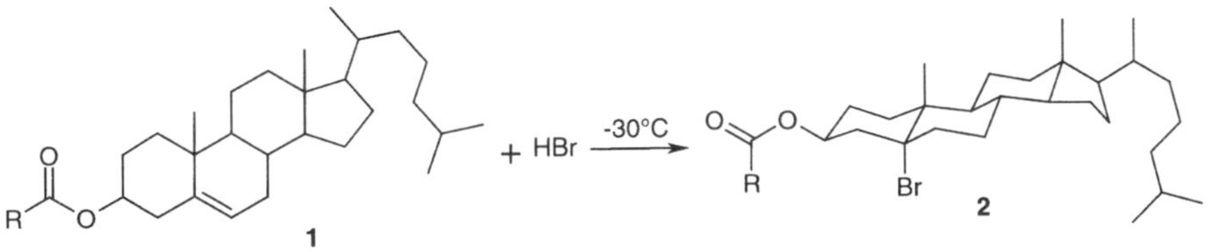

a: R= (CH$_2$)$_7$CH=CH(CH$_2$)$_7$CH$_3$ (*cis*) (10 h)
b: R= Me (6 days)

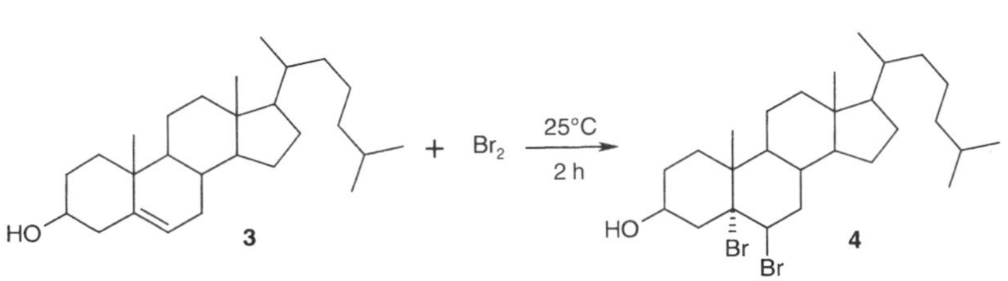

Experimental procedures:
Powdered cholesterol oleate **1a** (3.00 g, 4.6 mmol) was evacuated in a 500 mL flask and cooled to –30 °C. HBr gas (1 bar, 22 mmol) was let in. After 10 h, excess gas was pumped off and condensed in a cold trap for further use. The quantitatively obtained crystals **2a** were collected in pure form.

Similarly, quantitative yields of **2b** and **4** were obtained.

References: G. Kaupp, C. Seep, *Angew. Chem.*, **100**, 1568 (1988); *Angew. Chem. Int. Ed. Engl.*, **27**, 1511 (1988).

Type of reaction: C–halogen bond formation
Reaction condition: solid-state
Keywords: epoxide, stereoselective, addition, ring opening, gas-solid reaction

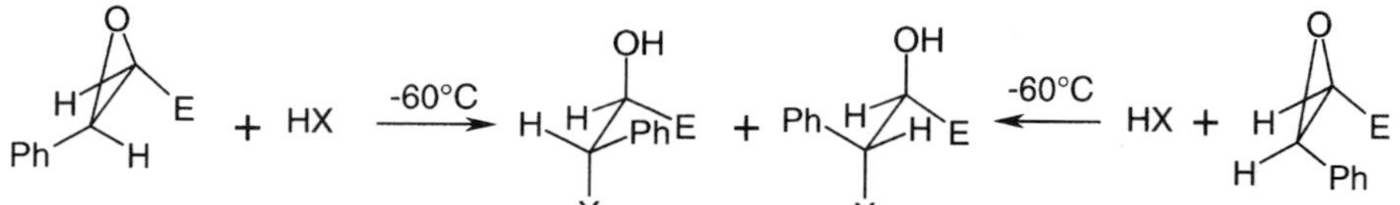

rac-**1** **2** (*threo*) **3** (*erythro*) *rac*-**4**

$E=CO_2Me$ a: X=Cl
b: X=Br

	front side	backside
1a	**2a** (73)	**3a** (27)
1b	**2b** (74)	**3b** (26)
4a	**3a** (71)	**2a** (29)
4b	**3b** (65)	**2b** (35)
5	**6** (74)	**7** (26)

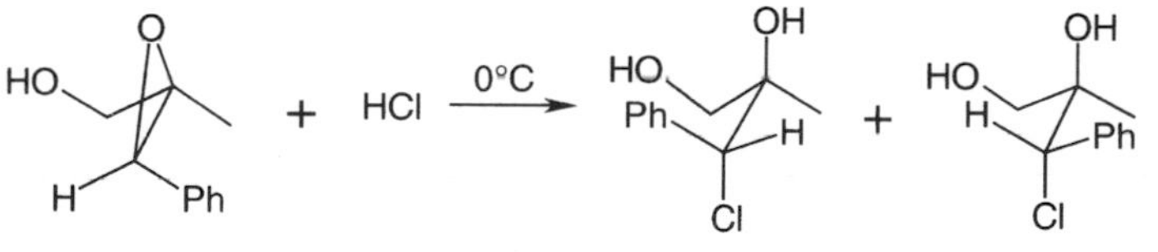

5 (2*R*, 3*R*) **6** (2*R*, 3*R*) **7** (2*R*,3*S*)

Experimental procedures:
Liquid *rac*-**1** (500 mg, 2.8 mmol) was crystallized by cooling to –60 °C in a 100-mL flask under vacuum. HCl gas (1 bar, 4.5 mmol) was let in through a vacuum line. After 15 h at –60 °C the excess gas was pumped off and 600 mg (100%) yellow crystals (73:27 mixture of **2a** and **3a**) *were obtained that melted at room temperature.* **2a** (mp 65–67 °C) was obtained pure by crystallization from *n*-hexane, though involving heavy losses.

Similarly, the reactions of *rac*-**1** and HBr, or *rac*-**4** and HCl or HBr, or **5** (2*R*,3*R*) and HCl gave the selectivities as indicated in complete conversions without side-products.

References: G. Kaupp, A. Ulrich, G. Sauer, *J. prakt. Chem.,* **334**, 383 (1992).

Type of reaction: C–halogen bond formation
Reaction condition: solid-state
Keywords: solid diazonium salt, potassium iodide, iodination, solid-solid reaction, aryl iodide

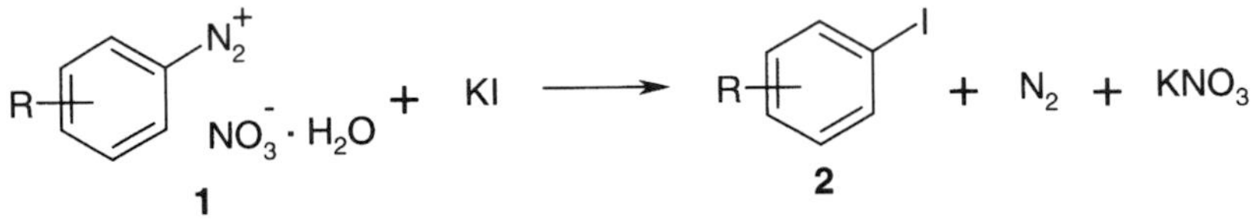

a: R= 4-NO$_3$
b: R= 4-COOH
c: R= 4-Br
d: R= 2,3-(CO-C$_6$H$_4$-CO)
e: R= 4-CN
f: R= 2-COOH

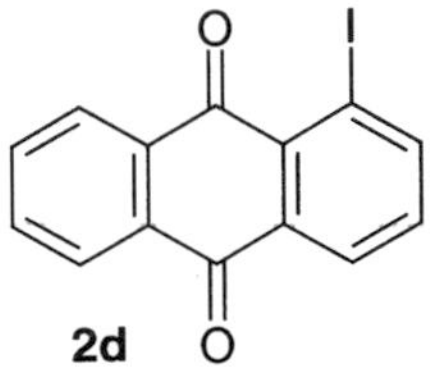

2d

Experimental procedures:
Caution: Solid diazonium salts are heat- and shock-sensitive; do not ball-mill!

Potassium iodide (830 mg, 5.0 mmol) was finely ground in an agate mortar and the diazonium salt **1** (0.50 mmol) added in five portions and co-ground for 5 min, each. After a 24 h rest with occasional grinding, the diazonium band in the IR spectra had completely disappeared. The potassium salts were removed by washings with cold water. The yield of pure aryl iodide **2** was 100% throughout.

References: G. Kaupp, A. Herrmann, J. Schmeyers, *Chem. Eur. J.,* **8**, 1395 (2002).

Type of reaction: C–halogen bond formation
Reaction condition: solid-state
Keywords: camphene, rearrangement, stereospecific, waste-free, gas-solid reaction, isobornyl halide

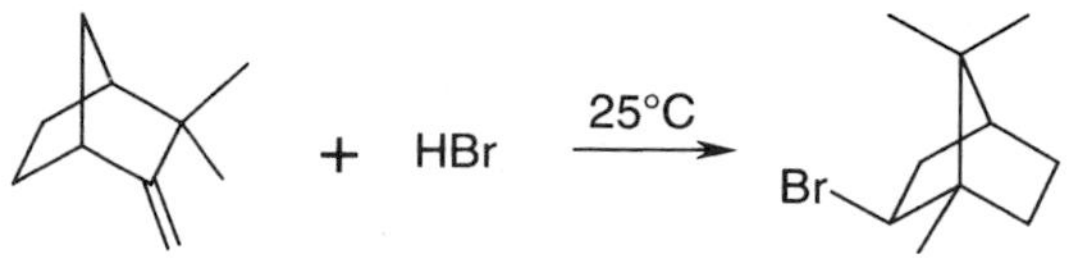

Experimental procedures:

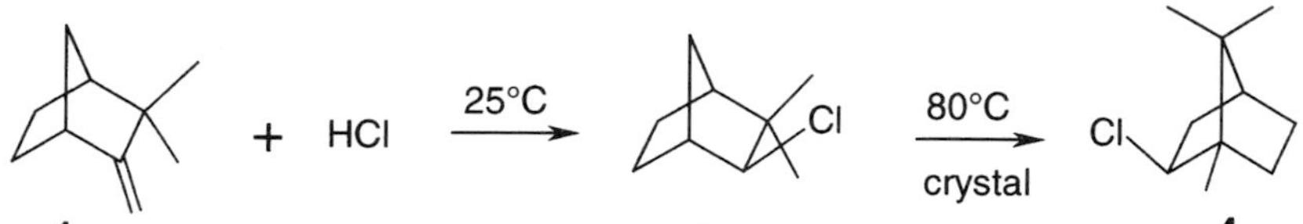

2: Camphene **1** (mp 51–52 °C; 1.09 g, 8.0 mmol) was treated with HBr gas (0.5 bar, 11.2 mmol) at 25 °C for 10 h in an evacuated 50-mL flask that was connected to a 500-mL flask. Excess gas was recovered in a cold trap at −196 °C. The solid isobornylbromide **2** was quantitatively obtained.

3: Camphene **1** (1.09 g, 8.0 mmol) was treated with HCl gas (0.5 bar, 11.2 mmol) at 25 °C for 10 h in an evacuated 50-mL flask that was connected to a 500-mL flask. Excess gas was recovered in a cold trap at −196 °C. The pure solid camphene hydrochloride **3** was quantitatively obtained.

4: The Wagner Meerwein rearrangement of **3** to **4** occurred rapidly in solution but slowly in the crystal where it was complete after three years at room temperature or in 6 h at 80 °C.

References: G. Kaupp, J. Schmeyers, J. Boy, *Chemosphere*, **43**, 55 (2001).

Type of reaction: C–halogen bond formation
Reaction condition: solid-state
Keywords: dimethylthiocarbamoyl compounds, chlorination, gas-solid reaction, solid-solid reaction, Viehe salt

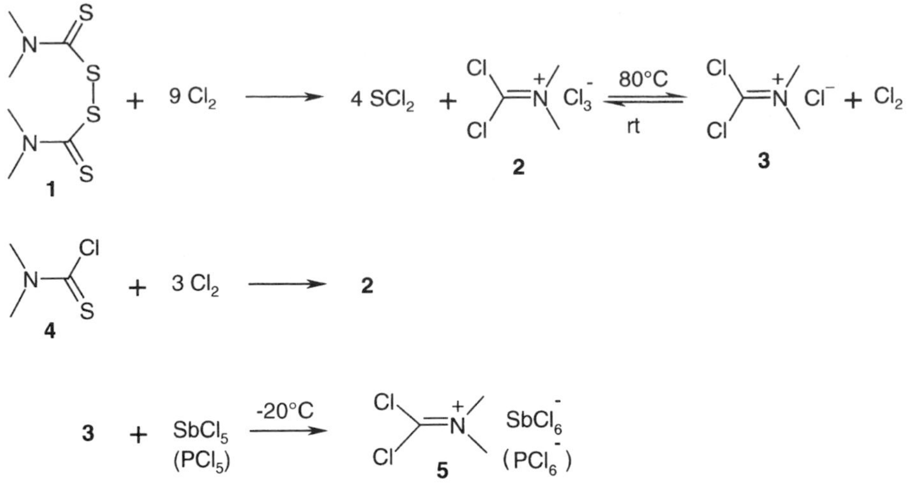

Experimental procedures:

3: Tetramethylthiocarbamoyl disulfide **1** (1.20 g, 5.00 mmol) or dimethyl-thiocarb-amoylchloride **4** (1.24 g, 10.0 mmol) were reacted in an evacuated 1-L flask with Cl_2 (1 bar, 45 mmol). After 10 h, all Cl_2 was consumed to form solid **2** with in-cluded SCl_2. This product was heated to 80 °C for 2 h in a vacuum with a cold trap (−196 °C) condensing the liberated Cl_2 and SCl_2. The yield of pure **3** was 1.62 g (100%) in both cases.

    5: Solid $SbCl_5$ (598 mg, 2.00 mmol) or PCl_5 (417 mg, 2.00 mmol) was weighed to a 10-mL steel beaker of a ball-mill under N_2 and the balls were added. The beaker was closed and cooled to −20 °C prior to the addition of Viehe salt **3** (325 mg, 2.00 mmol) that was previously cooled to −20 °C under N_2. The closed beaker was connected to the cooling system (−20 °C) and milling was performed at −20 °C for 1 h. The highly reactive **5** was quantitatively obtained and should be handled under protecting gas throughout.

References: G. Kaupp, J. Boy, J. Schmeyers, *J. Prakt. Chem., 340,* 346 (1998).

9 Nitrogen–Nitrogen Bond Formation

9.1 Solvent-Free N–N Bond Formation

Type of reaction: N–N bond formation
Reaction condition: solid-state
Keywords: solid diazonium salt, anilines, diphenylamine, gas-solid reaction, triazenes

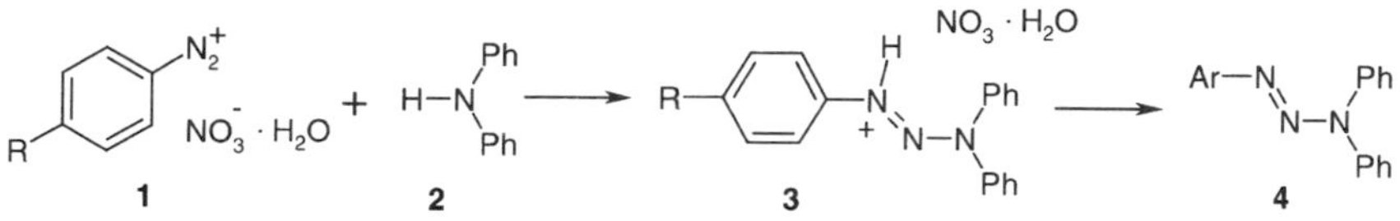

a: R=NO$_2$ (3 h)
b: R=COOH (6 h)
c: R=Cl   (3.5 h)
d: R=Br   (3 h)

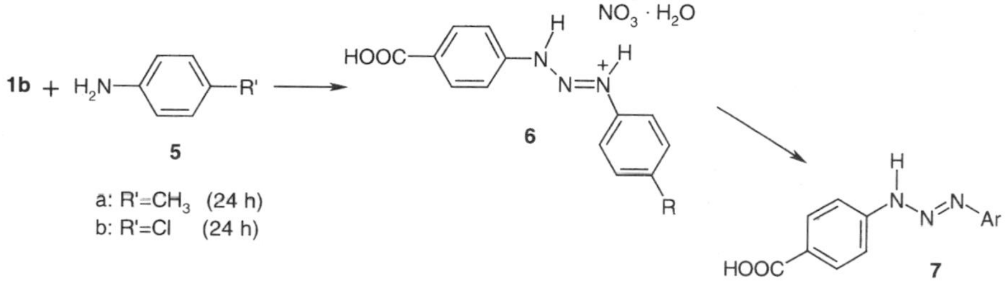

a: R'=CH$_3$ (24 h)
b: R'=Cl   (24 h)

Experimental procedures:

Caution: Solid diazonium salts are heat- and shock-sensitive; do not ball-mill solid diazonium salts!

Diphenylamine **2** (1.00 mmol) or substituted aniline **5a,b** (1.00 mmol) was ground in an agate mortar. The diazonium salt **1** (1.00 mmol) was added in five portions and co-ground for 5 min. To complete the reaction, the solid mixture was transferred to a test tube and then exposed to ultrasound in a cleaning bath, the temperature of which was maintained at 20–25 °C for the time given, when all of the diazonium band in the IR had disappeared. The triazenium salts **3** or **6** were obtained quantitatively. The free triazene bases **4** or **7** were obtained by trituration of their salts with 0.1 n NaOH (20 mL), filtering, washing (H$_2$O) and drying. The yield was >99% in all cases.

References: G. Kaupp, A. Herrmann, J. Schmeyers, *Chem. Eur. J.*, **8**, 1395 (2002).

Type of reaction: N–N bond formation
Reaction condition: solid-state
Keywords: solid diazonium salt, amine, gas-solid reaction, triazene

a: R=4-COOH
b: R=4-Br
c: R=4-Cl
d: R=4-NO$_2$
e: R=3-NO$_2$
f: R=2-COOH
g: R=4-CN
h: R=4-SO$_3$H (as zwitterion)
i: R=2-SO$_3$H (as zwitterion)
j: R=4-(4-C$_6$H$_4$N$_2^+$ NO$_3^-$·H$_2$O)

Experimental procedures:
Caution: These reactions may occur violently, use protecting shield!
    The diazonium nitrate (**1a–g**, 0.70 mmol, **1j**, 0.35 mmol) or the zwitterion (**1h, i**, 0.70 mmol) or 4-nitrophenyldiazonium tetrafluoroborate (**1d**·BF4; 0.70 mmol) in an evacuated 250-mL flask was cautiously treated with dimethylamine (**1a, f, h, i** and **1d**·BF4 at room temperature; **1b, c, d, e, g, j** at 0 °C): slow application was obtained by connecting to an evacuated flask with 70 mg (1.56 mmol) Me$_2$NH cooled to –196 °C and then removing the liquid nitrogen bath, for security reasons behind a protecting shield. After the thawing-up, the slight excess of gas was condensed back to the other flask and quantitative reaction secured by weighing. The triazenes **2** were extracted from the dimethylammonium salts with dry EtOAc and evaporated. The purity was checked with mp and by spectroscopic techniques.

References: G. Kaupp, A. Herrmann, *J. Prakt. Chem.*, **339**, 256 (1997).

Type of reaction: N–N bond formation
Reaction condition: solid-state
Keywords: anilines, nitrosobenzene, azomethine, nitrosyl chloride, nitrogen monoxide, diazotation, gas-solid reaction, solid diazonium salts

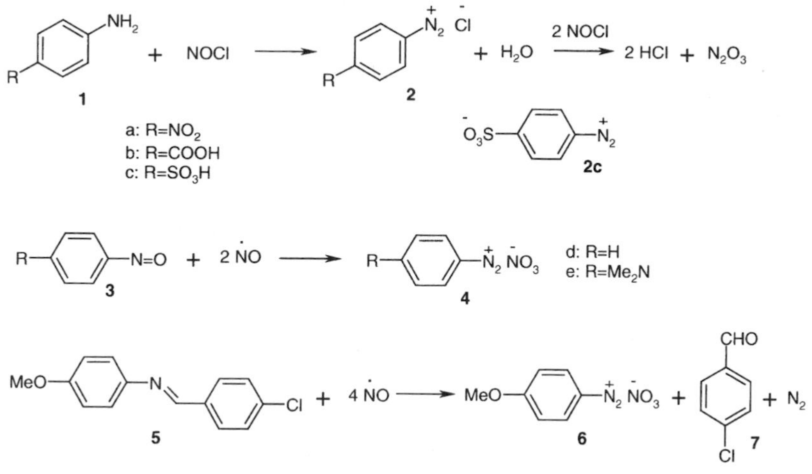

R–C$_6$H$_4$–NH$_2$ **(1)** + NOCl ⟶ R–C$_6$H$_4$–N$_2^+$ Cl$^-$ **(2)** + H$_2$O

2 NOCl + H$_2$O ⟶ 2 HCl + N$_2$O$_3$

a: R=NO$_2$
b: R=COOH
c: R=SO$_3$H

$^-$O$_3$S–C$_6$H$_4$–N$_2^+$ **2c**

R–C$_6$H$_4$–N=O **(3)** + 2 NO ⟶ R–C$_6$H$_4$–N$_2^+$ NO$_3^-$ **(4)**

d: R=H
e: R=Me$_2$N

MeO–C$_6$H$_4$–N=CH–C$_6$H$_4$–Cl **(5)** + 4 NO ⟶ MeO–C$_6$H$_4$–N$_2^+$ NO$_3^-$ **(6)** + Cl–C$_6$H$_4$–CHO **(7)** + N$_2$

Experimental procedures:

Caution: Solid diazonium salts explode upon heating, upon shock and upon grinding at sharp edges!

2: The solid aniline **1** (1.00 mmol) in an evacuated 250-mL flask was connected by a vacuum line to a 250-mL flask that was filled with NOCl (1 bar, 11 mmol). After 24 h, the gases were condensed back into the gas reservoir at −196 °C, absorbed in water and neutralized with NaOH for disposal. The yellow-orange crystals **2** were quantitatively obtained.

4: The nitrosobenzene **3d** or **3e** (2.00 mmol) in an evacuated 250-mL flask was connected by a vacuum line with a 250-mL flask containing NO (1 bar, 11 mmol) that had been freed from traces of NO$_2$ by storing over 4-chloroaniline. After storing the sample in a refrigerator at 4 °C for 24 h or 48 h, respectively, excess gas was recovered in a cold trap at −196 °C. Pure **4** was quantitatively obtained.

6: The imine **5** (491 mg, 2.00 mmol) in an evacuated 250-mL flask was connected by a vacuum line to a 250-mL flask containing NO (1 bar, 11 mmol) that was freed from traces of NO$_2$. After 24 h, excess gas was recovered as above and the solid material washed with dry ether (2×10 mL) in order to remove **7** and non-polar impurities. The product **6** (315 mg, 80%) was spectroscopically pure.

References: G. Kaupp, A. Herrmann, J. Schmeyers, *Chem. Eur. J.*, **8**, 1395 (2002).

Type of reaction: N–N bond formation
Reaction condition: solid-state
Keywords: anilines, nitrogen dioxide, diazotation, waste-free, gas-solid reaction, solid diazonium salt

a: R=4-COOH
b: R=4-Br
c: R=4-Cl
d: R=4-NO₂
e: R=3-NO₂
f: R=2-COOH
g: R=4-CN
h: R=4-SO₃H
i: R=2-SO₃H
j: R=4-(4-C₆H₄NH₂)
k: R=2,3-(C=O)₂C₆H₄

Experimental procedures:

Caution: Solid diazonium salts explode upon heating to the melting point and upon shock or upon grinding at sharp edges. Do not ball-mill!

The solid aniline derivatives **1a–k** (2.0 mmol, 1.0 mmol with **1j**) were treated with NO₂ gas in an evacuated 50-mL flask at 0 °C (**1a, k** at room temperature). NO₂ (460 mg, 10 mmol) from a 250-mL flask was applied through a stopcock in 5 small portions, each after the brown color of the previous portion had disappeared. Finally, the excess gas was let in and the reaction completed by 6 h rest. Excess gas was recovered by cooling the 250-mL flask to –196 °C. Quantitative conversion to the diazonium nitrate hydrates **2** (except with **2d** where not all of the water could be accomodated by the crystal: 92%) was secured by weight, spectroscopy and quantitative coupling with β-naphthol. **2h, i** were freed from HNO₃ and water at 5×10^{-4} Torr (12 h) and were obtained as zwitterions. **2d** was purified by washings with ethyl acetate in order to remove unreacted **1d**, a technique that should be applied in all cases where the aniline derivative 1 contained unpolar impurities that are most easily removed at that stage. Thus, the synthesis of **2k** started with **1k** of 97% purity. **2k** was obtained in pure form by two washings with EtOAc.

References: G. Kaupp, A. Herrmann, *J. Prakt. Chem.*, **339**, 256 (1997); G. Kaupp, A. Herrmann, J. Schmeyers, *Chem. Eur. J.*, **8**, 1395 (2002).

10 Rearrangement

10.1 Solvent-Free Rearrangement

Type of reaction: rearrangement
Reaction condition: solid-state
Keywords: 1,2-diol, *p*-TsOH, pinacol rearrangement, ketone

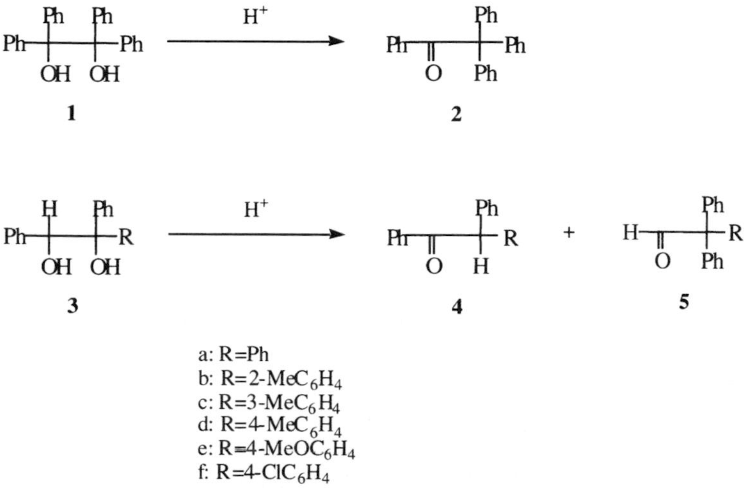

Ph Ph
Ph Ph
OH OH
1
H+
Ph
Ph Ph
O Ph
2
H Ph
Ph R
OH OH
3
H+
Ph
Ph R
O H
4
+
Ph
H R
O Ph
5
a: R=Ph
b: R=2-MeC6H4
c: R=3-MeC6H4
d: R=4-MeC6H4
e: R=4-MeOC6H4
f: R=4-ClC6H4

Exerimental procedures:
When a mixture of 1:3 molar ratio of powdered **3** and *p*-TsOH was kept at 60 °C for 2.5 h, products **4** and **5** were obtained in 89 and 8% yields, respectively.

References: F. Toda, T. Shigemasa, *Chem. Commun.*, 209 (1989).

Type of reaction: rearrangement
Reaction condition: solid-state
Keywords: benzil, alkyli metal hydroxide, benzylic acid rearrangement, benzylic acid

$$Ar-\overset{\overset{O}{\|}}{C}-\overset{\overset{O}{\|}}{C}-Ar' \xrightarrow{\text{KOH}} Ar-\overset{\overset{Ar'}{|}}{\underset{\underset{OH}{|}}{C}}-\overset{\overset{O}{\|}}{C}OH$$

1   **2**

a: Ar=Ph; Ar'=Ph
b: Ar=Ph; Ar'=4-ClC$_6$H$_4$
c: Ar=4-ClC$_6$H$_4$; Ar'=4-ClC$_6$H$_4$
d: Ar=Ph; Ar'=4-NO$_2$C$_6$H$_4$
e: Ar=3-NO$_2$C$_6$H$_4$; Ar'=3-NO$_2$C$_6$H$_4$
f: Ar=Ph; Ar'=4-MeOC$_6$H$_4$
g: Ar=Ph; Ar'=3,4-(OCH$_2$O)C$_6$H$_3$
h: Ar=4-MeOC$_6$H$_4$; Ar'=4-MeOC$_6$H$_4$
i: Ar=3,4-CH$_2$O$_2$C$_6$H$_3$; Ar'=3,4-(OCH$_2$O)C$_6$H$_3$

Experimental procedures:
A mixture of finely powdered benzil **1a** (0.5 g, 2.38 mmol) and KOH (0.26 g, 4.76 mmol) was heated at 80°C for 0.2 h, and the reaction product was mixed with 3N HCl (20 mL) to give benzylic acid **2a** as colorless needles (0.49 g, 90% yield).

References: F. Toda, K. Tanaka, Y. Kagawa, Y. Sakaino, *Chem. Lett.,* 373 (1990).

Type of reaction: rearrangement
Reaction condition: solid-state
Keywords: propargyl alcohol, TsOH, Meyer-Schuster rearrangement, cinnamic aldehyde

$$R_1-\overset{\overset{R_2}{|}}{\underset{\underset{OH}{|}}{C}}-\!\!\!\equiv\!\!\!-H \xrightarrow{\text{TsOH}} R_1R_2C\text{=CHCHO}$$

1   **2**

a: R$_1$= R$_2$=Ph
b: R$_1$=Ph; R$_2$=2-ClC$_6$H$_4$
c: R$_1$= R$_2$=2,4-Me$_2$C$_6$H$_3$

Experimental procedures:
The reaction was carried out by keeping a mixture of powdered **1** and an equimolar amount of TsOH at 50°C for 2–3 h.

References: F. Toda, H. Takumi, M. Akehi, *Chem. Commun.,* 1270 (1990).

Type of reaction: rearrangement
Reaction condition: solid-state
Keywords: methyl 2-(methylthio)benzenesulfonate, rearrangement, 2-(dimethyl-sulfonium)benzenesulfonate

Experimental procedures:
Powdered samples of **1** were kept at various temperatures between 36 and 49 °C (mp 54–55 °C) for various periods. The ratio of **2** to **1** was estimated by ^{1}H NMR measurement of the product dissolved in d_6-DMSO. In one experiment, ca. 12–13% transformation of **1** to **2** occurred after 28 days (14 days at 36 °C and 14 days at 49 °C); in another, 32% transformation occurred after 20 days at 45 °C.

References: P. Venugoplalan, K. Venkatesan, J. Klausen, E. Novotny-Bregger, C. Leumann, A. Eshenmoser, J.D. Dunitz, *Helv. Chim. Acta.*, **74**, 662 (1991).

Type of reaction: rearrangement
Reaction condition: solid-state
Keywords: 2-amino-2'-hydroxy-3'-(methoxycarbonyl)-1,1'-binaphthyl, S_N2 substitution, 2-methylamino-2'-hydroxy-3'-carboxy-1,1'-binaphthyl

Experimental procedures:
Crystalline (±)-**1** (343 mg) was heated under argon in a sealed tube at 150 ± 2 °C for 30 min (to reach full conversion). The product was chromatographed on silica gel (20 g) using a CHCl$_3$-MeOH mixture (1:1) as eluent to give pure **2** (325 mg, 95%).

References: M. Smrcina, S. Vyskocil, V. Hanus, M. Polasek, V. Langer, B.G.M. Chew, D.B. Zax, H. Verrier, K. Harper, T.A. Claxton, P. Kocovsky, *J. Am. Chem. Soc.*, **118**, 487 (1996).

Type of reaction: rearrangement
Reaction condition: solid-state
Keywords: hydrido complex, ruthenium, tautomerization, vinylidene complex

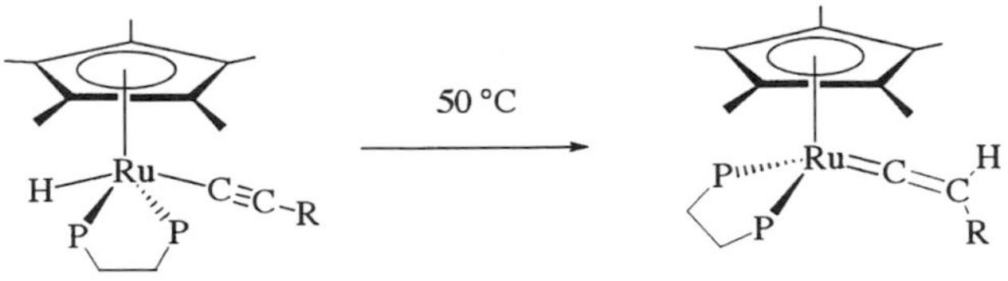

1   2

R=SiMe₃, Ph, H

Experimental procedures:
In the case of [(C₅Me₅RuH(C≡CH)(dippe)][BPh₄] **1**, which does not rearrange at an appreciable rate at 25 °C, the sample was kept in an oven at 50±3 °C. At regular time intervals, the sample was taken out, and its IR spectrum was recorded at room temperature. After recording each spectrum, the sample was returned back to the oven, and the cycle was repeated over a period of 10 h. After 21 h at 50±3 °C, conversion to [(C₅Me₅)RuH(C=CHR)(dippe)][BPh₄] **2** was found to be complete.

References: E. Bustelo, I. de los Rios, M.J. Tenorio, M.C. Puerta, P. Valerga, *Monatsh. Chem.,* **131**, 1311 (2000).

Type of reaction: rearrangement
Reaction condition: solvent-free
Keywords: ketoximes, Beckmann rearrangement, FeCl₃, amides

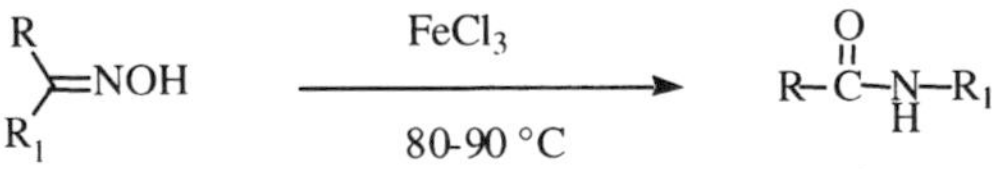

1   2

a: R, R₁= -(CH₂)₅-
b: R, R₁= -(CH₂)₇-
c: R, R₁= -(CH₂)₁₁-
d: R=R₁=Ph
e: R=R₁=CH(CH₃)₂
f: R=CH₃; R₁=CH₂CH(CH₃)₂
g: R=CH₃; R₁=Ph
h: R=CH₃; R₁=4-CH₃C₆H₄

Experimental procedures:
The oxime (10 mmol) and FeCl$_3$ (30–40 mmol) were mixed in a mortar. The reaction mixture was allowed to stand at 80–90 °C for 1.5–3 h. The progress of reaction was followed by dissolving a sample in acetone and monitoring on silica gel plates (SIL G/UV 254 plates). Water (15 mL) was added and the whole extracted with chloroform (3×20 mL). Drying and evaporation of the solvent gave the products in 81–95% yields which recrystallized from *n*-heptane.

References: M. M. Khodaei, F. A. Meybodi, N. Rezai, P. Salehi, *Synth. Commun.,* **31**, 2047 (2001).

Type of reaction: rearrangement
Reaction condition: solid-state
Keywords: oxime, Beckmann rearrangement, nitrile ylide, amide

a: R$_1$=R$_2$=Ph
b: R$_1$=Ph; R$_2$=Me
c: R$_1$=4-MeC$_6$H$_4$; R$_2$=Ph
d: R$_1$=R$_2$=PhCH$_2$
e: R$_1$,R$_2$=-(CH$_2$)$_5$-

Experimental procedures:
A stirred solution of the oxime **1** (1 g) in dry diethyl ether (20 mL), in a 100-mL two-necked round-bottomed flask protected by a CaCl$_2$ guard tube, was saturated with dry hydrogen chloride gas via an inlet. A white solid precipitated out, which was allowed to settle and collected by pipetting out the supernatant ether under nitrogen, and removing trace volatiles in vacuo. A pinch was transferred under nitrogen to a melting point capillary, which was then fully sealed before the determination of the mp. The rest was heated under nitrogen as indicated, the reaction being followed by TLC. The darkened solid was cooled and taken into ether; the resulting solution was washed successively with NaHCO$_3$ solution and water, dried (MgSO$_4$) and the solvent evaporated in vacuo. The products **2** were purified by recrystallization and identified by their mp's, IR and ^{1}H NMR spectra.

References: S. Chandrasekhar, K. Gopalaiah, *Tetrahedron Lett.,* **42**, 8123 (2001).

Type of reaction: rearrangement
Reaction condition: solid-state
Keywords: bisallene, ene-ine, cycloaddition, thermolysis, benzocyclobutene

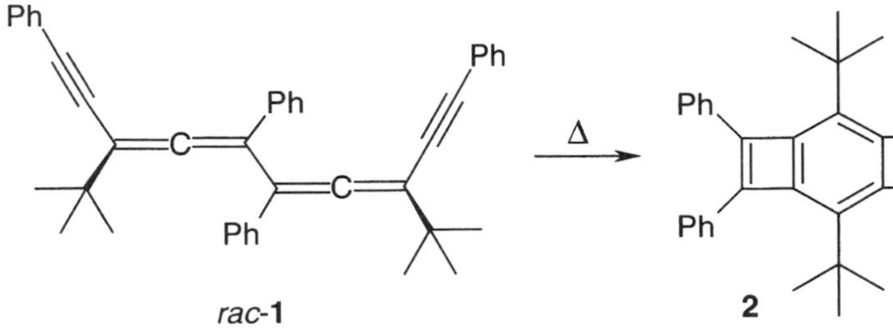

Experimental procedure:
The crystalline micronized bisallene *rac*-**1** was heated to 140–150 °C in a vacuum. The crystals turned dark-green and the reaction was completed within 4 h. Product **2** (20–28%) was very sensitive to oxygen. Supermicroscopy (AFM) revealed that the reaction occurred only at outer and inner surfaces.

References: F. Toda, *Eur. J. Org. Chem.,* 1377 (2000); G. Kaupp, J. Schmeyers, M. Kato, K. Tanaka, N. Harada, F. Toda, *J. Phys. Org. Chem.,* **14**, 444 (2001).

Type of reaction: rearrangement
Reaction condition: solid-state
Keywords: bisallene, cyclization, stereospecific, thermolysis, bismethylene-cyclobutene

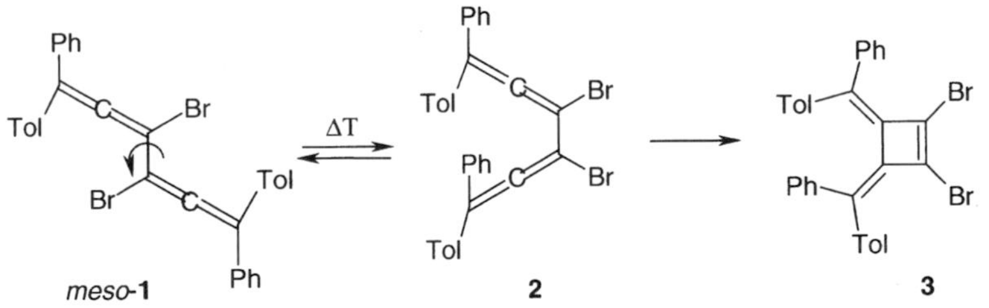

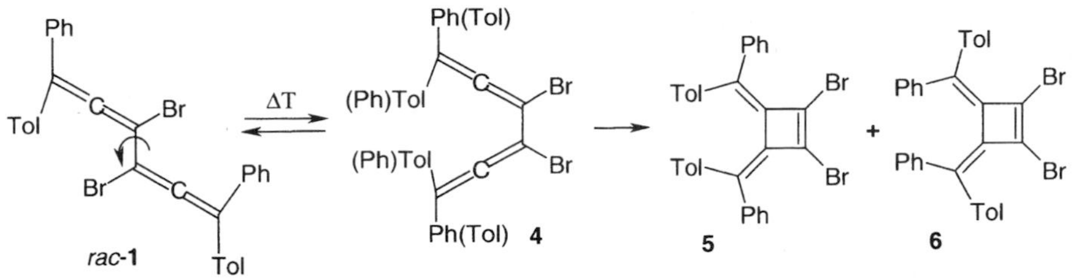

Experimental procedures:

The bisallene *meso*-**1** (0.20 g) was heated to 135 °C for 90 min. A quantitative yield of **3** was obtained (mp 214–215 °C, EtOAc).

Similarly, *rac*-**1** (0.40 g) was heated to 125 °C for 90 min. The quantitatively formed 1:1 mixture of **5** and **6** was fractionally crystallized to give pure **5** (mp 180–183 °C) and pure **6** (mp 215–218 °C).

The stereochemistry proves the occurrence of the intermediates **2** and **4**, respectively. These reactions were also studied by supermicroscopy (AFM) (Ref. 2).

References: (1) F. Toda, K. Tanaka, T. Tamashima, M. Kato, *Angew. Chem., 110,* 2852 (1998); *Angew. Chem. Int. Ed. Engl., 37,* 2724 (1998); (2) G. Kaupp, J. Schmeyers, M. Kato, K. Tanaka, F. Toda, *J. Phys. Org. Chem., 15,* 1 (2002).

Type of reaction: rearrangement
Reaction condition: solvent-free
Keywords: oxazepam, [1,3,2,3]-elimination, quinazoline aldehyde

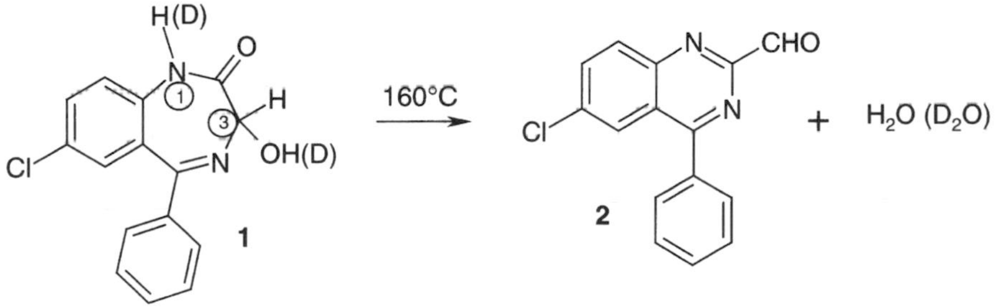

Experimental procedures:

The tranquillizer oxazepam **1** (mp 200–205 °C; 1.00 g, 3.5 mmol) was heated to 160 °C for 2 h. The water escaped and a quantitative yield of the aldehyde **2** (mp 176 °C) was obtained. The product **2** was sticky at 160 °C. Thus, the initial solid-state reaction changed to a (partial) melt reaction as it proceeded. It was not necessary to use solvent or a sealed tube. The mechanism was discussed on the basis of the deuteration experiment.

References: G. Kaupp, B. Knichala, *Chem. Ber.*, **118**, 462 (1985).

Type of reaction: rearrangement
Reaction condition: solid-state
Keywords: 17a-hydroxy-20-ketosteroid, acyloin rearrangement, substitutive rearrangement, stereoselective, gas-solid reaction

Experimental procedures:
17a-Hydroxyprogesterone **1** (1.98 g, 6.0 mmol) were co-ground with 4.0 g MgSO$_4$·2.3 H$_2$O. The mixture was treated with HCl (1 bar) in an evacuated 500-mL flask at room temperature for 16 h. Excess gas was recovered in a cold trap at −196 °C and the organic material extracted with CH$_2$Cl$_2$. The raw material (2.09 g) contained **2** and **3** in a 1:1 ratio and 8% of the dehydration product of **1** without ring enlargement. The sensitive products were separated by preparative TLC on 400 g SiO$_2$ with CH$_2$Cl$_2$-EtOAc (1:5) to give pure **2** (320 mg, 16%; mp 183–185 °C, diisopropyl ether) and **3** (820 mg, 44%; mp 189–193 °C, benzene).

Similarly, **4** (270 mg, 0.88 mmol) was treated with 0.9 bar HCl in a 100-mL flask at 4 °C for 4 days (no drying agent). Excess gas was pumped off and preparative TLC on 200 g SiO$_2$ with CH$_2$Cl$_2$-EtOAc (5:1) yielded 62% **5**, 20% unchanged **4** and **6** (2 stereoisomers). The latter stereoisomers **6** prevailed if the reaction was performed in the presence of 1.0 g MgSO$_4$·2.3 H$_2$O.

References: G. Kaupp, E. Jaskulska, G. Sauer, G. Michl, *J. Prakt. Chem.*, **336**, 686 (1994).

Type of reaction: rearrangement
Reaction condition: solvent-free
Keywords: benzimidazoline, melt reaction, tetrahydrobenzopyrazine

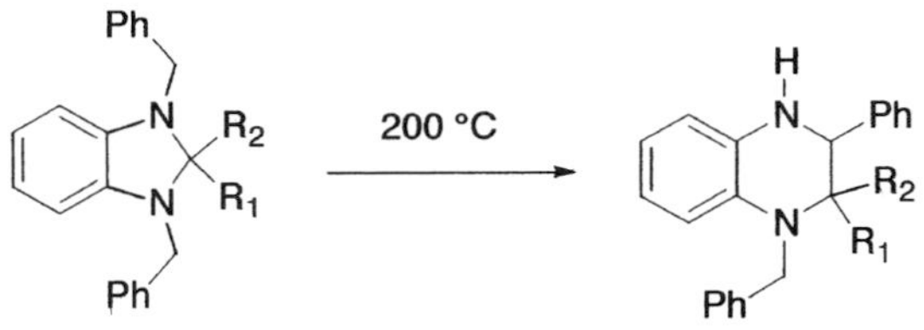

a: $R^1 = R^2 = Me$	(11%)
b: $R^1 = Me$, $R^2 = Et$	(13%)
c: $R^1 = Me$, $R^2 = Pr$	(10%)
d: R^1, $R^2 = (CH_2)_5$	(16%)
e: R^1, $R^2 = (CH_2)_4$	(14%)

Experimental procedures:
The 2,2-dialkylbenzimidazoline **1** was heated to 200 °C in a sealed tube and the rearranged product **2** was isolated by column chromotography with petroleum ether.

References: C. M. Reddy, P. L. Prasunamba, P. S. N. Reddy, *Tetrahedron Lett.*, **37**, 3355 (1996).

10.2 Solvent-Free Rearrangement under Photoirradiation

Type of reaction: rearrangement
Reaction condition: solid-state
Keywords: 1,4-pentadiene derivatives, 2-cyclohexenone derivatives, di-π-methane rearrangement, photoirradiation

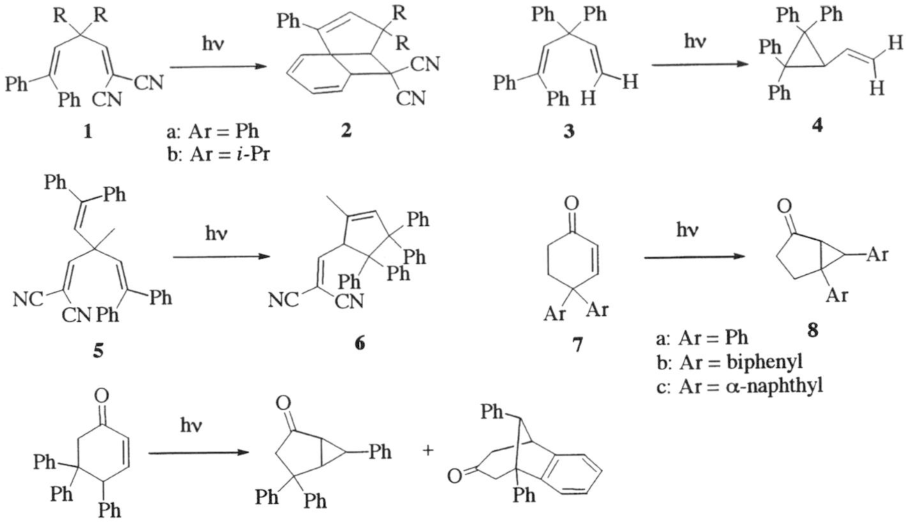

1
a: Ar = Ph
b: Ar = i-Pr

2

3

4

5

6

7

a: Ar = Ph
b: Ar = biphenyl
c: Ar = α-naphthyl

8

9

10

11

Experimental procedures:

A crystalline film of 314 mg (0.744 mmol) of 1,1,3,3-tetraphenyl-5,5-dicyano-1,4-pentadiene **1a**, deposited by slow evaporation of a 20% ether in hexane solution, was irradiated at 78 °C for 2.5 h through Pyrex. The resulting orange-red solid was subjected to preparative HPLC eluted with 8% ether and 0.5% acetonitrile in hexane to give 227.1 mg (72.3%) of starting diene and 74.8 mg (23.8%) of **2a** as a pale purple-red solid. Recrystallization from ether-hexane yielded 69.7 mg (21.6%) of **2a** as a white solid (mp 125–128 °C).

References: H. E. Zimmerman, M. J. Zuraw, *J. Am. Chem. Soc.,* **111**, 7974 (1989); H. E. Zimmerman, Z. Zhu, *J. Am. Chem. Soc.,* **117**, 5245 (1995).

Type of reaction: rearrangement
Reaction condition: solid-state
Keywords: benzyl phenyl sufone, β-cyclodextrin, photorearrangement, 2-methyl-phenyl phenyl sulfone

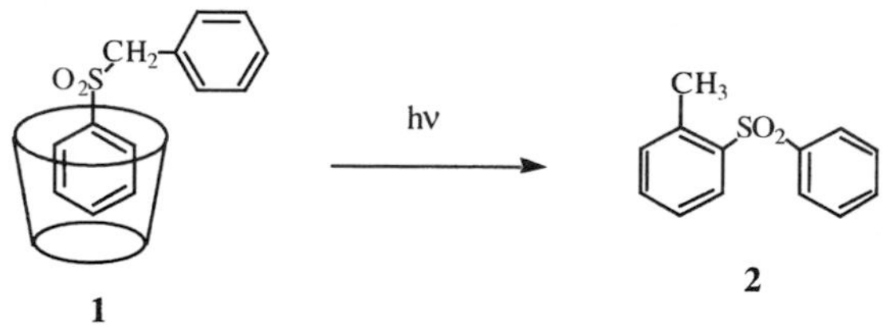

Experimental procedures:
The solid 1:1 complex of **1** with β-cyclodextrin was irradiated for 100 h. Photoproducts were identified using GC by coinjection of authentic samples synthesized separately and also by their ^{1}H NMR spectra.

References: K. Pitchumani, P. Velusamy, H.S. Banu, C. Srinivasan, *Tetrahedron Lett.*, **36**, 1149 (1995).

Type of reaction: rearrangement
Reaction condition: solid-state
Keywords: *cis*-1,2-dibenzoylalkene, photorearrangement, keten, enolether
Experimental procedures:

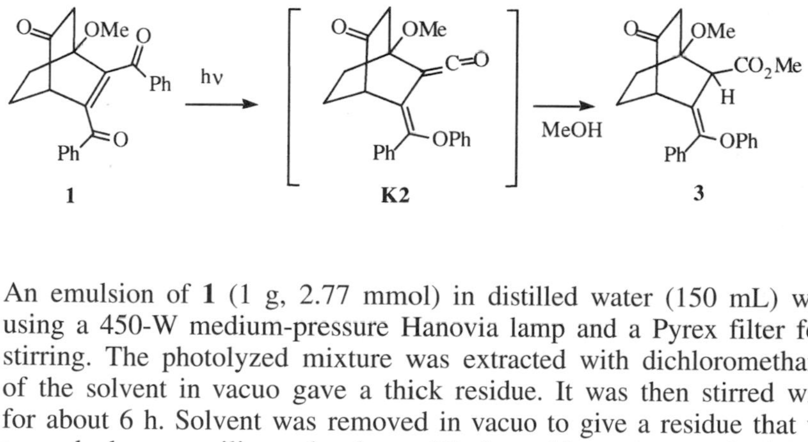

An emulsion of **1** (1 g, 2.77 mmol) in distilled water (150 mL) was irradiated using a 450-W medium-pressure Hanovia lamp and a Pyrex filter for 36 h with stirring. The photolyzed mixture was extracted with dichloromethane. Removal of the solvent in vacuo gave a thick residue. It was then stirred with methanol for about 6 h. Solvent was removed in vacuo to give a residue that was chromatographed over a silica gel column. Elution with a mixture of 15% ethyl acetate in PE gave **3** as a white solid, mp 41 °C (230 mg, 21%), which was crystallized with difficulty from a mixture of *n*-pentane and dichloromethane (4:1).

References: D. Maji, R. Singh, G. Mostafa, S. Ray, S. Lahiri, *J. Org. Chem.*, **61**, 5165 (1996).

Type of reaction: rearrangement
Reaction condition: solid-state
Keywords: bicyclo[3.1.0]hex-3-en-2-one, photorearrangement, phenol

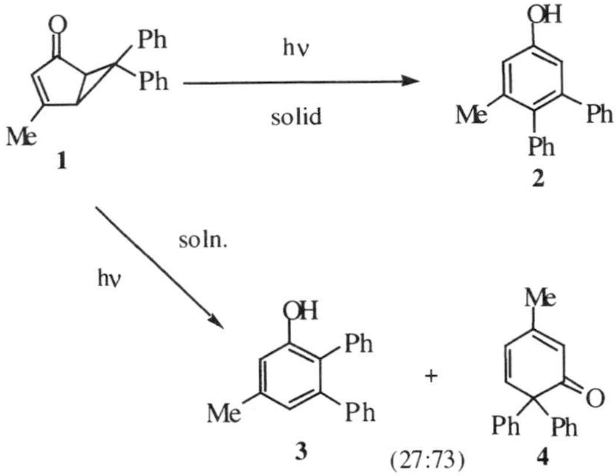

Experimental procedures:

Crystals were spread over a strip of "Parafilm". Both the photochemical reactor and the strip with crystals were placed in a drybox and purged with nitrogen for 1 h. A 40 mg portion of 4-methyl-6,6-diphenylbicyclo[3.1.0]hex-3-en-2-one **1** was irradiated under conditions described above for 10 min through a **5** mm filter solution of 10^{-3} M sodium metavanadate in 5% sodium hydroxide. The mixture was chromatographed on preparative TLC plate eluted with 7% ether in hexane. Band **1** (28.0 mg, 70%) contained 3-methyl-4,5-diphenylphenol **2**. Crystallization from hexane gave 14.1 mg (35%) of colorless crystals, mp 109–110 °C.

References: H.E. Zimmerman, P. Sebek, *J. Am. Chem. Soc.,* **119**, 3677 (1997).

Type of reaction: rearrangement
Reaction condition: solid-state
Keywords: S-aryl-o-arylbenzothioate, chiral crystal, photoirradiation, 1,4-aryl migration, phthalide

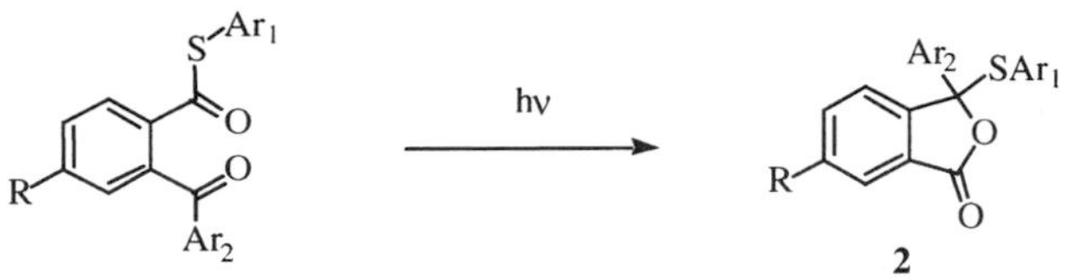

a: R=H, Ar$_1$=(*o*-tol), Ar$_2$=Ph
b: R=H, Ar$_1$=Ph, Ar$_2$=Ph
c: R=H, Ar$_1$=(*m*-tol), Ar$_2$=Ph
d: R=H, Ar$_1$=(*p*-tol), Ar$_2$=Ph
e: R=Me, Ar$_1$=Ph, Ar$_2$=Ph
f: R=H, Ar$_1$=Ph, Ar$_2$=(*p*-tol)
g: R=H, Ar$_1$=Ph, Ar$_2$=(*p*-ClPh)

Experimental procedures:

All samples were well ground and sandwiched between two Pyrex slides. The crystal samples packed in polyethylene bags in order to prevent soaking were purged by dried nitrogen, and were then closed in vacuo. They were placed in a cooling bath close to the light source. All crystal samples were irradiated with an Eikosha 500-W high-pressure mercury lamp. After irradiation, the photolysates were treated the same as that in solution photochemistry. The enantiomeric purities of **2a–2c** were determined by preparative HPLC using Chiralcell OJ (Daicel) with an eluent of EtOH-hexane (4:1).

References: M. Takahashi, N. Sekine, T. Fujita, S. Watanabe, K. Yamaguchi, M. Sakamoto, *J. Am. Chem. Soc.,* **120**, 12770 (1998); M. Sakamoto, M. Takahashi, S. Moriizumi, K. Yamaguchi, F. Fujita, S. Watanabe, *J. Am. Chem. Soc.,* **118**, 8138 (1996).

Type of reaction: rearrangement
Reaction condition: solid-state
Keywords: 2-cyanoethyl-isonicotinic acid-cobaltoxime complex, photoisomerization, 1-cyanoethyl-isonicotinic acid-cobaltoxime complex

hν
base
base

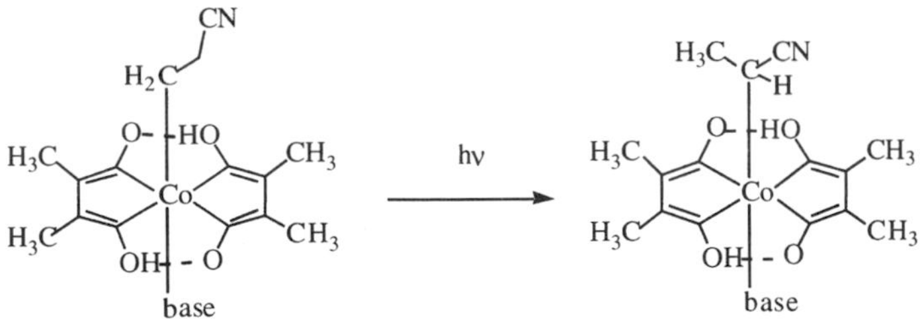

1 **2**

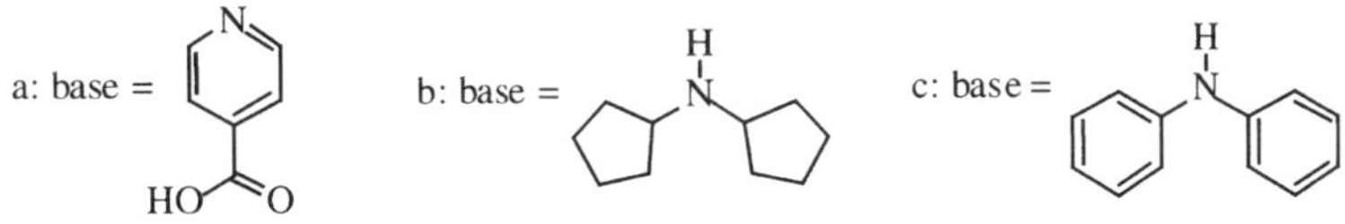

a: base =
b: base =
c: base =

Experimental procedures:
A KBr disk which contained 1% of the samples was exposed to a 500-W xenon lamp (Ushio UXL-5S+UI-501 C), the distance between the disk and the lamp being 20 cm. The absorption assigned to the streching vibrational mode of the cyano group of the 2-cyanoethyl complex, ν_{CN}, is at 2250 cm^{-1}, whereas ν_{CN} of the 1-cyanoethyl complex is at 2200 cm^{-1}. The infrared spectra of KBr disks including the powdered samples of crystals I, II and III were measured at a constant interval of 10 min, using a JASCO A-1000 IP spectrometer. The decrease of the absorption band at 2250 cm^{-1} within 40 min was explained by first-order kinetics for each sample and its rate constant was obtained by least-squares fitting. The rates for I, II and III were 1.88, 1.33 and $2.76\times10^{-4}s^{-1}$, respectively.

References: D. Hashizume, Y. Ohashi, *J. Chem. Soc., Perkin Trans. 2*, 1931 (1998).

Type of reaction: rearrangement
Reaction condition: solid-state
Keywords: *cis*-1,2-dibenzoylalkene, photoirradiation, vinylketene

hν
COPh
COPh
C=O
OPh
Ph

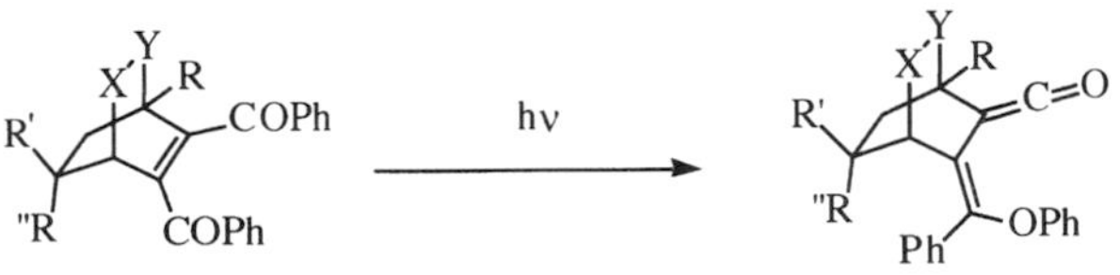

1 **2**

a: X=CO; Y= CH_2; R=OMe; R'=R"=H
b: X=CO; Y=CH_2; R=Me; R'=R"=H
c: X=CH_2; Y= CO; R=Me; R'=R"=H
d: X=CO; Y=CH_2; R=R'=R"=Me

Experimental procedures:
A suspension of the compound (0.5–1 g) in distilled water (150 mL) was irradiated with stirring using a Hanovia medium-pressure 450-W lamp and a Pyrex filter for ca. 35 h. The suspension was then extracted with CH_2Cl_2 and the residue obtained after removal of solvent was chromatographed on a silica gel column (60–120 mesh). The ketene was eluted with a mixture of ethyl acetate (5–10%) in petroleum ether (bp 60–80 °C).

References: M. Chanda, D. Maji, S. Lahiri, *Chem. Commun.*, 543 (2001).

10.3 Solvent-Free Rearrangement under Microwave Irradiation

Type of reaction: rearrangement
Reaction condition: solvent-free
Keywords: acetylenic alcohol, KSF clay, microwave irradiation, *α,β*-unsaturated ketone

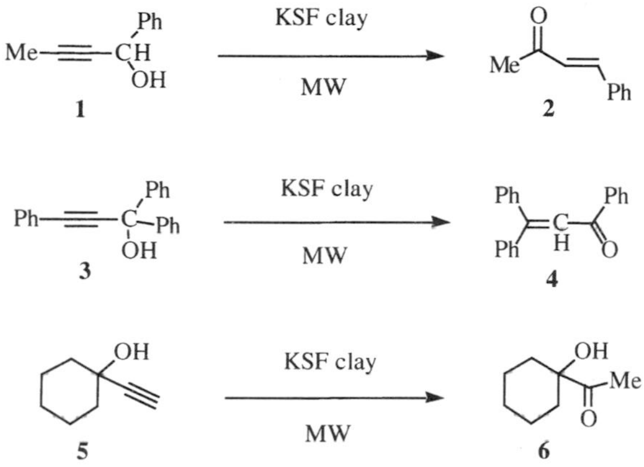

Experimental procedures:
The alcohol **1** (3 mmol) in solution in CH_2Cl_2 was absorbed onto KSF clay (1 g), solvent was evaporated in vacuo. The solid was irradiated by microwaves (270 W, 5 min) in a sealed Teflon vessel and after cooling at room temperature, was extracted by elution with MeCN. The product was recrystallized from ethanol after evaporation of the solvent (95%).

References: A.B. Alloum, B. Labiad, D. Villemin, *Chem. Commun.*, 386 (1989).

Type of reaction: rearrangement
Reaction condition: solvent-free
Keywords: eugenol KOH, phase transfer catalyst, microwave irradiation, isomerization, isoeugenol

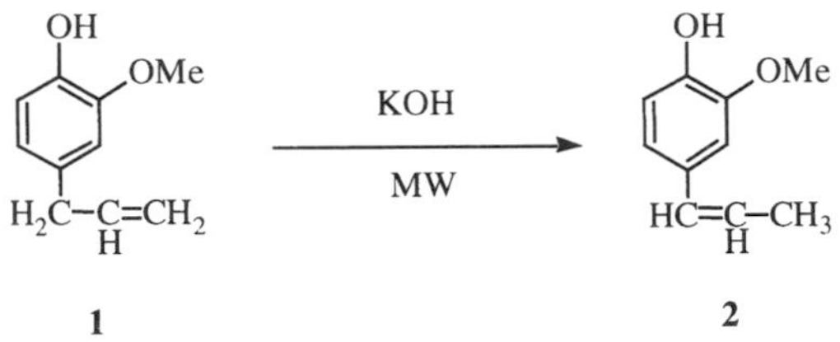

Experimental procedures:
A mixture of eugenol **1** (15 mmol), crushed potassium hydroxide or terbutoxide (33 mmol), the phase transfer catalyst (0.75 mmol) was placed either in a beaker in a domestic oven or in a Pyrex tube introduced into the Maxidigest MX 350 Prolabo microwave reactor filled with a rotational system. Microwave irradiation was carried out in the conditions described in Table 3 and 4. The mixture was cooled to ambient temperature. After elution with diethyl ether (50 mL) and subsequent filtration on Florisil, the organic products were analyzed by GC using an internal standard and characterized by ^{1}H NMR spectroscopy by comparison with authentic samples.

References: A. Loupy, L. N. Thach, *Synth. Commun.*, **23**, 2571 (1993).

Type of reaction: rearrangement
Reaction condition: solvent-free
Keywords: acyloxybenzene, Fries rearrangement, microwave irradiation, flavanone, *o*-acyl phenol

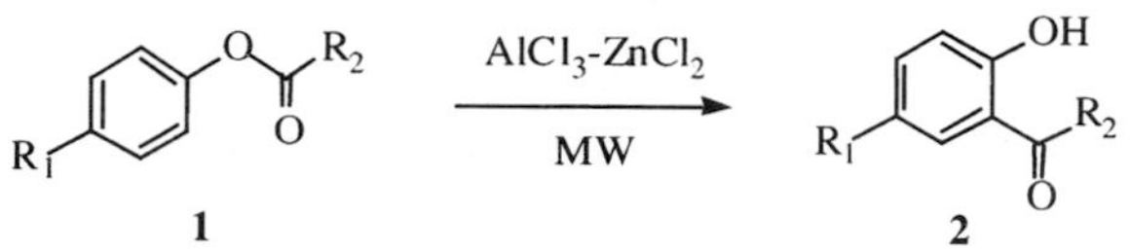

1

a: R_1=H; R_2=Me
b: R_1=p-Me; R_2=Me
c: R_1=p-Me; R_2=Ph
d: R_1=p-MeO; R_2=Me
e: R_1=p-MeCO$_2$; R_2=Me

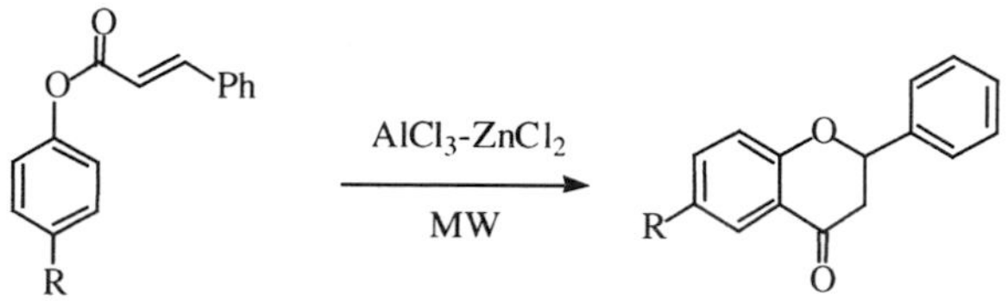

2

3

a: R=Me
b: R=OMe

4

Experimental procedures:

General Procedure for Microwave-Induced Fries Rearrangement. Neat substrate (0.01 mol) was mixed with the support (1:3, w/w) in a beaker and was exposed to microwave irradiation for 7 min. After cooling to room temperature the product was extracted with diethyl ether. Evaporation of the solvent gave almost pure product. Further purification was carried out by recrystallization or column chromatography on silica gel.

References: F. M. Moghaddam, M. Ghaffarzadeh, S. H. Abdi-Oskoui, *J. Chem. Res. (S),* 574 (1999).

Type of reaction: rearrangement
Reaction condition: solvent-free
Keywords: oximes, zinc chloride, microwave irradiation, Beckmann rearrangement, amide, nitrile

Ph–CH=N–OH $\xrightarrow[\text{MW}]{\text{ZnCl}_2}$ Ph–CONH$_2$ +   Ph–C≡N

1   2 (94%)   3 (6%)

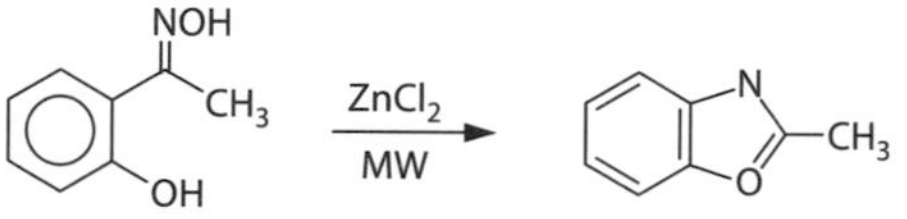

4

5 (86%)

Experimental procedures:
A mixture of *syn*-benzaldoxime **1** (10 mmol, 1.21 g) or 2-hydroxyacetophenone oxime **4** (10 mmol, 1.52 g) and one equivalent of 98% anhydrous zinc chloride (10 mmol, 1.36 g) was introduced in a Pyrex tube and submitted to microwave irradiation for 20 min in the Synthewave S402 monomode reactor at 140 °C. The mixture was cooled to room temperature and dissolved in methanol and then filtered through silica gel. Products were analyzed by capillary gas chromatography using methyl benzoate as an internal standard and compared by ^{1}H NMR with authentic samples.

References: A. Loupy, S. Regnier, *Tetrahedron Lett.*, **40**, 6221 (1999).

Type of reaction: rearrangement
Reaction condition: solvent-free
Keywords: aryl sulfonate, microwave irradiation, thia-Fries rearrangement, 2-arylsulfonyl phenol

a: R=H; Ar=4-MeC$_6$H$_4$
b: R=Me; Ar=4-MeC$_6$H$_4$
c: R=Cl; Ar=4-MeC$_6$H$_4$
d: R=Me; Ar=Ph
e: R=Cl; Ar=Ph

Experimental procedures:
To a solution of arylsulfonate **1** (1 g) in 5 mL of anhydrous chloroform, 3 g of support was added. After evaporation of the solvent, the mixture was subjected to microwave irradition for the given times. The cooled reaction mixture was extracted with ethyl acetate (3×50 mL) and the solvent was evaporated under vacuum. The product **2** was isolated by column chromatography of the crude reaction mixture on silica gel (elution: dichloromethane-petroleum ether).

References: F. M. Moghaddam, M. G. Dakamin, *Tetrahedron Lett.*, **41**, 3479 (2000).

11 Elimination

11.1 Solvent-Free Elimination

Type of reaction: elimination
Reaction condition: gas-solid
Keywords: alcohol, HCl gas, dehydration, alkene

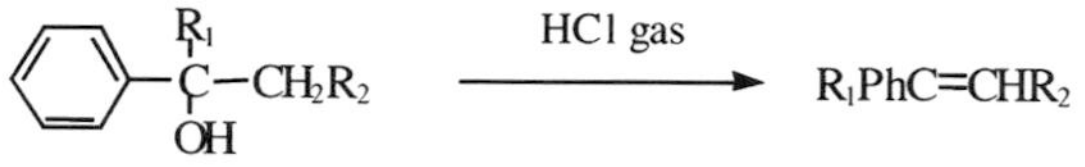

a: R_1=Ph; R_2=H
b: R_1=Ph; R_2=Me
c: R_1=Ph; R_2=Ph
d: R_1=Ph; R_2=2-$NO_2C_6H_4$
e: R_1=2-ClC_6H_4; R_2=Me

Experimental procedures:
Powdered 1,1-diphenylpropan-1-ol **1b** kept in a desiccator filled with HCl gas for 5.5 h gave pure 1,1-diphenylprop-1-ene **2b** in 99% yield.

References: F. Toda, H. Takumi, M. Akehi, *Chem. Commun.*, 1270 (1990).

Type of reaction: elimination
Reaction condition: solvent-free
Keywords: methanethiol, [2+2]-cycloreversion, [1,2,(3)4]-elimination, melt reaction

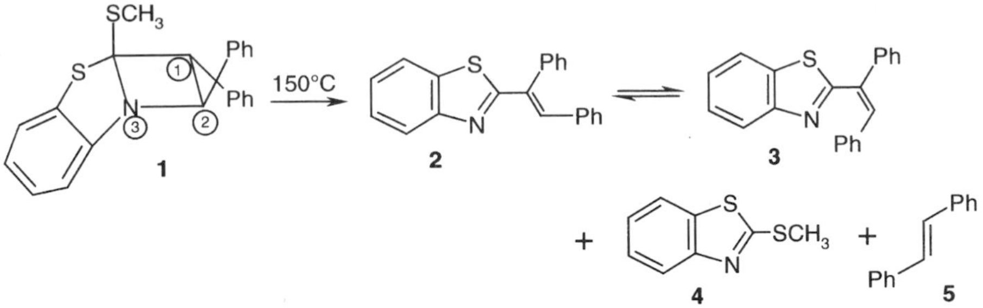

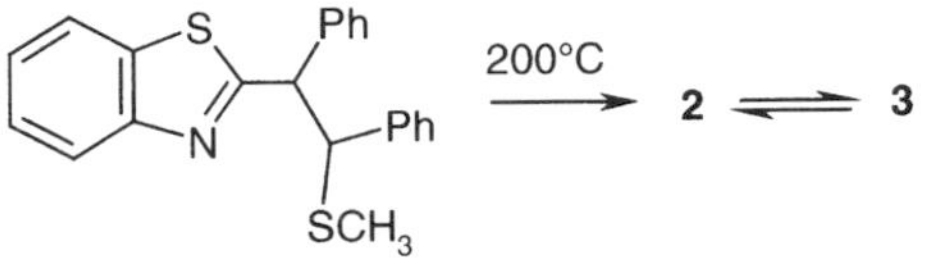

6 (threo + erythro)

Experimental procedures:

Compound **1** (150 mg, 0.41 mmol; from photolysis of stilbene with 2-methylmer-capto-benzothiazole) was sealed under vacuum in a 10-mL Pyrex tube that was previously heated to red glow in order to remove adhering water. The whole tube was heated to 150 °C for 20 h in the dark. Separation by preparative HPLC (RP18, 80% methanol) gave 62 mg (82%) **4** and 25 mg (34%) **5** as well as 18 mg (14%) **2** and 2 mg (2%) **3**. **2** and **3** equilibrated in the melt at 150 °C to give a 8.0:1.0 ratio (6.1:1.0 at 200 °C when similarly prepared from **6**).

References: G. Kaupp, *Chem. Ber.,* **118**, 4271 (1985).

Type of reaction: elimination
Reaction condition: solvent-free
Keywords: benzothiazepine, [2,1,3]-elimination, methanethiol

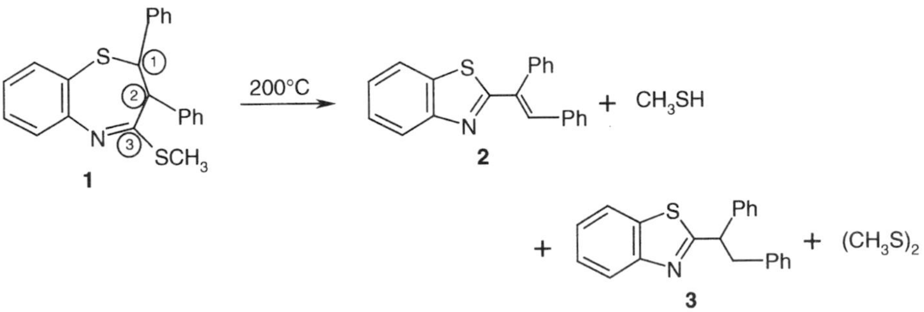

Experimental procedure:

Compound **1** (500 mg, 1.39 mmol; from photolysis of stilbene and 2-methylmer-capto-benzothiazole) was sealed under vacuum in a thick-walled Pyrex tube (the volume was about 100 mL) and heated to 200°C for 20 h in the dark. Methan-ethiol (from the [2,1,3]-elimination) and dimethyldisulfide were condensed to a cold trap (−196°C) at a vacuum line (65 mg, 97%). The residue was separated by preparative TLC on 200 g SiO_2 with benzene to give 340 mg (78%) of **2** (mp 108°C, methanol) and 96 mg (22%) of **3**.

References: G. Kaupp, *Chem. Ber.*, **117**, 1643 (1984).

Type of reaction: elimination
Reaction condition: solvent-free
Keywords: benzothiazepinone, [1,5,5′,(4)5]-elimination, ring contraction, complex elimination, benzothiazole

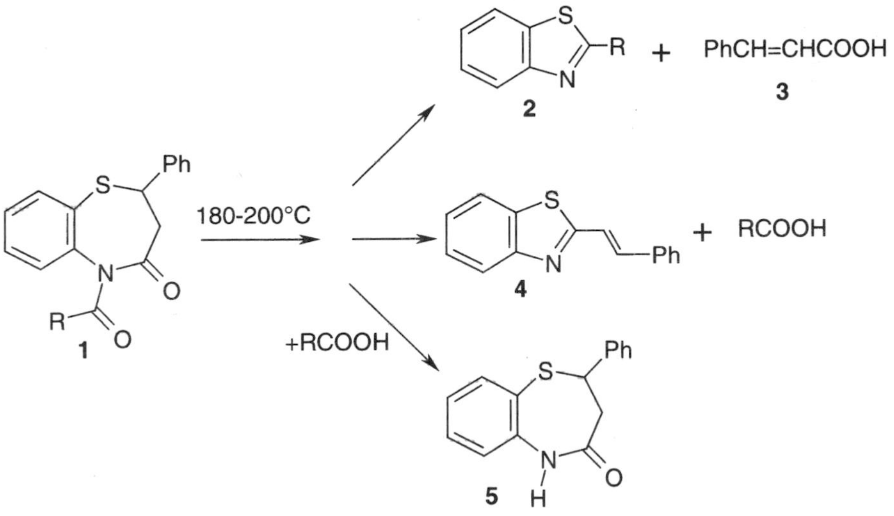

	R	T (°C)	t (h)	**2**(%)	**3**(%)	**4**(%)	**5**(%)
a	Me	200	5	35	27	40	12
b	Et	200	55	30	42	66	2
c	CH_2Ph	180	12	24	29	27	16
d	Ph	180	10	31	19	44	6

Experimental procedures:
Compound **1a** (500 mg, 1.61 mmol) was heated to 200 °C for 5 h in an evacuated Pyrex ampoule. The raw material contained 46 mg (39%) propionic acid. The other products **2a, 3, 4, 5** were separated by preparative TLC on 100 g SiO$_2$ with dichloromethane and obtained with the yields given. Similarly, **1b–d** gave **2b–d, 3–5** and RCOOH.

References: G. Kaupp, D. Matthies, *Chem. Ber.*, **120**, 1741 (1987).

Type of reaction: elimination
Reaction condition: solvent-free
Keywords: benzothiazepinone, [2,1,3]-elimination, [1,2,3,2(3)]-elimination, [1,1]-elimination, [1,2/2,1]-rearrangement, benzothiazole, benzothiazinone, chinolin-2-one

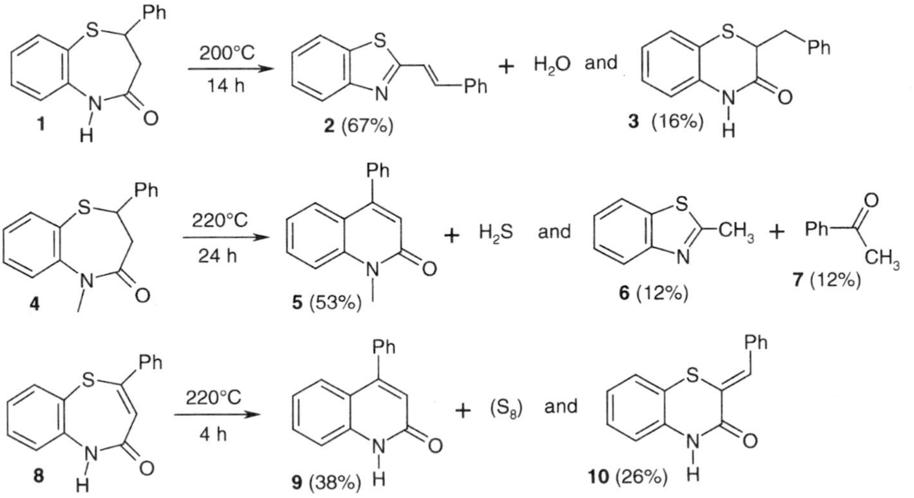

Experimental procedures:
Compound **1** (1.00 g; 3.9 mmol) was sealed in an evacuated Pyrex tube and heated to 200±2 °C for 14 h. Preparative TLC on 200 g SiO$_2$ with dichloromethane (two developments) gave 625 mg (67%) **2** ([2,1,3]-elimination from the iminol tautomer) and 160 mg (16%) of the [1,2/2,1]-rearrangement product **3**.

Similarly, compounds **4** and **8** were heated to 220 °C and the products of the [1,2,3,2(3)]-elimination **5**, of the [1,1]-elimination **9** and of the [1,2/2,1]-rearrangement **10** isolated by preparative TLC.

References: G. Kaupp, E. Gründken, D. Matthies, *Chem. Ber.*, **119**, 3109 (1986).

Type of reaction: elimination
Reaction condition: solvent-free
Keywords: chorohydrin, [1,2]-elimination, [1,2,3]-elimination, chlorotriphenyl-ethen, diphenylmethyl phenyl ketone

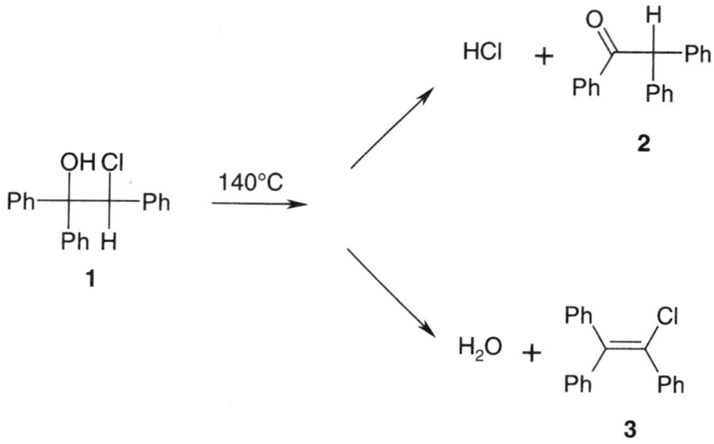

HCl +

O
‖
Ph—C—CH(Ph)—Ph

2

Ph—C(OH)(Ph)—C(Cl)(H)—Ph   →(140°C)

1

H₂O +

Ph(Ph)C=C(Cl)(Ph)

3

Experimental procedure:
The chlorohydrin **1** (mp 138–139 °C, decomp.; 400 mg) was heated to 140 °C. It melted with decomposition to give **2** (by [1,2,3]-elimination) and **3** (by [1,2]-elimination) in a 66:34 ratio. The products were separated by preparative TLC on SiO₂ with dichloromethane.

References: G. Kaupp, D. Matthies, *Chem. Ber.,* **120**, 1897 (1987).

Type of reaction: elimination
Reaction condition: solvent-free
Keywords: quinolizine, complex elimination, ring contraction, indolizine

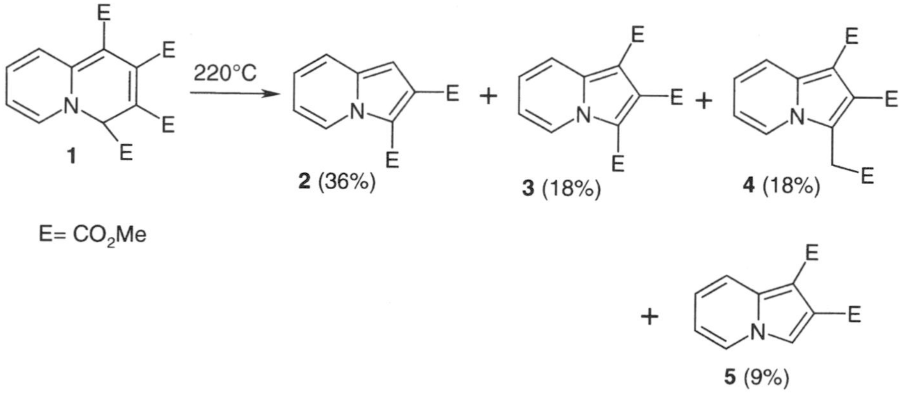

1 →(220°C) **2** (36%) + **3** (18%) + **4** (18%)

+ **5** (9%)

E= CO₂Me

Experimental procedure:
The quinolizine **1** (1.00 g, 2.75 mmol; from pyridine and dimethyl acetylenedicarboxylate) was heated to 220 °C for 10 h in an evacuated sealed tube or in a V2A-autoclave. Gaseous and liquid products were CO_2, CO, MeOH, MeOMe, CH_3CO_2Me, $OC(OMe)_2$ and $(MeO_2C)_2$. The solid residue contained the indolizines **2, 3, 4** and **5** in the ratio 10:5:5:1. These were separated and isolated by preparative TLC on 400 g SiO_2 with benzene-methylacetate 3:2 (v/v).

References: G. Kaupp, D. Hunkler, I. Zimmermann, *Chem. Ber.,* **115**, 2467 (1982).

Type of reaction: elimination
Reaction condition: solid-state
Keywords: hexahydrotriazine, [2+2+2]-cycloreversion, waste-free, gas-solid reaction, *N*-arylmethylenimine hydrochloride

a: Ar=phenyl
b: Ar=4-tolyl
c: Ar=4-anisyl

Experimental procedures:
The hexahydrotriazines **1a–c** (3.00 mmol) were cooled to –18 °C, –10 °C and –10 °C, respectively, in an evacuated 500-mL flask. HCl gas (500 mL, 1 bar) was let in through a vacuum line in 6 portions during 2 h or continuously through a stopcock in 3 h. Excess gas was evaporated after thawing to 4 °C and 8 h rest. A yellow (**2a**), orange (**2b**), and brown (**2c**) solid was obtained with quantitative yield. The versatile iminium salts **2** are only stable in the solid state and should be used for aminomethylations soon after their preparation.

References: G. Kaupp, J. Schmeyers, J. Boy, *J. Prakt. Chem.,* **342**, 269 (2000).

Type of reaction: elimination
Reaction condition: solvent-free
Keywords: ethyl acetoacetate, *o*-phenylendiamine, acid cleavage, decarboxylation, carboxylic ester

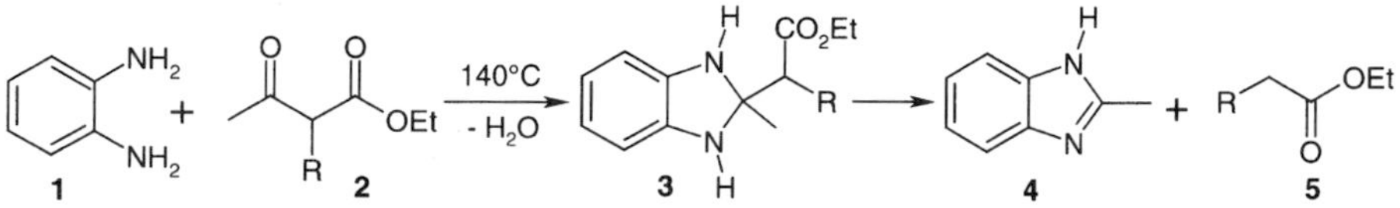

	R	yield (%)
a:	H	45
b:	Me	79
c:	Et	68
d:	*i*-Pr	61
e:	CH$_2$CH=CH$_2$	66
f:	COOEt	75
g:	CH$_2$COOEt	61
h:	CH$_2$CH$_2$COOEt	61
i:	CH$_2$Ph	67
j:	CH$_2$CH$_2$COOMe	65

Experimental procedures:

A mixture of 1,2-diaminobenzene **1** (0.20 mol) and *α*-substituted acetoacetate **2** (0.20 mol) was heated to 140 °C for 4 h, when water, small quantities of the volatile esters **5** and small quantities of ethanol (from a side reaction) distilled over. In the cases **a–e** the distillation of the ester was completed by raising the temperature to a maximum of 220 °C. Redistillation afforded the pure products **5a–e**. The esters **5f–i** were directly extracted from the cooled mixture with 4×60 mL CCl$_4$. The tetrachloromethane solutions were extracted with water, the organic phase dried, evaporated and filtered over SiO$_2$-60 (400 g) with CH$_2$Cl$_2$ followed by distillation under vacuum. **5j** was extracted with CH$_2$Cl$_2$ (3×50 mL) after addition of 20 mL 15% HCl and purified by distillation. The solid 2-methylbenzimidazol **4** was isolated from the residues of distillation (**a–e**) and extraction (**f–j**) by treatment with CCl$_4$ and filtration.

References: G. Kaupp, H. Frey, G. Behmann, *Synthesis*, 555 (1985).

11.2 Solvent-Free Elimination under Microwave Irradiation

Type of reaction: elimination
Reaction condition: solvent-free
Keywords: benzylamino alcohol, microwave irradiation, retro-Diels-Alder reaction, unsaturated benzylamino alcohol

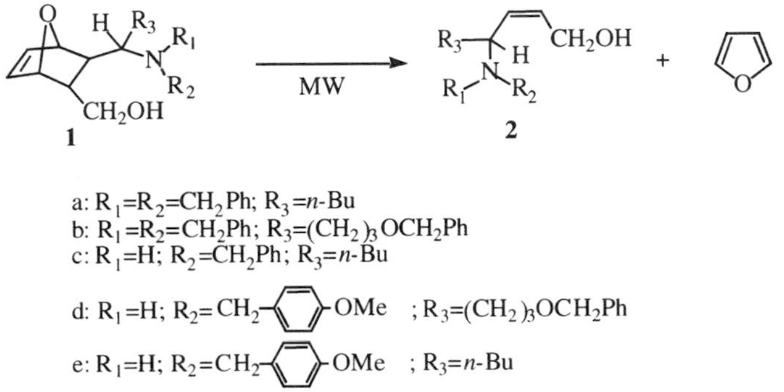

H R3
R1
N
R2
CH2OH
1
MW
R3
H
CH2OH
N
R1 R2
2
+
O
a: R1=R2=CH2Ph; R3=n-Bu
b: R1=R2=CH2Ph; R3=(CH2)3OCH2Ph
c: R1=H; R2=CH2Ph; R3=n-Bu
d: R1=H; R2=CH2 OMe ; R3=(CH2)3OCH2Ph
e: R1=H; R2=CH2 OMe ; R3=n-Bu

Experimental procedures:
Neat amino alcohols **1a–e** (10 mmol) were placed in a Pyrex open flask allowing the removal of furan. After microwave irradiation or conventional heating, the products were removed with methylene dichloride and analyzed by ^{1}H NMR spectroscopy. Unsaturated amino alcohols **2** obtained under microwaves were highly pure and did not need further purification.

References: M. Bortolussi, R. Bloch, A. Loupy, *J. Chem. Res. (S),* 34 (1998).

Type of reaction: elimination
Reaction condition: solvent-free
Keywords: α-halogen acetal, α-haogen thioacetal, KO*t*Bu, microwave irradiation, β-elimination, ketene *O,O*-acetal, ketene *S,S*-acetal

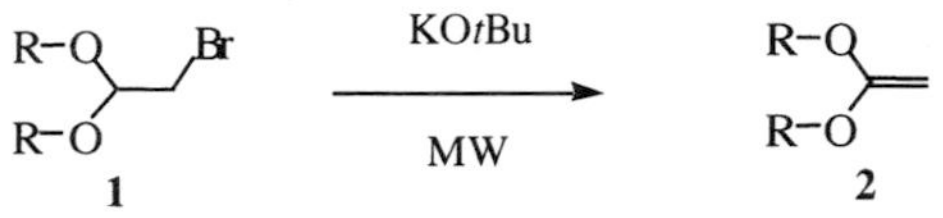

a: R,R=-(CH$_2$)$_4$-
b: R,R=-CH$_2$-CH=CH-CH$_2$-

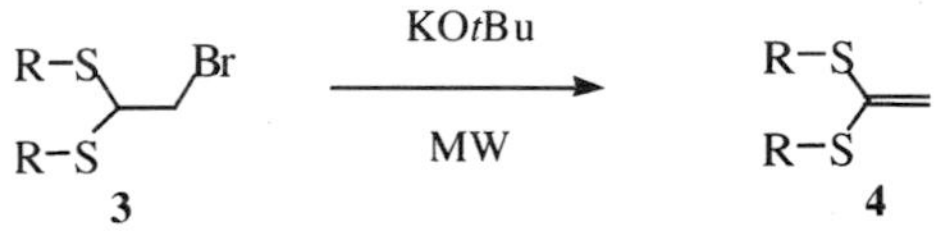

a: R,R=-(CH$_2$)$_2$-
b: R=n-Bu

Experimental procedures:

A cold mixture (0 °C) of the halogenated acetal (2 mmol), 5% of tetrabutylam-monium bromide (TBAB) (0.1 mmol), if it is necessary, and the appropriate base were introduced into a Pyrex tube which was then placed in a Synthewave 402 Prolabo microwave reactor fitted with a rotational system. Microwave irradiation was carried out for the time and at the power indicated. The crude reaction was purified by distillation or extraction to obtain the product.

References: A.D. Ortiz, P. Prieto, A. Loupy, D. Abenhaim, *Tetrahedron Lett.*, **37**, 1695 (1996).

12 Hydrolysis

12.1 Solvent-Free Hydrolysis

Type of reaction: hydrolysis
Reaction condition: solid-state
Keywords: phthalimide, substitution, gas-solid reaction

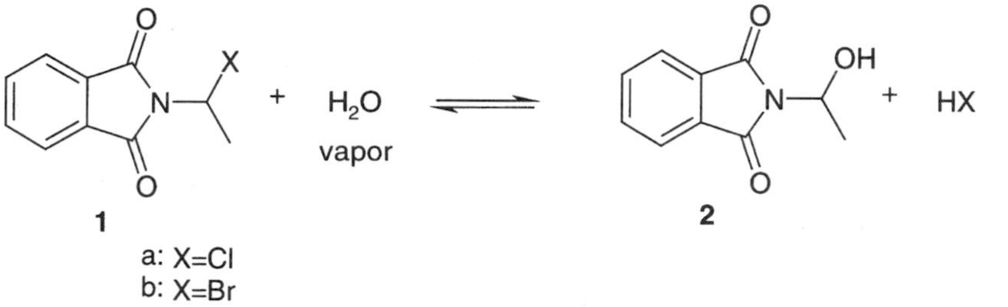

a: X=Cl
b: X=Br

Experimental procedures:
Compound **1a** (100 mg, 0.47 mmol) or **1b** (100 mg, 0.39 mmol) was placed on a glass frit and moist air was sucked through by an aspirator for 2 h. The slightly wettish crystals were dried in a high vacuum and consisted of pure **2** according to ^{1}H NMR analysis.

The reaction is reversible: If an excess of HX gas was applied to **2** the pure compound **1a** or **1b** was obtained.

References: G. Kaupp, D. Matthies, *Chem. Ber.,* **119**, 2387 (1986).

12.2 Solvent-Free Hydrolysis under Microwave Irradiation

Type of reaction: hydrolysis
Reaction condition: solvent-free
Keywords: nitrile, phthalic acid, microwave irradiation, hydrolysis, phthalimide

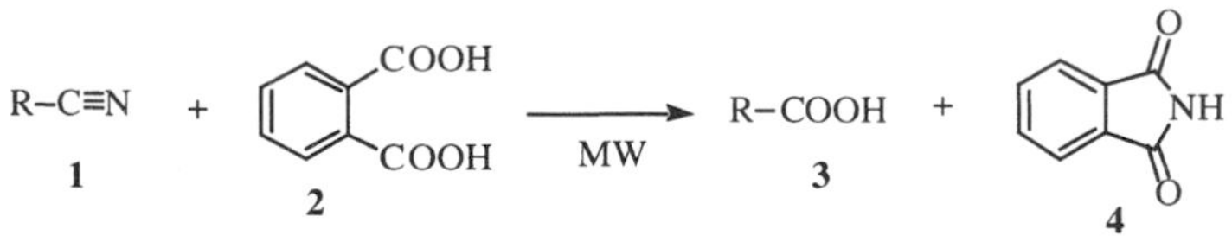

R–C≡N + [2, benzene ring with two COOH groups] $\xrightarrow{\text{MW}}$ R–COOH + [4, phthalimide structure with NH]

1   **2**   **3**   **4**

a: R=PhCH$_2$
b: R=Ph
c: R=Bu
d: R=(C$_3$H$_7$)$_2$CH
e: R=p-HOC$_6$H$_4$
f: R=ClC$_2$H$_4$
g: R= HOC$_2$H$_4$
i: R=C$_2$H$_5$OCOCH$_2$

Experimental procedures:

In a typical run, the nitrile **1** (30 mmol) and phthalic acid **2** (36 mmol) were introduced into the reactor, and heated under stirring. In the kinetic studies, time zero is taken at complete dissolution of the phthalic acid. At the desired reaction time, the reactor was rapidly cooled in a water-ice mixture and then chloroform (30 mL) was added. The mixture was stirred for 5 min and then the solid was filtered off. The chloroform solution contains the unchanged nitrile **1**, the amide and the carboxylic acid **3**. The residual solid contains unchanged phthalic acid **2**, phthalimide **4**, and as the major component, phthalic anhydride **5**. The volume of the chloroform solution was adjusted to 50 mL and naphthalene was added as an internal standard. The resulting solution was analyzed by GLC.

References: F. Chemat, M. Poux, J. Berlan, *J. Chem. Soc., Perkin Trans. 2*, 2597 (1994).

Type of reaction: hydrolysis
Reaction condition: solvent-free
Keywords: carbonic ester, hydrolysis, KF-Al$_2$O$_3$, microwave irradiation, carboxylic acid

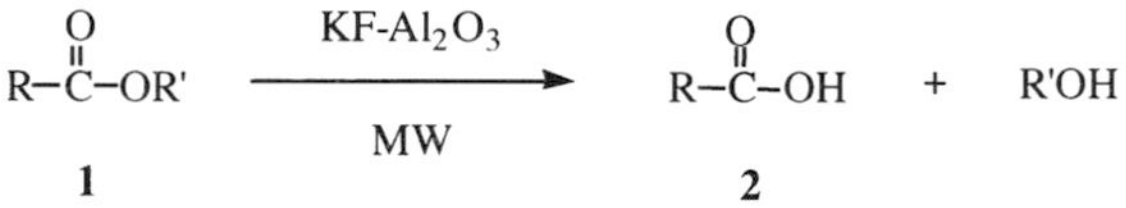

[R–C(=O)–OR'] $\xrightarrow[\text{MW}]{\text{KF-Al}_2\text{O}_3}$ [R–C(=O)–OH] + R'OH

1   **2**

a: R=Ph; R'=Me   h: R=p-NO$_2$C$_6$H$_4$; R'=Et
                                            i: R=Me(CH$_2$)$_{12}$; R'=Et
b: R=Ph; R'=n-Bu   j: R=Me(CH$_2$)$_{12}$; R'=i-Pr
c: R=Ph; R'=i-Pr   k: R=Me(CH$_2$)$_{12}$; R'=t-Bu
d: R=Ph; R'=t-Bu   l: R=Et(Me)$_2$; R'=n-Bu
e: R=o-IC$_6$H$_4$; R'=Et   m: R=CH$_2$=CH; R'=i-Pr
f: R=p-MeOC$_6$H$_4$; R'=Et

Experimental procedures:

Ethyl benzoate **1b** (0.300 g, 2.00 mmol) was added to KF/Al$_2$O$_3$ (1.00 g, 40 wt% KF) contained in a 10-mL round-bottomed flask. The mixture was stirred at room temperature to ensure efficient mixing. The flask was then fitted with a septum, placed in the microwave oven and irradiated at 100% power for 2 min (caution: heating volatile materials in commercial microwave ovens for extended periods can be hazardous). After cooling, water (5 mL) was added to the solid and stirred for 10 min (to ensure the potassium carboxylate was removed from the surface) and the mixture filtered. The filtrate was neutralized by addition of aqueous HCl. The product was filtered off and dried under vacuum to afford benzoic acid **2b** (0.240 g, 98%).

References: G.W. Kabalka, L. Wang, R.M. Pagni, *Green Chem.*, **3**, 261 (2001).

13 Protection

13.1 Solvent-Free Protection

Type of reaction: protection
Reaction condition: solid-state
Keywords: phenylboronic acid, cyclization, phenylboronic ester

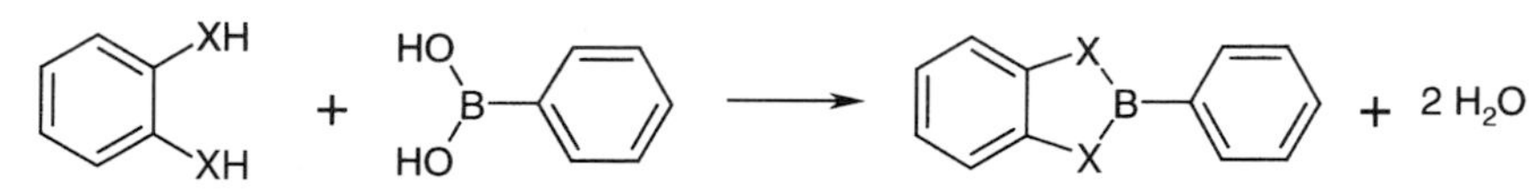

a: X=O   (80 °C)
b: X=NH   (40 °C)

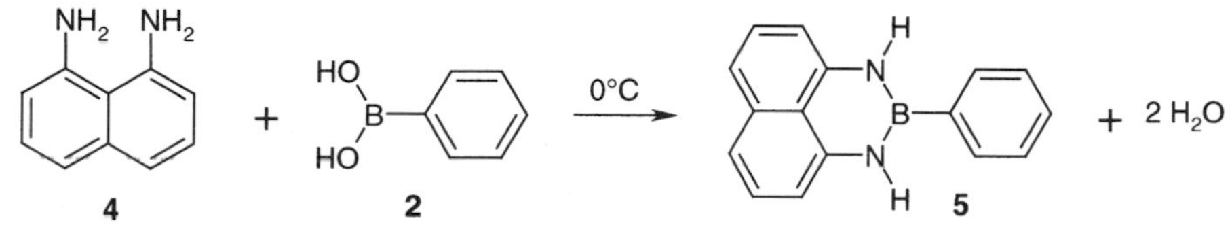

Experimental procedures:

Compound **1** or **4** (1.00 mmol) and phenylboronic acid **2** (1.00 mmol) were co-ground in a mortar at room temperature or 0 °C and heated for 1 h to the temperature given. The pure products **3a–d** or **5** were quantitatively obtained after drying in a vacuum at 80 °C.

The compounds **1** or **4** can be recovered by heating in aqueous NaHCO₃ to 40 °C.

References: G. Kaupp, V. A. Stepanenko, M. R. Naimi-Jamal, submitted to *Eur. J. Org. Chem.*

Type of reaction: protection
Reaction condition: solid-state
Keywords: glycols, pinacol, phenylboronic acid, phenylboronic ester

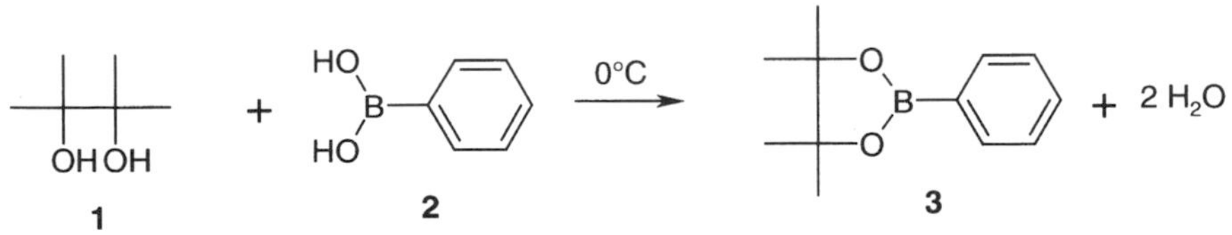

1
(mp 40-43°C)

2
(mp 217-220°C)

3
(mp 35-37°C)

4   HO

2

5

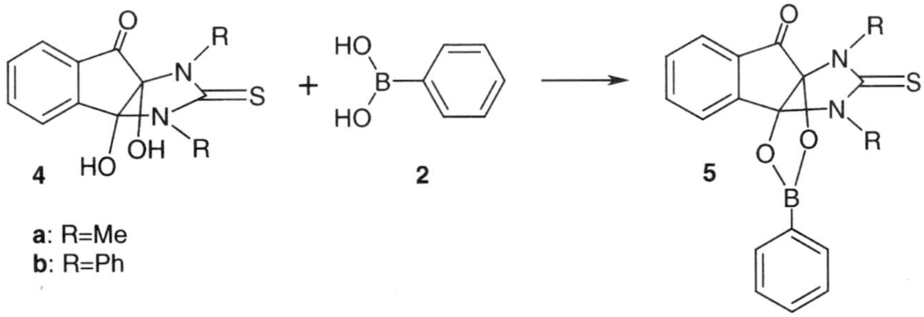

a: R=Me
b: R=Ph

Experimental procedures:

Pinacol **1** (1.00 mmol) and phenylboronic acid **2** were ball-milled at 0 °C for **1** h. The quantitatively obtained product **3** was dried in a vacuum. Mp 35–37 °C.

Compound **5a, b** was quantitatively obtained if an equimolar mixture of **4a, b** and **2** was ball-milled at 50 °C for 1 h or if the co-ground mixture was heated to 140 °C for 1 h.

References: G. Kaupp, V. A. Stepanenko, M. R. Naimi-Jamal, submitted to *Eur. J. Org. Chem.*

Type of reaction: protection
Reaction condition: solid-state
Keywords: phenylboronic acid, anthranilic acid, cyclization

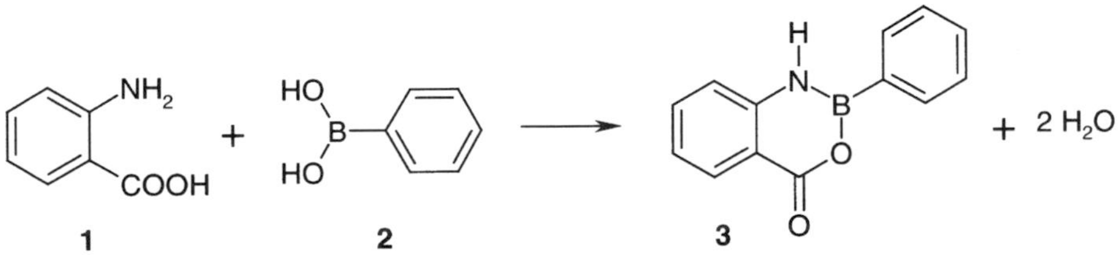

1

2

3

Experimental procedure:
Anthranilic acid **1** (274 mg, 2.00 mmol) and phenylboronic acid **2** (244 mg, 2.00 mmol) were ball-milled at room temperature for 1 h. The product **3** was quantitatively obtained after drying at 80 °C in a vacuum. Mp 223–224 °C.

References: G. Kaupp, V. A. Stepanenko, M. R. Naimi-Jamal, submitted to *Eur. J. Org. Chem.*

Type of reaction: protection
Reaction condition: solvent-free
Keywords: phenol, diethyl chlorophosphonate, magnesia, phosphorylation, aryl diethyl phosphate

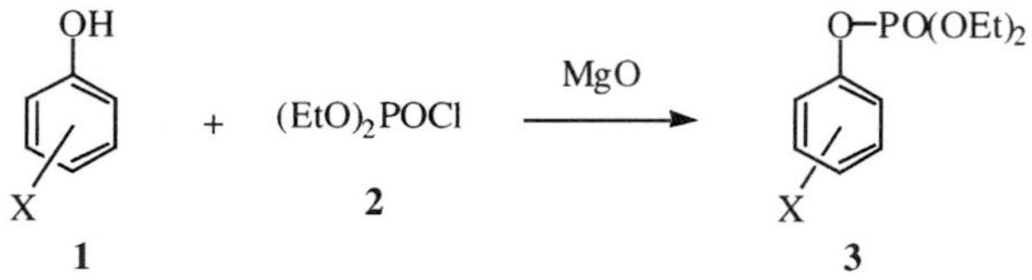

a: Ar=Ph
b: Ar=o-ClC$_6$H$_4$
c: Ar=p-MeC$_6$H$_4$
d: Ar=p-NO$_2$C$_6$H$_4$
e: Ar=m-BrC$_6$H$_4$

f: Ar=2,4-Cl$_2$C$_6$H$_3$
g: Ar=m-MeC$_6$H$_4$
h: Ar=o-NO$_2$C$_6$H$_4$
i: Ar=α-C$_{10}$H$_7$
j: Ar=β-C$_{10}$H$_7$

Experimental procedures:
General Procedure for the Phosphorylation of Phenol with Diethyl Chlorophosphonate on the Surface of Magnesia. Magnesia (0.3 g) was added to a mixture of diethyl chlorophosphonate (0.86 mL, 6 mmol) and the phenol (5 mmol). This mixture was stirred at room temperature or 60 °C for 0.25–1 h. The solid mixture was washed with dichloromethane (4×25 mL). The solution was then washed with a dilute NaOH solution and a saturated sodium chloride solution and dried over MgSO$_4$. After evaporating of solvent, the crude product was isolated in a pure state by distillation in vacuo in 85–95% yield.

References: B. Kaboudin, *J. Chem. Res. (S)*, 402 (1999).

13.2   Solvent-Free Protection under Microwave Irradiation

Type of reaction: protection
Reaction condition: solvent-free
Keywords: L-galactono-1,4-lactone, aldehyde, montmorillonite, microwave irradiation, acetals

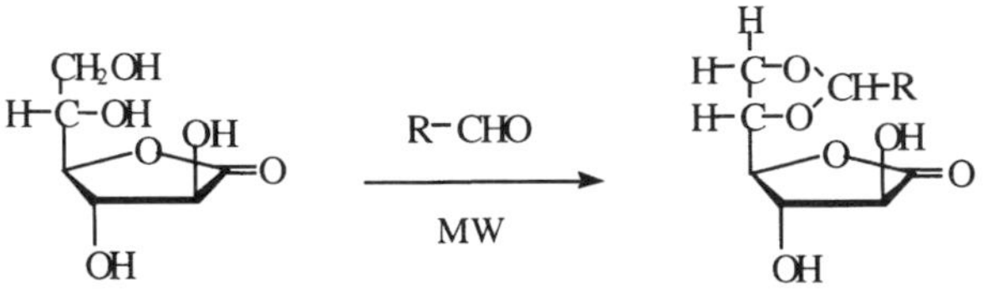

a: R=-(CH$_2$)$_4$-CH$_3$
b: R=-(CH$_2$)$_5$-CH$_3$
c: R=-(CH$_2$)$_6$-CH$_3$
d: R=-(CH$_2$)$_8$-CH$_3$
e: R=-(CH$_2$)$_{10}$-CH$_3$
f: R=-(CH$_2$)$_{12}$-CH$_3$

Experimental procedures:

The lactone (in aqueous solution) and the aldehyde (in CH$_2$Cl$_2$ solution) were impregnated separately on Fluka montmorillonite KSF, in the proportions 1:2 w/w, followed by subsequent removal of solvents under reduced pressure. One part of the supported lactone was mixed with 1.5 parts of the supported aldehyde. This mixture was placed in a tube, itself placed in another tube (in order to collect the water evaporated during the reaction on its walls), which was then introduced into the Maxidigest microwave reactor. Microwave irradiation was carried out at 60 W for 10 min. The final temperature (140–155 °C) was measured by introducing in the montmorillonite a digital thermometer at the end of the irradiation. The mixture was then cooled to ambient temperature. After elution with CH$_2$Cl$_2$ and subsequent filtration, the major organic products were crystallized (EtOAc-heptane).

References: M. Csiba, J. Cleophax, A. Loupy, J. Malthete, S. D. Gero, *Tetrahedron Lett.*, **34**, 1787 (1993).

Type of reaction: protection
Reaction condition: solvent-free
Keywords: aldehyde, ketone, orthoformate, KSF clay, microwave irradiation, acetals

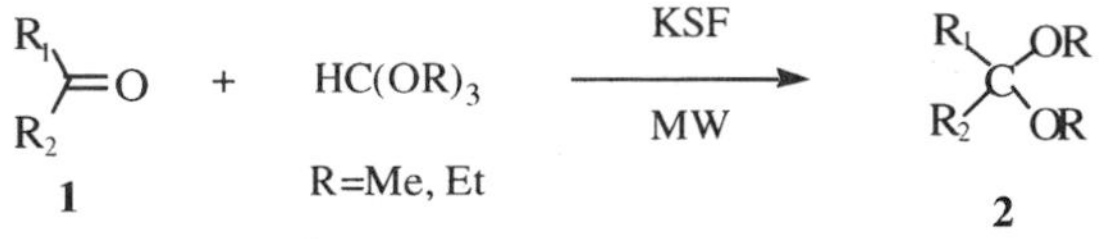

a: $R_1=C_6H_5(CH_2)_2$; $R_2=H$
b: $R_1=Pr$; $R_2=H$
c: $R_1=C_6H_5CH=CH$; $R_2=H$
d: $R_1=Ph$; $R_2=H$
e: $R_1=4\text{-}NO_2C_6H_4$; $R_2=H$
f: $R_1=3,4\text{-}(OCH_2O)C_6H_3$; $R_2=H$

g: $R_1=Ph$; $R_2=Me$
h: $R_1=R_2=Ph$
i: $R_1,R_2=cyclohexyl$
j: $R_1=Ph$; $CH=CH$; $R_2=Me$
k: $R_1=i\text{-}Bu$; $R_2=Me$

Experimental procedures:

The carbonyl compound (10 mmol) was introduced in the reactor of a Synthe-wave 402 single mode apparatus followed by the acidic catalyst, para-toluene sulfonic acid (PTSA) 10% or montmorillonite clay KSF 1 g (10 mmol) and the reagent (methyl or ethyl orthoformate, 2 equiv.). After irradiation, the product is filtered or washed by a solution of $NaHCO_3$ to remove the catalyst and the product is purified by distillation or recrystallization.

References: B. Perio, M.J. Dozias, P. Jacquault, J. Hamelin, *Tetrahedron Lett.,* **38**, 7867 (1997).

Type of reaction: protection
Reaction condition: solvent-free
Keywords: aldehyde, ethane-1,2-diol, SiO_2-$NaHSO_4$, microwave irradiation, acetalization, acetal

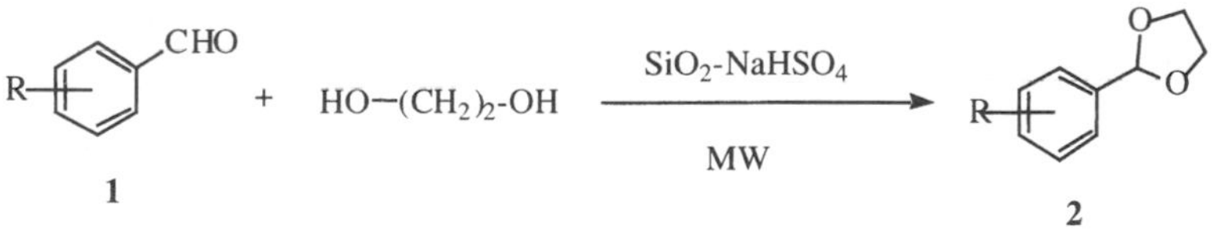

a: R=H
b: R=2-NO_2
c: R=2-MeO
d: R=2-Cl
e: R=3-CN
f: R=3-PhO
g: R=3-PhCH$_2$O

Experimental procedures:
Aldehyde (5 mmol), ethane 1,2-diol (5 mmol) and metal sulfate (5 mmol) supported on silica gel (1.65 g) were mixed in a Pyrex test tube and subjected to microwave irradiation for 36 min. After complete conversion, as indicated by TLC, the reaction mass was charged directly on small silica gel column (100–200 mesh) and eluted with ethylacetate-hexane (2:8) to afford pure acetal in 80–98% yield.

References: J. S. Yadav, B. V. S. Reddy, R. Srinivas, T. Ramalingam, *Synlett,* 701 (2000).

Type of reaction: protection
Reaction condition: solvent-free
Keywords: aldehyde, ketone, ethane-1,2-diol, cadmium iodide, microwave irradiation, acetalization, acetal

1: PhCHO, 4-ClC$_6$H$_4$CHO, 4-NO$_2$C$_6$H$_4$CHO, CH$_3$CH=CHCHO, PhCH=CHCHO, PhCOCHO

Experimental procedures:
A mixture of benzaldehyde (1.06 g, 10 mmol), ethane-1,2-diol (0.62 g, 10 mmol) and commercial grade cadmium iodide (1.85 g, 5 mmol) were thoroughly mixed at room temperature in an Erlenmeyer flask and placed in a commercial microwave oven operating at 2450 MHz frequency. After irradiation of the mixture for 1.5 min (monitored vide TLC) it was cooled to room temperature, extracted with dichloromethane, washed with sodium thiosulfate and dried over anhydrous Na$_2$SO$_4$. Evaporation of the solvent gave almost pure products and there was no evidence for the formation of any hydroxy ester or iodoester. Further purification was achieved by column chromatography on silica gel using 1:5 chloroform-petroleum ether as eluent.

References: J. A. Thakuria, M. Baruah, J. S. Sandhu, *Chem. Lett.,* 995 (1999).

Type of reaction: protection
Reaction condition: solvent-free
Keywords: aldehyde, acetic anhydride, montmorillonite K-10, microwave irradiation, 1,1-diacetate

$$R\text{-}CHO \quad + \quad Ac_2O \quad \xrightarrow[\text{MW}]{\text{K-10 clay}} \quad R\text{-}CH(OAc)_2$$

$$\mathbf{1} \qquad\qquad \mathbf{2} \qquad\qquad\qquad\qquad \mathbf{3}$$

a: R=*p*-ClC$_6$H$_4$
b: R=*m*-ClC$_6$H$_4$
c: R=*p*-MeOC$_6$H$_4$
d: R=*p*-NO$_2$C$_6$H$_4$
e: R=*p*-MeC$_6$H$_4$
f: R=Ph
g: R=PhCH=CH
h: R=MeCH=CH

Experimental procedures:
A mixture of *p*-chlorobenzaldehyde **1a** (1.41 g, 10 mmol), acetic anhydride **2** (2.04 g, 20 mmol) and montmorillonite K-10 clay (0.3 g) in an Erlenmeyer flask at room temperature was placed in a commercial microwave oven (operating at 2450 MHz frequency) and irradiated for 5 min. Upon completion, the reaction mixture was allowed to reach room temperature (monitored via TLC), water was added and the K-10 clay was filtered off. The catalyst was washed with dichloromethane (2×20 mL) and then the filtrate was extracted with dichloromethane (3×20 mL). The combined extract was washed with 10% HCl solution (20 mL) and brine (2×20 mL) dried over anhydrous sodium sulfate and distilled. The *p*-chlorobenzylidene diacetate **3a** thus obtained was purified by column chromatography using dichloromethane-light petroleum (1:1) as the eluent, mp 80–81 °C. Similarly other aldehydes gave the corresponding 1,1-diacetates in 75–96% yields.

References: D. Karmakar, D. Prajapati, J.S. Sandhu, *J. Chem. Res. (S)*, 382 (1998).

Type of reaction: protection
Reaction condition: solvent-free
Keywords: aldehyde, acetic anhydride, microwave irradiation, Envirocat EPZ10, 1,1-diacetate

$$\text{RCHO} \quad + \quad \text{Ac}_2\text{O} \quad \xrightarrow[\text{MW}]{\text{Envirocat EPZ10}} \quad \text{RCH(OAc)}_2$$

1   **2**

a: $R=Et$
b: $R=Pr$
c: $R=C_5H_{11}$
d: $R=C_8H_{17}$
e: $R=4\text{-Br}C_6H_4$
f: $R=4\text{-Cl}C_6H_4$

g: $R=2,4\text{-Cl}_2C_6H_3$
h: $R=2\text{-NO}_2C_6H_4$
i: $R=3\text{-NO}_2C_6H_4$
j: $R=4\text{-NO}_2C_6H_4$
k: $R=3\text{-NO}_2\text{-} 4\text{-Cl}C_6H_3$
l: $R=3,4,5\text{-(MeO)}_3C_6H_2$

Experimental procedures:

Envirocat EPZ10 was obtained from Contract Chemicals, England, and used without activation. A mixture of the aldehyde (5 mmol), acetic anhydride (10 mmol) and Envirocat EPZ10 (100 mg) was irradiated with microwaves. After completion of the reaction (TLC), diethyl ether (2×10 mL) was added to the reaction mixture, and EPZ10 was filtered off. After drying (Na_2SO_4), the ether was evaporated under reduced pressure to give the products in high yield and in almost pure form.

References: B. P. Bandger, S. S. Makone, S. R. Kulkarni, *Monatsh. Chem.,* **131**, 417 (2000).

14 Deprotection

14.1 Solvent-Free Deprotection

Type of reaction: deprotection
Reaction condition: solvent-free
Keywords: *N-tert*-butoxycarbonyl group, Yb(OTf)$_3$, silica gel, amide

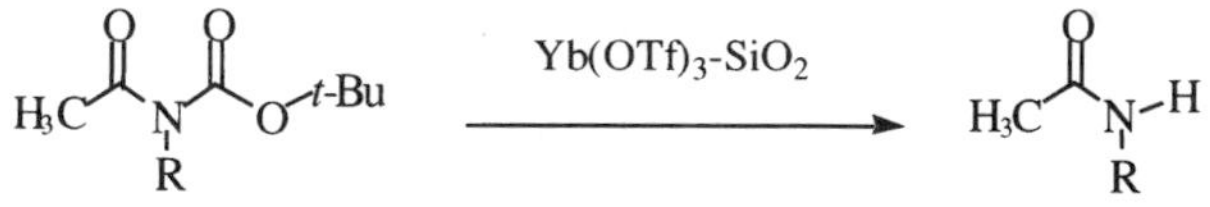

a: R=Ph
b: R=CH$_2$Ph

Experimental procedures:
A mixture of the starting material (2 mmol) and 9% Yb(OTf)$_3$-SiO$_2$ (800 mg) was suspended in 2 mL of CH$_2$Cl$_2$ and the solvent was removed in vacuo. After standing at either room temperature or at 40 °C, the product was extracted into ethyl acetate. Concentration and purification by silica gel column chromatography or preparative TLC gave the corresponding free amide.

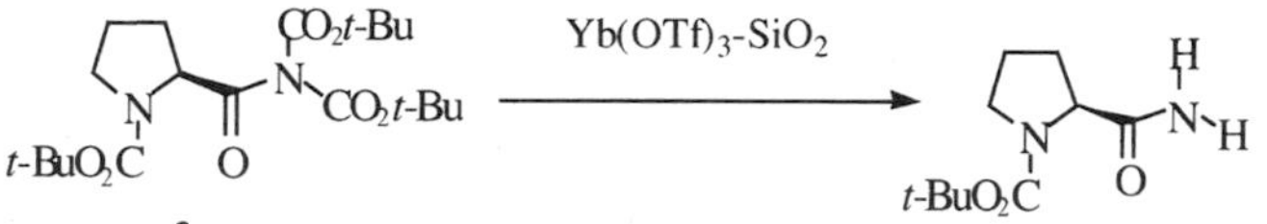

References: H. Kotsuki, T. Ohishi, T. Araki, K. Arimura, *Tetrahedron Lett.,* **39**, 4869 (1998).

Type of reaction: deprotection
Reaction condition: solvent-free
Keywords: oximes, semicabazone, 1-benzyl-4-aza-1-azoniabicyclo[2.2.2]octane, dichromate, aldehyde, ketone

a: X=OH; R=H; R$_1$= 5-Me-furyl
b: X=OH; R=H, R$_1$ = 3,4-MeO$_2$C$_6$H$_3$
c: X=OH; R=Me; R$_1$= 4-BrC$_6$H$_4$
d: X=OH; R=Me; R$_1$ = 4-ClC$_6$H$_4$
e: X=OH; R=CH$_2$Br; R$_1$= 4-BrC$_6$H$_4$
f: X=OH; R,R$_1$=-(CH$_2$)$_4$-
g: X=OH; R,R$_1$=-(CH$_2$)$_5$-
h: X=OH; R=Me; R$_1$ = Ph

i: X=OH; R=R$_1$=Ph
j: X=NHCONH$_2$; R=H; R$_1$= 5-Me-furyl
k X=NHCONH$_2$; R=H, R$_1$ = 2,5-MeO$_2$C$_6$H$_3$
l: X=NHCONH$_2$; R=Me; R$_1$= 4-BrC$_6$H$_4$
m:X=NHCONH$_2$; R=Me; R$_1$= 3-ClC$_6$H$_4$
n: X=NHCONH$_2$; R=CH$_2$Br; R$_1$= 4-BrC$_6$H$_4$
o: X=NHCONH$_2$; R=Me; R$_1$= 3-MeOC$_6$H$_4$
p: X=NHCONH$_2$; R=Me; R$_1$= 3-NO$_2$C$_6$H$_4$

Experimental procedures:
A mixture of benzophenone oxime (0.16 g, 0.803 mmol), 1-benzyl-4-aza-1-azo-niabicyclo[2.2.2]octane dichromate (0.50 g, 0.803 mmol), and aluminum chloride (11 mg, 0.08 mmol) was crushed with a mortar and pestle for 35 s until TLC showed complete disappearance of starting oximes. CCl$_4$ (10 mL) was added to the reaction mixture and, after vigorous stirring, the mixture was filtered off. The solvent was evaporated by rotary evaporator to give pure benzophenone (0.129 g, 88%).

References: A. R. Hajipour, S. E. Mallakpour, I. Mohammadpoor-Baltork, S. Khoee, *Synth. Commun.*, **31**, 1187 (2001).

14.2   Solvent-Free Deprotection under Microwave Irradiation

Type of reaction: deprotection
Reaction condition: solvent-free
Keywords: *tert*-butylsimethylsilyl ether, alumina, microwave irradiation, alcohol

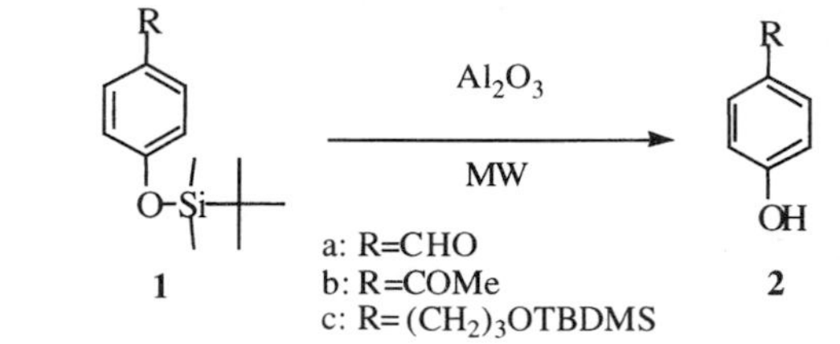

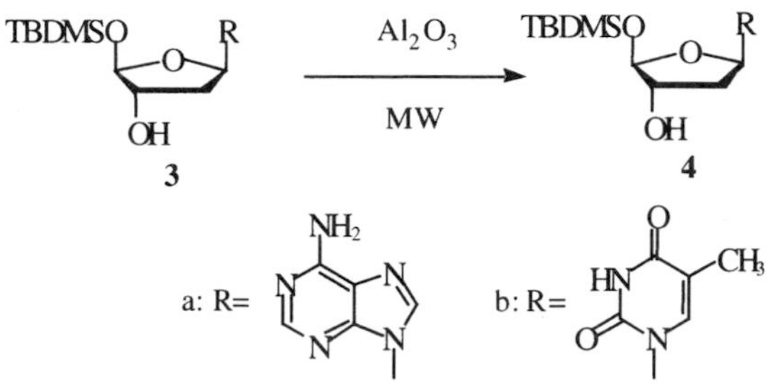

a: R=CHO
b: R=COMe
c: R=(CH₂)₃OTBDMS

a: R=

b: R=

Experimental procedures:

Neutral alumina (35 g) is added to a solution of **1c** (0.760 g, 2 mmol) dissolved in a minimum amount of dichloromethane (5 mL) at room temperature and the reaction mixture is thoroughly mixed using a vortex mixer. The adsorbed material is dried in air (beaker) and placed in an alumina bath inside the microwave oven. Upon completion of the reaction as followed by TLC examination (10 min), the product is extracted into dichloromethane (4×15 mL). Removal of the solvent, under reduced pressure yielded the product which is purified by crystallization from methanol-dichloromethane. Alternatively, the adsorbed material is charged directly on a silica gel column to afford the pure product, 3-(4-hydroxyphenyl)-1-propanol **2c** in 78% yield, mp 53–55 °C, in ethyl acetate-methanol (4:1, v/v) as an eluent.

References: R. S. Varma, J. B. Lamture, M. Varma, *Tetrahedron Lett.,* **34**, 3029 (1993).

Type of reaction: deprotection
Reaction condition: solvent-free
Keywords: benzyl esters, alumina, solid surface, microwave irradiation, carboxylic acid

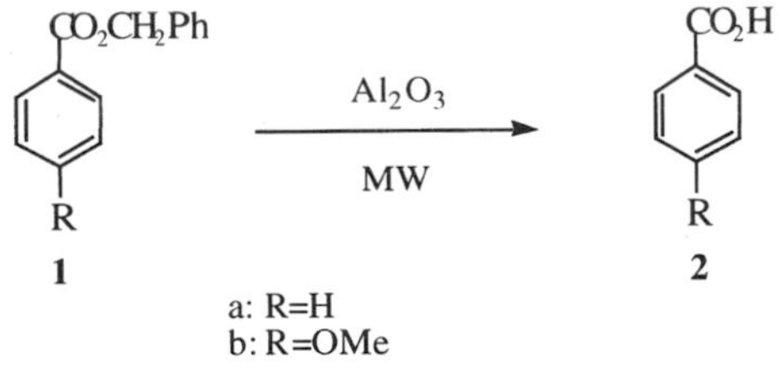

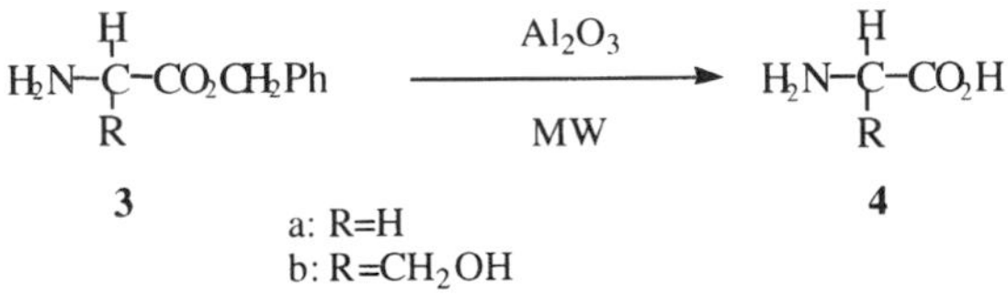

1
a: R=H
b: R=OMe

2

3
a: R=H
b: R=CH$_2$OH

4

Experimental procedures:

Acidic alumina (6.2 g) is added to a solution of benzyl ester (0.110 g, 0.52 mmol) dissolved in a minimum amount of dichloromethane (1×2 mL) at room temperature and the reaction mixture was thoroughly mixed using a vortex mixer. The adsorbed material is dried in air (beaker) and placed in an alumina bath inside the microwave oven. Upon completion of the reaction as followed by TLC examination (7 min), the products are extracted successively into hexane-ether (4:1, v/v; 2×5 mL) and methanol-dichloromethane (4:1, v/v; 4×5 mL). Removal of the methanol extract, under reduced pressure afforded pure benzoic acid (0.058 g, 92%). The hexane-ether extraction removes only the byproduct, benzyl alcohol.

References: R. S. Varma, A. K. Chatterjee, M. Varma, *Tetrahedron Lett.*, **34**, 4603 (1993).

Type of reaction: deprotection
Reaction condition: solvent-free
Keywords: methyl aryl ether, pyridine hydrochloride, microwave irradiation, phenol

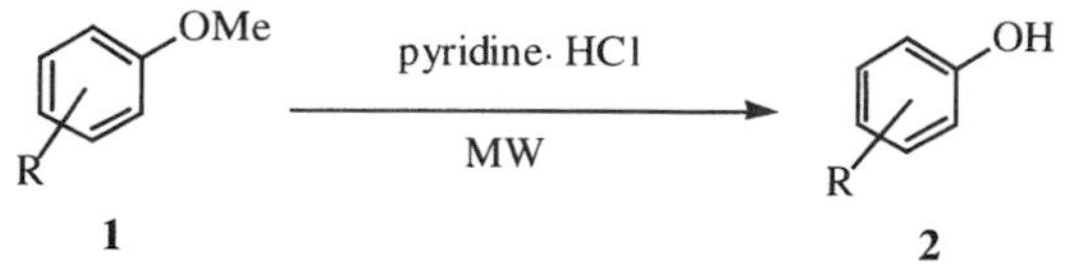

OMe → pyridine·HCl / MW → OH

1 **2**

a: R=H f: R=*p*-MeCO
b: R=*o*-Me g: R=*p*-Br
c: R=*m*-Me h: R=*p*-Cl
d: R=*p*-Me i: R=*o*-NO$_2$
e: R=*p*-CHO j: R=β-naphthyl

Experimental procedures:

A mixture of methyl aryl ether **1** (0.01 mol) and pyridine hydrochloride (0.05 mol) were placed in a stoppered round bottom flask and subjected to microwave irradiation at 215 W for various time intervals. After complete conversion the reaction mixture was decomposed using ice-water and extracted with diethyl ether. The ether extract was repeatedly washed with water, dried over anhydrous sodium sulfate and the ether removed to obtain the product **2**.

References: P.P. Kulkarni, A.J. Kadam, R.B. Mane, U.D. Desai, P.P. Wadgaonkar, *J. Chem. Res. (S)*, 394 (1999).

Type of reaction: deprotection
Reaction condition: solvent-free
Keywords: sulfonate, sulfonamide, KF-Al$_2$O$_3$, microwave irradiation, alcohol, amine

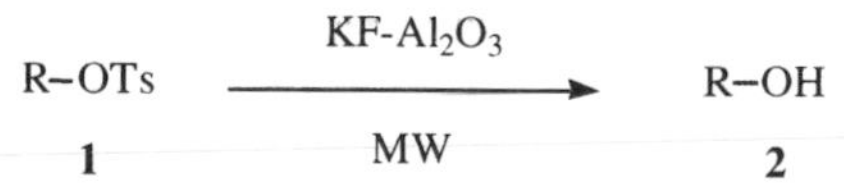

R–OTs → KF-Al$_2$O$_3$ / MW → R–OH

1 **2**

a: R=3,4-(CH$_2$O$_2$)C$_6$H$_3$ h: R=4-(CH$_2$=CH-CH$_2$O)C$_6$H$_4$
b: R=2-naphthyl i: R=4-(CH$_3$COOCH$_2$)C$_6$H$_4$
c: R=3,4-(CH$_3$)C$_6$H$_3$ j: R=4-(MeO)$_2$CHC$_6$H$_4$
d: R=4-NO$_2$C$_6$H$_4$ k: R=4-ClCH$_2$C$_6$H$_4$
e: R=4-ClC$_6$H$_4$ l: R=PhCH$_2$CH$_2$
f: R=4-CH$_3$OC$_6$H$_4$ m: R=C$_6$H$_{13}$
g: R=4-PhCH$_2$OC$_6$H$_4$ n: R=*c*-C$_6$H$_{11}$

R–N–R' (Ts) → KF-Al$_2$O$_3$ / MW → R–N–R' (H)

3 **4**

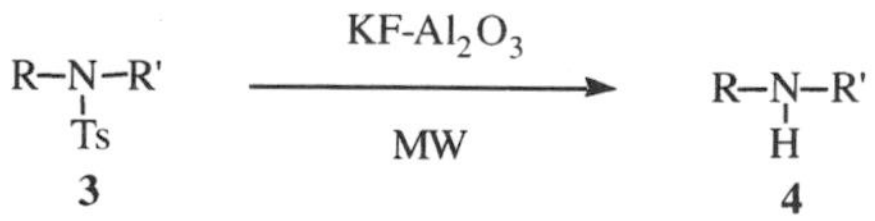

Experimental procedures:

Typical procedure for the cleavage of sulfonate: Piperonyl-toluenesulfonate **1a** (2.92 g, 10 mmol) and 37% KF on Al_2O_3 (3 weight equivalents of sulfonate) were admixed in a Pyrex test tube and subjected to microwave irradiation for 3 min. After cooling down to room temperature, the solid mass was extracted with ethyl acetate and concentrated in vacuo. The crude product was subsequently purified by column chromatography on silica gel (100–200 mesh, ethyl acetate-hexane, 2:8) to afford sesmol **2a** (1.22 g, 88% yield) as white solid.

Typical procedure for the cleavage of sulfonamides: Indole sulfonamide (2.71 g, 10 mmol) and 37% KF on Al_2O_3 (3 weight equivalents of sulfonamide) were admixed in a Pyrex test tube and subjected to microwave irradiation for 5 min. The solid mass was allowed to cool to room temperature, and was extacted with ethyl acetate (2×15 mL). The organic layer was concentrated in vacuo and was purified by column chromatography on silica gel (100–200 mesh, ethyl acetate-hexane 2:8) to afford indole (1.03 g, 88% yield) as a white solid.

References: G. Sabitha, S. Abraham, B.V.S. Reddy, J.S. Yadav, *Synlett,* 1745 (1999).

Type of reaction: deprotection
Reaction condition: solvent-free
Keywords: trimethylsilyl ether, montmorillonite K-10, microwave irradiation, phenol, alcohol

Experimental procedures:

Desilylation of phenols or alcohols in the presence of montmorillonite K-10 clay
1 mmol of silyl ether was placed in a 5-mL beaker, and 0.1 g of activated montmorillonite K-10 clay was added. The beaker was placed in a 50-mL Teflon container and irradiated in a microwave oven (900 W) for 1 to 5 min. The progress of the reaction was monitored by GLC. After completion of the reaction, the product was extracted with ether or CH_2Cl_2, filtered, and the solvent was evaporated under reduced pressure to yield the corresponding phenol.

Desilylation of phenols or alcohols in the presence of $PdCl_2(PhCN)_2$
1 mmol of silyl ether was placed in a 5-mL beaker, and 0.01 g of $PdCl_2(PhCN)_2$ (1 mol%) was added. Then one drop of water was added to the reaction mixture. The beaker was placed in a 50-mL Teflon container and irradiated in a micro-

wave oven (900 W) for 3 to 5 min. The progress of the reaction was monitored by GLC. After completion of the reaction, the product was extracted with ether or CH_2Cl_2, filtered, and the solvent was evaporated under reduced pressure to yield the corresponding phenol.

References: M.M. Mojtahedi, M.R. Saidi, M.M. Heravi, M. Bolourtchian, *Monatsh. Chem.,* **130**, 1175 (1999).

Type of reaction: deprotection
Reaction condition: solvent-free
Keywords: *tert*-butoxycarbonyl amide, microwave irradiation, amine, ester

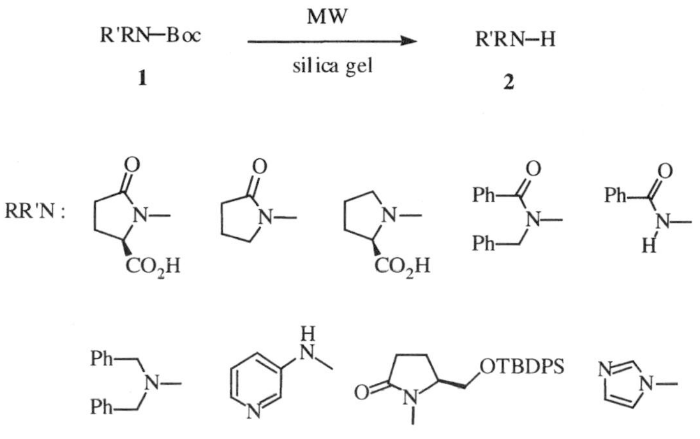

Experimental procedures:
N-Boc derivative (1 mmol) was dissolved in CH_2Cl_2 (50 mL) and silica gel (230–400 mesh, 10 g) was added. The solvent was taken off in vacuo and the powdered solid obtained was irradiated in the microwave oven, in an open Erlenmeyer flask at 450 W. The resulting solid was thoroughly washed with acetone or methanol, and then isolated to afford pure products.

References: J.G. Siro, J. Martin, J.L. Garcia-Navio, M.J. Remuinan, J.J. Vaquero, *Synlett,* 147 (1998).

Type of reaction: deprotection
Reaction condition: solvent-free
Keywords: *N-tert*-butoxycarbonyl group, aluminium chloride, microwave irradiation, amine

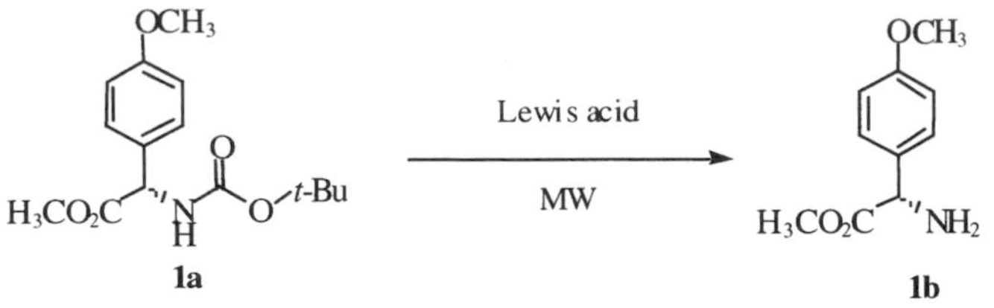

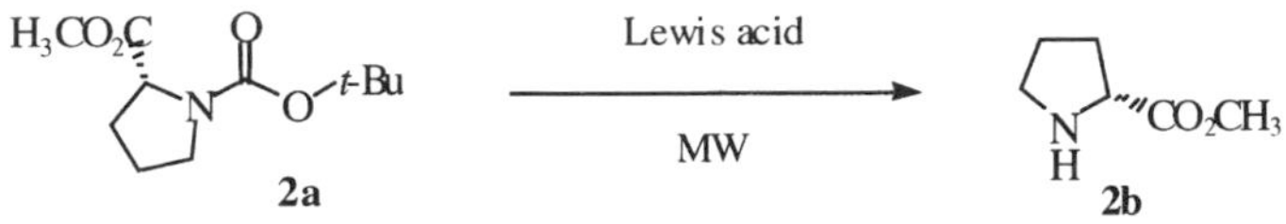

la → lb (Lewis acid, MW)

2a → 2b (Lewis acid, MW)

Experimental procedures:

The deprotection of *t*-Boc proline ester **2a** is representative of the general procedure employed. *tert*-Butyl carbamate (0.217 g, 1.0 mmol) and aluminium chloride (0.134 g, 1.0 mmol) doped on a neutral alumina (1.0 g) were mixed thoroughly on a vortex mixer. The reaction mixture was placed in an alumina bath inside an unmodified household microwave oven (operating at frequency 2450 MHz) and irradiated for a period of 1 min. After completion of the reaction (monitored by TLC, EtOAc-hexane, 9:1 v/v), it was neutralized with aqueous sodium bicarbonate solution and the product was extracted into ethyl acetate (2×15 mL). The ethyl acetate layer was separated, dried over magnesium sulfate, filtered, and the crude product thus obtained was purified by column chromatography to afford pure methyl ester **2b** in 88% yield.

References: D.S. Bose, V. Lakshminarayana, *Tetrahedron Lett.*, **39**, 5631 (1998).

Type of reaction: deprotection
Reaction condition: solvent-free
Keywords: benzaldehyde diracetate, alumina, microwave irradiation, deacetylation, benzaldehyde

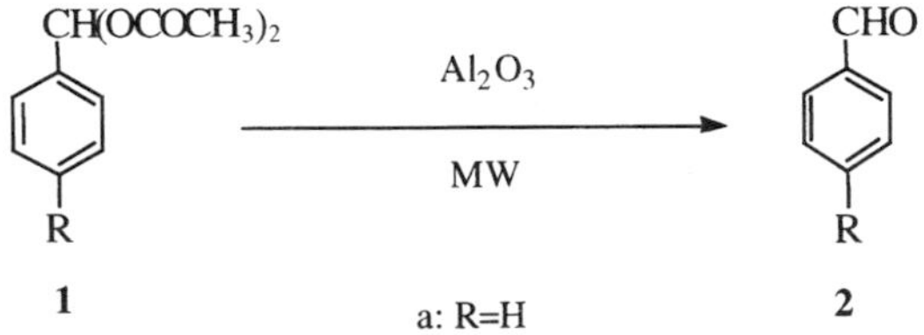

1 → 2 (Al_2O_3, MW)

a: R=H
b: R=Me
c: R=NO$_2$
d: R=CN

Experimental procedures:
Neutral alumina (3.6 g) is added to a solution of benzaldehyde diacetate **1a** (0.054 g, 0.284 mmol) dissolved in a minimum amount of dichloromethane (1–2 mL) at room temperature and the reaction mixture was thoroughly mixed using a vortex mixer. The adsorbed material is dried in air (beaker) and placed in an alumina bath inside the microwave oven. Upon completion of the reaction as followed by TLC examination (40 s), the product is extracted into dichloromethane (4×5 mL). Removal of the solvent, under reduced pressure afforded essentially quantitative yield of benzaldehyde **2a**.

References: R. S. Varma, A. K. Chatterjee, M. Varma, *Tetrahedron Lett.*, **34**, 3207 (1993).

Type of reaction: deprotection
Reaction condition: solvent-free
Keywords: benzaldehyde diacetate, deacetylation, zeolite HSZ-360, microwave irradiation, benzaldehyde

a: R=Ph
b: R=NO$_2$
c: R=OMe
d: R=Me
e: R=H

Experimental procedures:
0.5 g of 1,1-diacetate **1** and 0.5 g of zeolite HSZ-360 were placed, at 500 W, in a domestic oven, for 20 min. After cooling, the mixture was extracted with Et$_2$O and the catalyst was filtered off. Evaporation of the solvent followed by flash chromatography of the mixture produced the pure aldehyde **2**.

References: R. Ballini, M. Bordoni, G. Bosica, R. Maggi, G. Sartori, *Tetrahedron Lett.*, **39**, 7587 (1998).

Type of reaction: deprotection
Reaction condition: solvent-free
Keywords: tetrahydropyranyl ether, iron(III) nitrate, clay, microwave irradiation, ketone, aldehyde

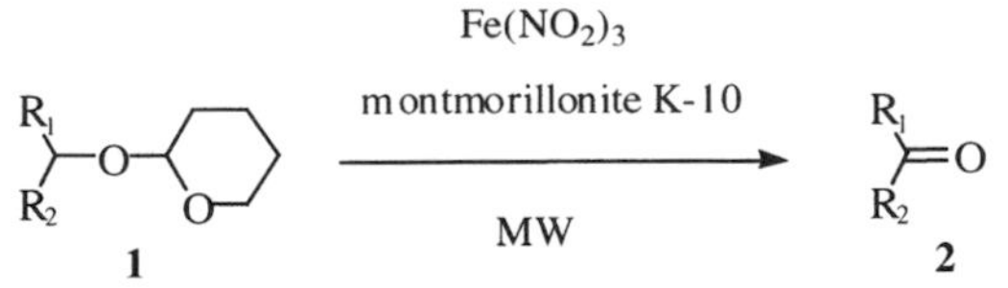

$$\text{1} \quad \xrightarrow[\text{MW}]{\substack{\text{Fe(NO}_2)_3 \\ \text{montmorillonite K-10}}} \quad \text{2}$$

Fe(NO$_2$)$_3$

montmorillonite K-10

MW

R$_1$, R$_2$ — O — (tetrahydropyranyl) **1**

R$_1$, R$_2$ =O **2**

a: R$_1$=Ph; R$_2$=H
b: R$_1$=4-MeC$_6$H$_4$; R$_2$=H
c: R$_1$=2-NO$_2$-5-MeC$_6$H$_3$; R$_2$=H
d: R$_1$=Ph; R$_2$=Me
e: R$_1$, R$_2$=Ph
f: R$_1$=C$_6$H$_5$CH=CH; R$_2$=H
g: R$_1$, R$_2$=-(CH$_2$)$_5$-

Experimental procedures:

Montmorillonite K-10 (1 g) was mixed thoroughly with Fe(NO$_2$)$_3\cdot$9H$_2$O (0.404 g, 1 mmol) using a pestle and mortar. This mixture was added to neat tetrahydropyranyl ether **1** (1 mmol) in a beaker and was placed into a microwave oven (900 W). The mixture was irradiated for the indicated time. After completion of the reaction, which was monitored by TLC or GC, the crude product was extracted with dichloromethane. GC analysis of the crude product showed that **2** were obtained in high to excellent yields. Final purification was achieved by column chromatography using hexane-ethyl acetate as eluent.

References: M. Heravi, D. Ajami, M.M. Mojtahedi, M. Ghassemzadeh, *Tetrahedron Lett.*, **40**, 561 (1999).

Type of reaction: deprotection
Reaction condition: solvent-free
Keywords: bisulfite, montmorillonite KSF clay, microwave irradiation, aldehyde

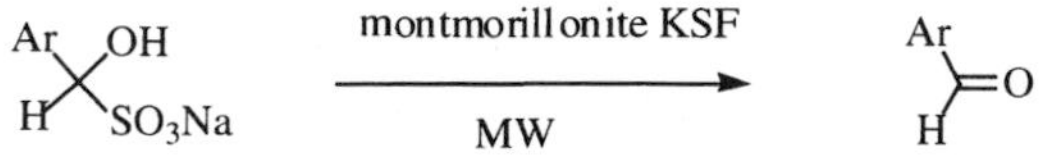

$$\text{Ar—C(OH)(H)(SO}_3\text{Na)} \xrightarrow[\text{MW}]{\text{montmorillonite KSF}} \text{Ar—C(H)=O}$$

1 → **2**

a: Ar=2-HOC$_6$H$_4$
b: Ar=4-MeOC$_6$H$_4$
c: Ar=3-MeOC$_6$H$_4$
d: Ar=4-ClC$_6$H$_4$
e: Ar=3-NO$_2$C$_6$H$_4$
f: Ar=4-NO$_2$C$_6$H$_4$
g: Ar=3-MeO-4-OHC$_6$H$_3$
h: Ar=3-HO-4-MeOC$_6$H$_3$
i: Ar=3,4-(MeO)$_2$C$_6$H$_3$
j: Ar=3,4-O-CH$_2$OCH$_2$-
k: Ar=3-MeO-4-PhCH$_2$O-C$_6$H$_3$
l: Ar=3-PhCH$_2$O-4-MeO-C$_6$H$_3$

Experimental procedures:

A mixture of bisulfite addition product (1 mmol) and montmorillonite KSF (300 mg) was taken in a 25-mL Erlenmeyer flask and kept over an alumina bath (heat sink) inside a domestic microwave oven and irradiated for 520 s and the reaction was monitored by TLC. The product was extracted with ethyl acetate (2×5 mL), washed with brine and dried over anhydrous sodium sulfate. Evaporation of the solvent afforded the products in excellent yield. All the products were characterized by ^{1}H NMR spectroscopy and by comparison with IR spectra of authentic samples.

References: A. K. Mitra, A. De, N. Karchaudhuri, *J. Chem. Res. (S)*, 560 (1999).

Type of reaction: deprotection
Reaction condition: solid-state
Keywords: thioacetal, dethioacetalization, clayfen, microwave irradiation, ketone, aldehyde

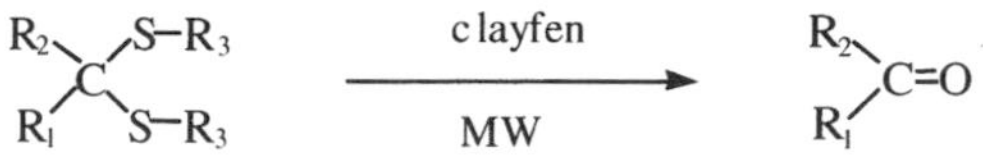

$$\text{R}_1\text{R}_2\text{C(S-R}_3)(\text{S-R}_3) \xrightarrow[\text{MW}]{\text{clayfen}} \text{R}_1\text{R}_2\text{C=O}$$

1 → **2**

a: R$_1$=C$_6$H$_5$; R$_2$=H; R$_3$,R$_3$=-(CH$_2$)$_2$-
b: R$_1$=4-NO$_2$C$_6$H$_4$; R$_2$=H; R$_3$,R$_3$=-(CH$_2$)$_2$-
c: R$_1$=4-NO$_2$C$_6$H$_4$; R$_2$=H; R$_3$,R$_3$=-(CH$_2$)$_3$-
d: R$_1$=C$_6$H$_5$; R$_2$=Me; R$_3$,R$_3$=-(CH$_2$)$_2$-
e: R$_1$=R$_2$=Ph; R$_3$=Et
f: R$_1$=R$_2$=Et; R$_3$=-(CH$_2$)$_2$-
g: R$_1$=4-MeOC$_6$H$_4$; R$_2$=H; R$_3$=-(CH$_2$)$_2$-

Experimental procedures:
Clayfen (1.13 g, 1.2 mmol of iron(III) nitrate) is thoroughly mixed with neat thioacetal **1b** (0.227 g, 1 mmol) in the solid state. The material is transferred in a test tube and placed in an alumina bath inside the microwave oven and irradiated (40 s). Upon completion of the reaction, monitored on TLC (hexane-EtOAc, 8:2, v/v), the product was extracted into ethylene chloride. The resulting solution is passed through a small bed of neutral alumina. Evaporation of the solvent delivers pure *p*-nitrobenzaldehyde **2b** in 97% yield. In the case of cyclic thio acetals and ketals, the liberated dithiols bind to the clay surface rather tightly and a simple washing of the clayfen affords clean products.

References: R. S. Varma, R. K. Saini, *Tetrahedron Lett.*, **38**, 2623 (1997).

Type of reaction: deprotection
Reaction condition: solvent-free
Keywords: oxime, sodium periodate, silica, microwave irradiation, ketone

$$\begin{array}{ccc} R_1 & \xrightarrow[\text{MW}]{\text{wet NaIO}_4\text{-silica}} & R_1 \\ R_2 \end{array} C=N-OH \qquad\qquad\qquad \begin{array}{c} R_1 \\ R_2 \end{array} C=O$$

1   2

a: R_1=Ph; R_2=Me
b: R_1=4-ClC$_6$H$_4$; R_2=Me
c: R_1=4-MeOC$_6$H$_4$; R_2=Me
d: R_1=4-MeC$_6$H$_4$; R_2=Me
e: R_1=4-NH$_2$C$_6$H$_4$; R_2=Me
f: R_1=R_2=Ph
g: R_1=*n*-Bu; R_2=Et
h: R_1, R_2=-(CH$_2$)$_5$

Experimental procedures:
The reagent is prepared as described earlier by adding silica gel (40 g, 230–400 mesh, Baxter) to a stirred solution of NaIO$_4$ (10 g, 46.7 mmol) in 60 mL of water. After removal of water the resulting white powder is dried in an oven at 120 °C for 12 h. The reagent (2.14 g, 2 mmol of NaIO$_4$) is wetted with water (0.6 mL) and is mixed with the neat ketoxime **1** (1 mmol) in a small beaker. The beaker is placed in an alumina bath (heat sink) inside a Kenmore microwave oven (2450 MHz) operating at full power (900 W) for the specific time. After completion of the reaction (monitored by TLC) the product **2** is extracted into dichloromethane (3×15 mL).

References: R. S. Varma, R. Dahiya, R. K. Saini, *Tetrahedron Lett.*, **38**, 7029 (1997).

Type of reaction: deprotection
Reaction condition: solvent-free
Keywords: oxime, hydrazone, deoximation, ammonium persulfate-silica gel, microwave irradiation, ketone, aldehyde

$$
\begin{array}{c}
R_1 \\
C{=}N{-}OH \\
R_2 \\
\mathbf{1}
\end{array}
\quad
\xrightarrow[\text{MW}]{(NH_4)_2S_2O_8\text{-silica}}
\quad
\begin{array}{c}
R_1 \\
C{=}O \\
R_2 \\
\mathbf{2}
\end{array}
$$

a: R_1=Ph; R_2=Me
b: R_1=4-ClC$_6$H$_4$; R_2=Me
c: R_1=4-MeOC$_6$H$_4$; R_2=Me
d: R_1=4-MeC$_6$H$_4$; R_2=Me
e: R_1=Ph; R_2=H
f: R_1=1-naphthyl; R_2=H
g: R_1=4-NO$_2$C$_6$H$_4$; R_2=H
h: R_1=3,4-(MeO)$_2$C$_6$H$_3$; R_2=H

$$
\begin{array}{c}
R_1 \\
C{=}N{-}N{-}R \\
R_2 \quad H \\
\mathbf{3}
\end{array}
\quad
\xrightarrow[\text{MW}]{(NH_4)_2S_2O_8\text{-clay}}
\quad
\begin{array}{c}
R_1 \\
C{=}O \\
R_2 \\
\mathbf{4}
\end{array}
$$

a: R_1=Ph; R_2=Me;   R=CONH$_2$
b: R_1=4-ClC$_6$H$_4$; R_2=Me; R=CONH$_2$
c: R_1=4-HOC$_6$H$_4$; R_2=Me; R=CONH$_2$
d: R_1=4-MeC$_6$H$_4$; R_2=Me; R=CONH$_2$
e: R_1=4-MeOC$_6$H$_4$; R_2=Me; R=CONH$_2$
f: R_1=4-NH$_2$C$_6$H$_4$; R_2=Me; R=CONH$_2$
g: R_1=4-MeOC$_6$H$_4$CH$_2$CH$_2$; R_2=Me; R=CONH$_2$
h: R_1=n-Bu; R_2=Et; R=CONH$_2$
i: R_1=Ph; R_2=Me; R=Ph
j: R_1=n-Bu; R_2=Et; R=Ph

Experimental procedures:
Neat oxime (1 mmol) or dissolved in dichloromethane (2–3 mL) is combined with silica gel (10 times, w/w) and the "dry" powder is mixed with ammonium persulfate (5 mmol) using a vortex mixer. The contents are irradiated at full power in an alumina bath inside a Kenmore microwave oven (2450 MHz, 800 W). After completion of the reaction (monitored by TLC) the product is extracted into dichloromethane (4×30 mL) and purified by column chromatography. In the case of aldehydes, the crude product is filtered through a small bed of neutral alumina with dichloromethane.

References: R.S. Varma, H.M. Meshram, *Tetrahedron Lett.,* **38**, 5427 (1997); R.S. Varma, H.M. Meshram, *Tetrahedron Lett.,* **38**, 7973 (1997).

Type of reaction: deprotection
Reaction condition: solid state
Keywords: oxime, semicarbazone, silica gel, sodium bismuthate, microwave irradiation, ketone

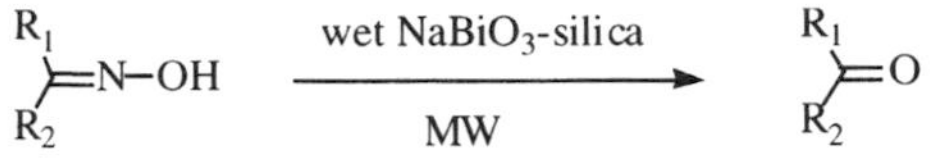

$$R_1R_2C=N-OH \xrightarrow[\text{MW}]{\text{wet NaBiO}_3\text{-silica}} R_1R_2C=O$$

1

a: R_1=Me; R_2=i-Bu
b: R_1,R_2=-(CH$_2$)$_5$-
c: R_1=Me; R_2=Ph
d: R_1=Me; R_2=1-naphthyl
e: R_1=Me; R_2=4-MeC$_6$H$_4$
f: R_1=Me; R_2=4-NO$_2$C$_6$H$_4$
g: R_1=Me; R_2=2-HOC$_6$H$_4$
h: R_1=R_2=Ph

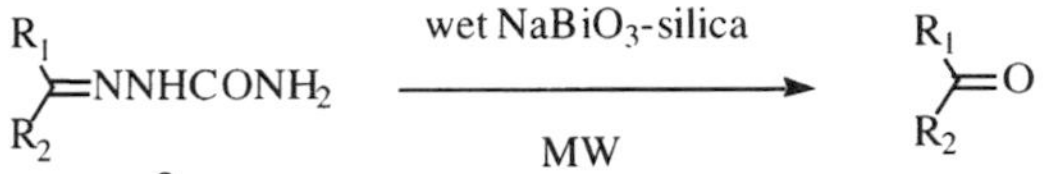

2

$$R_1R_2C=NNHCONH_2 \xrightarrow[\text{MW}]{\text{wet NaBiO}_3\text{-silica}} R_1R_2C=O$$

3

a: R_1=Me; R_2=i-Bu
b: R_1=R_2=-(CH$_2$)$_5$-
c: R_1=Me; R_2=Ph
d: R_1=Me; R_2=4-MeC$_6$H$_4$
e: R_1=Me; R_2=4-MeOC$_6$H$_4$
f: R_1=Me; R_2=2-NO$_2$C$_6$H$_4$
g: R_1=Me; R_2=2-HOC$_6$H$_4$
h: R_1=Me; R_2=2-naphthyl

4

Experimental procedures:

The reagent was prepared by adding silica gel (20 g, 230–400 mesh, SRL) to a stirred solution of NaBiO$_3$ (6.5 g, 23.35 mmol) in 30 mL of water. After removal of water, the resulting powder was dried in an oven at 120 °C for 12 h. The reagent (2.3 g) was moistened with water (0.5 mL) and was mixed with the neat ketoxime (1 mmol) in a 25-mL Erlenmeyer flask. The flask was then placed in an alumina bath (heat sink) inside a BPL-SANYO domestic microwave oven (2450 MHz) operating at full power (1200 W) for the specified time. After completion of the reaction (monitored by TLC) the product was extracted with dichloromethane (3×10 mL). All the compounds obtained were characterized by ^{1}H NMR spectroscopy and by comparison with infrared spectra of authentic samples.

References: A.K. Mitra, A. De, N. Karchaudhuri, *Synlett,* 1345 (1998); A.K. Mitra, A. De, N. Karchaudhuri, *J. Chem. Res. (S),* 320 (1999).

Type of reaction: deprotection
Reaction condition: solvent-free
Keywords: oxime, clayfen, microwave irradiation, aldehyde, ketone

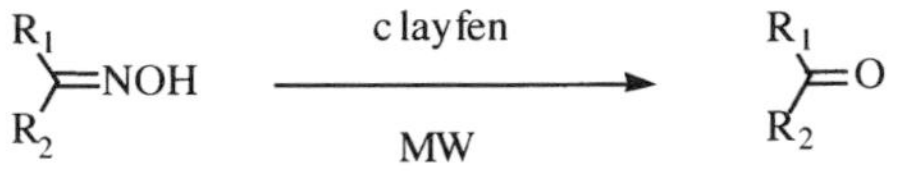

a: R_1=4-MeC$_6$H$_4$; R_2=H
b: R_1=4-ClC$_6$H$_4$; R_2=H
c: R_1=2-ClC$_6$H$_4$; R_2=H
d: R_1=Ph; R_2=Me
e: R_1=4-MeOC$_6$H$_4$; R_2=Me
f: R_1=Ph; R_2=Et
g: R_1=4-ClC$_6$H$_4$; R_2=Ph

Experimental procedures:
In a small beaker oxime (10 mmol) and freshly prepared clayfen reagent (6.6 mmol of iron(III) nitrate) were mixed together to make an intimate mixture. The beaker was placed in a household microwave oven for the specified time. The residue was washed with CH_2Cl_2 (10 mL) and filtered. The filtrate was evaporated to dryness to afford the corresponding carbonyl compound.

References: M. M. Heravi, D. Ajami, M. M. Mojtahedi, *J. Chem. Res. (S)*, 126 (2000).

Type of reaction: deprotection
Reaction condition: solvent-free
Keywords: oxime, deoximation, ammonium chlorochromate, montmorillonite K-10, microwave irradiation, aldehyde, ketone

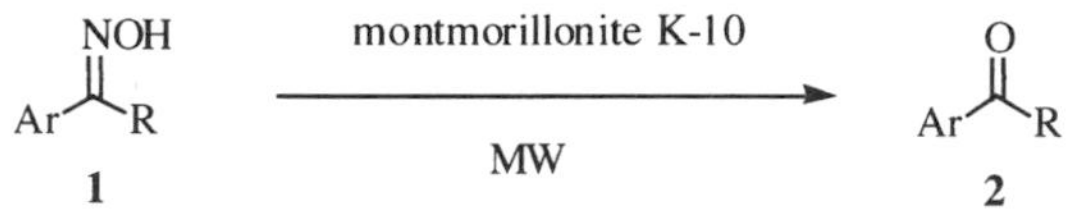

a: Ar=4-MeC$_6$H$_4$; R=H
b: Ar=4-ClC$_6$H$_4$; R=H
c: Ar=2-MeC$_6$H$_4$; R=H
d: Ar=Ph; R=Me
e: Ar=4-MeOC$_6$H$_4$; R=Me
f: Ar=Ph; R=Et
g: Ar=4-ClC$_6$H$_4$; R=Ph

Experimental procedures:

Ammonium chlorochromate supported on montmorillonite K-10 (2.6 mmol) was mixed with the neat oxime (1.6 mmol) in a small beaker. The beaker was placed inside a microwave oven operating at full power (900 W). After completion of the reaction (monitored by TLC), the product was extracted with CH$_2$Cl$_2$. The filtrate was evaporated to dryness, and the residue was passed through a small bed of silica gel to afford the corresponding carbonyl compounds after removal of the solvent.

References: M.M. Heravi, Y.S. Beheshtiha, M. Ghasemzadeh, R. Hekmatshoar, N. Sarmad, *Monatsh. Chem.*, **131**, 187 (2000).

Type of reaction: deprotection
Reaction condition: solvent-free
Keywords: *N,N*-dimethylhydrazone, CeCl$_3 \cdot$7H$_2$O-SiO$_2$, microwave irradiation, ketone

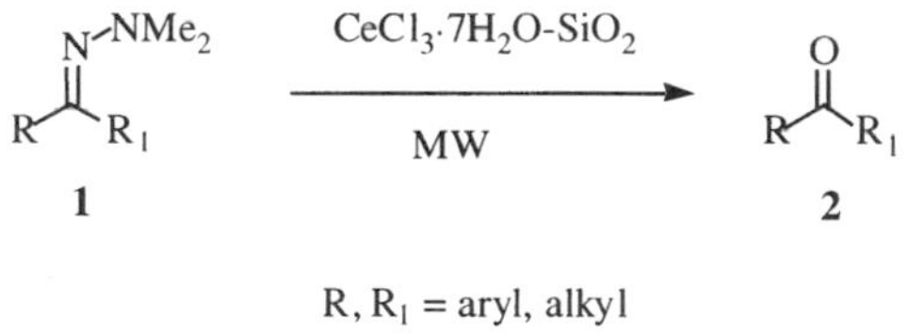

R, R$_l$ = aryl, alkyl

Experimental procedures:

Ketone *N,N*-dimethyl hydrazone (5 mmol) and CeCl$_3 \cdot$7H$_2$O (5 mmol) were admixed with silica gel (1.5 g Merck, finer than 200 mesh) and subjected to microwave irradiation at 450 W for 3–5 min. After completion of the reaction as indicated by TLC, the reaction mixture was filtered and washed with dichloromethane (2×20 mL). The combined organic layers were concentrated in vacuo to give the parent ketone (>95% purity).

References: J.S. Yadav, B.V.S. Reddy, M.S.K. Reddy, G. Sabitha, *Synlett,* 1134 (2001).

Type of reaction: deprotection
Reaction condition: solvent-free
Keywords: semicarbanzone, oxidation, montmorillonite K-10, microwave irradiation, aldehyde, ketone

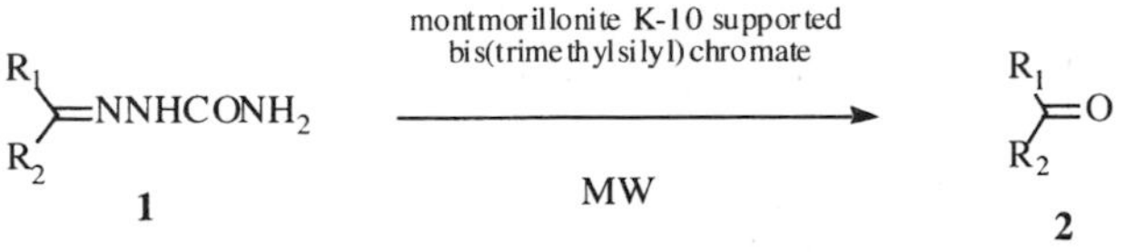

montmorillonite K-10 supported
bis(trimethylsilyl)chromate

$$R_1 \diagdown C{=}NNHCONH_2 \xrightarrow[\text{MW}]{} R_1 \diagdown C{=}O$$

1   MW   2

a: R_1=H; R_2=Ph   f: R_1=H; R_2=2-HOC$_6$H$_4$
b: R_1=H; R_2=4-ClC$_6$H$_4$ g: R_1=Me; R_2=Ph
c: R_1=H; R_2=2-ClC$_6$H$_4$ h: R_1=H; R_2=CH=CHMe
d: R_1=H; R_2=3-ClC$_6$H$_4$ i: R_1=Ph; R_2=Ph
e: R_1=H; R_2=2-NO$_2$C$_6$H$_4$ j: R_1,R_2=-(CH$_2$)$_5$-

Experimental procedures:

Montmorillonite K-10 supported bis(trimethylsilyl)chromate (0.75 g, equivalent to 1.2 mmol of chromium(VI)) was mixed thoroughly with 1 mmol of semicarbazone and irradiated by the microwaves (900 W). The progress of the reaction was monitored by TLC. After completion of the reaction, the solid phase was taken up in CH$_2$Cl$_2$, filtered, and washed with an excess of CH$_2$Cl$_2$. The filtrate was evaporated to dryness and purified by column chromatography using hexane-AcOEt, 8:2 as eluent to afford the corresponding carbonyl compound.

References: M.M. Heravi, M. Tajbakhsh, H. Bakooie, D. Ajami, *Monatsh. Chem.*, **130**, 933 (1999); M.M. Heravi, D. Ajami, B. Mohajerani, M. Tajbakhsh, M. Ghassemzadeh, K. Tabar-Hydar, *Monatsh. Chem.*, **132**, 881 (2001).

List of Journals

The following is a list of the journals scanned in the preparation of this volume. In most cases they were scanned from 1980 to 2001.

Angew. Chem.
Amgew. Chem. Int. Ed. Engl.
Bull. Chem. Soc. Jpn.
Chem. Ber.
Chem. Commun.
Chem. Exp.
Chem. Lett.
Chem. Eur. J.
Chem. Engin. Sci.
Chemosphere
Chemie Technik
Eur. J. Org. Chem.
Green Chem.
Heterocycles
Helv. Chim. Acta
Indian J. Chem. Sect. B
J. Am. Chem. Soc.
J. Chem. Soc., Perkin Trans. 1
J. Chem. Soc., Perkin Trans. 2
J. Chem. Res. (S)
J. Mol. Cat.
J. Fluorine Chem.
J. Org. Chem.

J. Organomet. Chem.
J. Photochem. Photobiol. A
J. Phys. Org. Chem.
J. Prakt. Chem.
Liebigs Ann.
Macromolecule
Monatsh. Chem.
Mendeleev Commun.
Merck Spectrum
Mol. Cryst. Liq. Cryst.
Nippon Kagaku Kaishi
Nature
New J. Chem.
Org. Lett.
Phosphorus, Sulfur and Silicon
Synlett
Synthesis
Supramol. Chem.
Synth. Commun.
Tetrahedron
Tetrahedron:Asym.
Tetrahedron Lett.

Subject Index